McGraw·Hill Companies

Congress Cataloging-in-Publication Data

omas.
Sigma handbook : a complete guide for green belts, black belts, and
s at all levels / Thomas Pyzdek, Paul A. Keller.—3rd ed.
cm.
es bibliographical references and index.
978-0-07-162338-4 (alk. paper)
duction management—Handbooks, manuals, etc. 2. Quality control—
l methods—Handbooks, manuals, etc. 3. Six sigma (Quality control
) I. Keller, Paul A. II. Title.
799 2009
—dc22

2009030973

Hill books are available at special quantity discounts to use as premiums and sales
ns, or for use in corporate training programs. To contact a representative please
at bulksales@mcgraw-hill.com.

The Six Sigma Handbook, Third Edition

DOC/DOC 1 4 3

3-0-07-162338-4
-07-162338-8

es within this book were printed on acid-free paper.

ring Editor
Bass

tions Coordinator
hael Mulcahy

l Supervisor
id E. Fogarty

Manager
a Dixit

Editor
al Chandel

Proofreader
Eina Malik

Indexer
Robert Swanson

Production Supervisor
Richard C. Ruzycka

Composition
Glyph International

Art Director, Cover
Jeff Weeks

A Complete Gu

Black Belts, and Ma

New Y
Lisbon Lo
M
Seoul S

The I

Library of C

Pyzdek, Th
The Si
managen
p.
Includ
ISBN
1. Pro
Statistica
standard
TS155.P
658.5'62

McGraw-
promotio
e-mail us

7 8 9 0

ISBN 97
MHID 0

The pag

Sponso
Judy

Acquisi
Mic

Editori
Da

Projec
Ek

Copy
So

Contents

PART II Six Sigma Tools and Techniques

Appendices

About the Authors

Thomas Pyzdek is author or coauthor of more than 50 books including *The Six Sigma Handbook, The Quality Engineering Handbook,* and *The Handbook of Quality Management*. His works are used by thousands of universities and organizations around the world to teach process excellence. Mr. Pyzdek has provided training and consulting to employers and clients in all industries since 1967. He provides consulting guidance from the executive suite to "Belts" working in the trenches. In his online and live public seminars and client classes Mr. Pyzdek has taught Six Sigma, Lean, Quality and other business process improvement methodologies to thousands.

Mr. Pyzdek is a Fellow of ASQ and recipient of the ASQ Edward's Medal and the Simon Collier Quality Award, both for outstanding contributions to the field of quality management, and the ASQ E.L. Grant Medal for outstanding contributions to Quality Education. He serves on numerous editorial boards, including *The Quality Management Journal, Quality Engineering*, and *International Journal of Six Sigma and Competitive Advantage*.

Paul Keller is vice president of and senior consultant with Quality America, Inc. He has developed and implemented successful Six Sigma and Quality Improvement programs in service and manufacturing environments. Mr. Keller (just for consistency!) has been with Quality America since 1992, where he has:

- Developed and managed systems for overall operations, including quality improvement, product development, partner relations, marketing, sales, order fulfillment, and technical support.
- Provided primary statistical expertise to customers, as well as to internal software development, sales, and technical support teams.
- Developed and implemented Six Sigma related courses, including Quality Management, Statistical Process Control (SPC), and Designed Experiments, to hundreds of companies in a wide variety of industries including Roche Pharmaceuticals, Core3 Inc. Business Process Outsourcing, U.S. Army, MacDermid Printing Solutions, Boeing Satellite, Dow Corning, Antec, Pfizer, Warner Lambert, and many others.

Preface

The Six Sigma approach has been adopted by a growing majority of the Fortune 500 companies, as well as many small and mid-sized organizations. Its application in both for-profit and non-profit organizations is a reflection of its broad objectives in improving processes at the core of an organization's mission. While initial perceptions often focus on quality improvements, successful deployments look beyond to profitability, sustainability, and long term growth.

As these words are written, what is now the longest and deepest recession since the Great Depression has upset a record period of global growth and expansion. During the expansion, Six Sigma proved a valuable strategy to meet the strong market demand for products and services through capacity and productivity improvements and focus on reduced time to market. Where competitive pressures from emerging global markets were especially strong, service improvement, cost of delivery and cost to manufacture strategies proved successful. This recession has been labeled a "game changer" by more than a few economists, upsetting supply chains and forcing entire industries to rethink their business model. There will certainly be many organizational casualties of this recession in a wide array of industries. Yet, there will undoubtedly be survivors, who will gain market share and become the pillars of this new century. Those organizations will focus first on core businesses, ensuring continued market share and profitability. They will apply structured Six Sigma efforts directed at key cost, quality and service objectives. This will demand a fresh look at their internal processes, from the eyes of their customer base, to maximize value and reduce cost. They will then seize new opportunities, left open by the weakened competition. Their ability to expand into these markets will depend on diligent planning and successful execution, hallmarks of a Six Sigma approach. The simplicity and adaptability of the DMAIC approach will provide the means towards achieving a strengthened competitive advantage.

The key benefits we sought to achieve in this third revision include:

- Clearly define the management responsibilities and actions necessary for successful deployment.
- Fully incorporate Lean, Problem Solving and Statistical techniques within the Six Sigma methodology.
- Create an easy to use reference guide written in easy-to-understand language.

- Provide examples using Minitab, Excel and other software to demonstrate application of problem-solving and statistical techniques in a variety of settings.
- Emphasize service applications of Six Sigma, since all organizations are at their core a service organization.

We direct this revision toward executive-level management, or those who aspire to those positions, as a means to discover the potential of a properly designed and deployed Lean Six Sigma effort. Operational-level practitioners will also value the detailed deployment plans, and structured approach to the tools and methods used by project teams. The core principles and tools of Lean, with the statistical validation, root-cause analysis and DMAIC problem-solving methodology, are integrated throughout this handbook. The presentation of this third edition is based on the implementation strategy for Six Sigma: initial topics cover the management responsibilities, with subsequent topics addressing the details of the Lean Six Sigma DMAIC problem solving methodology.

We hope you enjoy it.

PART I

Six Sigma Implementation and Management

CHAPTER 1

Building the Responsive Six Sigma Organization

What Is Six Sigma?

Six Sigma is a rigorous, focused, and highly effective implementation of proven quality principles and techniques. Incorporating elements from the work of many quality pioneers, Six Sigma aims for virtually error-free business performance. Sigma, σ, is a letter in the Greek alphabet used by statisticians to measure the variability in any process. A company's performance is measured by the sigma level of their business processes. Traditionally companies accepted three or four sigma performance levels as the norm, despite the fact that these processes created between 6,200 and 67,000 problems per million opportunities! The Six Sigma standard of 3.4 problems-per-million opportunities* is a response to the increasing expectations of customers and the increased complexity of modern products and processes.

Despite its name, Six Sigma's magic isn't in statistical or high-tech razzle-dazzle. Six Sigma relies on tried and true methods that have been used for decades. By some measures, Six Sigma discards a great deal of the complexity that characterized Total Quality Management (TQM). Six Sigma takes a handful of proven methods and trains a small cadre of in-house technical leaders, known as Six Sigma Black Belts, to a high level of proficiency in the application of these techniques. To be sure, some of the methods Black Belts use are highly advanced, including up-to-date computer technology. But the tools are applied within a simple performance improvement model known as Define-Measure-Analyze-Improve-Control, or DMAIC. DMAIC is described briefly as follows:

D Define the goals of the improvement activity.

M Measure the existing system.

A Analyze the system to identify ways to eliminate the gap between the current performance of the system or process and the desired goal.

I Improve the system.

C Control the new system.

---

*Statisticians note: The area under the normal curve beyond Six Sigma is 2 parts-per-billion. In calculating failure rates for Six Sigma purposes we assume that performance experienced by customers over the life of the product or process will be much worse than internal short-term estimates predict. To compensate, a "shift" of 1.5 sigma from the mean is added before calculating estimated long-term failures. Thus, you will find 3.4 parts-per-million as the area beyond 4.5 sigma on the normal curve.

Why Six Sigma?

When a Japanese firm took over a Motorola factory that manufactured Quasar television sets in the United States in the 1970s, they promptly set about making drastic changes in the way the factory operated. Under Japanese management, the factory was soon producing TV sets with 1/20th as many defects as they had produced under Motorola's management. They did this using the same workforce, technology, and designs, and did it while lowering costs, making it clear that the problem was Motorola's management. It took a while but, eventually, even Motorola's own executives finally admitted "Our quality stinks" (Main, 1994).

It took until nearly the mid-1980s before Motorola figured out what to do about it. Bob Galvin, Motorola's CEO at the time, started the company on the quality path known as Six Sigma and became a business icon largely as a result of what he accomplished in quality at Motorola. Using Six Sigma Motorola became known as a quality leader and a profit leader. After Motorola won the Malcolm Baldrige National Quality Award in 1988 the secret of their success became public knowledge and the Six Sigma revolution was on. Today it's hotter than ever. Even though Motorola has been struggling for the past few years, companies such as GE and AlliedSignal have taken up the Six Sigma banner and used it to lead themselves to new levels of customer service and productivity.

It would be a mistake to think that Six Sigma is about quality in the traditional sense. Quality, defined traditionally as conformance to internal requirements, has little to do with Six Sigma. Six Sigma focuses on helping the organization make more money by improving customer value and efficiency. To link this objective of Six Sigma with quality requires a new definition of quality: the value added by a productive endeavor. This quality may be expressed as potential quality and actual quality. Potential quality is the known maximum possible value added per unit of input. Actual quality is the current value added per unit of input. The difference between potential and actual quality is waste. Six Sigma focuses on improving quality (i.e., reducing waste) by helping organizations produce products and services better, faster, and cheaper. There is a direct correspondence between quality levels and "sigma levels" of performance. For example, a process operating at Six Sigma will fail to meet requirements about 3 times per million transactions. The typical company operates at roughly four sigma, equivalent to approximately 6,210 errors per million transactions. Six Sigma focuses on customer requirements, defect prevention, cycle time reduction, and cost savings. Thus, the benefits from Six Sigma go straight to the bottom line. Unlike mindless cost-cutting programs which also reduce value and quality, Six Sigma identifies and eliminates costs which provide no value to customers: waste costs.

For non-Six Sigma companies, these costs are often extremely high. Companies operating at three or four sigma typically spend between 25 and 40 percent of their revenues fixing problems. This is known as the cost of quality, or more accurately the cost of poor quality. Companies operating at Six Sigma typically spend less than 5 percent of their revenues fixing problems (Fig. 1.1). COPQ values shown in Fig. 1.1 are at the lower end of the range of results reported in various studies. The dollar cost of this gap can be huge. General Electric estimated that the gap between three or four sigma and Six Sigma was costing them between $8 billion and $12 billion per year.

One reason why costs are directly related to sigma levels is very simple: sigma levels are a measure of error rates, and it costs money to correct errors. Figure 1.2 shows the relationship between errors and sigma levels. Note that the error rate drops exponentially as the sigma level goes up, and that this correlates well to the empirical cost data shown in Fig. 1.1. Also note that the errors are shown as errors per million opportunities, not as

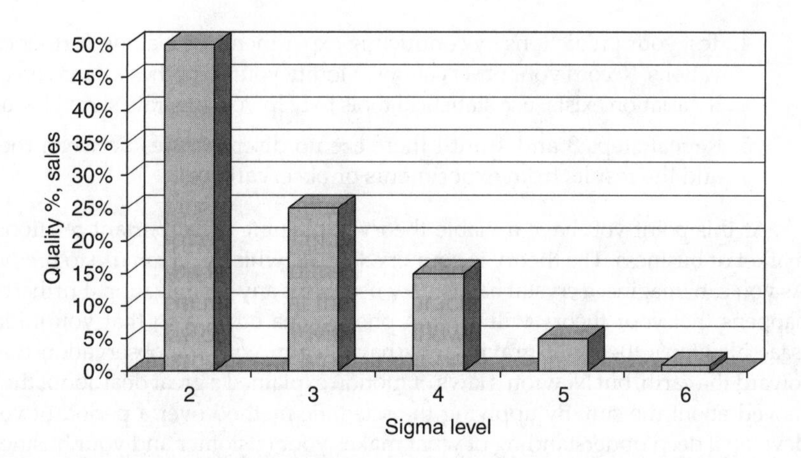

FIGURE 1.1 Cost of poor quality versus sigma level.

percentages. This is another convention introduced by Six Sigma. In the past we could tolerate percentage error rates (errors per hundred opportunities) today we cannot.

The Six Sigma Philosophy

Six Sigma is the application of the scientific method to the design and operation of management systems and business processes which enable employees to deliver the greatest value to customers and owners. The scientific method works as follows:

1. Observe some important aspect of the marketplace or your business.
2. Develop a tentative explanation, or hypothesis, consistent with your observations.
3. Based on your hypothesis, make predictions.

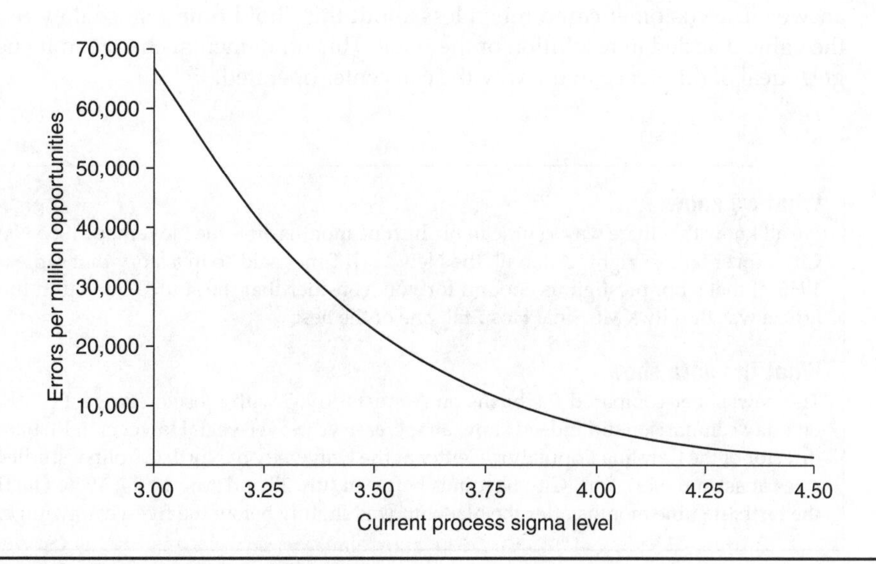

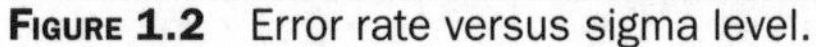
FIGURE 1.2 Error rate versus sigma level.

4. Test your predictions by conducting experiments or making further careful observations. Record your observations. Modify your hypothesis based on the new facts. If variation exists, use statistical tools to help you separate signal from noise.
5. Repeat steps 3 and 4 until there are no discrepancies between the hypothesis and the results from experiments or observations.

At this point you have a viable theory explaining an important relationship in your market or business. The theory is your crystal ball, which you can use to predict the future. As you can imagine, a crystal ball is very useful for any organization. Furthermore, it often happens that your theory will explain phenomena other than that you initially studied. Isaac Newton's theory of gravity may have begun with the observation that apples fell toward the earth, but Newton's laws of motion explained a great deal about the way planets moved about the sun. By applying the scientific method over a period of years you will develop a deep understanding of what makes your customer and your business tick.

When this approach is applied across the organization, the political influence that stalls organizations is minimized and a "show me the data" attitude prevails. While corporate politics can never be eliminated where human beings interact, politics is much less an influence in Six Sigma organizations than in traditional organizations. People are often quite surprised at the results of this seemingly simple shift in attitude. The essence of these results is stated quite succinctly by "Pyzdek's law":

Most of what you know is wrong!

Like all such "laws," this is an overstatement. However, you'll be stunned by how often people are unable to provide data supporting positions on basic issues when challenged. For example, the manager of a technical support call center was challenged by the CEO to show that customers cared deeply about hold time. Upon investigation, the manager determined that customers cared more about the time it took to reach a technician and whether or not their issue was resolved. The call center's information system measured hold time to include both the time until the technician first answered the phone and the time the customer was on hold while the technician researched the answer. The customer cared much less about this "hold time," since they recognized the value it added in resolution of the issue. This fundamental change in focus made a great deal of difference in the way the call center operated.

What we know

We all know that there was a surge in births nine months after the November 1965 New York City power failure, right? After all, the New York Times said so in a story that ran August 8, 1966. If that's not prestigious enough for you, consider that the source quoted in the Times article was the city's Mt. Sinai Hospital, one of the best.

What the data show

The newspaper compared the births on August 8, 1965 with those on August 8, 1966. This one-day comparison did indeed show an increase year-over-year. However, J. Richard Udry, director of the Carolina Population Center at the University of North Carolina, studied birthrates at several New York City hospitals between July 27 and August 14, 1966. His finding: the birthrate nine months after the blackout was slightly below the five-year average.

The Six Sigma philosophy focuses the attention on the stakeholders for whom the enterprise exists. It is a cause-and-effect mentality. Well-designed management systems and business processes operated by happy employees cause customers and owners to be satisfied or delighted. Of course, none of this is new. Most leaders of traditional organizations honestly believe that this is what they already do. What distinguishes the traditional approach from Six Sigma is the degree of rigor and commitment to the core principles.

Six Sigma versus Traditional Three Sigma Performance

The traditional quality model of process capability differed from Six Sigma in two fundamental respects:

1. It was applied only to manufacturing processes, while Six Sigma is applied to all important business processes.
2. It stipulated that a "capable" process was one that had a process standard deviation of no more than one-sixth of the total allowable spread, where Six Sigma requires the process standard deviation be no more than one-twelfth of the total allowable spread.

These differences are far more profound than one might realize. By addressing all business processes Six Sigma not only treats manufacturing as part of a larger system, it removes the narrow, inward focus of the traditional approach. Customers care about more than just how well a product is manufactured. Price, service, financing terms, style, availability, frequency of updates and enhancements, technical support, and a host of other items are also important. Also, Six Sigma benefits others besides customers. When operations become more cost-effective and the product design cycle shortens, owners or investors benefit too. When employees become more productive their pay can be increased. Six Sigma's broad scope means that it provides benefits to all stakeholders in the organization.

The second point also has implications that are not obvious. Six Sigma is, basically, a process quality goal, where sigma is a statistical measure of variability in a process. As such it falls into the category of a process capability technique. The traditional quality paradigm defined a process as capable if the process's natural spread, plus and minus three sigma, was less than the engineering tolerance. Under the assumption of normality, this three sigma quality level translates to a process yield of 99.73%. A later refinement considered the process location as well as its spread and tightened the minimum acceptance criterion so that the process mean was at least four sigma from the nearest engineering requirement. Six Sigma requires that processes operate such that the nearest engineering requirement is at least Six Sigma from the process mean.

One of Motorola's most significant contributions was to change the discussion of quality from one where quality levels were measured in percent (parts-per-hundred), to a discussion of parts-per-million (PPM) or even parts-per-billion. Motorola correctly pointed out that modern technology was so complex that old ideas about "acceptable quality levels" could no longer be tolerated. Modern business requires near perfect quality levels.

One puzzling aspect of the "official" Six Sigma literature is that it states that a process operating at Six Sigma will produce 3.4 parts-per-million nonconformances. However, if a special normal distribution table is consulted (very few go out to Six Sigma) one finds that the expected nonconformances are 0.002 PPM (2 parts-per-billion, or PPB). The difference occurs because Motorola presumes that the process mean can drift 1.5 sigma in either direction. The area of a normal distribution beyond 4.5 sigma from the mean is indeed 3.4 PPM. Since control charts will easily detect any process shift of this magnitude in a single sample, the 3.4 PPM represents a very conservative upper bound on the nonconformance rate.

In contrast to Six Sigma quality, the old three sigma quality standard of 99.73% translates to 2,700 PPM failures, even if we assume zero drift. For processes with a series of steps, the overall yield is the product of the yields of the different steps. For example, if we had a simple two-step process where step #1 had a yield of 80% and step #2 had a yield of 90%, then the overall yield would be $0.8 \times 0.9 = 0.72 = 72\%$. Note that the overall yield from processes involving a series of steps is always less than the yield of the step with the lowest yield. If three sigma quality levels (99.97% yield) are obtained from every step in a 10-step process, the quality level at the end of the process will contain 26,674 defects per million. Considering that the complexity of modern processes is usually far greater than 10 steps, it is easy to see that Six Sigma quality isn't optional, it's required if the organization is to remain viable.

The requirement of extremely high quality is not limited to multiple-stage manufacturing processes. Consider what three sigma quality would mean if applied to other processes:

- Virtually no modern computer would function
- 10,800,000 mishandled healthcare claims each year
- 18,900 lost U.S. savings bonds every month
- 54,000 checks lost each night by a single large bank
- 4,050 invoices sent out incorrectly each month by a modest-sized telecommunications company
- 540,000 erroneous call detail records each day from a regional telecommunications company
- 270,000,000 (270 million) erroneous credit card transactions each year in the United States

With numbers like these, it's easy to see that the modern world demands extremely high levels of error-free performance. Six Sigma arose in response to this realization.

Just Do It!

It's important to note that Six Sigma organizations are not academic institutions. They compete in the fast-paced world of business, and they don't have the luxury of taking years to study all aspects of a problem before deciding on a course of action. A valuable skill for the leader of a Six Sigma enterprise, or for the sponsor of a Six Sigma project, is to decide when enough information has been obtained to warrant taking a particular course of action. Six Sigma leadership should be conservative when spending the shareholders' dollars. As a result, project research tends to be tightly focused on delivering information useful for management decision-making. Once a level of confidence is achieved, management must direct the Black Belt to move the project from the Analyze phase to the Improve phase, or from the Improve phase to the Control phase. Projects are closed and resources moved to new projects as quickly as possible.

Six Sigma organizations are not infallible; they make their share of mistakes and miss opportunities. Yet, research has shown they make fewer mistakes than their traditional counterparts and perform significantly better in the long run. Their systems incorporate the ability to learn from these mistakes, with resulting systematic improvements.

What's Important?

While working with an aerospace client, I was helping an executive set up a system for identifying potential Six Sigma projects in his area. I asked "What are your most important metrics? What do you focus on?" "That's easy," he responded. "We just completed our monthly ops review so I can show you."

He then called his secretary and asked that she bring the ops review copies. Soon the secretary came in lugging three large, loose-leaf binders filled with copies of PowerPoint slides. This executive and his staff spend one very long day each month reviewing all of these metrics, hoping to glean some direction to help them plan for the future. This is not focusing, it's torture!

Sadly, this is not an isolated case. Over the years I've worked with thousands of people in hundreds of companies and this measurement nightmare is commonplace, even typical. The human mind isn't designed to make sense of such vast amounts of data. Crows can track three or four people, beyond that they lose count.* Like crows, we can only hold a limited number of facts in our minds at one time. We are simply overwhelmed when we try to retain too much information. One study of information overload found the following (Waddington, 1996):

- Two-thirds of managers report tension with work colleagues, and loss of job satisfaction because of stress associated with information overload.
- One-third of managers suffer from ill health, as a direct consequence of stress associated with information overload. This figure increases to 43% among senior managers.
- Almost two-thirds (62%) of managers testify that their personal relationships suffer as a direct result of information overload.
- 43% of managers think important decisions are delayed, and the ability to make decisions is affected as a result of having too much information.
- 44% believe the cost of collating information exceeds its value to business.

Clearly, more information isn't always better.

When pressed, nearly every executive or manager will admit that there are a half-dozen or so measurements that really matter. The rest are either derivatives or window dressing. When asked what really interested him, my client immediately turned to a single slide in the middle of one of the binders. There were two "Biggies" that he focused on. The second-level drill down involved a half-dozen major drivers. Tracking this number of metrics is well within the abilities of humans, if not crows! With this tighter focus the executive could put together a system for selecting good Six Sigma projects and team members.

Six Sigma activities focus on the few things that matter most to three key constituencies: customers, shareholders, and employees. The primary focus is on customers,

*See Joe Wortham, "Corvus brachyrhynchos," http://www.geocities.com/jswortham/corvus.html.

but shareholder interests are not far behind. The requirements of these two groups are determined using scientific methods, of course. Yet the science of identifying customer and shareholder desires is not fully mature, so the data are supplemented with a great deal of personal contact at all levels of the organization. Employee requirements are also aggressively sought. Well-treated employees stay longer and do a better job.

Focus comes from two perspectives: down from the top-level goals and up from problems and opportunities. The opportunities meet the goals at the Six Sigma project, whose selection and development become critical aspects of meeting organizational objectives. Six Sigma projects link the activities of the enterprise to its improvement goals. The linkage is so tight that in a well-run enterprise people working on Six Sigma projects can tell you which enterprise objectives will be impacted by their project, and senior leaders are able to measure the impact of Six Sigma on the enterprise in clear and meaningful terms. The costs and benefits of Six Sigma are monitored using enterprise-wide tracking systems that can slice and dice the data in many different ways. At any point in time an executive can determine if Six Sigma is pulling its weight. In many TQM programs of the past people were unable to point to specific bottom-line benefits, so interest gradually waned and the programs were shelved when times got tough. Six Sigma organizations know precisely what they're getting for their investment.

Six Sigma also has an indirect and seldom measured benefit to an enterprise: its impact on human behavior. Six Sigma doesn't operate in a vacuum. When employees observe Six Sigma's dramatic results, they naturally modify how they approach their work. Seat-of-the-pants management doesn't sit well (pardon the pun!) in Six Sigma organizations that have reached "critical mass." Critical mass occurs when the organization's culture has changed as a result of Six Sigma's successful deployment across a large segment of the organization. The initial clash of cultures has worked itself out, and those opposed to the Six Sigma way have either left, converted, or learned to keep quiet.

When deploying Six Sigma, it's important not to stifle creativity for the sake of operational efficiencies. For example, successful Research and development (R&D) involves a good deal of original creative thinking. Research may actually suffer from too much rigor and focus on error prevention. Cutting-edge research is necessarily trial and error and requires a high tolerance for failure. The chaos of exploring new ideas is not something to be managed out of the system; it is expected and encouraged. To the extent that it involves process design and product testing, including the concept of manufacturability, Six Sigma will certainly make a contribution to the development part of R&D. The objective is to selectively apply Six Sigma to those areas where it provides benefit.

Taking a broader view, a business is a complex undertaking, requiring creativity, innovation, and intuition for successful leadership. While it's good to be "data-driven," leaders need to question data effectively, especially since some of the most important components of success in business are unmeasured and perhaps immeasurable. Challenge counterintuitive data and subject it to a gut check. It may be that the counterintuitive result represents a startling breakthrough in knowledge, but it may simply be wrong.

Consider this example. A software client had a technical support call center to help their customers solve problems with the software. Customer surveys were collected and the statistician made an amazing discovery, hold time didn't matter! The data showed that customer satisfaction was the same for customers served immediately and for those on hold for an hour or more. Discussions began along the lines of how many fewer staff would be required due to this new information. Impressive savings were forecast.

Fortunately, the support center manager hadn't left his skepticism at the front door. He asked for additional data, which showed that the abandon rate increased steadily as people were kept on hold. The surveys were given only to those people who had waited for service. These people didn't mind waiting. Those who hung up the phone before being served apparently did. In fact, when a representative sample was obtained, excessive hold time was the number one complaint.

The Change Imperative

In traditional organizations the role of management is to design systems to create and deliver value to customers and shareholders. Unfortunately, however, too many of these organizations fail to recognize that this is a never-ending task. Competitors constantly innovate in an attempt to steal your customers. Customers continuously change their minds about what they want. Capital markets offer investors new ways to earn a return on their investment. The result is an imperative to constantly change management systems.

Despite the change imperative, most enterprises resist change until there are obvious signs that current systems are failing one or more stakeholder groups. Perhaps declining market share makes it clear that your products or services are not as competitive as they once were. Customers may remain loyal, but complaints have reached epidemic proportions. Or share price, the perceived market value of your business, may be trending ominously downward. Traditional organizations watch for such signs and react to them. Change occurs, as it must, but it does so in an atmosphere of crisis and confusion. Substantial loss may result before the needed redesign is complete. People may lose their jobs or even their careers. Many organizations that employ these reactionary tactics don't survive the shock.

Sadly, as this page is written, the U.S. automobile industry is reeling from the combined effects of global competition, a worldwide credit crisis, and an extended period of high fuel costs. While arguments can be made as to the predictability of these events, it is clear that the strength of their competitors lies primarily in their ability to adapt. A recent poll found that more than 60% of global respondents agreed that the ability to change is an organization's main competitive advantage (Blauth, 2008). The ability to respond to customer demand, whether that demand is stagnant or dynamic, is a key focus of Six Sigma projects. Applied at a process level, the Lean principles deployed within these projects stress reduced inventories with decreased cycle times to quickly satisfy shifts in customer demand. As an organizational strategy, these principles result in agile organizations that invest in adaptability rather than volume efficiencies. Resources are deployed only when needed, so they can be constantly refocused to meet the current customer value definitions.

In this way, the Six Sigma enterprise proactively embraces change by explicitly incorporating change into their management systems. Full- and part-time change agent positions are created with a supporting infrastructure designed to integrate change into the routine. Systems are implemented to monitor changing customer, shareholder, and employee inputs, and to rapidly integrate the new information into revised business processes. The approach may employ sophisticated computer modeling, or more basic statistical analysis, to minimize unneeded tampering by separating signal from noise.

These analytical techniques are applied to stakeholder inputs and to enterprise and process metrics at all levels.

The intended consequence of deploying Six Sigma is a change in behavior, as well as the more obvious organizational effectiveness and efficiencies. Conventional wisdom is respectfully questioned: the phrase "How do you know?" is heard repeatedly.

- "Nice report on on-time deliveries, Joan, but show me why you think this is important to the customer. If it is, I want to see a chart covering the last 52 weeks, and don't forget the control limits."
- "This budget variance report doesn't distinguish between expected variation and real changes to the system! I want to see performance across time, with control limits, so we know how to effectively respond."
- "Have these employee survey results been validated? What is the reliability of the questions? What are the main drivers of employee satisfaction? How do you know?"
- "How do these internal dashboards relate to the top-level dashboards that are important to shareholders?"

Yet, the act of challenging accepted practices poses risk. The challenger may feel isolated; those being challenged may feel threatened. These represent behavioral costs to the change effort. The net result of the challenge, ultimately, is the need for further information, which comes at a monetary cost and opportunity risk to the organization. These risks and costs must be effectively managed.

Managing Change

Three goals of change may be summarized as follows:

1. Change the way people in the organization think. Helping people modify their perspective is a fundamental activity of the change agent. All change begins with the individual, at a personal level. Unless the individual is willing to change his behavior, no real change is possible. Changing behavior requires a change in thinking. In an organization where people are expected to use their minds, people's actions are guided by their thoughts and conclusions. The change agent's job starts here.
2. Change the norms. Norms consist of standards, models, or patterns which guide behavior in a group. All organizations have norms or expectations of their members. Change cannot occur until the organization's norms change. In effective Six Sigma organizations, the desired norm is data-driven decision making focused on providing maximum value to key stakeholders.
3. Change the organization's systems or processes. This is the "meat" of the change. Ultimately, all work is a process and quality improvement requires change at the process and system level. However, this cannot occur on a sustained basis until individuals change their behavior and organizational norms are changed.

Change agents fundamentally accomplish these goals by building buy-in within the key stakeholder groups affected by the change. While this is challenging at the process level, it is considerably more so at the organizational level, as is discussed in the next section.

The press of day-to-day business, combined with the inherent difficulties of change, make it easy to let time slip by without significant progress. Keeping operations going is a full-time job, and current problems present themselves with an urgency that meeting a future goal can't match. Without the constant reminders from change agents that goals aren't being met, the leadership can simply forget about the transformation. It is the change agent's job to become the "conscience" of the leadership and to challenge them when progress falls short of goals.

Implementing Six Sigma

After nearly two decades of Six Sigma experience, there is now a solid body of scientific research that successful deployment involves focusing on a small number of high-leverage items. The activities and systems required to successfully implement Six Sigma are well documented.

1. Leadership. Leadership's primary role is to create a clear vision for Six Sigma success and to communicate their vision clearly, consistently, and repeatedly throughout the organization. In other words, leadership must lead the effort. Their primary responsibility is to ensure that Six Sigma goals, objectives, and progress are properly aligned with those of the enterprise as a whole. This is done by modifying the organization such that personnel naturally pursue Six Sigma as part of their normal routine. This requires the creation of new positions and departments, and modified reward, recognition, incentive, and compensation systems. These key issues are discussed throughout this chapter. The Six Sigma deployment will begin with senior leadership training in the philosophy, principles, and tools they need to prepare their organization for success.
2. Infrastructure. Using their newly acquired knowledge, senior leaders direct the development and training of an infrastructure to manage and support Six Sigma.
3. Communication and awareness. Simultaneously, steps are taken to "soft-wire" the organization and to cultivate a change-capable environment where innovation and creativity can flourish. A top-level DMAIC project is focused on the change initiative and the communication required to build buy-in of the initiative, as outlined later in this chapter.
4. Stakeholder feedback systems. Systems are developed for establishing close communication with customers, employees, and suppliers. This includes developing rigorous methods of obtaining and evaluating customer, owner, employee, and supplier input. Baseline studies are conducted to determine the starting point and to identify cultural, policy, and procedural obstacles to success. These systems are discussed in more detail later in this chapter.
5. Process feedback systems. A framework for continuous process improvement is developed, along with a system of indicators for monitoring progress and success. Six Sigma metrics focus on the organization's strategic goals, drivers, and key business processes, as discussed in Chap. 2.
6. Project selection. Six Sigma projects are proposed for improving business processes by people with process knowledge at various levels of the organization. Six Sigma projects are selected based on established protocol by senior

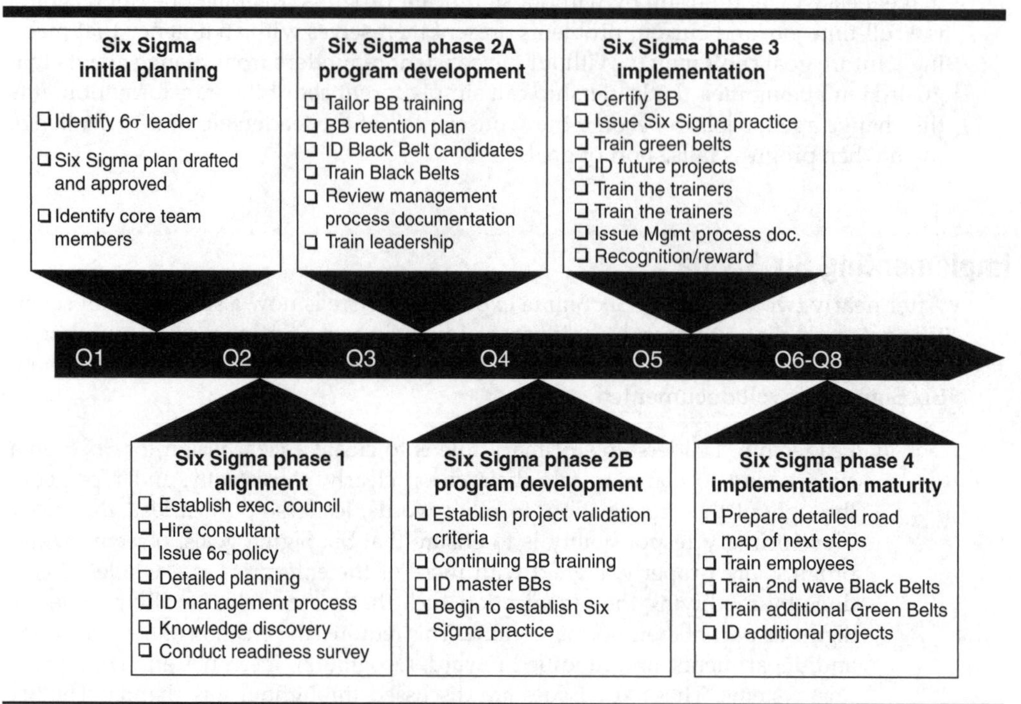

Figure 1.3 Typical deployment activities and timeline.

management to achieve business performance objectives linked to measurable financial results, as discussed in Chap. 4.

7. Project deployment. Six Sigma projects are conducted by project teams lead by Black Belts (or by Green Belts with the technical assistance of Black Belts). Project deployment is discussed in detail in Part II of this book.

Timetable

Figure 1.3 shows a typical set of deployment activities to reach system maturity within two years. The resulting benefits are dependent on the rate of project deployment and the organization's initial quality levels. A typical goal is an improvement rate of approximately 10 times every two years, measured in terms of errors (or defects) per million opportunities (DPMO)*. For example, an organization starting at a typical sigma level of 3.0 would seek to reduce their overall error rate from approximately 67,000 to about 6,700 (or about 4.0 sigma level) in two years time. Figure 1.4 provides a rough guideline for determining when you will reach Six Sigma based on the initial quality level, assuming the 10 times improvement every years. For the typical company starting at three sigma, Fig. 1.4 indicates they will reach Six Sigma levels of performance after approximately five years from the time they have deployed Six Sigma. Given the deployment timeline shown in Fig. 1.3,

*This is about twice the rate of improvement reported by companies using TQM. For example, Baldrige winner Milliken & Co. implemented a "ten-four" improvement program requiring reductions in key adverse measures by a factor of ten every four years.

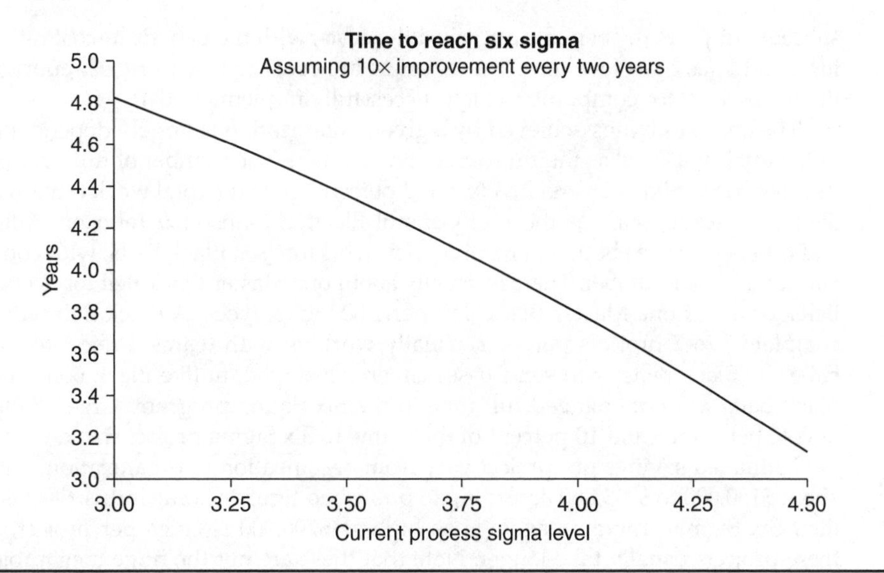

FIGURE 1.4 Time to reach Six Sigma performance levels.

it will be approximately seven years from date of program initiation. Of course, results will begin to appear within a year of starting the deployment.

Yet, even when the enterprise reaches a performance level of five or Six Sigma overall, there may still be processes operating at poor sigma levels, demonstrating the fallibility of the DPMO metric, especially when interpreted across an entire organization. Individual customers judge your organization based on their individual experiences, and customer expectations are a moving target, as previously discussed.

Figure 1.5 shows General Electric's published data on their Six Sigma program. Note there was sufficient savings to cover costs during the first year. In the second and

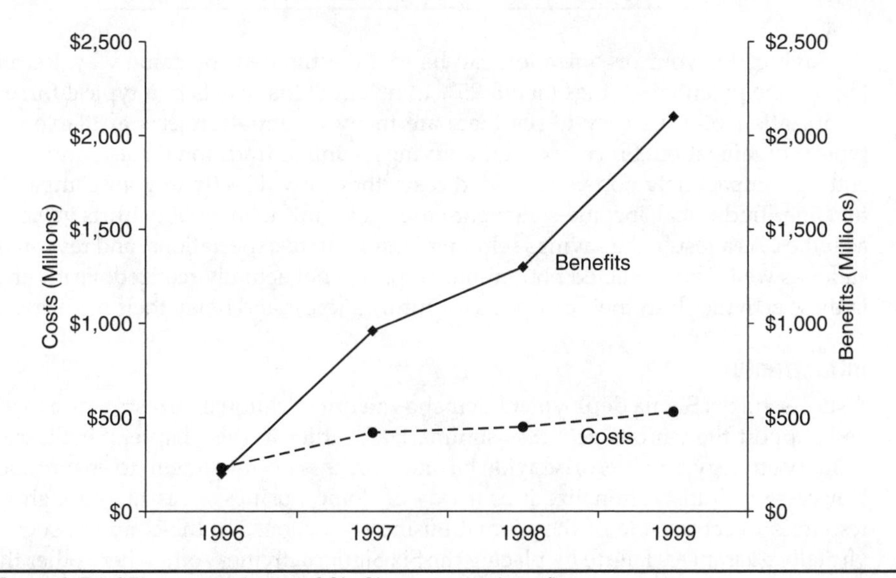

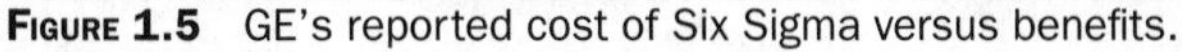

FIGURE 1.5 GE's reported cost of Six Sigma versus benefits.

subsequent years the benefits outpaced the costs, with the benefit-to-cost ratio improving steadily as costs level out. These results are consistent with those reported by academic research for companies which successfully implemented TQM.

The annual savings achieved by a given organization is largely dependent on their initial quality, as well as their resource commitment. The number of full-time personnel devoted to Six Sigma is a relatively small percentage of the total work force. Mature Six Sigma programs, such as those of General Electric, Johnson & Johnson, AlliedSignal, and others average about 1 percent of their workforce as Black Belts, with considerable variation in that number. There is usually about one Master Black Belt for every 10 Black Belts, or about one Master Black Belt per 1,000 employees. A Black Belt will typically complete 5 to 7 projects per year, usually working with teams. Project teams are led either by Black Belts or in some cases Green Belts, who, unlike Black Belts and Master Black Belts, are not engaged full time in the Six Sigma program. Green Belts usually devote between 5 and 10 percent of their time to Six Sigma project work.

Estimated savings per project vary from organization to organization, but average about $150,000 to $243,000 according to published figures. Some industries just starting their Six Sigma programs average as high as $700,000 savings per project, although these projects usually take longer. Note that these are not the huge megaprojects such as pursued by reengineering. Yet, by completing 5 to 7 projects per year per Black Belt the company will add in excess of $1 million per year per Black Belt to its bottom line. For a company with 1,000 employees the resource requirement and estimated savings are shown in the following table:

Master Black Belts:	1
Black Belts:	10
Projects:	50 to 70 (5 to 7 per Black Belt)
Estimated saving:	$9 million to $14.6 million (i.e., $14,580 savings per employee)

Savings for your organization can be easily estimated the same way. Recall from Fig. 1.1 the potential savings (about 25% of revenue) that exists in a typical three sigma organization, and it's easy to see there are many potential projects available within a typical organization. Since Six Sigma savings—unlike traditional slash and burn cost cutting—impact only non-value-added costs, they flow directly to your company's bottom line. Traditional, income-statement-based cost cutting inevitably hurts value-adding activities. As a result, the savings seldom measure up to expectations, and revenues often suffer as well. The predicted bottom-line impact is not actually realized. Firms engaging in these activities hurt their prospects for future success and delay their recovery.

Infrastructure

A successful Six Sigma deployment demands an organizational infrastructure to manage and support the various activities summarized earlier in this chapter. Six Sigma is the primary strategy for enterprise-wide business process improvement; to ensure success it is necessary to institutionalize it as a way of doing business. It is not enough to train resources to act outside of the normal business functions. To the contrary, such a plan virtually guarantees failure by placing the Six Sigma activities somewhere other than the mainstream. Instead, process improvement must become an ongoing part of the business to meet the ever-changing market conditions and customer value definitions.

It's interesting to note that companies institutionalizing the principles of TQM obtained excellent results, which are comparable to the results reported by companies implementing Six Sigma. Those that didn't invariably failed to achieve lasting results. Six Sigma provides a quasi-standardized set of guidelines for deployment, resulting in a much higher success rate. Although each organization will develop its own unique approach to Six Sigma, it is helpful to review the practices of successful companies.

Most importantly, successful Six Sigma deployment is always a top-down affair. For Six Sigma to have a major impact on overall enterprise performance, it must be fully embraced and actively led by top management. Isolated efforts at division or department levels are doomed from the outset. Like flower gardens in a desert, they may flourish and produce a few beautiful results for a time, but sustaining the results requires immense effort by local heroes in constant conflict with the mainstream culture, placing themselves at risk. Sooner or later, the desert will reclaim the garden. Six Sigma shouldn't require heroic effort—there are never enough heroes to go around. Once top management has accepted its leadership responsibility the organizational transformation process can begin.

A key decision is whether Black Belts will report to a central Six Sigma organization or to managers located elsewhere in the organization. The experience of most successful Six Sigma enterprises is that centralized reporting is best. Internal studies by one company that experimented with both types of reporting revealed the results shown in Table 1.1. The major reason for problems with the decentralized approach was disengaging people from routine work and firefighting. Six Sigma is devoted to change, and it seems change tends to take a back seat to current problems. To be sure, the Black Belt possesses a skill set that can be very useful in putting out fires. Black Belts also tend to excel at whatever they do. This combination makes it difficult to resist the urge to pull the Black Belt off of his or her projects "just for a while." In fact, some organizations have trouble getting the Black Belt out of their current department and into the central organization. In one case the CEO intervened personally on behalf of the Black Belts to break them loose. Such stories are testimony to the difficulties encountered in making drastic cultural changes.

The transformation process involves new roles and responsibilities on the part of many individuals in the organization. In addition, new change agent positions must be created. Table 1.2 lists some typical roles and responsibilities. In a Six Sigma organization, improvement and change are the full-time job of a small but critical percentage of the organization's personnel. These full-time change agents are the catalyst that institutionalizes change.

Where Black Belt Reported	Black Belts Successfully Certified
Local organization	40%
Centralized Six Sigma organization	80%

Table 1.1 Black Belt Certification versus Reporting Arrangement

Responsible Entity	Roles	Responsibilities
Executive Six Sigma Council	Strategic leadership	• Ensures Six Sigma goals are linked to enterprise goals • Develops new policies as required • Aligns process excellence efforts across the organization • Suggests high-impact projects • Approves project selection strategy
	Ensures progress	• Provides resources • Tracks and controls progress toward goals • Reviews improvement teams' results (BB, GB, Lean, Supply Chain, other) • Reviews effectiveness of Six Sigma deployment: systems, processes, infrastructure, etc.
	Cultural transformation	• Communicates vision • Removes formal and informal barriers • Commissions modification of compensation, incentive, reward and recognition systems
Director, Six Sigma	Manages Six Sigma infrastructure and resources	• Six Sigma champion for ACME • Develops Enterprise Six Sigma deployment • Owns the Six Sigma project selection and prioritization process for ACME • Ensures Six Sigma strategies and projects are linked through quality function deployment to business plans • Achieves defect reduction and cost take-out targets through Six Sigma activities • Member of Executive Six Sigma Council • Leads and evaluates the performance of Black Belts and Master Black Belts • Communicates Six Sigma progress with customers, suppliers and the enterprise • Champions Six Sigma reward and recognition, as appropriate
Six Sigma Certification Board	Certifies Black Belts Board representatives include Master Black Belts and key Six Sigma leaders	• Works with local units to customize Black Belt and Green Belt requirements to fit business needs • Develops and implements systems for certifying Black Belts and Green Belts • Certifies Black Belts

Six Sigma Core Team	Cross-functional Six Sigma team Part-time change agent	• Provides input into policies and procedures for successful implementation of Six Sigma across ACME • Facilitates Six Sigma activities such as training, special recognition events, Black Belt testing, etc.
Master Black Belt	Enterprise Six Sigma expert Permanent full-time change agent Certified Black Belt with additional specialized skills or experience especially useful in deployment of Six Sigma across the enterprise	• Highly proficient in using Six Sigma methodology to achieve tangible business results • Technical expert beyond Black Belt level on one or more aspects of process improvement (e.g., advanced statistical analysis, project management, communications, program administration, teaching, project coaching) • Identifies high-leverage opportunities for applying the Six Sigma approach across the enterprise • Basic Black Belt training • Green Belt training • Coach/Mentor Black Belts • Participates on ACME Six Sigma Certification Board to certify Black Belts and Green Belts
Black Belt	Six Sigma technical expert Temporary, full-time change agent (will return to other duties after completing a two to three year tour of duty as a Black Belt)	• Leads business process improvement projects where Six Sigma approach is indicated • Successfully completes high-impact projects that result in tangible benefits to the enterprise • Demonstrated mastery of Black Belt body of knowledge • Demonstrated proficiency at achieving results through the application of the Six Sigma approach • Internal Process Improvement Consultant for functional areas • Coach/Mentor Green Belts • Recommends Green Belts for Certification

Table 1.2 Six Sigma Roles and Responsibilities

Responsible Entity	Roles	Responsibilities
Green Belt	Six Sigma project originator Six Sigma project leader Part-time Six Sigma change agent. Continues to perform normal duties while participating on Six Sigma project teams Six Sigma champion in local area	• Demonstrated mastery of Green Belt body of knowledge • Demonstrated proficiency at achieving results through the application of the Six Sigma approach • Recommends Six Sigma projects • Participates on Six Sigma project teams • Leads Six Sigma teams in local improvement projects • Works closely with other continuous improvement leaders to apply formal data analysis approaches to projects • Teaches local teams, shares knowledge of Six Sigma • Successful completion of at least one Six Sigma project every 12 months to maintain their Green Belt certification
Six Sigma Improvement Team	Primary ACME vehicle for achieving Six Sigma improvements	• Completes chartered Six Sigma projects that deliver tangible results • Identifies Six Sigma project candidates
ACME Leaders and Managers	Champions for Six Sigma	• Ensures flow-down and follow-through on goals and strategies within their organizations • Plans improvement projects • Charters or champions chartering process • Identifies teams or individuals required to facilitate Six Sigma deployment • Integrates Six Sigma with performance appraisal process by identifying measurable Six Sigma goals/objectives/results • Identifies, sponsors and directs Six Sigma projects • Holds regular project reviews in accordance with project charters • Includes Six Sigma requirements in expense and capital budgets • Identifies and removes organizational and cultural barriers to Six Sigma success • Rewards and recognizes team and individual accomplishments (formally and informally) • Communicates leadership vision • Monitors and reports Six Sigma progress • Validates Six Sigma project results • Nominates highly qualified Black Belt and/or Green Belt candidates

Project Sponsor	Charters and supports Six Sigma project teams	• Sponsor is ultimately responsible for the success of sponsored projects • Actively participates in projects • Ensures adequate resources are provided for project • Personal review of progress • Identifies and overcomes barriers and issues • Evaluates and accepts deliverable
"Matrixed" Project Manager	Manages Six Sigma resources dedicated to a particular area (e.g., teams of Black Belts on special assignment) Champions Six Sigma Black Belt team	• Provides day-to-day direction for Six Sigma project Black Belt and team activities • Provides local administrative support, facilities, and materials • Conducts periodic reviews of projects • Provides input on Black Belt performance appraisals • Makes/implements decisions based on recommendations of Six Sigma Black Belts
Six Sigma Improvement Team Member	Learns and applies Six Sigma tools to projects	• Actively participates in team tasks • Communicates well with other team members • Demonstrates basic improvement tool knowledge • Accepts and executes assignments as determined by team

Table 1.2 Six Sigma Roles and Responsibilities (*Continued*)

Education and training are important means of changing individual perceptions and behaviors. In this discussion, a distinction is made between training and education. *Training* refers to instruction and practice designed to teach a person how to perform one or more tasks. Training focuses on concrete tasks to be completed. *Education* refers to instruction in thinking. Education focuses on integrating abstract concepts into one's knowledge of the world. An educated person will view the world differently after being educated. This is an essential part of the process of change.

Six Sigma training is a subproject of the Six Sigma deployment plan, whose timetables must be tightly linked. Training provided too early or too late is a mistake. When training is provided too early, the recipient will forget much of what he has learned before it is needed. When it is provided too late, the quality of the employee's work will suffer. When it comes to training, just-in-time delivery is the goal.

The cost of Six Sigma training should be included in the previously discussed estimates of Six Sigma cost-benefit ratios and include:

- Trainer salaries
- Consulting fees
- Classroom space and materials
- Lost time from the job
- Staff salaries
- Office space of training staff

The estimated benefits of the training include the subsequent project deliverables, often on an annualized basis. Since trained Black Belts and Green Belts will often work on multiple projects during the year, it's best to consider these costs and benefits on a program-wide basis, rather than a per-class or per-project basis.

Champions and Sponsors

Six Sigma champions are high-level individuals who understand Six Sigma and are committed to its success. In larger organizations Six Sigma will be led by a full-time, high-level champion, such as an executive vice president. In all organizations, champions also include informal leaders who use Six Sigma in their day-to-day work and communicate the Six Sigma message at every opportunity. Sponsors are owners of processes and systems who help initiate and coordinate Six Sigma improvement activities in their areas of responsibilities.

Leaders should receive guidance in the art of "visioning." Visioning involves the ability to develop a mental image of the organization at a future time; without a vision, there can be no strategy. The future organization will more closely approximate the ideal organization, where "ideal" is defined as that organization which completely achieves the organization's values. Will the organizational structure change? What roles and responsibilities will change? Who will be its key customers? How will it behave toward its customers, employees, and suppliers? Developing a lucid image of this organization will help the leader see how she should proceed with her primary duty of transforming the present organization. Without such an image in her mind, the executive will lead the organization through a maze with a thousand dead ends. Conversely, with a vision as a guide, the transformation will proceed on course. This is not to say that the transformation is ever "easy." But when there is a leader with a vision, it's as if the organization is following an expert scout through hostile territory. The destination is clear, but the journey is still difficult.

Leaders need to be masters of communication. Fortunately, most leaders already possess outstanding communication skills; few rise to the top without them. However, training in effective communication is still wise, even if it is only refresher training. When large organizations are involved, communications training should include mass communication media, such as video, radio broadcasts and print media. Communicating with customers, investors, and suppliers differs from communicating with employees and colleagues, and special training is often required.

Communicating vision is very different from communicating instructions or concrete ideas. Visions of organizations that embody abstract values are necessarily abstract in nature. To effectively convey the vision, the leader must convert the abstractions to concretes. One way to do this is by living the vision. The leader demonstrates her values in every action she takes, every decision she makes, which meetings she attends or ignores, when she pays rapt attention and when she doodles absentmindedly on her notepad. Employees who are trying to understand the leader's vision will pay close attention to the behavior of the leader.

Another way to communicate abstract ideas is through stories. In organizations there is a constant flow of events involving customers, employees, and suppliers. From time to time an event occurs that captures the essence of the leader's vision. A clerk provides exceptional customer service, an engineer takes a risk and makes a mistake, a supplier keeps the line running through a mighty effort. These are concrete examples of what the leader wants the future organization to become. She should repeat these stories to others and publicly recognize the people who made the stories. She should also create stories of her own, even if it requires staging an event. There is nothing dishonest about creating a situation with powerful symbolic meaning and using it to communicate a vision. For example, Nordstrom has a story about a sales clerk who accepted a customer return of a defective tire. This story has tremendous symbolic meaning because Nordstrom doesn't sell tires! The story illustrates Nordstrom's policy of allowing employees to use their own best judgment in all situations, even if they make "mistakes," and of going the extra mile to satisfy customers. However, it is doubtful that the event ever occurred. This is irrelevant. When employees hear this story during their orientation training, the message is clear. The story serves its purpose of clearly communicating an otherwise confusing abstraction.

Leaders need training in conflict resolution. In their role as process owners in a traditional organization, leaders preside over a report-based hierarchy trying to deliver value through processes that cut across several functional areas. The inevitable result is competition for limited resources, which creates conflict. Of course, the ideal solution is to resolve the conflict by designing organizations where there is no such destructive competition. Until then, the leader can expect to find a brisk demand for his conflict-resolution services.

Finally, leaders should demonstrate strict adherence to ethical principles. Leadership involves trust, and trust isn't granted to one who violates a moral code that allows people to live and work together. Honesty, integrity, and other moral virtues should be second nature to the leader.

Black Belts

Candidates for Black Belt status are technically oriented individuals held in high regard by their peers. They should be actively involved in the process of organizational change and development. Candidates may come from a wide range of disciplines and need not be formally trained statisticians or analysts. However, because they are expected to master a wide variety of technical tools in a relatively short period of time, Black Belt candidates will probably possess a background in college-level mathematics, the basic tool of quantitative analysis. Coursework in statistical methods should be considered a strong plus or even a prerequisite. Black Belts receive from three to six weeks of training in the technical tools of Six Sigma. Three-week curricula are usually given to Black Belts working in service or transaction-based businesses, administrative areas, or finance. Four-week programs are common for manufacturing environments. Six weeks of training are provided for Black Belts working in R&D or similar environments. Figure 1.6 shows the curriculum used for courses in General Electric for personnel with finance backgrounds who will be applying Six Sigma to financial, general business, and e-commerce processes. Figure 1.7 shows GE's curriculum for the more traditional manufacturing areas.

Week 1
The DMAIC and DFSS (design for Six Sigma) improvement strategies
Project selection and "scoping" (define)
QFD (quality function deployment)
Sampling principles (quality and quantity)
Measurement system analysis (also called "Gage R&R")
Process capability
Basic graphs
Hypothesis testing
Regression

Week 2
Design of experiments (DOE) (focus on two-level factorials)
Design for Six Sigma tools
Requirements flowdown
Capability flowup (prediction)
Piloting
Simulation
FMEA (failure mode and effects analysis)
Developing control plans
Control charts

Week 3
Power (impact of sample size)
Impact of process instability on capability analysis
Confidence intervals (vs. hypothesis tests)
Implications of the Central Limit Theorem
Transformations
How to detect "lying with statistics"
General linear models
Fractional factorial DOEs

FIGURE 1.6 Sample curriculum for finance Black Belts. From Hoerl (2001). P. 395. Reprinted by permission of ASQ.

Context[1]
- Why Six Sigma
- DMAIC and DFSS processes (sequential case studies)
- Project management fundamentals
- Team effectiveness fundamentals

Define[1]
- Project selection
- Scoping projects
- Developing a project plan
- Multigenerational projects
- Process identification (SIPOC)

Measure[1]
- QFD
- Developing measurable CTQs
- Sampling (data quantity and data quality)
- Measurement system analysis (not just gage R&R)
- SPC Part I
    - The concept of statistical control (process stability)
    - The implications of instability on capability measures
- Capability analysis

Analyze[2]
- Basic graphical improvement tools ("Magnificent 7")
- Management and planning tools (Affinity, ID, etc.)
- Confidence intervals (emphasized)
- Hypothesis testing (de-emphasized)
- ANOVA (de-emphasized)
- Regression
- Developing conceptual designs in DFSS

Improve[3,4]
- DOE (focus on two-level factorials, screening designs, and RSM)
- Piloting (of DMAIC improvements)
- FMEA
- Mistake-proofing
- DFSS design tools
    - CTQ flowdown
    - Capability flowup
    - Simulation

Control[4]
- Developing control plans
- SPC Part II
    - Control charts
- Piloting new designs in DFSS

FIGURE 1.7 Sample curriculum for manufacturing Black Belts. (The week in which the material appears is noted as a superscript.) From Hoerl (2001). P. 399. Reprinted by permission of ASQ.

Although some training companies offer highly compressed two-week training courses, these are not recommended. Even in a six-week course, students receive the equivalent of two semesters of college-level applied statistics in just a few days. Humans require a certain "gestation period" to grasp challenging new concepts; providing too much material in too short a time period is counterproductive. Successful candidates will be comfortable with computers. At a minimum, they should be proficient with one or more operating systems, spreadsheets, database managers, presentation programs, and word processors. As part of their training they will also be required to become proficient in the use of one or more advanced statistical analysis software packages and probably simulation software. Six Sigma Black Belts work to extract actionable knowledge from an organization's information warehouse. To ensure access to the needed information, Six Sigma activities should be closely integrated with the information systems of the organization. Obviously, the skills and training of Six Sigma Black Belts must be enabled by an investment in software and hardware. It makes no sense to hamstring these experts by saving a few dollars on computers or software.

As a full-time change agent, the Black Belt needs excellent interpersonal skills. In addition to mastering a body of technical knowledge, Black Belts must

- Communicate effectively verbally and in writing
- Communicate effectively in both public and private forums
- Work effectively in small group settings as both a participant and a leader
- Work effectively in one-on-one settings
- Understand and carry out instructions from leaders and sponsors

A change agent deficient in these soft skills will nearly always be ineffective. They are usually frustrated and unhappy souls who don't understand why their technically brilliant case for change doesn't cause instantaneous compliance by all parties. The good news is that if the person is willing to apply as much time and effort to soft-skill acquisition and mastery as they applied to honing their technical skills, they will be able to develop proficiency.

In general, Black Belts are hands-on oriented people selected primarily for their ability to get things done. Tools and techniques are provided to help them do this. The training emphasis is on application, not theory. In addition, many Black Belts will work on projects in an area where they possess a high degree of subject-matter expertise. Therefore, Black Belt training is designed around projects related to their specific work areas. This requires Master Black Belts or trainers with very broad project experience to answer application-specific questions. When these personnel aren't available, examples are selected to match the Black Belt's work as closely as possible. For example, if no trainer with human resource experience is available, the examples might be from another service environment; manufacturing examples would be avoided. Another common alternative is to use consultants to conduct the training. Consultants with broad experience within the enterprise as well as with other organizations can sometimes offer insights.

Black Belts must work on projects while they are being trained. Typically, the training classes are conducted at monthly intervals and project work is pursued between classes. One of the critical differences between Six Sigma and other initiatives is the emphasis on using the new skills to get tangible results. It is relatively easy to sit in a classroom and absorb the concepts well enough to pass an exam. It's another thing entirely to apply the new approach to a real-world problem. The Black Belt has to be able to use change agent skills to recruit sponsors and team members and to get these people to work together on a project with a challenging goal and a tight timetable. While the instructors can provide coaching and project-specific training and advice, there's no better time to initiate the process than during the training.

The process for selecting Black Belts should be clearly defined. This ensures consistency and minimizes the possibility of bias and favoritism. Figure 1.8 provides a list of seven success factors, with their relative importance weights, that can be used to compare Black Belt candidates.

The weights are, of course, subjective and only approximate, and are based on an exercise with a group of consultants and Master Black Belts. Organizations can easily identify their own set of criteria and weights, such as shown by Keller (2005). The important thing is to determine the criteria and then develop a method of evaluating candidates on each criterion. The sum of the candidate's criterion score times the criterion weight will give you an overall numerical assessment for ranking the Black Belt candidates. Of course, the numerical assessment is not the only input into the selection decision, but it is a very useful one.

Notice the relatively low weight given to math skills. The rationale is that Black Belts will receive 200 hours of training, much of it focused on the practical application of statistical techniques using computer software and requiring very little actual mathematics. Software automates the analysis, making math skills less necessary.

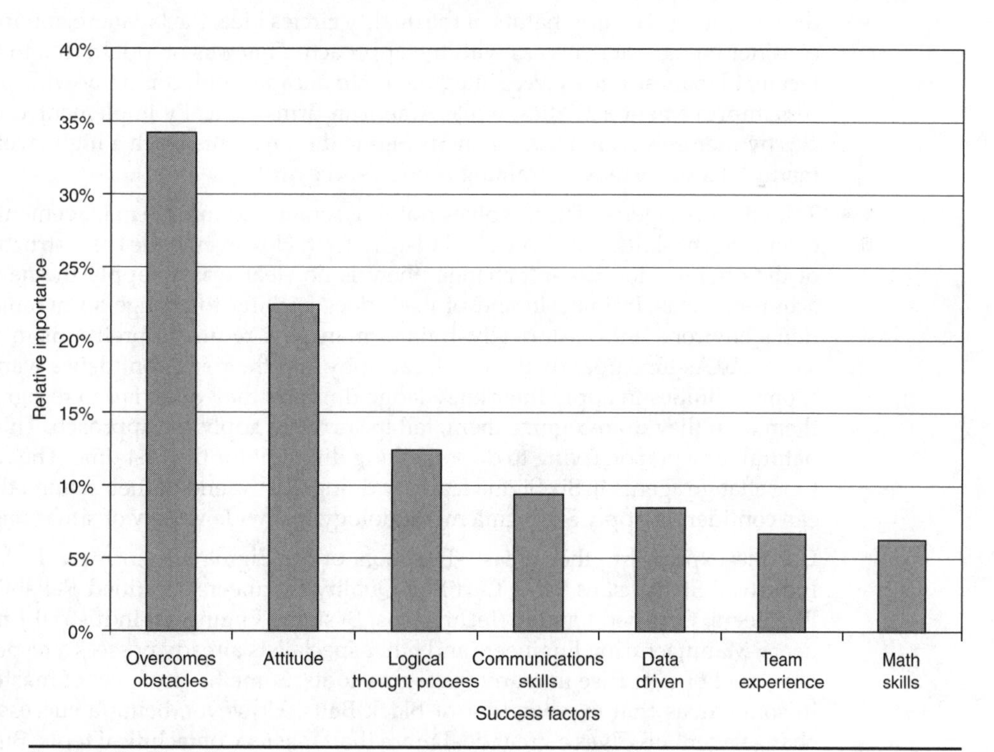

FIGURE 1.8 Black Belt success factors and importance weights.

The mathematical theory underlying a technique is not discussed beyond the level necessary to help the Black Belt properly apply the tool. Black Belts who need help with a particular tool have access to Master Black Belts, other Black Belts, consultants, professors, and a wealth of other resources. Most statistical techniques used in Six Sigma are relatively straightforward and often graphical; spotting obvious errors is usually not too difficult for trained Black Belts. Projects seldom fail due to a lack of mathematical expertise. In contrast, the Black Belts will often have to rely on their own abilities to deal with the obstacles they will inevitably encounter. Failure to overcome the obstacle will often spell failure of the entire project.

Figure 1.9 provides an overview of a process for the selection of Black Belt candidates.

Past improvement initiatives, such as TQM, shared much in common with Six Sigma. TQM also had management champions, improvement projects, sponsors, etc. One of the main differences in the Six Sigma infrastructure is the creation of more formally defined change agent positions. Some observers criticize this practice as creating corps of "elites," especially Black Belts and Master Black Belts. Let's examine the commonly proposed alternatives to creating a relatively small group of highly trained technical experts:

- Train the masses—This is the "quality circles" approach, where people in the lowest level of the organizational hierarchy are trained in the use of basic tools and set loose to solve problems without explicit direction from leadership. When this approach was actually tried in America in the 1970s the results were disappointing. The originators of the quality circles idea, the Japanese, reported considerably greater success with the approach. This was no doubt due to the fact that Japanese circles were integrated into decades- old, company-wide process improvement activities, while American firms typically implemented circles by themselves. Indeed, when Six Sigma deployments reach a high level of maturity, more extensive training is often successful.
- Train the managers—This involves training senior and middle management in change agent skills. This isn't a bad idea in itself. However, if the basic structure of the organization doesn't change, there is no clear way to apply the newly acquired skills. Training in and of itself does nothing to change an organization's environment. Historically, trained managers return to pretty much the same job. As time goes by their skills atrophy and their self-confidence wanes. If opportunities to apply their knowledge do arise, they often fail to recognize them or, if they do recognize them, fail to correctly apply the approach. This is natural for a person trying to do something different for the first time. The full-time change agents in Six Sigma learn by doing. By the end of their tenure, they can confidently apply Six Sigma methodology to a wide variety of situations.
- Use the experts in other areas—The tools of Six Sigma are not new. In fact, Industrial Statisticians, ASQ Certified Quality Engineers, Certified Reliability Engineers, Certified Quality Technicians, Systems Engineers, Industrial Engineers, Manufacturing Engineers and other specialists already possess a respectable level of expertise in many Six Sigma tools. Some have a level of mastery in some areas that exceeds that of Black Belts. However, being a successful change agent involves a great deal more than mastery of technical tools. Black Belts, Green Belts, and Master Black Belts learn tools and techniques in the context

Minimum Criteria

Education—Bachelors Degree, minimum.

Work Experience—At least 3 years of business, technical, or managerial experience plus technical application of education and experience as a member or leader of functional and cross-functional project teams.

Technical Capability—Project management experience is highly desired. Understanding of basic principles of process management. Basic college algebra proficiency as demonstrated by exam.

Computer Proficiency—MS Office Software Suite.

Communication—Demonstrate excellent oral and written communication skills.

Team Skills—Ability to conduct meetings, facilitate small groups and successfully resolve conflicts. Ability to mentor and motivate people.

Final Candidate Selection

To ensure that the Black Belts will be able to address enterprise-wide issues and processes, the Director of Six Sigma and the Executive Six Sigma Council will determine the number of Black Belts to be trained in each functional area, division, department, etc. Black Belt candidates are ranked using a system of points assigned during the screening process. Rank-ordered lists of Black Belt candidates are prepared for designated areas and presented to the senior management of the area for final selection. Area management nominates candidates from their list in numbers sufficient to fill the spaces allocated by the Director of Six Sigma and the Executive Six Sigma Council.

Commitment to Black Belt Assignment

Selected candidates are required to attend 200 hours of Black Belt training (see Chap. 4 for the training content). Within one year of completing training, the Black Belt candidate is required to become certified by passing a written examination and successfully completing at least two major projects. (See Appendix 15 for detailed Black Belt certification process information.) The Black Belt is assigned to Six Sigma full time as a Black Belt for a minimum period of 2 full years, measured from the time he or she is certified as a Black Belt.

Reintegration of Black Belts into the Organization

Black Belts are employed in the Black Belt role for two or three years. After that time they leave the Six Sigma organization and return to other duties. Accomplishing this transition is the joint responsibility of the Black Belt, the Director of Six Sigma, and the management of the Black Belt's former department. Collectively this group comprises the "Transition Team" for the Black Belt. However, senior leadership must accept ultimate responsibility for assuring that Black Belts are not "homeless" after completing their Black Belt tour of duty.

The Director of Six Sigma will inform the Black Belt at least six months prior to the scheduled return. The Black Belt should maintain contact with their "home" organization during his tenure in Six Sigma. If it appears that there will be a suitable position available at approximately the time the Black Belt is scheduled to return, arrangements should be made to complete or hand-off the Black Belt's Six Sigma projects in preparation for his return. If no suitable openings will be available, the Transition Team needs to develop alternative plans. Alternatives might include extending the Black Belt's term of service in Six Sigma, looking for openings in other areas, or making temporary arrangements.

Figure 1.9 Black Belt candidate selection process and criteria.

of following the DMAIC approach to drive organizational change. This is very different than using the same techniques in routine daily work. Quality analysts, for example, generally work in the quality department as permanent, full-time employees. They report to a single boss and have well-defined areas of responsibility. Black Belts, in contrast, go out and seek projects rather than work on anything routine. They report to many different people, who use different criteria to evaluate the Black Belt's performance. They are accountable for delivering measurable, bottom-line results. Obviously, the type of person who is good at one job may not be suitable for the other.

- Create permanent change agent positions. Another option to the Black Belt position is to make the job permanent. After all, why not make maximum use of the training by keeping the person in the Black Belt job indefinitely? Furthermore, as Black Belts gain experience they become more proficient at completing projects. There are, however, arguments against this approach. Having temporary Black Belts allows more people to go through the position, thus increasing the number of people in management with Black Belt experience. Since Black Belts work on projects that impact many different areas of the enterprise, they have a broad, process-oriented perspective that is extremely valuable in top management positions. The continuous influx of new blood into Black Belt and Green Belt positions keeps the thinking fresh and prevents the "them-versus-us" mentality that often develops within functional units. New Black Belts have different networks of contacts throughout the organization, which leads to projects in areas that might otherwise be missed. Permanent Black Belts would almost certainly be more heavily influenced by their full-time boss than temporary Black Belts, thus leading to a more provincial focus.

Green Belts

Green Belts are Six Sigma project leaders capable of forming and facilitating Six Sigma teams and managing Six Sigma projects from concept to completion. Green Belt training consists of five days of classroom training and is conducted in conjunction with Six Sigma projects. (In some cases a 10 day course is offered to increase the time allotted for software training and exercises). Training covers project management, quality management tools, quality control tools, problem solving, and descriptive data analysis. Six Sigma champions should attend Green Belt training. Usually, Six Sigma Black Belts help Green Belts define their projects prior to the training, attend training with their Green Belts, and assist them with their projects after the training.

Green Belts are change agents who work part time on process improvement. The bulk of the Green Belt's time is spent performing their normal work duties. Although most experts advocate that the Green Belt spend 10 to 20% of their time on projects, in most cases it is only 2 to 5%. A Green Belt will usually complete one or two major projects per year, usually as a team member rather than a team leader. Since a Green Belt is not trained in all the tools needed in the DMAIC cycle, when they lead projects they must be actively supported by a Black Belt. Few Green Belt projects cover enterprise-wide processes. However, since there are usually more Green Belts than Black Belts (by a factor of 2 to 5), Green Belt projects can have a tremendous impact on the enterprise. Figure 1.10 provides an overview of a process for the selection of Green Belt candidates.

Master Black Belts

This is the highest level of technical and organizational proficiency. Master Black Belts provide technical leadership of the Six Sigma program. They must be thoroughly familiar with the Black Belts Body of Knowledge, as well as additional skills including the mathematical theory that forms the basis of the statistical methods, project management, coaching, teaching, and program organization at the enterprise level. Master Black Belts must be able to assist Black Belts in applying the methods correctly in unusual situations. Whenever possible, statistical training should be conducted only by

Minimum Criteria

Education—High school or equivalent.

Work Experience—At least 3 years of business, technical, or managerial experience.

Technical Capability—High school algebra proficiency as demonstrated by a passing grade in an algebra course.

Computer Proficiency—Word processing, presentation and spreadsheet software.

Team Skills—Willingness to lead meetings, facilitate small groups and successfully resolve conflicts. Ability to mentor and motivate people.

Final Candidate Selection

Based on the organizational need for Green Belts, as determined by the Director of Six Sigma and the Executive Six Sigma Council, Green Belt training allotments areprovided to Master Black Belts, Black Belts and/or General Managers. Green Beltcandidacy requires the consent of the candidate's management.

Commitment

Each Green Belt candidate selected will be required to complete a 40 hour Green Belt training course, and to lead at least one successful Six Sigma project every 12 months, or participate on at least two successful Six Sigma projects every 12 months. Green Belt certification is accomplished as described in the Appendix 16.

FIGURE 1.10 Green Belt candidate selection process and criteria.

qualified Master Black Belts or equivalently skilled consultants. If it becomes necessary for Black Belts and Green Belts to provide training, they should only do so under the guidance of Master Black Belts. Otherwise the familiar "propagation of error" phenomenon will occur; that is., Black Belt trainers pass on errors to Black Belt trainees who pass them on to Green Belts, who pass on greater errors to team members. Because of the nature of the Master's duties, all Master Black Belts must possess excellent communication and teaching skills.

Master Black Belts are recruited from the ranks of Black Belts. The process is usually less formal and less well defined than that for Black Belts or Green Belts and there is a great deal of variability between companies. Master Black Belt candidates usually make their interest known to Six Sigma leadership. Leadership selects candidates based on the needs of the enterprise and Six Sigma's role in meeting those needs. For example, in the early stages of deployment Master Black Belt candidates with excellent organizational skills and the ability to communicate the leadership's Six Sigma vision may be preferred. Intermediate deployments might favor candidates who excel at project selection and Black Belt coaching. Mature Six Sigma programs might look for Master Black Belts with training ability and advanced statistical know-how. Master Black Belts often have advanced technical degrees and extensive Black Belt experience. Many organizations provide Master Black Belts with additional training. Certification requirements for Master Black Belts vary with the organization. Many organizations do not certify Master Black Belts.

Change Agent Compensation and Retention

Experienced Black Belts and Master Black Belts are in great demand throughout the manufacturing and services sectors.* Given their proven talent for effecting meaningful

*Although Green Belts are also highly trained change agents, they are not full-time change agents and we will not discuss their compensation here.

change in a complex environment, this is no surprise. Since organizations exist in a competitive world, steps must be taken to protect the investment in these skilled change agents, or they will be lured away by other organizations, perhaps even competitors. The most common (and effective) actions involve compensation and other financial incentives, such as:

- Bonuses
- Stock options
- Results sharing
- Payment of dues to professional societies
- Pay increases

There are also numerous nonfinancial and quasi-financial rewards. For example, Black Belts reentering the workforce after their tour of duty often enter positions that pay significantly higher than the ones they left when becoming Black Belts. In fact, in some companies the Black Belt position is viewed as a step on the fast track to upper management positions. Also, change is "news" and it is only natural that the names of Master Black Belts and Black Belts involved in major change initiatives receive considerable publicity on company Web sites as well as in newsletters, recognition events, project fairs, etc. Even if they don't receive formal recognition, Six Sigma projects often generate a great deal of internal excitement and discussion. The successful Black Belt usually finds that his work has earned him a reputation that makes him a hot commodity when it's time to end his Black Belt career.

There are, of course, innumerable complexities and details to be decided and worked out. Usually these issues are worked out by a team of individuals with members from Human Resources, the Six Sigma Core Team, and other areas of the organization. The team will address such issues as:

- What pay grade is to be assigned to the Black Belt and Master Black Belt positions?
- Should the pay grade be determined by the pay grade of the candidate's job prior to becoming a Black Belt?
- Should the Black Belt pay grade be guaranteed when the Black Belt leaves the Black Belt position to return to the organization?
- How do we determine eligibility for the various rewards? For example, are there key events such as acceptance as a Black Belt candidate, completion of training, completion of first project, successful certification, and so forth?
- What about Black Belts who were certified by other organizations or third parties?
- Do we provide benefits to Green Belts as well? If so, what and how?
- Who will administer the benefits package?

The plan will be of great interest to Black Belt candidates. If not done properly, the organization will find it difficult to recruit the best people.

Integrating Six Sigma and Related Initiatives

At any given time most companies have numerous activities underway to improve their operations. For example, the company may have functional areas devoted to Lean

Implementation, Continuous Improvement, or Business Process Reengineering, to name just a few.

Leadership must give careful thought as to how the various overlapping activities can best be organized to optimize their impact on performance and minimize confusion over jurisdiction, resources and authority. An "umbrella concept" often provides the needed guidance to successfully integrate the different but related efforts. One concept that is particularly useful is that of "Process Excellence" (PE).

Organizations are typically designed along functional lines, with functions such as engineering, marketing, accounting, and manufacturing assigned responsibility for specific tasks often corresponding closely to university degree programs. Persons with higher education in a specific discipline specialize in the work assigned to that function. Resources are allocated to each function based on the needs of the enterprise.

If the enterprise is to be successful, the "needs of the enterprise" must be based on the needs of its customers. However, customers obtain value from products or services created by the cooperative efforts and resources of many different functional areas. Most customers couldn't care less about how the enterprise creates the values they are purchasing.* A similar discussion applies to owners and shareholders. There is a substantial body of opinion among management experts that focusing internally on functional concerns can be detrimental to the enterprise as a whole. An alternative is to focus on the process or value stream that creates and delivers value.

A process focus means that stakeholder values are determined and activities are classified as either relating to the creation of the final value (value-added activity) or not (non-value-added activity). Processes are evaluated on how effectively and efficiently they create value. *Effectiveness* is defined as delivering what the customer requires, or exceeding the requirements; it encompasses quality, price, delivery, timeliness and everything else that goes into perceived value. *Efficiency* is defined as being effective using a minimum of resources; more of an owner's perspective. Excellent processes are those that are both effective and efficient.

Table 1.3 illustrates the contrast between the way that staff functions used to operate under the traditional system of management, and the way they can operate more effectively.

PE is the set of activities specifically designed to create excellent processes. PE is change-oriented and cross-functional. It includes Six Sigma, all of the initiatives listed earlier, and many more as well. By creating a top-level position for PE, leadership assigns clear responsibility for this important work. The PE leader, usually a Vice President, leads a Process Excellence Leadership Team (PELT) which includes functional leaders as well as full-time PE personnel such as the Director of Six Sigma. The VP of PE isn't responsible for particular processes, but she has the authority to identify key processes and nominate owners for approval by the CEO or the PELT. Examples of processes include:

- Order fulfillment
- Coordinating improvement activities of Six Sigma, Lean, and so forth.
- Customer contact with the company
- Handling public relations emergencies

---

*There are exceptions to this. Many large customers, such as the Department of Defense or automobile or aircraft manufacturers, take a very active interest in the internal operations of their key suppliers.

	From	To
Role	Customer—for information, evidence, and reports from others	Supplier—of information, expertise, and other services
Strategy	Control—by imposition of policies and procedures, and by audit and inspection	Support—by gearing efforts to the needs of others Self-control by client
Goal	Departmental—achievement of departmental objectives	Collective achievement of the organization's objectives
Style of working with others	Competitive, adversarial	Integrating, collaborative
Focus of attention	Some aspects of outcomes; for example, product quality, financial results Some pieces of the process; for example, adherence to policy and procedure	The relationship between the entire underlying process and the achievement of all the desired outcomes
Image	Regulator, inspector, policeman	Educator, helper, guide

Hutton (1994). P. 220. Reprinted with permission.

TABLE 1.3 How Staff Functions Are Changing

- Getting ideas for improvement projects
- Matching improvement projects with customer needs
- Innovating
- Communicating with the outside world
- Communicating internally
- Identifying talent
- Handling customer problems
- Avoiding legal disputes

In other words, the VP of PE has a "meta-process" responsibility. She is responsible for the process of identifying and improving processes. PE activities such as Six Sigma, Lean, etc. provide PE with resources to direct toward the organization's goal of developing internal processes that give it a competitive advantage in securing the best employees, delivering superior customer value, and earning a premium return for its investors.

Deployment to the Supply Chain

In the early part of the twentieth century Henry Ford pursued a great vision by building the Ford River Rouge Complex. By 1927 the Rouge was handling all production of Ford automobiles. It was truly a marvel. The Rouge was the largest single manufacturing complex in the United States, with peak employment of about 120,000. Here Henry Ford achieved self-sufficiency and vertical integration in automobile production, a continuous work flow from iron ore and other raw materials to finished automobiles. The complex included dock facilities, blast furnaces, open-hearth steel mills, foundries, a rolling mill, metal stamping facilities, an engine plant, a glass manufacturing building, a tire plant, and its own power house supplying steam and electricity.

On June 2, 1978, the Rouge was listed a National Historic Landmark: From state-of-the-art wonder to historical curiosity in just 50 years.

A related historical artifact is the idea that a firm can produce quality products or services by themselves. This may have been the case in the heyday of the Rouge, when the entire "supply chain" was a single, vertically integrated behemoth entity, but it is certainly no longer true. In today's world fully 50 to 80% of the cost of a manufactured product is in purchased parts and materials. When the customer forks over her good money for your product, she doesn't differentiate between you and your suppliers.

You say you're not in manufacturing? The situation is likely the same, regardless of industry. Consider personal finance software. Your customer runs your software on a computer you didn't design with an operating system you have no control over. They're using your software to access their account at their financial institution to complete a tax return, which they will file electronically with the Internal Revenue Service (IRS). When your customers click the icon to run your product, they consider all of these intermediaries to be part of the value they are paying to receive.

The service industry is no different. Consider a discount brokerage company, whose customers want to use your service to buy common stocks, fixed income instruments, derivatives, etc. They also want debit cards, check writing, bill paying, pension plans, and a variety of other services, including financial advice, investment portfolio analysis, and annuities. When your customers put their money into their account at your firm, they expect you to be responsible for making all of the "third parties" work together seamlessly.

In short, you'll never reach Six Sigma quality levels with three sigma suppliers.

A primary objective with regard to suppliers is to obtain Six Sigma levels of supplier quality with minimal costs through projects that involve suppliers. The organization responsible for supply chain management (SCM) will take the lead in developing the supplier Six Sigma program, including preparation of a Supplier Six Sigma Deployment Plan with the following attributes:

- Policies on supplier Six Sigma
- Goals and deliverables of the supplier Six Sigma program
- Supplier communication plan
- Timetable for deployment, including phases (e.g., accelerated deployment to most critical suppliers)
- Procedures defining supplier contact protocols, supplier project charter, supplier project reporting and tracking, etc.
- Training requirements and timeline
- Methods of assessing supplier Six Sigma effectiveness
- Integration of the supplier Six Sigma program and in-house activities

SCM receives guidance from the Executive Six Sigma Council and the Six Sigma organization. The Six Sigma organization often provides expertise and other resources to the supplier Six Sigma effort.

SCM should sponsor or cosponsor supplier Six Sigma projects. In some cases SCM will lead the projects, often with supplier personnel taking a coleadership role. In others they will assist Black Belts or Green Belts working on other projects that involve suppliers.

Full SCM sponsorship is usually required when the project's primary focus is on the supplier's product or process, such as to reduce the number of late deliveries of a key product. Projects involving suppliers, but not focused on them, can be co-sponsored by SCM, such as a project involving the redesign of an order fulfillment process requiring only minor changes to the supplier's web ordering form. SCM assistance can take a number of different forms, for example

- Acting as a liaison between the internal team members and suppliers
- Negotiating funding and budget authority for supplier Six Sigma projects
- Estimating and reporting supplier project savings
- Renegotiating contract terms
- Resolving conflicts
- Defining responsibility for action items
- Scheduling supplier visits
- Defining procedures for handling of proprietary supplier information
- Responding to supplier requests for assistance with Six Sigma

In addition to SCM, other elements within your organization play important supporting roles. Usually Black Belts will come from the Six Sigma organization, although some larger enterprises assign a team of Black Belts to work on SCM projects full time. Green Belts often come from organizations sponsoring supplier-related projects. Team members are assigned from various areas, as with any Six Sigma project.

The customer certainly has the final say in the process and project requirements, but ultimate responsibility for the process itself should remain with the supplier, who owns and controls their processes, and may have liability and warranty obligations. Six Sigma teams must be clear that only SCM has the authority to make official requests for change. It can be embarrassing if a Black Belt makes a suggestion that the supplier believes to be a formal requirement to change. SCM may receive a new bid, price change, complaint letter, etc. from the supplier over such misunderstandings. Supplier relationships are often quite fragile, "Handle with care" is a good motto for the entire Six Sigma team to follow.

In addition to accepting responsibility for their processes, suppliers must often take the lead role in Six Sigma teams operating in supplier facilities. Supplier leadership must support Six Sigma efforts within their organizations. Suppliers must agree to commit the resources necessary to successfully complete projects, including personnel and funding.

Communications and Awareness

Top-level Six Sigma projects using the DMAIC methodology can be defined to build buy-in for the change initiative and build awareness through communication, as follows: (Keller, 2005).

Define

DEFINE the scope and objectives for the Six Sigma change initiative, which is usually an enterprise undertaking.

Define the key stakeholder groups that will be impacted by the change. The key stakeholders are those groups whose involvement is key to the success of the change initiative, which can include:

- Key customers
- Shareholders or other owners
- Senior leadership
- Middle management
- Six Sigma change agents
- The general employee population
- Suppliers

Define one or more metrics that can be used to track the current organizational culture on quality, which is discussed in the Measure description in the following section.

Measure

Measure the baseline level of buy-in for the change initiative among these key stakeholder groups, as well as the baseline quality culture.

Buy-in can be measured according to the following scale (Forum, 1996), from lowest to highest: Hostility, Dissent, Acceptance, Support, Buy-In. Note that the desired level of buy-in surpasses mere support; enthusiasm is required for complete buy-in. Surveys and focus groups, further discussed in Chap. 2, are often used to measure buy-in as well as perceptions on quality.

Juran and Gryna (1993) define the company quality culture as the opinions, beliefs, traditions, and practices concerning quality. While sometimes difficult to quantify, an organization's culture has a profound effect on the quality produced by that organization. Without an understanding of the cultural aspects of quality, significant and lasting improvements in quality levels are unlikely.

Two of the most common means of assessing organization culture are the focus group and the written questionnaire. These two techniques are discussed in greater detail below. The areas addressed generally cover attitudes, perceptions, and activities within the organization that impact quality. Because of the sensitive nature of cultural assessment, anonymity is usually necessary. The author believes that it is necessary for each organization to develop its own set of questions. The process of getting the questions is an education in itself. One method for getting the right questions that has produced favorable results in the past is known as the critical-incident technique. This involves selecting a small representative sample ($n \approx 20$) from the group you wish to survey and asking open-ended questions, such as:

> "Which of our organization's beliefs, traditions and practices have a beneficial impact on quality?"
>
> "Which of our organization's beliefs, traditions and practices have a detrimental impact on quality?"

The questions are asked by interviewers who are unbiased and the respondents are guaranteed anonymity. Although usually conducted in person or by phone, written responses are sometimes obtained. The order in which the questions are asked (beneficial/detrimental) is randomized to avoid bias in the answer. Interviewers are instructed not to prompt the respondent in any way. It is important that the responses be recorded verbatim, using the respondent's own words. Participants are urged to provide as many responses as they can; a group of 20 participants will typically produce 80 to 100 responses.

The responses themselves are of great interest and always provide a great deal of information. In addition, the responses can be grouped into categories and the categories examined to glean additional insight into the dimensions of the organization's quality culture. The responses and categories can be used to develop valid survey items and to prepare focus-group questions. The follow-up activity is why so few people are needed at this stage—statistical validity is obtained during the survey stage.

Analyze

Analyze the primary causes of buy-in resistance, which can include issues and resolutions such as (Forum, 1996; resolutions by Keller, 2005):

- Unclear goals—Goals need to be clearly communicated throughout the stakeholder groups.
- No personal benefit—Goals should be stated in terms that provide a clear link to personal benefits for stakeholders, such as decreased hassles or improved working conditions.
- Predetermined solutions—When teams are given the solution without chance for analysis of alternatives, they will likely be skeptical of the result. The root cause of this practice is often management resistance to free thinking or experimentation by process personnel or a lack of customer focus, as further described in Chap. 2.
- Lack of communication—Analyses and results should be communicated throughout the stakeholder groups.
- Too many priorities—Teams need to be focused on achievable results.
- Short-term focus—Goals should provide clear benefits over short and longer terms.
- No accountability—Clearly defined Project Sponsors, stakeholders and team members provide accountability.
- Disagreement on the definition of customer—Clearly defined stakeholder groups are needed for project success. This can also be associated with so-called turf wars between various functional areas within an organization.
- Low probability of implementation—Formal project sponsorship and approvals provide a clear implementation channel.
- Insufficient resources—Stakeholder groups need to understand that the project is sufficiently funded and resources allocated—Training project teams is essential.
- Midstream change in direction or scope—Changes in project scope or direction provide a potential for a loss of buy-in. Changes must be properly communicated to stakeholder groups to prevent this reduction in buy-in.

Improve

Improve buy-in by addressing the causes of resistance, such as suggested by the resolutions noted above. Communication is the primary method of building buy-in, and can be effectively improved by developing and managing a Six Sigma communication plan. Successful implementation of Six Sigma will only happen if the leadership's vision and implementation plans are clearly understood and embraced by employees, shareholders, customers, and suppliers. Because it involves cultural change, Six Sigma frightens many people, and good communication is an antidote to fear: without it rumors run rampant and morale suffers. The commitment to Six Sigma must be clearly and unambiguously understood throughout the organization. This doesn't happen by accident; it is the result of careful planning and execution.

Communicating the Six Sigma message is a multimedia undertaking. The modern organization has numerous communications technologies at its disposal. Keep in mind that communication is a two-way affair; be sure to provide numerous opportunities for upward and lateral as well as downward communication. Here are some suggestions to accomplish the communications mission:

- All-hands launch event, with suitable pomp and circumstance
- Mandatory staff meeting agenda item
- Internal newsletters, magazines, and Web sites, with high profile links to enterprise Six Sigma Web site on home page
- Six Sigma updates in annual report
- Stock analyst updates on publicly announced Six Sigma goals
- Intranet discussion forums
- Two-way mail communications
- Surveys
- Suggestion boxes
- Videotape or DVD presentations
- Closed circuit satellite broadcasts by executives, with questions and answers
- All-hands discussion forums
- Posters
- Logo shirts, gear bags, keychains, coffee mugs, and other accessories
- Speeches and presentations
- Memoranda
- Recognition events
- Lobby displays
- Letters

Promoting Six Sigma awareness is, in fact, an internal marketing campaign. A marketing expert, perhaps from your company's marketing organization, should be consulted. If your organization is small, a good book on marketing can provide guidance [e.g., Levinson et al. (1995)].

For each stakeholder group, the key concerns include:

1. Who is primarily responsible for communication with this group?
2. What are the communication needs for this group? For example, key customers may need to know how Six Sigma will benefit them; employees may need to understand the process for applying for a change agent position such as Black Belt.
3. What communication tools, techniques and methods will be used? These include meetings, newsletters, email, one-on-one communications, and Web sites.
4. What will be the frequency of communication? Remember, repetition will usually be necessary to be certain the message is received and understood.
5. Who is accountable for meeting the communication requirement?
6. How will we measure the degree of success? Who will do this?

The requirements and responsibilities can be organized using tables, such as Table 1.4.

Group		Method	Frequency	Accountability
Senior Leadership				
Requirement	Program strategy, goals and high-level program plan	• Senior staff meetings • Senior leadership training	• At least monthly • Start of program	• CEO • Six Sigma Director • Training department
	Metrics/status performance to program plan	• Senior staff meetings	• At least monthly	• Six Sigma Director
Middle Management				
Program strategy, goals and management-level program plan		• Regular flow down of upper level staff meeting notes/flow down; newsletter • Management training	• At least monthly for staff meetings; newsletter piece every 2 weeks during program rollout, as needed thereafter • Prior to 1st wave of Six Sigma projects	• Senior Leadership for staff meeting flow down • Internal communications via core team for company newsletter • Training department

Etc. for customers, owners, stock analysts, change agents, bargaining unit, exempt employees, suppliers, or other stakeholder group.

Table 1.4 Six Sigma Communications Plan and Requirements Matrix

When buy-in is reduced because of a perceived lack of management support, action is necessary to increase the leadership involvement. Senior managers' time is in great demand from a large number of people inside and outside of the organization. It is all too easy to schedule a weekly meeting to discuss "Six Sigma" for an hour, and then think you've done your part. In fact, transforming an organization, large or small, requires a prodigious commitment of the time of senior leadership, not just a few hours a month for update meetings.

One way to maximize the value of an executive's time investment is the use of symbolic events, or stories, that capture the essence of management's commitment (or lack of it) to the change effort. Stakeholders repeat and retain stories far better than proclamations and statements. For example, there's a story told by employees of a large U.S. automotive firm that goes as follows:

In the early 1980s the company was just starting their quality improvement effort. At a meeting between upper management and a famous quality consultant, someone casually mentioned that quality levels were seasonal—quality was worse in the summer months. The consultant asked why this should be so. Were different designs used? Were the machines different? How about the suppliers of raw materials? The answer to each of these questions was "No." An investigation revealed that the problem was vacations. When one worker went on vacation, someone else did her job, but not quite as well. And that "someone" also vacated a job, which was done by a replacement, etc. It turned out that the one person going on vacation led to six people doing jobs they did not do routinely. The solution was to have a vacation shutdown of two weeks. This greatly reduced the number of people on new jobs and brought summer quality levels up to the quality levels experienced the rest of the year. This worked fine for a couple of years, given an auto industry recession and excess capacity. One summer, however, the senior executives were asked by the finance department to reconsider their shutdown policy. Demand had picked up and the company could sell every car it could produce. The accountants pointed out that the shutdown would cost $100 million per day in lost sales.

When the vice president of the truck division asked if anything had been done to address the cause of the quality slippage in the summer, the answer was "No, nothing had been done." The president asked the staff "If we go back to the old policy, would quality levels fall like they did before?" Yes, he was told, they would. "Then we stay with our current policy and shut down the plants for vacations," the president announced.

The president was challenged by the vice president of finance. "I know we're committed to quality, but are you sure you want to lose $1.4 billion in sales just to demonstrate our commitment?" The president replied, "Frank, I'm not doing this to 'demonstrate' anything. We almost lost our company a few years back because our quality levels didn't match our overseas competition. Looking at this as a $1.4 billion loss is just the kind of short-term thinking that got us in trouble back then. I'm making this decision to save money."

This story had tremendous impact on the managers who heard it, and it spread like wildfire throughout the organization. It demonstrated many things simultaneously: senior leadership's commitment to quality, political parity between operations and finance, how seemingly harmless policies can have devastating effects, an illustration of how short-term thinking had damaged the company in the past, and how long-term thinking worked in a specific instance, etc. It is a story worth 100 speeches and mission statements.

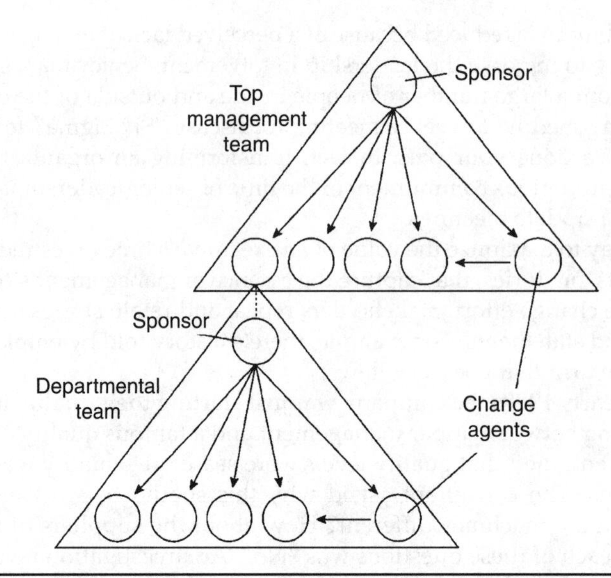

Figure 1.11 Cascading of sponsorship. From Hutton (1994). Copyright © 1994 by David W. Hutton.

Control

Control the change effort with a plan to maintain buy-in. Personnel trained as change agents can be placed in strategic positions throughout the organization, as shown in Fig. 1.11. This makes it possible for them to assist in the development and implementation (including sponsorship) of future quality improvement projects. Quality improvement of any significance nearly always involves multiple departments and levels in the organization. Change agents help organize an assessment of the organization to identify its strengths and weaknesses. Change is usually undertaken to either reduce areas of weakness, or exploit areas of strength. The assessment is part of the education process. Knowing one's specific strengths and weaknesses is useful in mapping the process for change.

CHAPTER 2

Recognizing Opportunity

Chapter 1 focused on the need to become a responsive, change-capable organization, and proposed a basic implementation plan to meet that need. The infrastructure described in Chap. 1 allows the organization to define and deploy Six Sigma projects focused on specific needs of key stakeholder groups, including customers, shareholders, and employees. In this way, the Six Sigma projects serve as the means of achieving these broad organizational objectives. Chapter 2 describes the organizational mechanisms required to recognize the key stakeholder needs, which are the opportunities for organizational growth and in some case, survival.

A core principle of Six Sigma is the recognition of customer value, a term derived from Lean practices. *Value* (as discussed in Chap. 10) is something the customer is willing to pay for; the remainder of our activities is *waste*. While this chapter describes a number of approaches for gathering customer input on their value definition, the optimal approach incorporates customer feedback regarding value into the organization's daily operations. In this way, the organization collects and disseminates information on value-definition as part of its intrinsic function. Each and every customer contact provides that opportunity, and at the heart of the responsive organization is the inherent motivation to capture that data as the first step in satisfying customer needs. It becomes the lifeblood of the organization, whose products and services exist only to satisfy the customers' need for value.

A practice recently popularized in sales departments is *Solution Selling*, the title of a best-selling book (Eades, 2004). In *Solution Selling*, Eades discusses the need to identify the customer's pain: their critical business issue, challenge or, source of dissatisfaction. In *Solution Selling*, pain that is identified and admitted by a prospective buyer provides the first level of qualification for a sale. The seller's product or service is then, where appropriate, positioned as the solution to that pain. The relevance of this technique is its simplicity: engaging in conversation with informed stakeholders is often sufficient to define the pain that exists in their system. The solutions to alleviate that pain provide value that the customer is willing to pay for.

An organization in the software sector has successfully deployed a similar model implemented (and improved) over the course of the last 15 years. Their sales and technical support areas have access to a common database that maintains every suggestion or complaint voiced by a customer. This Quality Management System (QMS) database is integrated with statistical process control analysis tools to provide trend analysis of issues, resolution times, etc. Most notable is their response to these Improvement Opportunities (IO, as they call them). As the IO is entered into the system, it is circulated throughout the technical support, sales and development areas, up to the vice president level. This provides immediate VISIBILITY (another lean concept) to all internal stakeholders. High priority issues, such as errors that significantly impact customers or the potential for a significant sale based on a critical customer need, are immediately investigated and scheduled for deployment. Of course, this requires a number of back and forth discussions with the customer to fully understand their needs, sometimes at a fairly high level in the organization to ensure the proper information is received. While this takes resource and time, it serves the organization's long-term vision to build relationships with their customer base. Their relatively flat organizational structure is designed to provide customer access to nearly all aspects of the organization, including subject matter experts. Their ability to rapidly integrate these new ideas into their products is heavily influenced by deployment of lean concepts for rapid turnaround of new products. They have effectively removed non-value added bureaucracies, while focusing their limited resources on creating marketable value. Their business continues to grow based on new customers as well as the continued upgrades and growth opportunities from long term customers.

The practices described by example previously are further discussed in principle in the following section.

Becoming a Customer and Market-Driven Enterprise

The proper place of the customer in the organization's hierarchy is illustrated in Fig. 2.1. Note that this perspective is precisely the opposite of the traditional view of the organization. The difficulties involved in making such a radical change should not be underestimated.

Edosomwan (1993) defines a customer-and market-driven enterprise as one that is committed to providing excellent quality and competitive products and services to satisfy the needs and wants of a well-defined market segment. This approach is in contrast to that of the traditional organization, as shown in Table 2.1.

The journey from a traditional to a customer-driven organization has been made by enough organizations to allow us to identify a number of distinct milestones that mark the path to success. Generally, the journey begins with recognition that a crisis is either upon the organization, or imminent. This wrenches the organization's leadership out of denial and forces them to abandon the status quo.

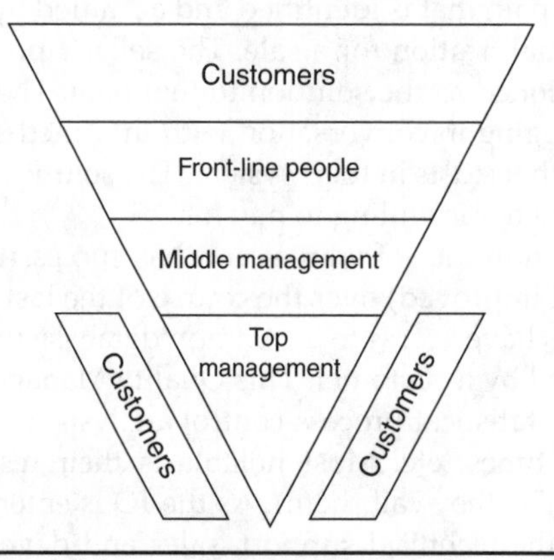

FIGURE 2.1. The "correct" view of the company organization chart. (*From Kotler (1991). P 21. Copyright © 1991 by Prentice-Hall, Inc. Reprinted by permission.*)

	Traditional Organizations	Customer-Driven Organizations
Product and service planning	– Short-term focus – Reactionary management – Management by objectives planning process	– Long-term focus – Prevention-based management – Customer-driven strategic planning process
Measures of performance	– Bottom-line financial results – Quick return on investment	– Customer satisfaction – Market share – Long-term profitability – Quality orientation – Total productivity
Attitudes toward customers	– Customers are irrational and a pain – Customers are a bottleneck to profitability – Hostile and careless – "Take it or leave it" attitude	– Voice of the customer is important – Professional treatment and attention to customers is required – Courteous and responsive – Empathy and respectful attitude
Quality of products and services	– Provided according to organizational requirements	– Provided according to customer requirements and needs
Marketing focus	– Seller's market – Careless about lost customers due to poor customer satisfaction	– Increased market share and financial growth achieved
Process management approach	– Focus on error and defect detection	– Focus on error and defect prevention
Product and service delivery attitude	– It is OK for customers to wait for products and services	– It is best to provide fast time to market products and services
People orientation	– People are the source of problems and are burdens on the organization	– People are an organization's greatest resource
Basis for decision-making	– Product-driven – Management by opinion	– Customer-driven – Management by data
Improvement strategy	– Crisis management – Management by fear and intimidation	– Continuous process improvement – Total process management
Mode of operation	– Career-driven and independent work – Customers, suppliers, and process owners have nothing in common	– Management-supported improvement – Teamwork between suppliers, process owners, and customers practiced

TABLE 2.1 Traditional Organizations versus Customer-Driven Organizations (*From Johnson. Copyright © 1993 by ASQ. Reprinted by permission.*)

When the familiar ways of the past are no longer acceptable, the result is a feeling of confusion among the leaders. At this stage the leadership must answer some very basic questions:

- What is the organization's purpose?
- What are our values?
- What does an organization with these values look like?

A "value" is that which one acts to gain and/or keep. It presupposes an entity capable of acting to achieve a goal in the face of an alternative. Values are not simply nice-sounding platitudes, they represent *goals*. Pursuing the organization's values implies building an organization which embodies these values. This is the leadership's *vision*, to create a reality where their values have been achieved

After the vision has been clearly developed, the next step is to develop a strategy for building the new organization (see Chap. 1).

Elements of the Transformed Organization

Customer-driven organizations share certain common features:

- **Flattened hierarchies**—Getting everyone closer to the customer involves reducing the number of bureaucratic "layers" in the organization structure. It also involves the "upside-down" perspective of the organization structure shown in Fig. 2.1. The customer comes first, not the boss. Everyone serves the customer.
- **Risk-taking**—Customers' demands tend to be unpredictable. Responsiveness requires that organizations be willing to change quickly, which involves uncertainty and risk Customer-driven organizations encourage risk-taking in a variety of ways. One important aspect is to celebrate mistakes made by individuals who engage in risky behavior. Bureaucratic impediments such as excessive dependence on written procedures are minimized or eliminated. Employees are encouraged to act on their own best judgments and not to rely on formal approval mechanisms.
- **Effective communication**—During transformation the primary task of the leadership team is the clear, consistent, and unambiguous transmission of their vision to others in the organization. One way this is done is through "internal marketing" which involves using the principles of marketing to get the message to the target "market": the employees. It is vital that the leaders' actions are completely consistent with their words. The assistance of outside consultants may be helpful in identifying inconsistencies.
    - Leaders should realize that their behavior carries tremendous symbolic meaning. This can contribute to the failure of convincing employees; a single action which is inconsistent with the stated message is sufficient to destroy all credibility. On the plus side, an action that clearly shows a commitment to the vision can help spread the word that "They're serious this time." The leadership should seek out stories that capture the essence of the new organization and repeat these stories often. For example, Nordstrom employees

all hear the story of the sales clerk who allowed the customer to return a tire (Nordstrom's doesn't sell tires). This story captures the essence of the Nordstrom "rule book" which states:

- Rule 1—Use your own best judgment.
- Rule 2—There are no other rules.

Leaders should expect to devote a minimum of 50% of their time to communication during the transition.

- **Supportive boards of directors**—It is vital to obtain the enthusiastic endorsement of the new strategy by the board. Management cannot focus their attention until this support has been received. This will require that management educate their board and ask them for their approval. However, boards are responsible for governance, not management. Don't ask the board to approve tactics. This bogs down the board, stifles creativity in the ranks, and slows the organization down.
- **Partnered trade unions**—In the transformed organization, everyone's job changes. If the organization's employees are unionized, changing jobs requires that the union become management's partner in the transformation process. In the flat organization union employees will have greater authority. Union representatives should be involved in all phases of the transformation, including planning and strategy development. By getting union input, the organization can be ensured that during collective bargaining the union won't undermine the company's ability to compete or sabotage the strategic plan. Unions also play a role in auditing the company's activities to ensure that they comply with contracts and labor laws.
- **Measured results**—It is important that the right things be measured. The "right things" are measurements that determine that you are delivering on your promises to customers, investors, employees, and other stakeholders. You must also measure for the right reasons. This means that measurements are used to learn about how to improve, not for judgment. Finally, you must measure the right way. Measurements should cover processes as well as outcomes. Data must be available quickly to the people who use them. Measurements must be easy to understand. These topics are discussed in more detail in Chap. 3.
- **Rewarded employees**—Care must be taken to avoid punishing with rewards. Rewarding individuals with financial incentives for simply doing their jobs well implies that the employee wouldn't do the job without the reward. It is inherently manipulative. The result is to destroy the very behavior you seek to encourage (Kohn, 1993). The message is that rewards should not be used as control mechanisms. Employees should be treated like adults and provided with adequate and fair compensation for doing their jobs. Recognizing exceptional performance or effort should be done in a way that encourages cooperation and team spirit, such as parties and public expressions of appreciation. Leaders should ensure fairness, for example, management bonuses and worker pay cuts don't mix.

Strategies for Communicating with Customers and Employees

There are several primary strategies commonly used to obtain information from or about customers and employees:

- Operational feedback systems
- Sample surveys
- Focus groups
- Field experiments

Operational Feedback Systems

As with the software company described at the beginning of this chapter, many organizations have complaint and suggestion systems which provide customers an easy-to-use avenue for both favorable and unfavorable feedback to management. Due to selection bias, these methods do not provide statistically valid information. However, because they are a census rather than a sample, they provide opportunities for individual customers to have their say. These are moments of truth that can be used to increase customer loyalty. They also provide anecdotes that have high face validity and are often a source of ideas for improvement.

When a customer complaint has been received it represents an opportunity to increase customer loyalty, as well as a risk of losing the customer. The importance of complaint handling is illustrated in Fig. 2.2. These data illustrate that the decision to repurchase is highly dependent on whether their complaint was handled to their satisfaction. Considering that customers who complain are likely to tell as many as 14 others of their experience, the importance of complaint handling in customer relations becomes obvious.

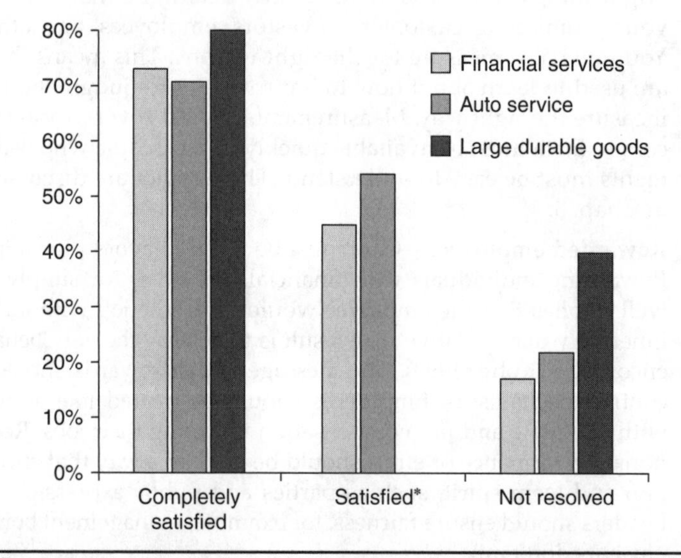

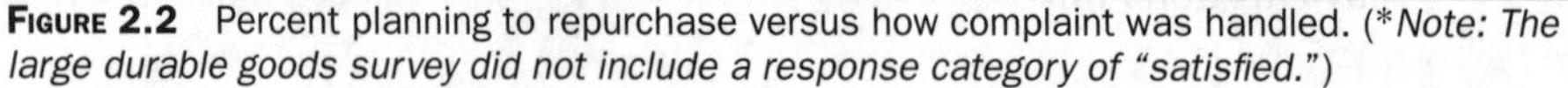

FIGURE 2.2 Percent planning to repurchase versus how complaint was handled. (**Note: The large durable goods survey did not include a response category of "satisfied."*)

Despite the impressive nature of Fig. 2.2, even these figures dramatically understate the true extent of the problem. Complaints represent people who were not only unhappy, they were unhappy enough to report that dissatisfaction to the company. Research indicates that up to 96% of unhappy customers never tell the company. This is especially unfortunate since it has been shown that customer loyalty is *increased* by proper resolution of complaints. Given the dramatic impact of a lost customer, it makes sense to maximize the opportunity of the customer to complain. Complaints should be actively sought, an activity decidedly against human nature. This suggests that a system must be developed and implemented to motivate employees to seek out customer complaints, ideas, and suggestions; to feel their pain. The system should also provide every conceivable way for an unhappy customer to contact the company on their own, including toll-free hotlines, email, comment cards, and so forth.

Once customer feedback has been obtained, it must be used to improve process and product quality. A system for utilizing customer feedback is described as follows:

1. Local managers and employees serve customers' needs on a daily basis, using locally modified procedures along with general corporate policies and procedures.
2. By means of a standardized and locally sensitive questionnaire, determine the needs and attitudes of customers on a regular basis.
3. Comparing financial data, expectations, and past attitude information, determine strengths and weaknesses and their probable causes.
4. Determine where and how effort should be applied to correct weaknesses and preserve strengths. Repeat the process by taking action—step 1—and maintain it to attain a steady state or to evolve in terms of customer changes.
5. A similar process can take place at higher levels, using aggregated data from the field and existing policy flows of the organization.

Surveys

In sample surveys, data are collected from a sample of a universe to estimate the characteristics of the universe, such as their range or dispersion, the frequency of occurrence of events, or the expected values of important universe parameters. The reader should note that these terms are consistent with the definition of enumerative statistical studies as described in Chap. 7. This is the traditional approach to such surveys. However, if survey results are collected at regular intervals, the results can be analyzed using the statistical control charts as described in Chap. 8 to obtain information on the underlying process. The process excellence leader should not be reticent in recommending that survey budgets be allocated for conducting small, routine, periodic surveys rather than infrequent "big studies." Without the information available from time-ordered series of data, it will not be possible to learn about processes which produce changes in customer satisfaction or perceptions of quality, or to verify progress towards improvement.

Survey development consists of the following major tasks (GAO, 1986, p.15):

1. Initial planning of the questionnaire
2. Developing the measures
3. Designing the sample

4. Developing and testing the questionnaire
5. Producing the questionnaire
6. Preparing and distributing mailing materials
7. Collecting data
8. Reducing the data to forms that can be analyzed
9. Analyzing the data

Guidelines for Developing Questions

The axiom that underlies the guidelines shown below is that the question-writer(s) must be thoroughly familiar with the respondent group and must understand the subject matter from the perspective of the respondent group.

This is often problematic for the employee when the respondent group is the customer; methods for dealing with this situation are discussed below. There are eight basic guidelines for writing good questions:

1. Ask questions in a format that is appropriate to the questions' purpose and the information required.
2. Make sure the questions are relevant, proper, and qualified as needed.
3. Write clear, concise questions at the respondent's language level.
4. Give the respondent a chance to answer by providing a comprehensive list of relevant, mutually exclusive responses from which to choose.
5. Ask unbiased questions by using appropriate formats and item constructions and by presenting all important factors in the proper sequence.
6. Get unbiased answers by anticipating and accounting for various respondent tendencies.
7. Quantify the response measures where possible.
8. Provide a logical and unbiased line of inquiry to keep the reader's attention and make the response task easier.

The above guidelines apply to the *form* of the question. Using the critical incident technique to develop good question *content* is described in the following section.

Response Types

There are several commonly used types of survey responses.

- **Open-ended questions**—These are questions that allow the respondents to frame their own response without any restrictions placed on the response. The primary advantage is that such questions are easy to form and ask using natural language, even if the question writer has little knowledge of the subject matter. Unfortunately, there are many problems with analyzing the answers received to this type of question. This type of question is most useful in determining the scope and content of the survey, not in producing results for analysis or process improvement.
- **Fill-in-the-blank questions**—Here the respondent is provided with directions that specify the units in which the respondent is to answer. The instructions

should be explicit and should specify the answer units. This type of question should be reserved for very specific requests, e.g., "What was your age on your last birthday?—(age in years)."

- **Yes/No questions**—Unfortunately, yes/no questions are very popular. Although they have some advantages, they have many problems and few uses. Yes/no questions are ideal for dichotomous variables, such as defective or not defective. However, too often this format is used when the measure spans a range of values and conditions, for example, "Were you satisfied with the quality of your new car (yes/no)?" A yes/no response to such questions contain little useful information.
- **Ranking questions**—The ranking format is used to rank options according to some criterion, for example, importance. Ranking formats are difficult to write and difficult to answer. They give very little real information and are very prone to errors that can invalidate all the responses. They should be avoided whenever possible in favor of more powerful formats and formats less prone to error, such as rating. When used, the number of ranking categories should not exceed five.
- **Rating questions**—With this type of response, a rating is assigned on the basis of the score's absolute position within a range of possible values. Rating scales are easy to write, easy to answer, and provide a level of quantification that is adequate for most purposes. They tend to produce reasonably valid measures. Here is an example of a rating format:

For the following statement, check the appropriate box: The workmanship standards provided by the purchaser are:

☐	Clear
☐	Marginally adequate
☐	Unclear

- **Guttman format**—In the Guttman format, the alternatives increase in comprehensiveness; that is, the higher-valued alternatives include the lower-valued alternatives. For example,

Regarding the benefit received from training in quality improvement:

☐	No benefit identified
☐	Identified benefit
☐	Measured benefit
☐	Assessed benefit value in dollar terms
☐	Performed cost/benefit analysis

- **Likert and other intensity scale formats**—These formats are usually used to measure the strength of an attitude or an opinion. For example,

Please check the appropriate box in response to the following statement: "The customer service representative was knowledgeable."

☐	Strongly disagree
☐	Disagree
☐	Neutral
☐	Agree
☐	Strongly agree

Intensity scales are very easy to construct. They are best used when respondents can agree or disagree with a statement. A problem is that statements must be worded to present a single side of an argument. We know that the respondent agrees, but we must infer what he believes. To compensate for the natural tendency of people to agree, statements are usually presented using the converse as well, for example, "The customer service representative was not knowledgeable."

When using intensity scales, use an odd-numbered scale, preferably with five or seven categories. If there is a possibility of bias, order the scale in a way that favors the hypothesis you want to disprove and handicaps the hypothesis you want to confirm. In this way you will confirm the hypothesis with the bias against you—a stronger result. If there is no bias, put the most undesirable choices first.

- **Semantic differential format**—In this format, the values that span the range of possible choices are not completely identified; only the end points are labeled. For example,

Indicate the number of times you initiated communication with your customer in the past month.

few	☐	☐	☐	☐	☐	☐	☐	many

The respondent must infer that the range is divided into equal intervals. The range seems to work well with seven categories.

Semantic differentials are very useful when we do not have enough information to anchor the intervals between the poles. However, they are very difficult to write well and if not written well the results are ambiguous.

Survey Development Case Study*

This actual case study involves the development of a mail survey at a community hospital. The same process has been successfully used by the author to develop customer surveys for clientele in a variety of industries.

The study of service quality and patient satisfaction was performed at a 213 bed community hospital in the southwestern United States. The hospital is a nonprofit, publicly funded institution providing services to the adult community; pediatric services are not provided. The purpose of the study was to:

1. Identify the determinants of patient quality judgments.
2. Identify internal service delivery processes that impacted patient quality judgments.

3. Determine the linkage between patient quality judgments and intent-to-patronize the hospital in the future or to recommend the hospital to others.

To conduct the study, the author worked closely with a core team of hospital employees, and with several ad hoc teams of hospital employees. The core team included the Nursing Administrator, the head of the Quality Management Department, and the head of Nutrition Services.[†]

The team decided to develop their criteria independently. It was agreed that the best method of getting information was directly from the target group, in-patients. Due to the nature of hospital care services, focus groups were not deemed feasible for this study. Frequently, patients must spend a considerable period of time convalescing after being released from a hospital, making it impossible for them to participate in a focus group soon after discharge. While the patients are in the hospital, they are usually too sick to participate. Some patients have communicable diseases, which makes their participation in focus groups inadvisable.

Since memories of events tend to fade quickly (Flanagan, 1954), the team decided that patients should be interviewed within 72 hours of discharge. The target patient population was, therefore, all adults treated as in-patients and discharged to their homes. The following groups were not part of the study: families of patients who died while in the hospital, patients discharged to nursing homes, patients admitted for psychiatric care.[‡]

The team used the Critical Incident Technique (CIT) to obtain patient comments. The CIT was first used to study procedures for selection and classification of pilot candidates in World War II (Flanagan, 1954). A bibliography assembled in 1980 listed over seven hundred studies about or using the CIT (Fivars, 1980). Given its popularity, it is not surprising that the CIT has also been used to evaluate service quality.

CIT consists of a set of specifically defined procedures for collecting observations of human behavior in such a way as to make them useful in addressing practical problems. Its strength lies in carefully structured data collection and data classification procedures that produce detailed information not available through other research methods. The technique, using either direct observation or recalled information collected via interviews, enables researchers to gather firsthand patient-perspective information. This kind of self-report preserves the richness of detail and the authenticity of personal experience of those closest to the activity being studied. Researchers have concluded that the CIT produces information that is both reliable and valid.

This study attempted to follow closely the five steps described by Flanagan as crucial to the CIT: (1) establishment of the general aim of the activity studied; (2) development of a plan for observers or interviewers; (3) collection of data; (4) analysis (classification) of data; and (5) interpretation of data.

Establishment of the general aim of the activity studied

The general aim is the purpose of the activity. In this case the activity involves the whole range of services provided to in-patients in the hospital. This includes every service activity between admission and discharge[¶] From the service provider's perspective the general aim is to create and manage service delivery processes in such a way as to produce a willingness by the patient to utilize the provider's services in the future. To do this, the service provider must know which particular aspects of the service are remembered by the patient.

Our general aim was to provide the service provider with information on what patients remembered about their hospital stay, both pleasant and unpleasant. This information was to be used to construct a new patient survey instrument that would be sent to recently discharged patients on a periodic basis. The information obtained would be used by the managers of the various service processes as feedback on their performance, from the patient's perspective.

Interview plan

Interviewers were provided with a list of patients discharged within the past 3 days. The discharge list included all patients. Nonpsychiatric patients who were discharged to "home" were candidates for the interview. Home was defined as any location other than the morgue or a nursing home. Interviewers were instructed to read a set of predetermined statements. Patients to be called were selected at random from the discharge list. If a patient could not be reached, the interviewer would try again later in the day. One interview form was prepared per patient. To avoid bias, 50% of the interview forms asked the patient to recall unpleasant incidents first and 50% asked for pleasant incidents first. Interviewers were instructed to record the patient responses using the patient's own words.

Collection of data

Four interviewers participated in the data collection activity; all were management level employees of the hospital. Three of the interviewers were female, one was male. The interviews were conducted when time permitted during the interviewer's normal busy work day. The interviews took place during the September 1993 time period. Interviewers were given the instructions recommended by Hayes (1992) for generating critical incidents.

A total of 36 telephone attempts were made and 23 patients were reached. Of those reached, three spoke only Spanish. In the case of one of the Spanish-speaking patients a family member was interviewed. Thus, 21 interviews were conducted, which is slightly greater than the 10 to 20 interviews recommended by Hayes (1992),. The 21 interviews produced 93 critical incidents.

Classification of data

The Incident Classification System required by CIT is a rigorous, carefully designed procedure with the end goal being to make the data useful to the problem at hand while sacrificing as little detail as possible (Flanagan, 1954). There are three issues in doing so: (1) identification of a general framework of reference that will account for all incidents; (2) inductive development of major area and sub-area categories that will be useful in sorting the incidents; and (3) selection of the most appropriate level of specificity for reporting the data.

The critical incidents were classified as follows:

1. Each critical incident was written on a 3 × 5 card, using the patient's own words.
2. The cards were thoroughly shuffled.
3. Ten percent of the cards (10 cards) were selected at random, removed from the deck and set aside.
4. Two of the four team members left the room while the other two grouped the remaining 83 cards and named the categories.
5. The ten cards originally set aside were placed into the categories found in step 4.
6. Finally, the two members not involved in the initial classification were told the names of the categories. They then took the reshuffled 93 cards and placed them into the previously determined categories.

The above process produced the following dimensions of critical incidents:

- Accommodations (5 critical incidents)
- Quality of physician (14 critical incidents)
- Care provided by staff (20 critical incidents)
- Food (26 critical incidents)
- Discharge process (1 critical incident)

- Attitude of staff (16 critical incidents)
- General (11 critical incidents)

Interpretation of data

Interjudge agreement, the percentage of critical incidents placed in the same category by both groups of judges, was 93.5%. This is well above the 80% cutoff value recommended by experts. The setting aside of a random sample and trying to place them in established categories is designed to test the comprehensiveness of the categories. If any of the withheld items were not classifiable it would be an indication that the categories do not adequately span the patient satisfaction space. However, the team experienced no problem in placing the withheld critical incidents into the categories.

Ideally, a critical incident has two characteristics: (1) it is specific and (2) it describes the service provider in behavior al terms or the service product with specific adjectives (Hayes, 1992). Upon reviewing the critical incidents in the General category, the team determined that these items failed to have one or both of these characteristics. Thus, the 11 critical incidents in the General category were dropped. The team also decided to merge the two categories "Care provided by staff" and "Attitude of staff" into the single category "Quality of staff care." Thus, the final result was a five dimension model of patient satisfaction judgments: Food, Quality of physician, Quality of staff care, Accommodations, and Discharge process.

A rather obvious omission in the above list is billing. This occurred because the patients had not yet received their bill within the 72 hour time frame. However, the patient's bill was explained to the patient prior to discharge. This item is included in the Discharge process dimension. The team discussed the billing issue and it was determined that billing complaints do arise after the bills are sent, suggesting that billing probably is a satisfaction dimension. However, the team decided not to include billing as a survey dimension because (1) the time lag was so long that waiting until bills had been received would significantly reduce the ability of the patient to recall the details of their stay; (2) fear that the patient's judgments would be overwhelmed by the recent receipt of the bill; and (3) a system already existed for identifying patient billing issues and adjusting the billing process accordingly.

Survey item development

As stated earlier, the general aim was to provide the service provider with information on what patients remembered about their hospital stay, both pleasant and unpleasant. This information was then to be used to construct a new patient survey instrument that would be sent to recently discharged patients on a periodic basis. The information obtained would be used by the managers of the various service processes as feedback on their performance, from the patient's perspective.

The core team believed that accomplishing these goals required that the managers of key service processes be actively involved in the creation of the survey instrument. Thus, ad hoc teams were formed to develop survey items for each of the dimensions determined by the critical incident study. The teams were given brief instruction by the author in the characteristics of good survey items. Teams were required to develop items that, in the opinion of the core team, met five criteria: (1) relevance to the dimension being measured; (2) concise; (3) unambiguous; (4) one thought per item; and (5) no double negatives. Teams were also shown the specific patient comments that were used as the basis for the categories and informed that these comments could be used as the basis for developing survey items.

Writing items for the questionnaire can be difficult. The process of developing the survey items involved an average of three meetings per dimension, with each meeting lasting approximately two hours. Ad hoc teams ranged in size from four to eleven members. The process was often quite tedious, with considerable debate over the precise wording of each item.

The core team discussed the scale to be used with each ad hoc team. The core team's recommended response format was a five point Likert-type scale. The consensus was to use a five point agree-disagree continuum as the response format. Item wording was done in such a way that agreement represented better performance from the hospital's perspective.

In addition to the response items, it was felt that patients should have an opportunity to respond to open-ended questions. Thus, the survey also included general questions that invited patients to comment in their own words. The benefits of having such questions are well known. In addition, it was felt that these questions might generate additional critical incidents that would be useful in validating the survey.

The resulting survey instrument contained 50 items and three open-ended questions and is included in the Appendix.

Survey administration and pilot study

The survey was to be tested on a small sample. It was decided to use the total design method (TDM) to administer the survey (Dillman, 1983). Although the total design method is exacting and tedious, Dillman indicated that its use would ensure a high rate of response. Survey administration would be handled by the Nursing Department.

TDM involves rather onerous administrative processing. Each survey form is accompanied by a cover letter, which must be hand-signed in blue ink. Follow up mailings are done 1, 3 and 7 weeks after the initial mailing. The 3 and 7 week follow ups are accompanied by another survey and another cover letter. No "bulk processing" is allowed, such as the use of computer-generated letters or mailing labels. Dillman's research emphasizes the importance of viewing the TDM as a completely integrated approach (Dillman, 1983).

Because the hospital in the study is small, the author was interested in obtaining maximum response rates. In addition to following the TDM guidelines, he recommended that a $1 incentive be included with each survey. However, the hospital administrator was not convinced that the additional $1 per survey was worthwhile. It was finally agreed that to test the effect of the incentive on the return rate $1 would be included in 50% of the mailings, randomly selected.

The hospital decided to perform a pilot study of 100 patients. The patients selected were the first 100 patients discharged to home starting April 1, 1994. The return information is shown in Table 2.2.

Although the overall return rate of 49% is excellent for normal mail-survey procedures, it is substantially below the 77% average and the 60% "minimum" reported by Dillman. As possible explanations, the author conjectures that there may be a large Spanish-speaking constituency for this hospital. As mentioned above, the hospital is planning a Spanish version of the survey for the future.

The survey respondent demographics were analyzed and compared to the demographics of the nonrespondents to ensure that the sample group was representative. A sophisticated statistical analysis was performed on the responses to evaluate the reliability and validity of each item. Items with low reliability coefficients or questionable validity were reworded or dropped.

---

*The survey referenced by this case study is located in the Appendix.

†The nutrition services manager was very concerned that she gets sufficient detail on her particular service. Thus, the critical incident interview instrument she used included special questions relating to food service.

‡The team was unable to obtain a Spanish-speaking interviewer, which meant that some patients that were candidates were not able to participate in the survey.

¶Billing was not covered in the CIT phase of the study because patients had not received their bills within 72 hours.

A. Numbers	B. Survey Responses by Mailing
Surveys mailed: 100	Number of surveys returned after:
Surveys delivered: 92	Initial mailing: 12
Surveys returned as undeliverable: 8	One week follow up: 16
Survey returned, needed Spanish version: 1	Three week follow up: 8
	Seven week follow up: 9
Total surveys returned: 45	
Percentage of surveys delivered returned: 49%	
Number delivered that had $1 incentive: 47	
Number returned that had $1 incentive: 26	
Percentage returned that had $1 incentive: 55%	
Number delivered that had no $1 incentive: 45	
Number returned that had no $1 incentive: 19	
Percentage returned that had no $1 incentive: 42%	

TABLE 2.2 Pilot Patient Survey Return Information

Focus Groups

The focus group is a special type of group in terms of purpose, size, composition, and procedures. A focus group is typically composed of seven to ten participants who are unfamiliar with each other. These participants are selected because they have certain characteristic(s) in common that relate to the topic of the focus group.

The researcher creates a permissive environment in the focus group that nurtures different perceptions and points of view, without pressuring participants to vote, plan, or reach consensus. The group discussion is conducted several times with similar types of participants to identify trends and patterns in perceptions. Careful and systematic analyses of the discussions provide clues and insights as to how a product, service, or opportunity is perceived.

A focus group can thus be defined as a carefully planned discussion designed to obtain perceptions on a defined area of interest in a permissive, nonthreatening environment. The discussion is relaxed, comfortable, and often enjoyable for participants as they share their ideas and perceptions. Group members influence each other by responding to ideas and comments in the discussion.

In Six Sigma, focus groups are useful in a variety of situations:

- Prior to starting the strategic planning process
- Generate information for survey questionnaires

- Needs assessment, for example, training needs
- Test new program ideas
- Determine customer decision criteria
- Recruit new customers

The focus group is a socially oriented research procedure. The advantage of this approach is that members stimulate one another, which may produce a greater number of comments than would individual interviews. If necessary, the researcher can probe for additional information or clarification. Focus groups produce results that have high face validity, that is, the results are in the participant's own words rather than in statistical jargon. The information is obtained at a relatively low cost, and can be obtained very quickly.

There is less control in a group setting than with individual interviews. When group members interact, it is often difficult to analyze the resulting dialogue. The quality of focus group research is highly dependent on the qualifications of the interviewer. Trained and skilled interviewers are hard to find. Group-to-group variation can be considerable, further complicating the analysis. Finally, focus groups are often difficult to schedule.

Another method somewhat related to focus groups is that of customer panels. Customer panels are composed of a representative group of customers who agree to communicate their attitudes periodically via phone calls or mail questionnaires. These panels are more representative of the range of customer attitudes than customer complaint and suggestion systems, yet more easily managed than focus groups. To be effective, the identity of customers on the panel must be withheld from the employees serving them.

Calculating the Value of Customer Retention

Customers have value. This simple fact is obvious when one looks at a customer making a single purchase. The transaction provides revenue and profit to the firm. However, when the customer places a demand on the firm, such as a return of a previous purchase or a call for technical support, there is a natural tendency to see this as a loss. At these times it is important to understand that customer value must not be viewed on a short-term transaction-by-transaction basis. Customer value must be measured *over the lifetime of the relationship*. One method of calculating the lifetime value of a loyal customer, based on work by Frederick Reichheld of Bain and Co. and the University of Michigan's Claes Fornell, is as follows (Stewart, 1995):

1. Decide on a meaningful period of time over which to do the calculations. This will vary depending on your planning cycles and your business: A life insurer should track customers for decades, a disposable diaper maker for just a few years, for example.
2. Calculate the profit (net cash flow) customers generate each year. Track several samples—some newcomers, some old-timers—to find out how much business they gave you each year, and how much it cost to serve them. If possible, segment them by age, income, sales channel, and so on. For the first year, be sure to subtract the cost of acquiring the pool of customers, such as advertising, commissions, back-office costs of setting up a new account. Get specific

numbers—profit per customer in year one, year two, and so on—not averages for all customers or all years. Long-term customers tend to buy more, pay more (newcomers are often lured by discounts), and create less bad debt.

3. Chart the customer "life expectancy," using the samples to find out how much your customer base erodes each year. Again, specific figures are better than an average like "10% a year"; old customers are much less likely to leave than freshmen. In retail banking, 26% of account holders defect in the first year; in the ninth year, the rate drops to 9%.
4. Once you know the profit per customer per year and the customer-retention figures, it's simple to calculate net present value (NPV). Pick a discount rate—if you want a 15% annual return on assets, use that. In year one, the NPV will be profit ÷ 1.15. Next year, NPV = (year-two profit × retention rate) ÷ $(1.15)^2$. In year n, the last year in your figures, the NPV is the n year's adjusted profit ÷ $(1.15)^n$. The sum of the years one through n is how much your customer is worth—the net present value of all the profits you can expect from his tenure.

This is very valuable information. It can be used to find out how much to spend to attract new customers, and which ones. Better still, you can exploit the leverage customer satisfaction offers. Take your figures and calculate how much more customers would be worth if you increased retention by 5%. Figure 2.3 shows the increase in customer NPV for a 5% increase in retention for three industries.

Once the lifetime value of the customer is known, it forms the basis of loyalty-based management℠ of the customer relationship. According to Reichheld (1996), loyalty-based management is the practice of carefully selecting customers, employees, and investors, and then working hard to retain them. There is a tight, cause-and-effect connection between investor, employee, and customer loyalty. These are the human assets of the firm.

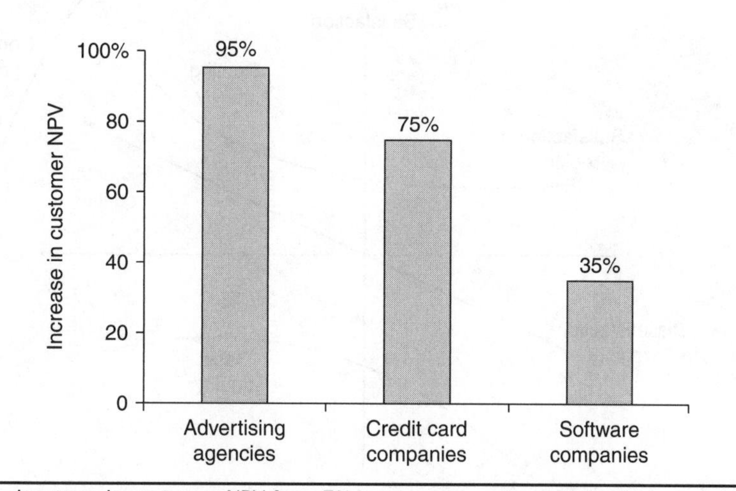

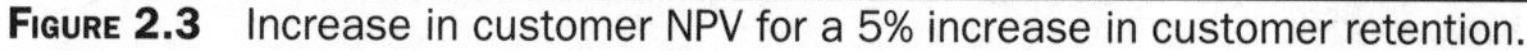

FIGURE 2.3 Increase in customer NPV for a 5% increase in customer retention.

Customer Expectations, Priorities, Needs, and "Voice"

Although customers seldom spark true innovation (for example, they are usually unaware of state-of-the art developments), their input is extremely valuable. Obtaining valid customer input is a science itself. Market research firms use scientific methods such as critical incident analysis, focus groups, content analysis and surveys to identify the "voice of the customer." Noritaki Kano developed the following model of the relationship between customer satisfaction and quality (Fig. 2.4).

The Kano model shows that there is a basic level of quality that customers assume the product will have. For example, all automobiles have windows and tires. If asked, customers don't even mention the basic quality items, they take them for granted. However, if this quality level *isn't* met the customer will be dissatisfied; note that the entire "Basic quality" curve lies in the lower half of the chart, representing dissatisfaction. However, providing basic quality isn't enough to create a satisfied customer.

The "Expected quality" line represents those expectations which customers explicitly consider. For example, the length of time spent waiting in line at a checkout counter. The model shows that customers will be dissatisfied if their quality expectations are not met; satisfaction increases as more expectations are met.

The "Exciting quality" curve lies entirely in the satisfaction region. This is the effect of innovation. Exciting quality represents *unexpected* quality items. The customer receives more than they expected. For example, Cadillac pioneered a system where the headlights stay on long enough for the owner to walk safely to the door. When first introduced, the feature excited people.

Competitive pressure will constantly raise customer expectations. Today's exciting quality is tomorrow's basic quality. Firms that seek to lead the market must innovate constantly. Conversely, firms that seek to offer standard quality must constantly research customer expectations to determine the currently accepted quality levels. It is not enough to track competitors since expectations are influenced by outside factors as

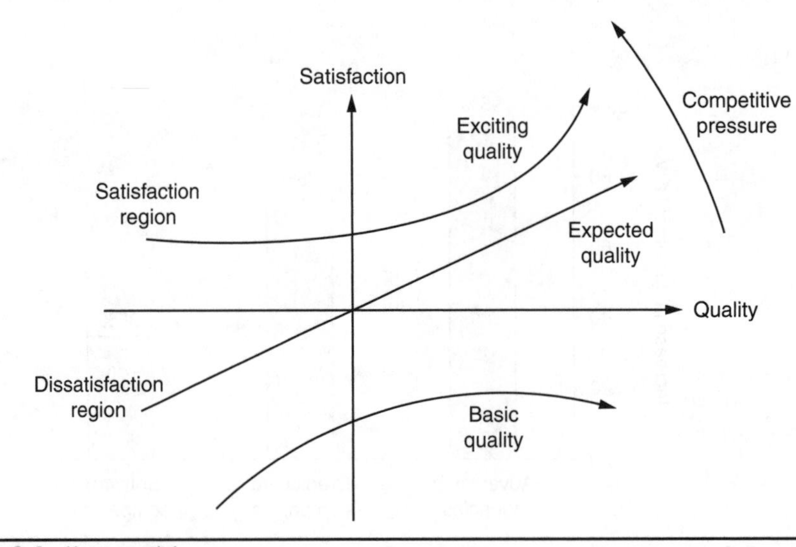

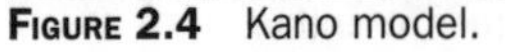
FIGURE 2.4 Kano model.

well. For example, the quality revolution in manufacturing has raised expectations for service quality as well.

Garden Variety Six Sigma Only Addresses Half of the Kano Customer Satisfaction Model

Some people, including your author, believe that even Six Sigma doesn't go far enough. In fact, even "zero defects" falls short. Defining quality as only the lack of nonconforming product reflects a limited view of quality. Motorola, of course, never intended to define quality as merely the absence of defects. However, some have misinterpreted the Six Sigma program in this way.

One problem with "garden variety" Six Sigma is that it addresses only half of the Kano model. By focusing on customer expectations and prevention of nonconformances and defects, Six Sigma addresses the portion of the Kano model on and below the line labeled "Expected Quality." While there is nothing wrong with improving these aspects of business performance, it will not ensure that the organization remains viable in the long term. Long-term success requires that the organization innovate. Innovation is the result of creative activity, not analysis. Creativity is not something that can be done "by the numbers." In fact, excessive attention to a rigorous process such as Six Sigma can detract from creative activities if not handled carefully. As discussed previously, the creative organization is one which exhibits variability, redundancy, quirky design, and slack. It is vital that the organization keep this paradox in mind.

Quality Function Deployment

Once information about customer expectations has been obtained, techniques such as quality function deployment (QFD) can be used to link the voice of the customer directly to internal processes.

Tactical quality planning involves developing an approach to implementing the strategic quality plan. One of the most promising developments in this area has been policy deployment. Sheridan (1993) describes policy deployment as the development of a measurement-based system as a means of planning for continuous quality improvement throughout all levels of an organization. Originally developed by the Japanese, American companies also use policy deployment because it clearly defines the long-range direction of company development, as opposed to short-term.

QFD is a customer-driven process for planning products and services. It starts with the voice of the customer, which becomes the basis for setting requirements. QFD matrices, sometimes called "the house of quality," are graphical displays of the result of the planning process. QFD matrices vary a great deal and may show such things as competitive targets and process priorities. The matrices are created by interdepartmental teams, thus overcoming some of the barriers which exist in functionally organized systems.

QFD is also a system for design of a product or service based on customer demands, a system that moves methodically from customer requirements to specifications for the product or service. QFD involves the entire company in the design and control activity. Finally, QFD provides documentation for the decision-making process. The QFD approach involves four distinct phases (King 1987):

1. **Organization phase**. Management selects the product or service to be improved, the appropriate interdepartmental team, and defines the focus of the QFD study.

2. **Descriptive phase**. The team defines the product or service from several different directions such as customer demands, functions, parts, reliability, cost, and so on.
3. **Breakthrough phase**. The team selects areas for improvement and finds ways to make them better through new technology, new concepts, better reliability, cost reduction, etc., and monitors the bottleneck process.
4. **Implementation phase**. The team defines the new product and how it will be manufactured.

QFD is implemented through the development of a series of matrices. In its simplest form QFD involves a matrix that presents customer requirements as rows and product or service features as columns. The cell, where the row and column intersect, shows the correlation between the individual customer requirement and the product or service requirement. This matrix is sometimes called the "requirement matrix." When the requirement matrix is enhanced by showing the correlation of the columns with one another, the result is called the "house of quality." Figure 2.5 shows one commonly used house of quality layout.

The house of quality relates, in a simple graphical format, customer requirements, product characteristics, and competitive analysis. It is crucial that this matrix be developed carefully since it becomes the basis of the entire QFD process. By using the QFD approach, the customer's demands are "deployed" to the final process and product requirements.

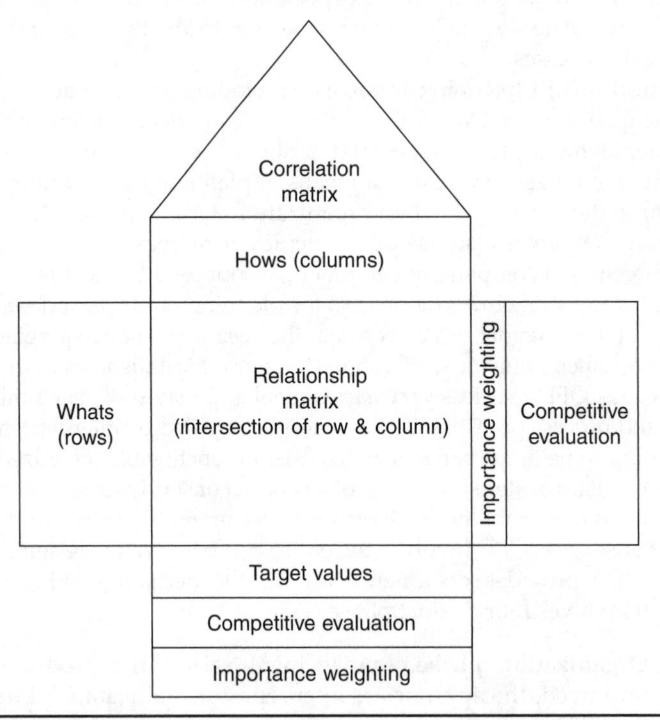

FIGURE 2.5 The house of quality.

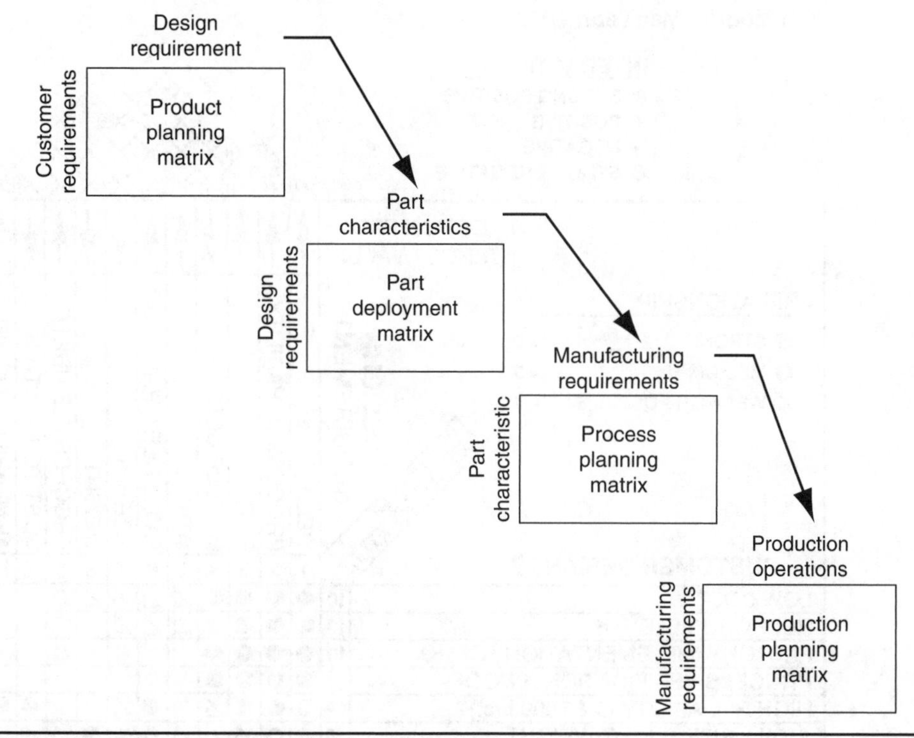

FIGURE 2.6 QFD methodology: Macabe approach.

One rendition of QFD, called the Macabe approach, proceeds by developing a series of four related matrices (King, 1987): product planning matrix, part deployment matrix, process planning matrix, and production planning matrix. Each matrix is related to the previous matrix as shown in Fig. 2.6.

Figure 2.7 shows an example of an actual QFD matrix.

Data Collection and Review of Customer Expectations, Needs, Requirements, and Specifications

Another approach to QFD is based on work done by Yoji Akao. Akao (1990, pp. 7–8) presents the following 11-step plan for developing the quality plan and quality design, using QFD.

1. First, survey both the expressed and latent quality demands of consumers in your target marketplace. Then decide what kinds of "things" to make.
2. Study the other important characteristics of your target market and make a demanded quality function deployment chart that reflects both the demands and characteristics of that market.
3. Conduct an analysis of competing products on the market, which we call a competitive analysis. Develop a quality plan and determine the selling features (sales points).
4. Determine the degree of importance of each demanded quality.

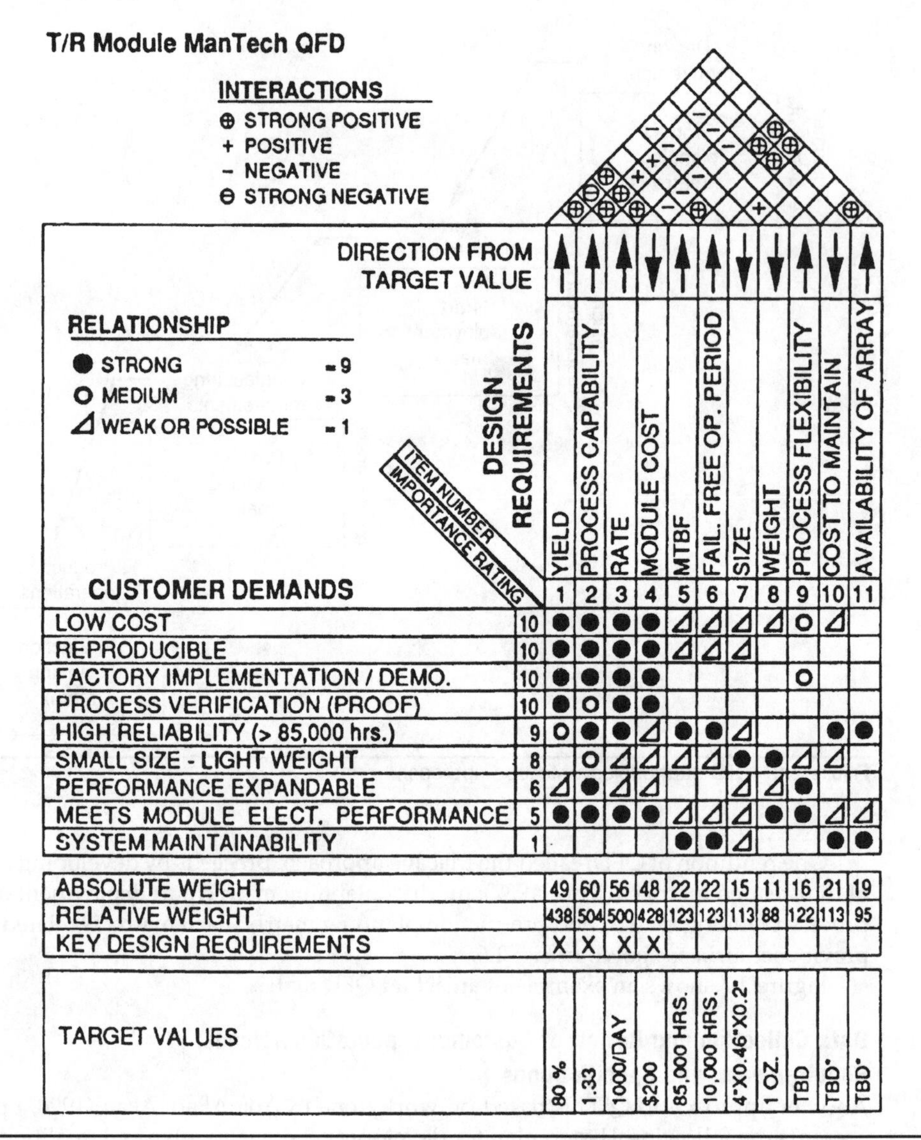

FIGURE 2.7 QFD matrix for an aerospace firm. (From Wahl and Bersbach (1991). Reprinted with permission.)

5. List the quality elements and make a quality elements deployment chart.
6. Make a quality chart by combining the demanded quality deployment chart and the quality elements deployment chart.
7. Conduct an analysis of competing products to see how other companies perform in relation to each of these quality elements.
8. Analyze customer complaints.
9. Determine the most important quality elements as indicated by customer quality demands and complaints.

10. Determine the specific design quality by studying the quality characteristics and converting them into quality elements.
11. Determine the quality assurance method and the test methods.

The Six Sigma Process Enterprise

I am often asked "Will Six Sigma work for…" where the blank is "health care," "oil exploration," "custom-built homes," etc. The list is unending. My typical response is that, if a process is involved, Six Sigma may be able to help you improve it. Personally, I don't believe that *everything* will benefit from the application of Six Sigma rigor. There are some things that aren't processes, such as pure creativity, love and unstructured play. I don't believe a chess grand master would benefit from the advice of a Black Belt in applying DMAIC to his moves, nor would his equivalent in the R&D area. Practices such as research, social relationships, criminal behavior, or curing substance abuse are process oriented, but so poorly understood that we would have difficulty using the Six Sigma approach to improve them. Nonetheless the vast majority of processes encountered in business, nonprofit organizations, and government services fall into the category of processes that can be improved by the application of Six Sigma methods.

But what exactly is a "process"? There is a tendency to narrowly interpret the term process to refer to a manufacturing operation that converts raw materials into finished products. That's true, of course. But as the term is used throughout this book it has a much broader meaning. It refers to any activity or set of activities that transform inputs to create values for stakeholders. The inputs can be labor, expertise, raw materials, products, transactions, or services that someone is willing to pay more for than they cost to create. In other words, the process adds value to the inputs. Said another way, *the process is the act of creating value*. The value can be a cured disease, a tasty banana split, a great movie, a successfully completed credit card transaction, or a cold soda purchased at a convenience store.

Reengineering, the process redesign fad so popular in the early 1990s, has become associated in the minds of many with brutal downsizing. Many academics condemned it as heartless and cold. But the problem wasn't caused by reengineering. Reengineering (and Six Sigma) focuses attention on broken and inefficient processes. The truth is this focus enabled companies to operate faster and more efficiently and to use information technology more productively. It gave employees more authority and a clearer view of how their work fit into the broader scheme of things. Customers benefited from lower prices, higher quality and better services, and investors enjoyed a higher rate of return. And, more germane to our discussion of processes, reengineering taught business leaders to see their organizations not as control structures, but as processes that deliver value to customers in a way that creates profits for shareholders.

Examples of Processes

Many business leaders think of their organizations as extremely complex. From a process perspective, this is seldom the case, at least at the high levels. For example, Texas Instruments was able to break its $4 billion semiconductor business into six core processes:

1. Strategy development
2. Product development
3. Customer design and support

4. Manufacturing capability development
5. Customer communication
6. Order fulfillment

A large financial software company described its four core processes in plain English as:

1. Provide good products at good prices.
2. Acquire customers and maintain good relations with them.
3. Make it easy to buy from us.
4. Provide excellent service and support after the sale.

Both of these companies have thousands of employees and generate billions of dollars in sales. Yet what they do for customers is really very simple. Once the basic (core) processes have been identified, the relationship between them should be determined and drawn on a *process map*. (Process mapping is discussed in greater detail in Part II of this handbook.) The process map presents employees with a simple picture that illustrates how the enterprise serves its customers. It is the basis for identifying subprocesses and, eventually, Six Sigma projects. Table 2.3 gives some examples of high-level processes and subprocesses.

Core Process	Subprocess
Product development	• R&D • Design creation • Prototype development • Design production support
Marketing	• Inspiration, concept discovery • Customer identification • Developing market strategies • Concept production support • Customer acquisition and maintenance
Product creation	• Manufacturing • Procurement • Installation
Sales and service	• Fulfillment (order ▸payment) • Pre-sale customer support • Installation and front-line service • Usage
Meta-processes	• Process excellence (Six Sigma) • Voice of customer • Voice of shareholder • Voice of employee

Table 2.3 Examples of High-Level Processes and Subprocesses

The truth is, it's the organizational structure that's complicated, not the business itself. The belief that the business is complicated results from a misplaced internal perspective by its leaders and employees. In a traditional organization tremendous effort is wasted trying to understand what needs to be done when goals are not well defined and people don't know how their work relates to the organization's purpose. A process focus is often the first real "focus" an employee experiences, other than pleasing one's superiors.

The Source of Conflict

Management structures, since the time of Alfred P. Sloan in the 1920s and 1930s, are designed to divide work into discrete units with clear lines of responsibility and authority. While this approach produced excellent results for a time, it has inherent flaws that became quite apparent by 1980. Organizations put leadership atop a pyramid-shaped control system designed to carry out their strategies. Control of the resources needed to accomplish this resided in the vertical pillars, known as "functions" or "divisions." This command-and-control approach is depicted in Fig. 2.8.

This arrangement creates "turf" where, much like caste systems, activities within a given area are the exclusive domain of that area. Personnel in engineering, for example, are not allowed to engage in activities reserved to the finance group, nor is finance allowed to "meddle" in engineering activities. These turfs are jealously guarded. In such a structure employees look to the leadership to tell them what to do and to obtain the resources needed to do it. This upward-inward focus is the antithesis of an external-customer focus. As Fig. 2.8 also shows, customer value is created by processes that draw resources from several different parts of the organization and end at a customer contact point. If an organization wants to be customer-focused, then it must change the traditional structure so its employees look *across* the organization at processes. As you might expect, this calls for a radical rethinking of the way the enterprise operates.

As long as control of resources and turf remain entirely with the functional units, the enterprise will remain focused inwardly. Goals will be unit-based, rather than process-based. In short, Six Sigma (or any other process-oriented initiative) will not work. Functional department leaders have both the incentive and the ability to thwart cross-functional process improvement efforts. This doesn't mean that these people are "bad." It's simply that their missions are defined in such a way that they are faced with a dilemma: pursue the mission assigned to my area to the best of my ability, or support an initiative that detracts from it but benefits the enterprise as a whole. Social scientists call this "the tragedy of the commons." It is in the best interest of all fishermen not to overharvest the fishing grounds, but it is in the interest of each individual fisherman to get all he can from this "common resource." Similarly, it is in the best interest of the enterprise as a whole to focus on customers, but it is in each functional leader's best interest to pursue his or her provincial self-interest. After all, if every other functional manager tries to maximize the resources devoted to their area and I don't, I will lose my department's share of the resources. Self-interest wins hands down.

A Resolution to the Conflict

Some companies—such as IBM, Texas Instruments, Owens Corning, and Duke Power—have successfully made the transition from the traditional organizational structure to an alternative system called the "Process Enterprise" (Hammer and Stanton, 1999). In these companies the primary organizational unit is not the functional department, but the process development team. These cross-functional teams, like the reengineering

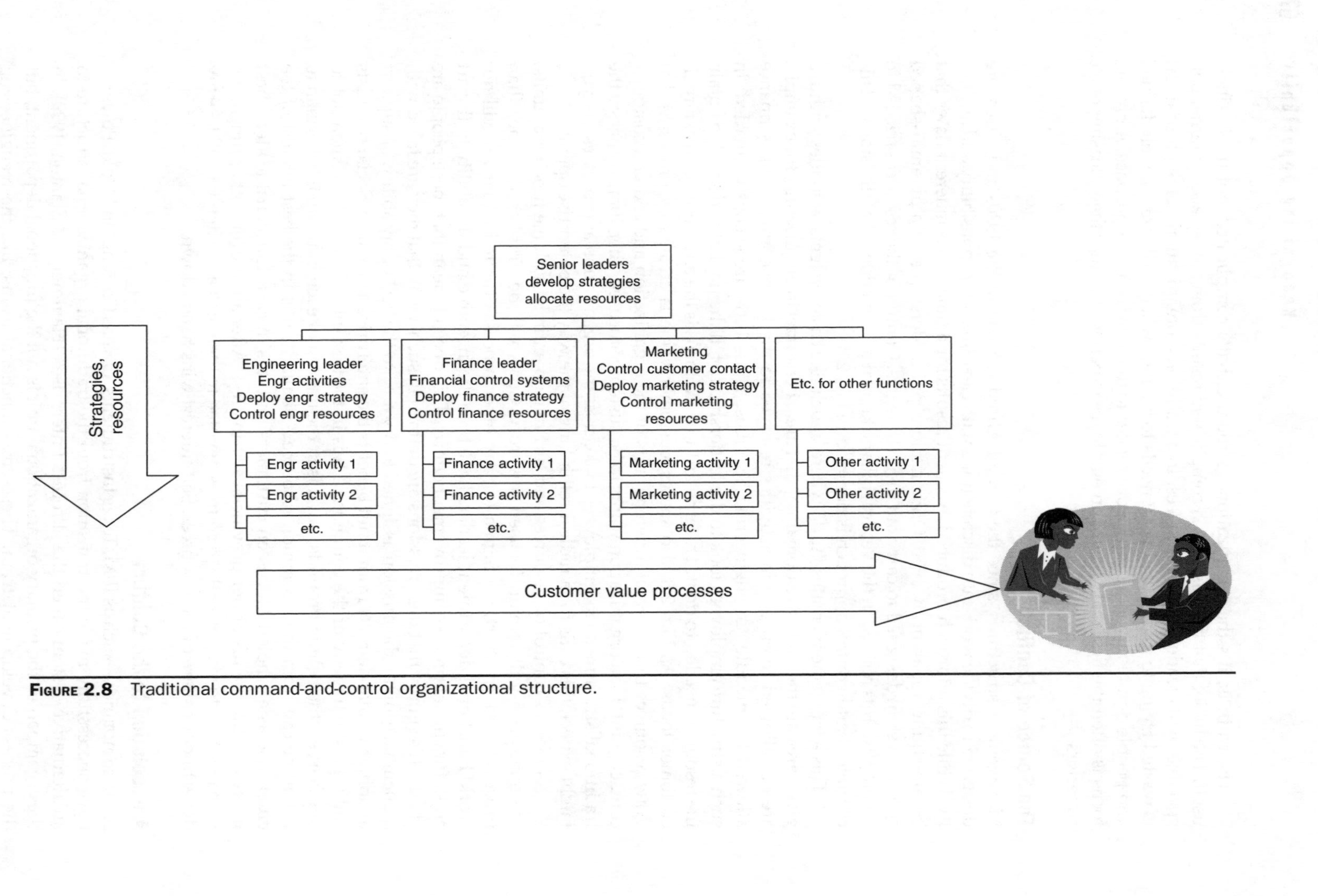

FIGURE 2.8 Traditional command-and-control organizational structure.

teams of old, have full responsibility for a major business process. For example, a product development team would work together in the same location to build the product development process from concept to launch. They would produce the design, documentation, training materials, advertising, and so on. In a Process Enterprise authority and control of resources is redistributed in a manner that achieves a balance of power between the process-focused and structure-focused aspects of the enterprise.

The differences between Process Enterprises and traditional organizations are fundamental. In the Process Enterprise a new position is created, that of Process Owner or Business Process Executive (BPE). The BPE position is permanent. BPEs are assigned from the senior-most executive body and given responsibility for designing and deploying the process, as well as control over all expenditures and supporting technology. They establish performance metrics, set and distribute budgets, and train the frontline workers who perform the process work. However, the people who perform the process work report to unit heads, not BPEs. In the Process Enterprise process goals are emphasized over unit goals. Process performance is used as a basis for compensation and advancement.

In a Process Enterprise lines of authority are less well defined. BPEs and functional unit managers are expected to work together to resolve disagreements. The BPE doesn't exert direct control over the workers, but because he controls budgets and sets goals by which unit managers will be evaluated, he does have a good deal of influence. The unit managers have to see to it that the process designs are sound, the resource allocation sufficient, and the goals clear and fair. In short, managing in a Process Enterprise places a premium on collaboration and cooperation.

One tool that has been developed to help clarify the different roles and responsibilities is the Decision Rights Matrix (Hammer and Stanton, 1999). This matrix specifies the roles the different managers play for each major decision, such as process changes, personnel hiring, setting budgets, and so on. For example, on a given decision must a given manager:

- Make the decision?
- Be notified in advance?
- Be consulted beforehand?
- Be informed after the fact?

The Decision Rights Matrix serves as a roadmap for the management team, especially in the early stages of the transition from traditional organization to Process Enterprise. Eventually team members will internalize the matrix rules.

BPEs must also work together. Processes overlap and process handoffs are critical. Often the same worker works with different processes. To avoid "horizontal turf wars" senior leadership needs to set enterprise goals and develop compensation and incentive systems that promote teamwork and cooperation between process owners.

The need for interprocess cooperation highlights the fact that no process is an island. From the customer's perspective, it's all one process. Overall excellence requires that the entire business be viewed as the customer sees it. One way to accomplish this is to set up a separate process with a focus of overall process excellence. For the sake of discussion, let's call this Process Excellence (PEX). PEX will have a BPE and it will be considered another core business process. The mission of PEX is to see to it that all business processes accomplish the enterprise goals as they relate to customers, shareholders, and

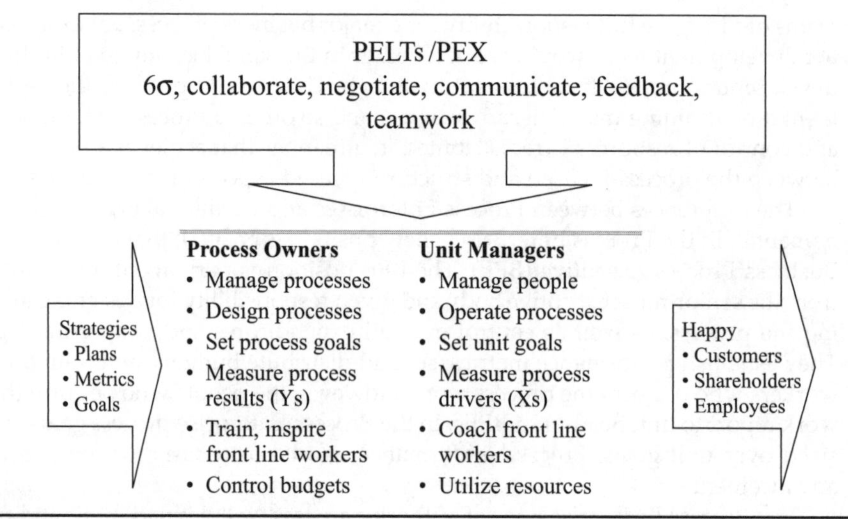

FIGURE 2.9 Process enterprise roles and responsibilities.

employees. PEX is also concerned with helping BPEs improve their processes, both internally and across the process boundaries. In other words, PEX is a *meta-process*, a process of processes. BPEs, unit managers, and Process Excellence leaders work together through Process Excellence Leadership Teams (PELTs) to ensure that the needs of the major stakeholder groups are met (Fig. 2.9).

Six Sigma and the Process Enterprise

Once the decision is made to become a Six Sigma Process Enterprise, the question of how to integrate the Six Sigma infrastructure will arise. Here are my recommendations:

1. Designate Process Excellence (PEX) as one of the enterprise's core processes and select a BPE.
2. Master Black Belts will report to PEX. The Master Black Belts will have an enterprise-wide perspective. They will be concerned with the internal processes in PEX, as well as the overall value creation and delivery produced by the cooperative efforts of the core processes.
3. Black Belts will report to the BPEs, but the budget for the Black Belts comes from Process Excellence. This gives PEX influence which helps maintain the enterprise perspective, but leaves day-to-day management and control with the Black Belt's customers, the BPEs.
4. BPEs have PEX goals, tied to incentives. PEX incentives are in the PEX budget.
5. Unit managers have process-based incentives. Process incentives are in the BPE's budgets.
6. The PEX leader and BPEs should collaboratively create a Decision Rights Matrix identifying:
    - The roles and responsibilities of PEX, BPEs, and unit managers. For example, hiring, budgets, project selection.

- Who makes the decision in the areas just described?
- Who must be consulted in decision-making?
- What is the communication plan?

7. PEX helps develop a BPE Coordination Plan addressing such interprocess issues as:
   - Where does the core processes overlap?
   - How will cross-process Six Sigma projects be chartered and coordinated?
   - Who will ensure that overlapping activities and handoffs are coordinated? (PEX plays a facilitation role here.)
   - When is standardization across processes best and when isn't it? The process intersections should be invisible to customers (e.g., customers shouldn't have to provide the same information more than once; single form information for ordering of products, support plans, and registration). However, diversity may be necessary to serve unique customer needs.

You may have noticed that having Black Belts reporting to BPEs instead of to PEX seems to contradict the advice given in the first chapter where I strongly recommended having the Black Belts report centrally. However, there is a critical difference. The traditional organizational structure was assumed in Chap. 1, so if the Black Belts didn't report to the Six Sigma organization (referred to here as PEX) they would have been reporting to the *unit managers*. I am not recommending that they report to unit managers, but to BPEs. BPEs are process owners, which gives them a much different perspective than the unit manager. This perspective, unlike that of unit managers, meshes very well with the Six Sigma focus on process improvement.

Linking Six Sigma Projects to Strategies

A common problem with Six Sigma is that there is a cognitive disconnect between the Six Sigma projects and top leadership's strategic goals. In Chap 3 we will discuss the development of Strategy Deployment Plans. Strategy Deployment Plans are simple maps showing the linkage between stakeholder satisfaction, strategies, and metrics. However, these maps are inadequate guides to operational personnel trying to relate their activities to the vision of their leadership. Unfortunately, more complexity is required to communicate the strategic message throughout the organization. We will use QFD for this purpose in an example, based on a Strategy Deployment Plan.

The Strategy Deployment Matrix

The process for developing the Strategy Deployment Matrix is:

1. Create a matrix of the strategies and metrics.
2. Determine the strength of the relationship between each strategy and metric.
3. Calculate a weight indicating the relative importance of the metric.

To begin we create a matrix where the rows are the strategies (what we want to accomplish) and the columns are the dashboard metrics (how we will operationalize the strategies and monitor progress). Note that this is the typical what-how QFD matrix layout, just with a different spin. In each cell (intersection of a row and a column) we will place a symbol assigning a weight to the relationship between the row and the column. A completed Phase I Strategy Deployment Matrix is shown in Fig. 2.10. The weights and symbols used are shown in Fig. 2.11.

The weights are somewhat arbitrary and you can choose others if you desire. These particular values increase more-or-less exponentially, which places a high emphasis on strong relationships, the idea being that we are looking for clear priorities. Weights of 1-2-3 would treat the different relationship strengths as increasing linearly. Choose the weighting scheme you feel best fits your strategy.

After the relationships have been determined for each cell, we are ready to calculate scores for each row. Remember, the rows represent strategies. For example, the first row represents our productivity strategy. The QFD analysis shows a strong relationship between the productivity strategy and inventory turns, which affects asset utilization. Critical to quality (CTQ) and profit per customer are somewhat related to this strategy. To get an overall score for the productivity strategy just sum the weights across the first row; the answer is 29. These row (strategy) weights provide information on how well the dashboards measure the strategies. A zero would indicate that the strategy isn't measured at all. However, a relatively low score doesn't necessarily

Sorted strategy matrix		Dashboard metric: New product introductions	Revenue from new sources	Customer relationship	R & D deployment time	Inventory turns	Fast service	New product revenues	Fast delivery	Product functionality	Skills audit gaps	CTQs	Asset utilization	Profit-per-customer	Price	Cost-per-unit	Compliance audit score	Employee feedback	Product quality	Shipping & handling costs	Product selection	Area score (row sum)	Relative area performance
Strategy: Financial performance	Productivity					◉						△	◉	△		◉						29	
	Revenue growth	◉	◉	△	○			◉						◉								40	
Customer value	Operational excellence					◉			◉		△	○	○	△	◉	△			◉	△	◉	55	
	Customer intimacy		△	◉				△			△	○										15	
	Product attributes	◉					◉			◉					○	○						33	
Internal process excellence	Innovation	◉	◉		◉			◉		△												37	
	Customer management process			◉			○				○	◉		△								25	
	Operations and logistics					◉	◉		◉				○	△	△					◉		41	
	Regulatory compliance																◉					9	
Learning and growth	Employee competencies		○		○		△		△		◉							○	△			21	
	Technology	◉	◉		◉			△		◉				△	△							39	
	Corporate culture		○	◉							○						○	◉				27	
Criteria performance target		+50%	20% of total revenues	VOC average > 6.5	−30%	+20%	Top 25%	25% of total	Above industry	All-weather capability	3.5 sigma	4.5 sigma	15% RONA	10% increase	No price increase	−6%	4 sigma	1Avg > 6.2	Top 20%	−10%	5% improvement		
Criteria score		36	34	28	24	27	22	20	19	19	17	16	15	14	14	13	12	12	10	10	9		
Strategic importance score		●	●	●	●	✓	✓	✓	✓	✓	✓	✓	✓	✓	✓	✓	✓	✓	✓	✓	✓		
Relative metric weight																							

FIGURE 2.10 Strategy deployment matrix.

Relationship description	Weight	Symbol
Strong relationship	9	⊙
Moderate relationship	3	○
Some relationship	1	△
Differentiator metric	5	●
Key requirement metric	1	✓

Figure 2.11 Weights and symbols in strategy deployment matrix.

indicate a problem. For example, the regulatory compliance strategy has a score of 9, but that comes from a strong relationship between the regulatory compliance audit and the strategy. Since the audit covers all major compliance issues, it's entirely possible that this single metric is sufficient.

The columns represent the metrics on the top-level dashboard, although only the differentiator metrics will be monitored on an ongoing basis. The metric's target is shown at the bottom of each column in the "how" portion of the matrix. QFD will provide a reality check on the targets. As you will see, QFD will link the targets to specific Six Sigma activities designed to achieve them. At the project phase it is far easier to estimate the impact the projects will have on the metric. If the sum of the project impacts isn't enough to reach the target, either more effort is needed or the target must be revised. Don't forget, there's more to achieving goals than Six Sigma. Don't hesitate to use QFD to link the organization's other activities to the goals.

In this example, leadership's vision for the hypothetical company is that they be the supplier of choice for customers who want state-of-the-art products customized to their demanding requirements. To achieve this vision they will focus their strategy on four key differentiators: new product introductions, revenue from new sources, intimate customer relationship, and R&D deployment time. With our chosen weighting scheme differentiator columns have a strategic importance score of 5, indicated with a • symbol in the row labeled Strategic Importance Score. These are the metrics that leadership will focus on throughout the year, and the goals for them are set very high. Other metrics must meet less demanding standards and will be brought to the attention of leadership only on an exception basis. The row labeled Relative Metric Weight is the product of the criteria score times the strategic importance score as a percentage for each column. The four differentiator metrics have the highest relative scores, while product selection (i.e., having a wide variety of standard products for the customer to choose from) is the lowest.

It is vital when using QFD to focus on only the most important columns!

Columns identified with a ▶ in the row labeled Strategic Importance Score are not part of the organization's differentiation strategy. This isn't to say that they are unimportant. What it does mean is that targets for these metrics will probably be set at or near their historical levels as indicated by process behavior charts. The goals will be to maintain these metrics, rather than to drive them to new heights. An organization has only limited resources to devote to change, and these resources must be focused if they are to make a difference that will be noticed by customers and shareholders. This organization's complete dashboard has twenty metrics, which can hardly be considered a "focus." By limiting attention to the four differentiators, the organization can pursue the strategy that their leadership believes will make them stand out in the marketplace for customer and shareholder dollars.*

Deploying Differentiators to Operations

QFD most often fails because the matrices grow until the analysis becomes burdensome. As the matrix grows like Topsy and becomes unwieldy, the team performing QFD begins to sense the lack of focus being documented by the QFD matrix. Soon, interest begins to wane and eventually the effort grinds to a halt. This too, is avoided by eliminating ▶ key requirements from the strategy deployment matrix. We will create a second-level matrix linked only to the differentiators. This matrix relates the differentiator dashboard metrics to departmental support strategies and it is shown in Fig. 2.12.

To keep things simple, we only show the strategy linkage for three departments: engineering, manufacturing, and marketing; each department can prepare its own QFD matrix. Notice that the four differentiator metric columns now appear as rows in the matrix shown in Fig. 2.12. These are the QFD "whats." The software automatically brings over the criteria performance target, criteria scores, and relative criteria scores for each row. This information is used to evaluate the strategy support plans for the three departments.

The support plans for the three departments are shown as columns, the QFD "hows," or how these three departments plan to implement the strategies. The relationship between the whats and hows is determined as described above. For each column the sum of the relationship times the row criteria score is calculated and shown in the score row near the bottom of the chart. This information will be used to select and prioritize Six Sigma projects in the next phase of the QFD.

Figure 2.12 also has a roof, which shows correlations between the whats. This is useful in identifying related Six Sigma projects, either within the same department or in different departments. For example, there is a strong relationship between the two engineering activities: faster prototype development and improve concept-to-design cycle time. Perhaps faster prototype development should be a subproject under the broad heading of improve concept-to-design cycle time. This also suggests that "improve concept-to-design cycle time" may have too large a scope. The marketing strategy of "improve ability to respond to changing customer needs" is correlated with three projects in engineering and manufacturing. When a strategy support plan involves many cross-functional projects it may indicate the existence of a core process. This suggests a need for high-level sponsorship, or the designation of a process owner to coordinate projects.

---

*The key requirements probably won't require explicit support plans. However, if they do QFD can be used to evaluate the plans. Key requirements QFD should be handled separately.

Deployment to operations	Departments: Engineering — Faster prototype development	Engineering — Reduce Old Design Support	Engineering — Improve Concept-to-Design cycle time	Manufacturing — Faster ramp-up to full production	Manufacturing — Increase flexibility to handle product mix	Marketing — Improve ability to respond to changing customer needs	Marketing — Improve direct marketing response rate	Marketing — Identify target markets for new products	Criteria performance target	Criteria score	Relative criteria score
Dashboard Metric: New product introductions	⊙	○	⊙	⊙	△	○		⊙	+50%	36	
Dashboard Metric: Revenue from new sources	⊙		⊙	○	○	⊙	△	⊙	20% of total revenues	34	
Dashboard Metric: Customer relationship						⊙			VOC average > 6.5	28	
Dashboard Metric: R&D deployment time	○	⊙	⊙			○			-30%	24	
Numeric Relative Score	0.18	0.08	0.22	0.11	0.04	0.19	0.01	0.16			
Target	10 weeks	15% of time	12 weeks	6 weeks	-50% on setup time	+0.5 on VOC	+50%	20% of sales			
Score	702	324	846	426	138	738	34	630			
Relative Score											

FIGURE 2.12 Phase II matrix: differentiators. (*Chart produced using QFD Designer software. Qualsoft, LLC. www.qualisoft.com.*)

Deploying Operations Plans to Projects

Figure 2.13 is a QFD matrix that links the department plans to Six Sigma projects. (In reality this may require additional flow down steps, but the number of steps should be kept as small as possible.) The rows are the department plans. The software also carried over the numeric relative score from the bottom row of the previous matrix, which is a measure of the relative impact of the department plan on the overall differentiator strategy. The far right column, labeled "Goal Score" is the sum of the relationships for the row.

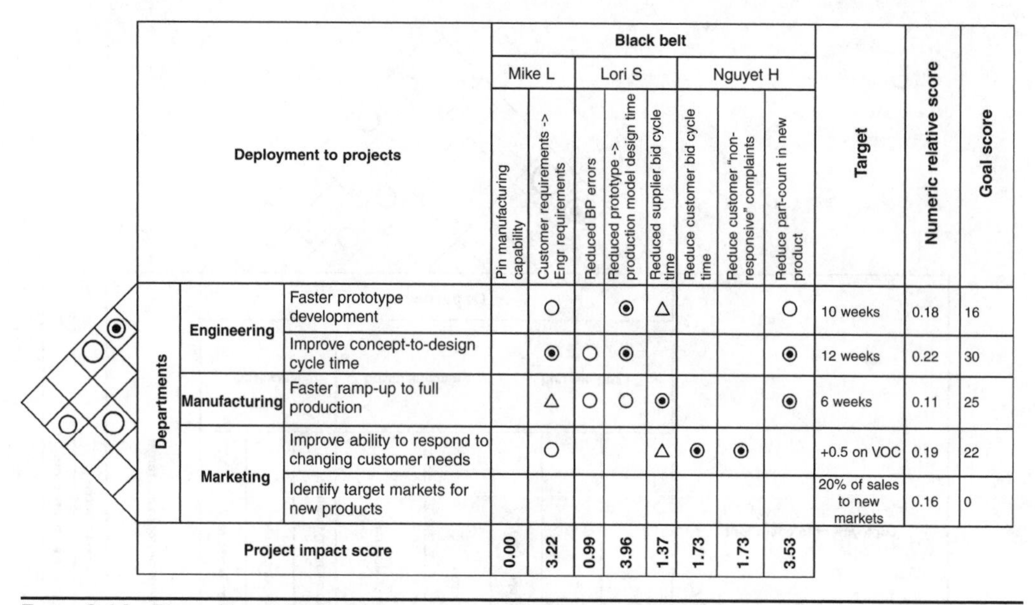

FIGURE 2.13 Phase III matrix: Six Sigma projects. (*Chart produced using QFD Designer software. Qualsoft, LLC. www.qualisoft.com.*)

For this example only the top five department plans are deployed to Six Sigma projects. By summing the numeric relative scores we can determine that these five plans account for 86% of the impact. In reality you will also only capture the biggest hitters, although there's nothing magic about the number five.

There are three Black Belts shown, and eight projects. Each project is shown as a column in the matrix. The relationship between the project and each departmental plan is shown in the matrix. The bottom row shows the "Project Impact Score," which is the sum of the relationships for the project's column times the row's numeric relative score.

Interpretation

Since the numeric relative scores are linked to department plans, which are linked to differentiator metrics, which are linked to strategies, the project impact score measures the project's impact on the strategy. The validity of these "carry-over scores" has been questioned (Burke *et al.*, 2002). Through the Strategy Deployment Plan we can trace the need for the project all the way back to stakeholders. This logical thread provides those engaged in Six Sigma projects with an anchor to reality and the meaning behind their activities.

The Goal Score column can also be used to determine the support Six Sigma provides for each department plan. Note that the marketing plan to "Identify target markets for new products" isn't receiving any support at all from Six Sigma projects (assuming that these eight projects are all of the Six Sigma projects). This may be okay, or it may not be. It all depends on how important the plan is to the strategic objectives, and what other activities are being pursued to implement the plan. The Executive Six Sigma Council may wish to examine project QFD matrices to determine if action is necessary to reallocate Six Sigma resources.

The Project Impact Score row is useful in much the same way. This row can be rank-ordered to see which projects have the greatest impact on the strategy. It is also useful in identifying irrelevant projects. The project Mike L is pursuing to improve "Pin manufacturing capability" has no impact on any of the departmental plans. Unless it impacts some other strategy support plan that isn't shown in the QFD matrix, it should be abandoned as a Six Sigma project. The project may still be something the manufacturing department wants to pursue, perhaps to meet a goal for a key requirement. However, as a general rule Six Sigma projects requiring a Black Belt should focus on plans that have a direct linkage to differentiator strategies.

Linking Customer Demands to Budgets

Once customers have made their demands known, it is important that these be translated into internal requirements and specifications. The term "translation" is used to describe this process because the activity literally involves interpreting the words in one language (the customer's) into those of another (the employee). For example, regarding the door of her automobile the customer might say "I want the door to close completely when I push it, but I don't want it swinging closed from just the wind." The engineer working with this requirement must convert it into engineering terminology such as pounds of force required to move the door from an open to a closed position, the angle of the door when it's opened, and so on. Care must be taken to maintain the customer's intent throughout the development of internal requirements. The purpose of specifications is to transmit the voice of the customer throughout the organization.

In addition to the issue of maintaining the voice of the customer, there is the related issue of the importance assigned to each demand by the customer. Design of products and services always involves tradeoffs: gasoline economy suffers as vehicle weight increases, but safety improves as weight increases. The importance of each criterion must be determined by the customer. When different customers assign different importance to criteria, design decisions are further complicated.

It becomes difficult to choose from competing designs in the face of ambiguity and customer-to-customer variation. Add to this the differences between internal personnel and objectives—department versus department, designer versus designer, cost versus quality, etc.—and the problem of choosing a design alternative quickly becomes complex. A rigorous process for deciding which alternative to settle on is helpful in dealing with the complexity.

Structured Decision-Making

The first step in deciding upon a course of action is to identify the goal. For example, let's say you're the owner of the Product Development process for a company that sells software to help individuals manage their personal finances. The product, let's call it DollarWise, is dominant in its market and your company is well respected by its customers and competitors, in large part because of this product's reputation. The business is profitable and the leadership naturally wants to maintain this pleasant set of circumstances and to build on it for the future. The organization has committed itself to a strategy of keeping DollarWise the leader in its market segment so it can capitalize on its reputation by launching additional new products directed towards other financially oriented customer groups, such as small businesses. They have determined that Product Development is a core process for deploying this strategy.

As the process owner, or Business Process Executive, you have control of the budget for product development, including the resources to upgrade the existing product. Although it is still considered the best personal financial software available, DollarWise is getting a little long in the tooth and the competition has steadily closed the technical gap. You believe that a major product upgrade is necessary and want to focus your resources on those things that matter most to customers. Thus, your goal is:

GOAL: Determine where to focus product upgrade resources

Through a variety of "listening posts" (focus groups, user laboratories, internet forums, trade show interviews, conference hospitality suites, surveys, letters, and technical support feedback.), you have determined that customers ask questions and make comments like the following:

- Can I link a DollarWise total to a report in my word processor?
- I have a high speed connection and I'd like to be able to download big databases of stock information to analyze with DollarWise.
- I like shortcut keys so I don't have to always click around in menus.
- I only have a 56K connection and DollarWise is slow on it.
- I use the Internet to pay bills through my bank. I'd like to do this using DollarWise instead of going to my bank's Web site.
- I want an interactive tutorial to help me get started.
- I want printed documentation.
- I want the installation to be simple.
- I want the user interface to be intuitive.
- I want to be able to download and reconcile my bank statements.
- I want to be able to upgrade over the Internet.
- I want to manage my stock portfolio and track my ROI.
- I'd like to have the reports I run every month saved and easy to update.
- It's a pain to set up the different drill downs every time I want to analyze my spending.
- It's clunky to transfer information between DollarWise and Excel.
- When I have a minor problem, I'd like to have easy-to-use self-help available on the Internet or in the help file.
- When it's a problem I can't solve myself, I want reasonably priced, easy to reach technical support.
- You should be making patches and bug-fixes available free on the Internet.

The first step in using this laundry list of comments is to see if there's an underlying structure embedded in them. If these many comments address only a few issues, it will simplify the understanding of what the customer actually wants from the product. While there are statistical tools to help accomplish this task (e.g., structural equation modeling, principal components analysis, factor analysis), they are quite advanced and require that substantial data be collected using well-designed survey instruments.

An alternative is to create an "affinity diagram," which is a simple procedure described elsewhere in this text. After creating the affinity diagram, the following structure was identified:

1. Easy to learn.
   1.1. I want the installation to be simple.
   1.2. I want an interactive tutorial to help me get started.
   1.3. I want printed documentation.
   1.4. I want the user interface to be intuitive.
2. Easy to use quickly after I've learned it well.
   2.1. I like shortcut keys so I don't have to always click around in menus.
   2.2. I'd like to have the reports I run every month saved and easy to update.
   2.3. It's a pain to set up the different drill downs every time I want to analyze my spending.
3. Internet connectivity.
   3.1. I use the Internet to pay bills through my bank. I'd like to do this using DollarWise instead of going to my bank's web site.
   3.2. I only have a 56K connection and DollarWise is slow on it.
   3.3. I have a high speed connection and I'd like to be able to download big databases of stock information to analyze with DollarWise.
   3.4. I want to be able to download and reconcile my bank statements.
   3.5. I want to manage my stock portfolio and track my ROI.
4. Works well with other software I own.
   4.1. It's clunky to transfer information between DollarWise and Excel.
   4.2. Can I link a DollarWise total to a report in my word processor?
5. Easy to maintain
   5.1. I want to be able to upgrade over the Internet.
   5.2. You should be making patches and bug-fixes available free on the Internet.
   5.3. When I have a minor problem, I'd like to have easy-to-use self-help available on the Internet or in the help file.
   5.4. When it's a problem I can't solve myself, I want reasonably priced, easy to reach technical support.

The reduced model shows that five key factors are operationalized by the many different customer comments (Fig. 2.14).

Next, we must determine importance placed on each item by customers. There are a number of ways to do this.

- Have customers assign importance weights using a numerical scale (e.g., "How important is 'Easy self-help' on a scale between 1 and 10?").

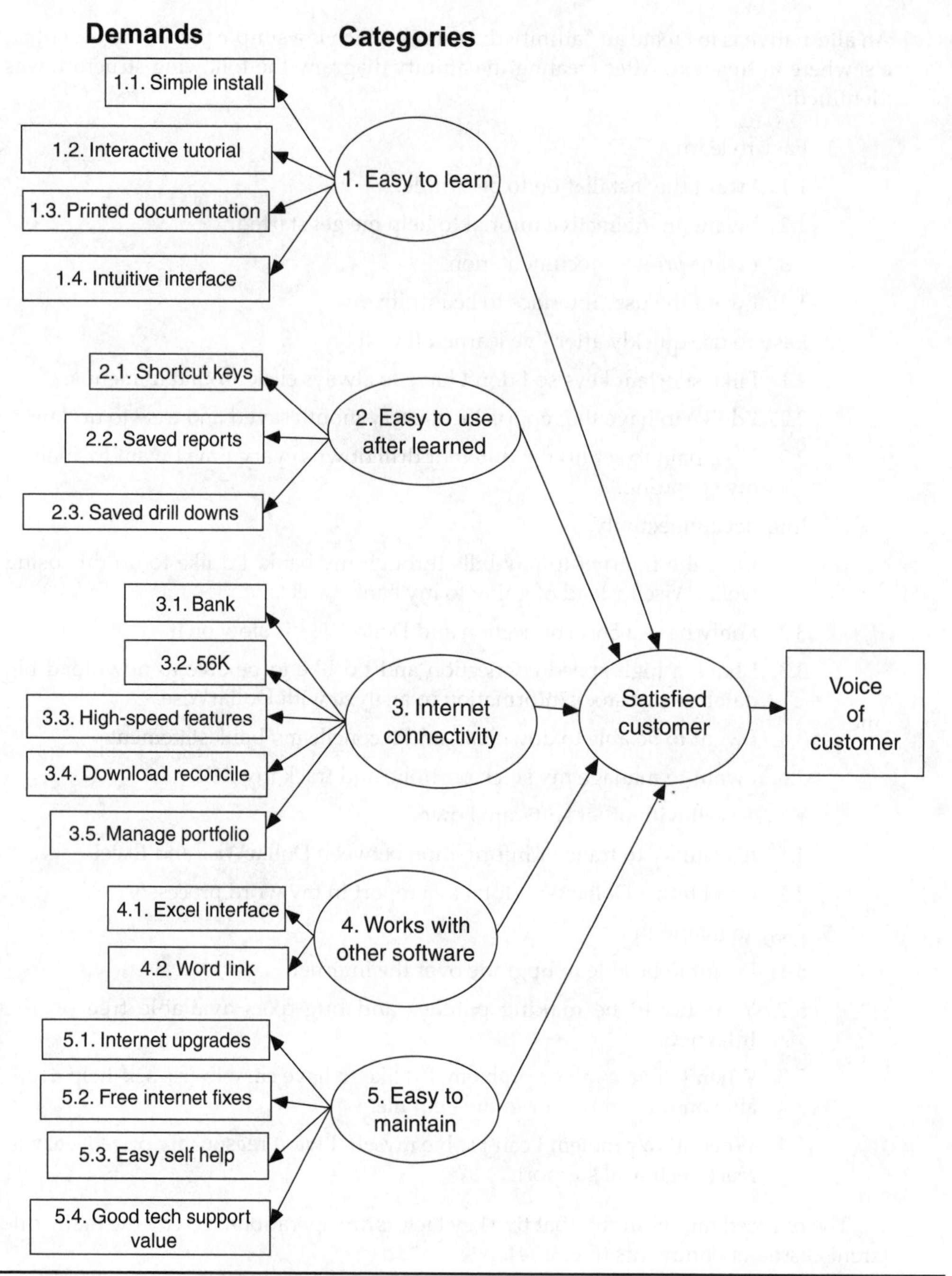

FIGURE 2.14 Customer demand model.

- Have customers assign importance using a subjective scale (e.g., unimportant, important, very important, etc.).
- Have customers "spend" $100 by allocating it among the various items. In these cases it is generally easier for the customer to first allocate the $100 to the major categories, and then allocate another $100 to items within each category.
- Have customers evaluate a set of hypothetical product offerings and indicate their preference for each product by ranking the offerings, assigning a "likely to buy" rating, etc. The product offerings include a carefully selected mix of items chosen from the list of customer demands. The list is selected in such a way that the relative value the customer places *on each item in the offering* can be determined from the preference values. This is known as *conjoint analysis*, an advanced technique that is covered in most texts on marketing statistics.
- Have customers evaluate the items in pairs, assigning a preference rating to one of the items in each pair, or deciding that both items in a pair are equally important. This is less tedious if the major categories are evaluated first, then the items within each category. The evaluation can use either numeric values or descriptive labels that are converted to numeric values. The pairwise comparisons can be analyzed to derive item weights using a method known as the Analytic Hierarchical Process (AHP) to determine the relative importance assigned to all of the items.

All of the above methods have their advantages and disadvantages. The simple methods are easy to use but less powerful (i.e., the assigned weights are less likely to reflect actual weights). The more advanced conjoint and AHP methods require special skills to analyze and interpret properly. We will illustrate the use of AHP for our hypothetical software product. AHP is a powerful technique that has been proven in a wide variety of applications. In addition to its use in determining customer importance values, it is useful for decision-making in general.

Category Importance Weights

We begin our analysis by making pairwise comparisons at the top level. The affinity diagram analysis identified five categories: easy to learn, easy to use quickly after I've learned it, internet connectivity, works well with other software I own, and easy to maintain. Arrange these items in a matrix as shown in Fig. 2.15.

For our analysis we will assign verbal labels to our pairwise comparisons; the verbal responses will be converted into numerical values for analysis. Customers usually find it easier to assign verbal labels than numeric labels. All comparisons are made relative to the customer's goal of determining which product he will buy, which we assume is synonymous with our goal of determining where to focus product upgrade efforts. The highlighted cell in the matrix compares the "easy to learn" attribute and the "easy to use quickly after I've learned it" attribute. The customer must determine which is more important to him, or if the two attributes are of equal importance. In Fig. 2.15 this customer indicates that "easy to learn" is moderately to strongly preferred over "easy to use quickly after I've learned it" and the software has placed a +4 in the cell comparing these two attributes. (The scale goes from −9 to +9, with "equal" being identified as a +1.) The remaining attributes are compared one by one, resulting in the matrix shown

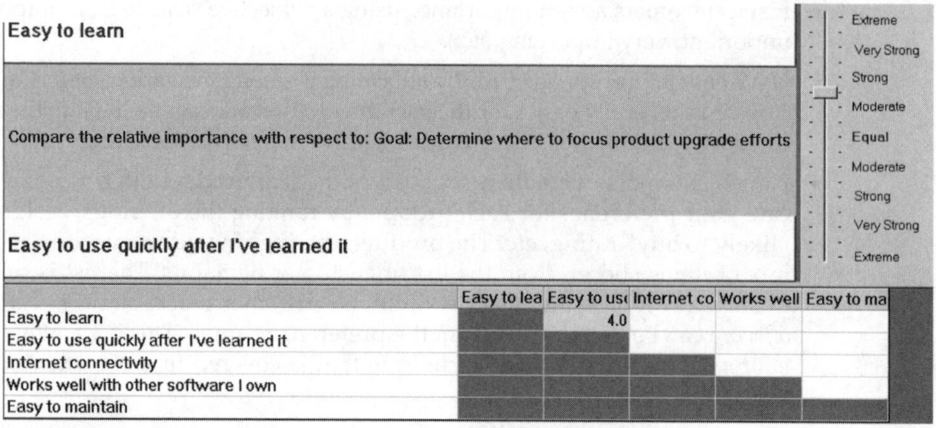

Easy to learn

Compare the relative importance with respect to: Goal: Determine where to focus product upgrade efforts

Easy to use quickly after I've learned it

Extreme
Very Strong
Strong
Moderate
Equal
Moderate
Strong
Very Strong
Extreme

	Easy to lea	Easy to us(	Internet co	Works well	Easy to ma
Easy to learn		4.0			
Easy to use quickly after I've learned it					
Internet connectivity					
Works well with other software I own					
Easy to maintain					

FIGURE 2.15 Matrix of categories for pairwise comparisons. (*Created using Expert Choice 2000 Software, www.expertchoice.com.**)

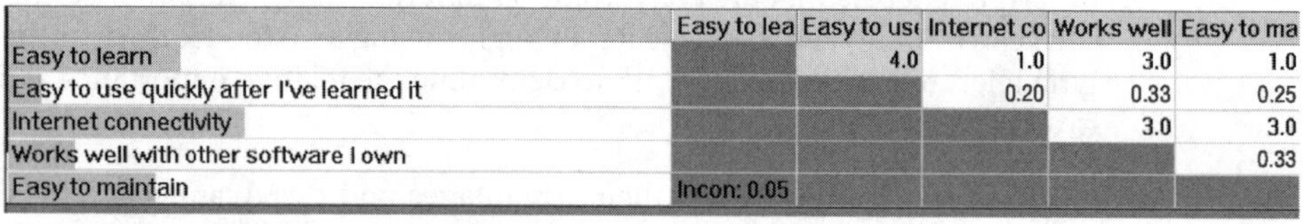

	Easy to lea	Easy to us(	Internet co	Works well	Easy to ma
Easy to learn		4.0	1.0	3.0	1.0
Easy to use quickly after I've learned it			0.20	0.33	0.25
Internet connectivity				3.0	3.0
Works well with other software I own					0.33
Easy to maintain	Incon: 0.05				

FIGURE 2.16 Completed top-level comparison matrix.

in Fig. 2.16. The shaded bars over the attribute labels provide a visual display of the relative importance of each major item to the customer. Numerically, the importance weights are:†

- Easy to learn: 0.264 (26.4%)
- Easy to use quickly after I've learned it: 0.054 (5.4%)
- Internet connectivity: 0.358 (35.8%)
- Works well with other software I own: 0.105 (10.5%)
- Easy to maintain: 0.218 (21.8%)

These relative importance weights can be used in QFD as well as in the AHP process that we are illustrating here. In our allocation of effort, we will want to emphasize those attributes with high-importance weights over those with lower weights.

---

*Although the analysis is easier with special software, you can obtain a good approximation using a spreadsheet. See the Appendix for details.

†See the Appendix for an example of how to derive approximate importance weights using Microsoft Excel.

Subcategory Importance Weights

The process used for obtaining category importance weights is repeated for the items within each category. For example, the items interactive tutorial, good printed documentation, and intuitive interface are compared pairwise within the category "easy to learn." This provides weights that indicate the importance of each item on the category. For example, within the "easy to learn" category, the customer weights might be:

- Interactive tutorial: 11.7%
- Good documentation: 20.0%
- Intuitive interface: 68.3%

If there were additional levels below these subcategories, the process would be repeated for them. For example, the intuitive interface subcategory might be subdivided into "number of menus," "number of submenus," "menu items easily understood," etc. The greater the level of detail, the easier the translation of the customers' demands into internal specifications. The tradeoff is that the process quickly becomes tedious and may end up with the customer being asked for input he isn't qualified to provide. In the case of this example, we'd probably stop at the second level.

Global Importance Weights

The subcategory weights just obtained tell us how much importance the item has with respect to the *category*, not with respect to the ultimate goal. Thus, they are often called *local importance weights*. However, the subcategory weights don't tell us the impact of the item on the overall goal, which is called its *global impact*. This is determined by multiplying the subcategory item weight by the weight of the category in which the item resides. The global weights for our example are shown in Table 2.4 in descending order.

The global importance weights are most useful for the purpose of allocating resources to the overall goal: *Determine where to focus product upgrade efforts*. For our example, Internet connectivity obviously has a huge customer impact. "Easy to use quickly after I've learned it" has relatively low impact. "Easy to learn" is dominated by one item: the user interface. These weights will be used to assess different proposed upgrade plans. Each plan will be evaluated on each subcategory item and assigned a value depending on how well it addresses the item. The values will be multiplied by the global weights to arrive at an overall score for the plan. The scores can be rank-ordered to provide a list that the process owner can use when making resource allocation decisions. Or, more proactively, the information can be used to develop a plan that emphasizes the most important customer demands. Table 2.5 shows part of a table that assesses project plans using the global weights. The numerical rating used in the table is 0=No Impact, 1=Some Impact, 3=Moderate Impact, 5=High Impact. Since the global weights sum to 1 (100%), the highest possible score is 5. Of the five plans evaluated, plan C has the highest score. It can be seen that plan C has a high impact on the six most important customer demands. It has at least a moderate impact on 10 of the top 11 items, with the exception of "Reasonably priced advanced technical support." These items account for almost 90% of the customer demands.

Category	Subcategory	Local Weight	Global Weight
Easy to learn	Intuitive interface	68.3%	18.0%
Internet connectivity	Online billpay	43.4%	15.5%
Internet connectivity	Download statements	23.9%	8.6%
Internet connectivity	Download investment information	23.9%	8.6%
Works well with other software	Hotlinks to spreadsheet	75.0%	7.9%
Easy to maintain	Free internet patches	35.7%	7.8%
Easy to maintain	Great, free self-help technical assistance on the internet	30.8%	6.7%
Easy to learn	Good documentation	20.0%	5.3%
Easy to maintain	Reasonably priced advanced technical support	20.0%	4.4%
Internet connectivity	Works well at 56K	8.9%	3.2%
Easy to learn	Interactive tutorial	11.7%	3.1%
Easy to maintain	Automatic internet upgrades	13.5%	2.9%
Works well with other software	Edit reports in word processor	25.0%	2.6%
Easy to use quickly after I've learned it	Savable frequently used reports	43.4%	2.3%
Easy to use quickly after I've learned it	Shortcut keys	23.9%	1.3%
Easy to use quickly after I've learned it	Short menus showing only frequently used commands	23.9%	1.3%
Easy to use quickly after I've learned it	Macro capability	8.9%	0.5%

Table 2.4 Local and Global Importance Weights

The plan's customer impact score is, of course, only one input into the decision-making process. The rigor involved usually makes the score a very valuable piece of information. It is also possible to use the same approach to incorporate other information, such as cost, timetable, feasibility, etc. into the final decision. The process owner would make pairwise comparisons of the different inputs (customer impact score, cost, feasibility, etc.) to assign weights to them, and use the weights to determine an overall plan score. Note that this process is a mixture of AHP and QFD.

Item	Plan Customer Impact Score	Intuitive Interface	Online Billpay	Download Statements	Download Investment Information	Hotlinks to Spreadsheet	Free Internet Patches	Great, Free Self-help Technical Assistance on the Internet	Good Documentation	Reasonably Priced Advanced Technical Support	Works Well at 56K	Interactive Tutorial
Global weight		18.0%	15.5%	8.6%	8.6%	7.9%	7.8%	6.7%	5.30%	4.4%	3.2%	3.1%
Plan A	3.57	3	5	1	1	3	3	4	5	5	5	5
Plan B	2.99	1	1	1	3	3	5	5	5	5	5	5
Plan C	4.15	5	5	5	5	5	5	3	3	1	3	3
Plan D	3.36	3	3	3	3	3	3	3	5	5	5	5
Plan E	2.30	5	0	0	0	5	5	1	1	0	1	1

TABLE 2.5 Example of Using Global Weights in Assessing Alternatives

CHAPTER 3

Data-Driven Management

Management decisions based on objective data is a key component of a Six Sigma initiative. On a project-by-project basis, Six Sigma projects provide a means of analyzing process data to achieve process improvements. In the larger organizational view, these process improvements are initiated so that the organization can achieve its organizational priorities. The priorities are developed based on analysis of key stakeholder needs and wants, including the customer, shareholder, and employee groups. In this way, data-driven management provides a means of achieving organizational objectives by quantifying needs or wants of stakeholder groups relative to current baselines, and acting upon the data to reduce those critical gaps in performance.

Attributes of Good Metrics

The choice of what to measure is crucial to the success of the organization. Improperly chosen metrics lead to suboptimal behavior and can lead people away from the organization's goals instead of toward them. Joiner (1994) suggests three systemwide measures of performance—overall customer satisfaction, total cycle time, and first-pass quality. An effective metric for quantifying first-pass quality is total cost of poor quality (later described in this chapter). Once chosen, the metrics must be communicated to the members of the organization. To be useful, the employee must be able to influence the metric through his performance, and it must be clear precisely how the employee's performance influences the metric.

Rose (1995) lists the following attributes of good metrics:

- They are customer centered and focused on indicators that provide value to customers, such as product quality, service dependability, and timeliness of delivery, or are associated with internal work processes that address system cost reduction, waste reduction, coordination and team work, innovation, and customer satisfaction.
- They measure performance across time, which shows trends rather than snapshots.
- They provide direct information at the level at which they are applied. No further processing or analysis is required to determine meaning.
- They are linked with the organization's mission, strategies, and actions. They contribute to organizational direction and control.
- They are collaboratively developed by teams of people who provide, collect, process, and use the data.

Rose also presents a performance measurement model consisting of eight steps:

- **Step 1: Performance category**—This category is the fundamental division of organizational performance that answers the question: What do we do? Sources for determining performance categories include an organization's strategic vision, core competencies, or mission statement. An organization will probably identify several performance categories. These categories define the organization at the level at which it is being measured.
- **Step 2: Performance goal**—The goal statement is an operational definition of the desired state of the performance category. It provides the target for the performance category and, therefore, should be expressed in explicit, action-oriented terms. An initial goal statement might be right on the mark, so complex that it needs further division of the performance category, or so narrowly drawn that it needs some combination of performance categories. It might be necessary to go back and forth between the performance goals in this step and the performance categories in step 1 before a satisfactory result is found for both.
- **Step 3: Performance indicator**—This is the most important step in the model because this is where progress toward the performance goal is disclosed. Here irrelevant measures are swept aside if they do not respond to an organizational goal. This is where the critical measures—those that communicate what is important and set the course toward organizational success—are established. Each goal will have one or more indicators, and each indicator must include an operational definition that prescribes the indicator's intent and makes its role in achieving the performance goal clear. The scope of the indicator might be viewed differently at various levels in the organization.
- **Step 4: Elements of measure**—These elements are the basic components that determine how well the organization meets the performance indicator. They are the measurement data sources—what is actually measured—and are controlled by the organization. Attempting to measure things that are beyond organizational control is a futile diversion of resources and energy because the organization is not in a position to respond to the information collected. This would be best handled in the next step.
- **Step 5: Parameters**—These are the external considerations that influence the elements of measure in some way, such as context, constraint, and boundary. They are not controlled by the organization but are powerful factors in determining how the elements of measure will be used. If measurement data analysis indicates that these external considerations present serious roadblocks for organizational progress, a policy change action could be generated.
- **Step 6: Means of measurement**—This step makes sense out of the preceding pieces. A general, how-to action statement is written that describes how the elements of measure and their associated parameters will be applied to determine the achievement level in the performance indicator. This statement can be brief, but clarifying intent is more important than the length.
- **Step 7: Notional metrics**—In this step, conceptual descriptions of possible metrics resulting from the previous steps are put in writing. This step allows everyone to agree on the concept of how the information compiled in the previous

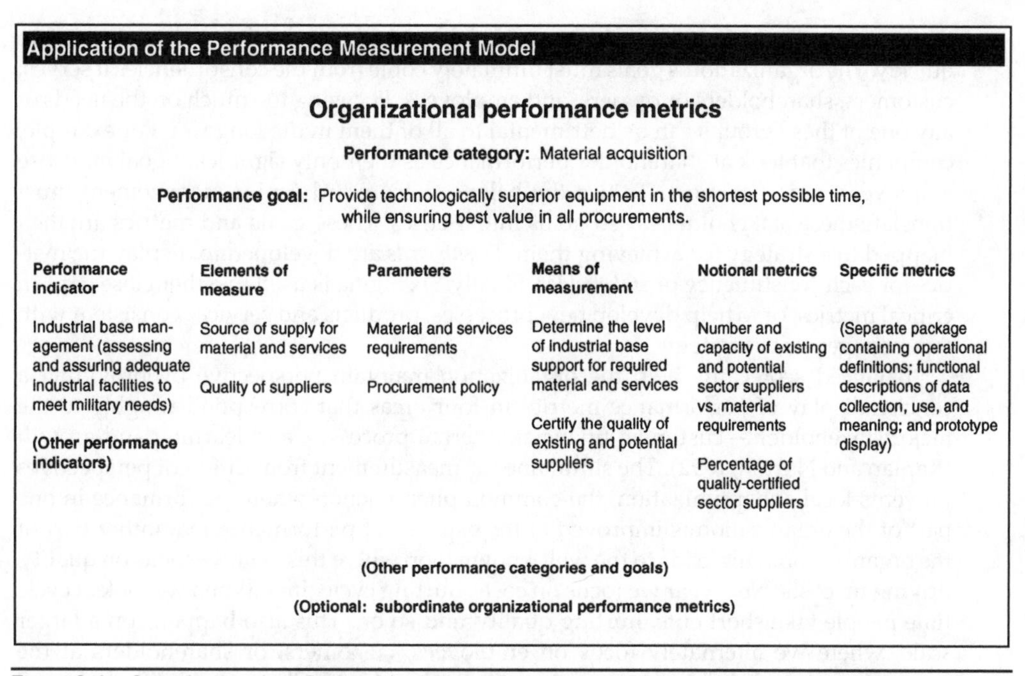
Application of the Performance Measurement Model

Organizational performance metrics

Performance category: Material acquisition

Performance goal: Provide technologically superior equipment in the shortest possible time, while ensuring best value in all procurements.

Performance indicator	Elements of measure	Parameters	Means of measurement	Notional metrics	Specific metrics
Industrial base management (assessing and assuring adequate industrial facilities to meet military needs)	Source of supply for material and services	Material and services requirements	Determine the level of industrial base support for required material and services	Number and capacity of existing and potential sector suppliers vs. material requirements	(Separate package containing operational definitions; functional descriptions of data collection, use, and meaning; and prototype display)
	Quality of suppliers	Procurement policy	Certify the quality of existing and potential suppliers	Percentage of quality-certified sector suppliers	
(Other performance indicators)					

(Other performance categories and goals)

(Optional: subordinate organizational performance metrics)

FIGURE 3.1 Organizational performance metrics. [*From Rose (1995). Reprinted by permission.*]

steps will be applied to measuring organizational performance. It provides a basis for validating the process and for subsequently developing specific metrics.

- **Step 8: Specific metrics**—In this final step, an operational definition and a functional description of the metrics to be applied are written. The definition and description describe the data, how they are collected, how they are used, and, most importantly, what the data mean or how they affect organizational performance. A prototype display of real or imaginary data and a descriptive scenario that shows what actions might be taken as a result of the measurement are also made. This last step is the real test of any metric. It must identify what things need to be done and disclose conditions in sufficient detail to enable subsequent improvement actions.

Rose presents an application of his model used by the U.S. Army Materiel Command, which is shown in Fig. 3.1.

The Balanced Scorecard

Given the magnitude of the difference between Six Sigma and the traditional three sigma performance levels, the decision to pursue Six Sigma performance obviously requires a radical change in the way things are done. The organization that makes this commitment will never be the same. Since the expenditure of time and resources will be huge, it is crucial that Six Sigma projects and activities are linked to the organization's top-level goals. It is even more important that these be the right goals. An organization that uses Six Sigma to pursue the wrong goals will just get to the wrong place more quickly. The organization's goals must ultimately come from the constituencies it serves: customers, shareholders or owners, and employees. Focusing too much on the needs of any one of these groups can be detrimental to all of them in the long run. For example, companies that look at shareholder performance as their only significant goal may lose employees and customers. To use the balanced scorecard senior management must translate these stakeholder-based goals into metrics. These goals and metrics are then mapped to a strategy for achieving them. *Dashboards* are developed to display the metrics for each constituency or *stakeholder*. Finally, Six Sigma is used to either close gaps in critical metrics, or to help develop new processes, products and services consistent with top management's strategy.

Balanced scorecards help the organization maintain perspective by providing a concise display of performance metrics in four areas that correspond roughly to the major stakeholders—customer, financial, internal processes, and learning and growth (Kaplan and Norton, 1992). The simultaneous measurement from different perspectives prevents local suboptimization, the common phenomenon where performance in one part of the organization is improved at the expense of performance in another part of the organization. This leads to the well-known loop where this year we focus on quality, driving up costs. Next year we focus on costs, hurting cycle time. When we look at cycle time people take short cuts, hurting quality, and so on. This also happens on a larger scale, where we alternately focus on employees, customers, or shareholders at the expense of the stakeholders who are not the current focus. Clearly, such "firefighting" doesn't make anyone happy. We truly need the "balance" in balanced scorecards.

Well-designed dashboards include statistical guidance to aid in interpreting the metrics. These guidelines most commonly take the form of statistical control limits, which are introduced in Chap. 7 and discussed in detail in Chap. 8. Limits are statistically calculated guidelines that operationally define when intervention is needed. Generally, when metrics fall within the limits, the process should be left alone. However, when a metric falls outside of the limits, it indicates that something important has changed that requires attention. An exception to these general rules occurs when a deliberate intervention is made to achieve a goal. In this case the metric is *supposed to* respond to the intervention by moving in a positive direction. The limits will tell leadership if the intervention produced the desired result. If so, the metric will go beyond the proper control limit indicating improvement. Once the metric stabilizes at the new and improved level, the limits should be recalculated so they can detect slippage.

Measuring Causes and Effects

Dashboard metrics are measurements of the results delivered by complex processes and systems. These results are, in a sense, "effects" caused by things taking place within the processes. For example, "cost per unit" might be a metric on a top-level dashboard. This is, in turn, composed of the cost of materials, overhead costs, labor, and so on. Cost of materials is a "cause" of the cost per unit. Cost of materials can be further decomposed into, say, cost of raw materials, cost of purchased sub-assemblies, and so on. At some level we reach a "root cause," or most basic reason behind an effect. Black Belts and Green Belts learn numerous tools and techniques to help them identify these root causes. However, the dashboard is the starting point for the quest.

In Six Sigma work, results are known as "*Ys*" and root causes are known as "*Xs*." Six Sigma's historical roots are technical and its originators generally came from engineering and scientific backgrounds. In the mathematics taught to engineers and scientists equations are used that often express a relationship in the form:

$$Y = f(X) \tag{3.1}$$

This equation simply means that the value identified by the letter Y is determined as a function of some other value X. The equation $Y = 2X$ means that if we know what X is, we can find Y if we multiply X by 2. If X is the temperature of a solution, then Y might be the time it takes the solution to evaporate. Equations can become more complicated. For example, $Y = f(X_1, X_2)$ indicates that the value Y depends on the value of two different X variables. You should think of the X in Eq. (3.1) as including any number of X variables. There can be many levels of dashboards encountered between the top-level Y, called the "Big Y," and the root cause Xs. In Six Sigma work some special notation has evolved to identify whether a root cause is being encountered, or an intermediate result. Intermediate results are sometimes called "Little Ys."

In these equations think of Y as the output of a process and the Xs as inputs. The process itself is symbolized by the f (). The process can be thought of as a *transfer function* that converts inputs into outputs in some way. An analogy is a recipe. Here's an example:

Corn Crisp Recipe
12 servings

3/4 cup yellow stone-ground cornmeal
1 cup boiling water
1/2 teaspoon salt
3 tablespoons melted butter

Preheat the oven to 400°F. Stir the cornmeal and boiling water together in a large glass measuring cup. Add the salt and melted butter. Mix well and pour onto a cookie sheet. Using a spatula, spread the batter out as thin as you possibly can—the thinner the crisper. Bake the cornmeal for half an hour or until crisp and golden brown. Break into 12 roughly equal pieces.

Here the Big Y is the customer's overall satisfaction with the finished corn crisp. Little Ys would include flavor ratings, "crunchiness" rating, smell, freshness, and other customer-derived metrics that drive the Big Y. Xs that drive the little Ys might include thinness of the chips, the evenness of the salt, the size of each chip, the color of the chip, and other measurements on the finished product. Xs could also be determined at each major step, for example, actual measurement of the ingredients, the oven temperature, the thoroughness of stirring, how much the water cools before it is stirred with the cornmeal, actual bake time, etc. Xs would also include the oven used, the cookware, utensils, etc.

Finally, the way different cooks follow the recipe is the *transfer function* or actual process that converts the ingredients into corn crisps. Numerous sources of variation (more Xs) can probably be identified by observing the cooks in action. Clearly, even such a simple process can generate some very interesting discussions. If you haven't developed dashboards it might be worthwhile to do so for the corn crisps as a practice exercise.

Figure 3.2 illustrates how dashboard metrics flow down until eventually linking with Six Sigma projects.

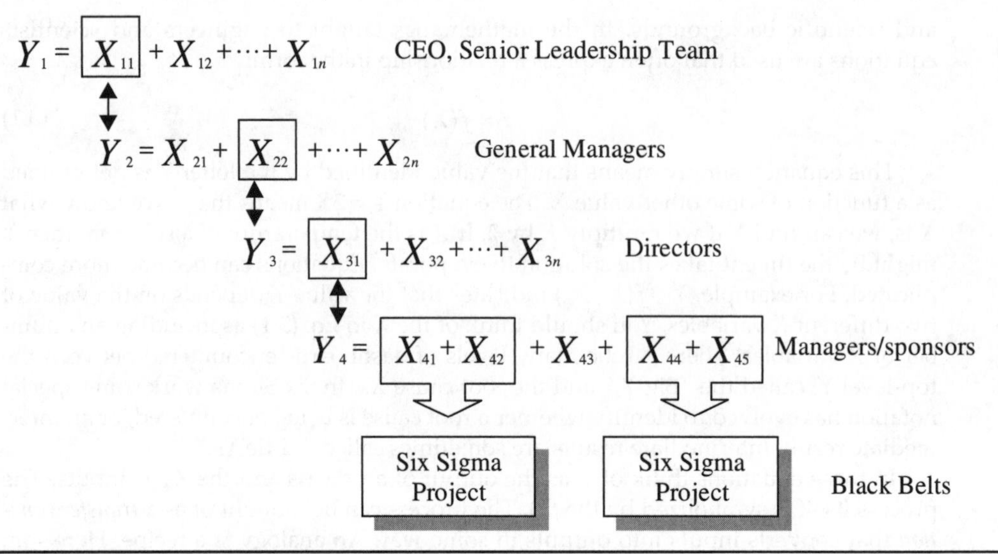

Figure 3.2 Flowdown of strategies to drivers and projects.

Customer Perspective

Let's take a closer look at each of the major perspectives on the balanced scorecard, starting with the customer. The balanced scorecard requires that management translate their vague corporate mission ("Acme will be #1 in providing customer value") into specific measures of factors that matter to customers. The customer scorecard answers the question: "How do our customers view us?"

To answer this, you must ask yourself two related questions: What things do customers consider when evaluating us? How do we know? While the only true way to answer these questions is to communicate with real customers, it is well established that customers in general tend to consider four broad categories of factors when evaluating an organization:

- **Quality**—How well do you keep your promises by delivering error free service or defect free product? Did I receive what I ordered? Was it undamaged? Are your promised delivery times accurate? Do you honor your warranty or pay your claims without a hassle?
- **Timeliness**—How fast is your service? How long does it take to have my order delivered? Do improvements appear in a timely manner?
- **Performance and service**—How do your products and services help me? Are they dependable?
- **Value**—What is the cost of buying and owning your product or service? Is it worth it?

The first step in the translation is to determine precisely *what* customers consider when evaluating your organization. This can be done by communicating with customers via one-on-one contacts, focus groups, questionnaires, chat rooms, forums, and so on. Management should see the actual, unvarnished words used by customers to describe what they think about the company, its products, and its services. Once management is thoroughly familiar with their target customer, they need to articulate their customer goals in words meaningful to them. For example, management might say:

- We will cut the time required to introduce a new product from 9 months to 3 months.
- We will be the best in the industry for on-time delivery.
- We will intimately involve our customers in the design of our next major product.

These goals must be operationalized by designating metrics to act as surrogates for the goals. Think of the goals themselves as *latent* or hidden constructs. The objective is to identify observable things directly related to the goals that can be measured. These are *indicators* that guide you toward your goals. Table 3.1 shows examples of how the goals mentioned above might be operationalized.

Plans, budgets, goals and targets are *key requirements* set by the leadership for the organization. If not done properly, the behavior driven by these key requirements may not be anywhere close to what the leadership desires, or expects. Key requirements are used to assess employee performance, which is linked to promotions, pay increases, bonuses and many other things that people care about a great deal.

The most common flaw in goal setting is the tendency to set goals that are mere wishes and hopes. The leadership looks at a metric and pontificates on what a "good" level of performance would be for it. If enough heads nod around the conference table, this becomes that metric's target.

A better way to arrive at goals for key requirements is to examine the actual history of the metric over time. This information should be plotted on a process behavior chart (see Chaps. 7 and 8). If the metric falls within the calculated limits the bulk of the time, then the process is considered predictable. Typically, unless the metric is operationalizing a differentiation strategy, the goal for predictable processes will be to maintain the historical levels. These metrics will not appear on the dashboards that the leadership reviews on a routine basis. However, their performance is monitored by process owners and the leadership is informed on an exception basis if the process behavior.

Goal	Candidate Metrics
We will cut the time required to introduce a new product from 9 months to 3 months.	• Average time to introduce a new product for most recent month or quarter • Number of new products introduced in most recent quarter
We will be the best in the industry for on-time delivery.	• Percentage of on-time deliveries • Best in industry on-time delivery percentage divided by our on-time delivery percentage • Percentage of late deliveries
We will intimately involve our customers in the design of our next major product.	• Number of customers on design team(s) • Number of customer suggestions incorporated in new design

Table 3.1 Operationalizing Goals

Internal Process Perspective

In the Internal Process section of the balanced scorecard we develop metrics that help answer the question: Which internal processes are critical to meet customer and shareholder goals? Internal process excellence is linked to customer perceived value, but the linkage is indirect and imperfect. It is often possible to hide internal problems from customers by throwing resources at problems; for example, increased inspection and testing. Also, customer perceived value is affected by factors other than internal processes such as price, competitive offerings, etc. Similarly, internal operations consume resources so they impact the shareholders. Here again, the linkage is indirect and imperfect. For example, sometimes it is in the organization's strategic interest to drive up costs in order to meet critical short-term customer demands or to head off competitive moves in the market. Thus, simply watching the shareholder or customer dashboards won't always give leadership a good idea of how well internal processes are performing. A separate dashboard is needed for this purpose.

This section of the scorecard gives operational managers the internal direction they need to focus on customer needs. Internal metrics should be chosen to support the leadership's customer strategy, with knowledge of what customers need from internal operations. Process maps should be created that show the linkage between suppliers, inputs, process activities, outputs and customers (SIPOC). SIPOC is a flowcharting technique that helps identify those processes that have the greatest impact on customer satisfaction, as discussed in greater detail in Chap. 7.

Companies need to identify and measure their core competencies. These are areas where the company must excel. It is the source of their competitive advantage. Goals in these areas must be ambitious and challenging. This is where you "Wow" your customer. Other key areas will pursue goals designed to satisfy customers, perhaps by maintaining competitive performance levels. Table 3.2 shows how core competencies might drive customer value propositions. The metrics may be similar for the different companies, but the goals will differ significantly. For example, Company A would place greater emphasis on the time required to develop and introduce new services. Companies B and C would not ignore this aspect of their internal operations, but their goals would be less ambitious in this area than Company A's. Company A is the industry *benchmark* for innovation.

Of course, it is possible that your competitor will try to leapfrog you in your core competency, becoming the new benchmark and stealing your customers. Or you may find that your customer base is dwindling and the market for your particular competency

Internal Process	Company A	Company B	Company C
Innovation	X		
Customer relationship management		X	
Operations and logistics			X
Customer value proposition	Product or service attributes	Flexibility, customization	Cost, dependability

"X" indicates the company's core competency.

TABLE 3.2 Customer Value Proposition versus Core Competency

is decreasing. Leadership must stay on the alert for such developments and be prepared to react quickly. Most companies will fight to maintain their position of industry leadership as long as there is an adequate market. Six Sigma can help in this battle because Six Sigma projects are usually of short duration strategically speaking, and Black Belts offer a resource that can be redeployed quickly to where they are most needed.

Innovation and Learning Perspective

In the Innovation and Learning Perspective section of the balanced scorecard we develop metrics that help answer the question: Can we continue to improve and create value? Success is a moving target. What worked yesterday may fail miserably tomorrow. Previous sections of the balanced scorecard have identified the metrics the leadership considers to be most important for success in the near future. But the organization must be prepared to meet the new and changing demands that the more distant future will surely bring. Building shareholder value is especially dependent on the company's ability to innovate, improve, and learn. The intrinsic value of a business is the discounted value of the cash that can be taken out of the business during its remaining life (Buffett, 1996). Intrinsic value is directly related to a company's ability to create new products and processes, to improve operating efficiency, to discover and develop new markets, and to increase revenues and margins. Companies able to do this well will throw off more cash over the long term than companies that do it poorly. The cash generated can be withdrawn by the owners, or reinvested in the business.

Innovation and learning were the areas addressed by the *continuous improvement* (CI) initiatives of the past. Devotees of CI will be happy to learn that it's alive and well in the Six Sigma world. However, CI projects were often local in scope, while most Black Belt Six Sigma projects are cross-functional. Many so-called Green Belt projects are reminiscent of the CI projects in the past. Also, CI tended to focus narrowly on work processes, while Green Belt projects cover a broader range of business processes, products, and services. A well-designed Six Sigma program will have a mix of Green Belt and Black Belt projects addressing a range of enterprise and local process improvement issues.

Dashboards designed to measure performance in the area of Innovation and Learning often address three major areas: employee competencies, technology, and corporate culture. These are operationalized in a wide variety of ways. One metric is the average rate of improvement in the sigma level of an organizational unit. As described in Chapter 1, an organizational Six Sigma strategy to reduce mistakes, errors, and defects may target for a factor of 10 improvements every two years, which translates to about 17% per month. This breakthrough rate of improvement is usually not attained instantly and a metric of the actual rate is a good candidate for including on the Innovation and Learning dashboard. The rate of improvement is a measure of the overall maturity of the Six Sigma initiative. Other Innovation and Learning metric candidates might include such things as:

- Results of employee feedback
- R&D cycle time
- Closure of gaps identified in the training needs audit

Financial Perspective

Obsession with financial metrics has been the undoing of many improvement initiatives. When senior leaders look only at results they miss the fact that these results come from a complex chain of interacting processes that effectively and efficiently produce value for customers. Only by providing value that customers are willing to pay for can an enterprise generate sales, and only by creating these values at a cost less than their price can it produce profits for owners. For many companies the consequence of looking only at short-term financial results has been a long-term decline in business performance. Many companies have gone out of business altogether.

The result of this unfortunate history is that many critics have advocated the complete abandonment of the practice of using financial metrics to guide leadership action. The argument goes something like this: since financial results are determined by a combination of customer satisfaction and the way the organization runs its internal operations, if we focus on these factors the financial performance will follow in due course. This is throwing the baby out with the bathwater. The flaw in the logic is that it assumes that leaders and managers *know* precisely *how* customer satisfaction and internal operational excellence lead to financial results. This arrogance is unjustified. Too often we learn in retrospect that we are focusing on the wrong things and the financial results fail to materialize. For example, we may busily set about improving the throughput of a process that already has plenty of excess capacity. All we get from this effort is more excess capacity. Even well-managed improvement efforts won't result in bottom-line impact because management fails to take the necessary steps such as reducing excess inventory, downsizing extra personnel, selling off unneeded equipment, etc. As Toyota's Taiichi Ohno says:

> If, as a result of labor saving, 0.9 of a worker is saved, it means nothing. At least one person must be saved before a cost reduction results. Therefore, we must attain worker saving.
>
> Taiichi Ohno
> Toyota Production System: Beyond Large-Scale Production

The truth is, it's very difficult to lay people off and a poor reward for people who may have participated in creating the improvement. Most managers agree that this is the worst part of their job. However, simply ignoring the issue isn't the best way to deal with it. Plans must be made before starting a project for adjusting to the consequences of success. If there will be no bottom-line impact because there are to be no plans to convert the savings into actual reductions in resource requirements, the project shouldn't be undertaken in the first place. On the other hand, plans can often be made at the enterprise level for dealing with the positive results of Six Sigma by such means as hiring moratoriums, early retirement packages, etc. Better still are plans to increase sales or to grow the business to absorb the new capacity. This can often be accomplished by modifying the customer value proposition through more reliable products, lower prices, faster delivery time, lower cycle times, etc. These enhancements are made possible as a result of the Six Sigma improvements.

There are many dangers in failing to monitor a financial results. We may blindly pour resources into improving customer satisfaction inaccurately measured by a faulty or incomplete survey. In other cases, the competition may discover a new technology that makes ours obsolete. The list of issues that potentially disrupt the link between internal strategies and financial performance is endless. Financial performance metrics provide the feedback needed to test assumptions.

Actual metrics for monitoring financial performance are numerous. The top-level dashboard will often include metrics in the areas of improved efficiency (e.g., cost per unit, asset utilization) or improved effectiveness (e.g., revenue growth, market share increase, profit per customer). These costs can often be broken down into operational quality costs, sometimes referred to as Cost of Quality (or more appropriately Cost of Poor Quality) categories.

Cost of Poor Quality

The history of quality costs dates back to the first edition of *Juran's QC Handbook* in 1951. Today, quality cost accounting systems are part of every modern organization's quality improvement strategy, as well as many quality standards. Quality cost systems identify internal opportunities for return on investment. As such, quality costs stress avoiding defects and other behaviors that cause customer dissatisfaction, yet provide little understanding of the product or service features that satisfy or delight customers. It is conceivable that a firm could drive quality costs to zero and still go out of business.

Cost of quality includes any cost that would not be expended if quality were perfect. This includes such obvious costs as scrap and rework, but it also includes less obvious costs, such as the cost to replace defective material, expedite shipments for replacement material, the staff and equipment to process the replacement order, etc. Service businesses also incur quality costs; for example, a hotel incurs a quality cost when room service delivers a missing item to a guest. Specifically, quality costs are a measure of the costs associated with the achievement or nonachievement of product or service quality—including all product or service requirements established by the company and its contracts with customers and society. Requirements include marketing specifications, end-product and process specifications, purchase orders, engineering drawings, company procedures, operating instructions, professional or industry standards, government regulations, and any other document or customer needs that can affect the definition of product or service. More specifically, quality costs are the total of the cost incurred by (a) investing in the *prevention* of nonconformances to requirements; (b) *appraising* a product or service for conformance to requirements; and (c) *failure* to meet requirements (Fig. 3.3).

For most organizations, quality costs are hidden costs. Unless specific quality cost identification efforts have been undertaken, few accounting systems include provision for identifying quality costs. Because of this, unmeasured quality costs tend to increase. Poor quality impacts companies in two ways: higher cost and lower customer satisfaction. The lower customer satisfaction creates price pressure and lost sales, which results in lower revenues. The combination of higher cost and lower revenues eventually brings on a crisis that may threaten the very existence of the company. Rigorous cost of quality measurement is one technique for preventing such a crisis from occurring. Figure 3.4 illustrates the hidden cost concept.

As a general rule, quality costs increase as the detection point moves further up the production and distribution chain. The lowest cost is generally obtained when errors are prevented in the first place. If nonconformances occur, it is generally least expensive to detect them as soon as possible after their occurrence. Beyond that point there is loss incurred from additional work that may be lost. The most expensive quality costs are from nonconformances detected by customers. In addition to the replacement or repair loss, a company loses customer goodwill and their reputation is damaged when the customer relates his experience to others. In extreme cases, litigation may result, adding even more cost and loss of goodwill.

Another advantage of early detection is that it provides more meaningful feedback to help identify root causes. The time lag between production and field failure makes it very difficult to trace the occurrence back to the process state that produced it. While

PREVENTION COSTS

The costs of all activities specifically designed to prevent poor quality in products or services. Examples are the costs of new product review, quality planning, supplier capability surveys, process capability evaluations, quality improvement team meetings, quality improvement projects, quality education and training.

APPRAISAL COSTS

The costs associated with measuring, evaluating or auditing products or services to ensure conformance to quality standards and performance requirements. These include the costs of incoming and source inspection/test of purchased material, in process and final inspection/test, product, process, or service audits, calibration of measuring and test equipment, and the costs of associated supplies and materials.

FAILURE COSTS

The costs resulting from products or services not conforming to requirements or customer/user needs. Failure costs are divided into internal and external failure cost categories.

INTERNAL FAILURE COSTS

Failure costs occurring prior to delivery or shipment of the product, or the furnishing of a service, to the customer. Examples are the costs of scrap, rework, reinspection, retesting, material review, and down grading.

EXTERNAL FAILURE COSTS

Failure costs occurring after delivery or shipment of the product, and during or after furnishing of a service, to the customer. Examples are the costs of processing customer complaints, customer returns, warranty claims, and product recalls.

TOTAL QUALITY COSTS

The sum of the above costs. It represents the difference between the actual cost of a product or service, and what the reduced cost would be if there was no possibility of substandard service, failure of products, or defects in their manufacture.

FIGURE 3.3 Quality costs general—description. [*From Campanella (1999). Copyright © 1999 by ASQ Quality Press.*]

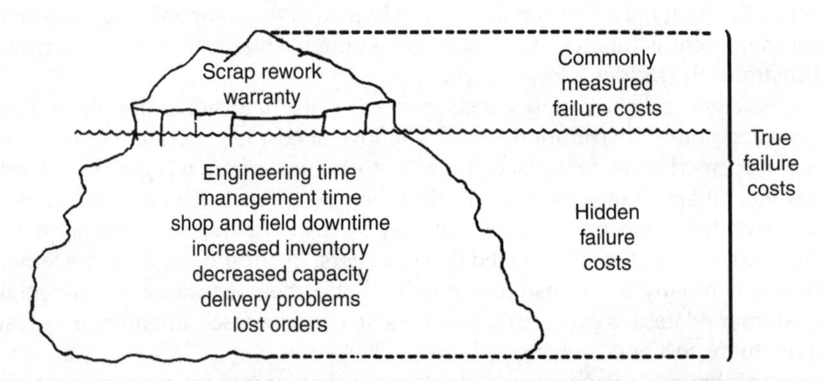

FIGURE 3.4 Hidden cost of quality and the multiplier effect. [*From Campanella (1990). Copyright © 1990 by ASQ Quality Press.*]

field failure tracking is useful in *prospectively* evaluating a "fix," it is usually of little value in *retrospectively* evaluating a problem.

Quality cost measurement need not be accurate to the penny to be effective. The purpose of measuring such costs is to provide broad guidelines for management decision-making and action. The very nature of cost of quality makes such accuracy impossible. In some instances it will only be possible to obtain periodic rough estimates of such costs as lost customer goodwill, cost of damage to the company's reputation, etc. These estimates can be obtained using special audits, statistical sampling, and other market studies. These activities can be jointly conducted by teams of marketing, accounting, and Six Sigma personnel. Since these costs are often huge, these estimates must be obtained. However, they need not be obtained every month. Annual studies are usually sufficient to indicate trends in these measures.

Cost of Quality Examples

I. Prevention costs—Costs incurred to prevent the occurrence of nonconformances in the future, such as:*

A. Marketing/customer/user
- Marketing research
- Customer/user perception surveys/clinics
- Contract/document review

B. Product/service/design development
- Design quality progress reviews
- Design support activities
- Product design qualification test
- Service design qualification
- Field tests

C. Purchasing
- Supplier reviews
- Supplier rating
- Purchase order tech data reviews
- Supplier quality planning

D. Operations (manufacturing or service)
- Operations process validation
- Operations quality planning
    - Design and development of quality measurement and control equipment
- Operations support quality planning
- Operator quality education
- Operator SPC/process control

*All detailed quality cost descriptions are from *Principles of Quality Costs*, John T. Hagan, editor. Milwaukee, WI: ASQ Quality Press, Appendix B.

E. Six Sigma, lean, and quality administration
    - Administrative salaries
    - Administrative expenses
    - Program planning
    - Performance reporting
    - Education and training
    - Improvement projects
    - Process and product audits
    - Other prevention costs

II. **Appraisal costs**—Costs incurred in measuring and controlling current production to ensure conformance to requirements, such as:

A. Purchasing appraisal costs
    - Receiving or incoming inspections and tests
    - Measurement equipment
    - Qualification of supplier product
    - Source inspection and control programs

B. Operations (manufacturing or service) appraisal costs
    - Planned operations inspections, tests, audits
        - Checking labor
        - Product or service quality audits
        - Inspection and test materials
    - Setup inspections and tests
    - Special tests (manufacturing)
    - Process control measurements
    - Laboratory support
    - Measurement equipment
        - Depreciation allowances
        - Measurement equipment expenses
        - Maintenance and calibration labor
    - Outside endorsements and certifications

C. External appraisal costs
    - Field performance evaluation
    - Special product evaluations
    - Evaluation of field stock and spare parts

D. Review of tests and inspection data

E. Miscellaneous quality evaluations

III. **Internal failure costs**—Costs generated before a product is shipped as a result of nonconformance to requirements, such as:

A. Product/service design failure costs (internal)
   - Design corrective action
   - Rework due to design changes
   - Scrap due to design changes

B. Purchasing failure costs
   - Purchased material reject disposition costs
   - Purchased material replacement costs
   - Supplier corrective action
   - Rework of supplier rejects
   - Uncontrolled material losses

C. Operations (product or service) failure costs
   - Material review and corrective action costs
     - Disposition costs
     - Troubleshooting or failure analysis costs (operations)
     - Investigation support costs
     - Operations corrective action
   - Operations rework and repair costs
     - Rework
     - Repair
   - Reinspection/retest costs
   - Extra operations
   - Scrap costs (operations)
   - Downgraded end product or service
   - Internal failure labor losses

D. Other internal failure costs

IV. **External failure costs**—Costs generated after a product is shipped as a result of nonconformance to requirements, such as:

A. Complaint investigation/customer or user service
B. Returned goods
C. Retrofit costs
D. Recall costs
E. Warranty claims
F. Liability costs
G. Penalties
H. Customer/user goodwill
I. Lost sales
J. Other external failure costs

Quality Cost Bases

The guidelines for selecting a base for analyzing quality costs are:

- The base should be related to quality costs in a meaningful way.
- The base should be well-known to the managers who will receive the quality cost reports.
- The base should be a measure of business volume in the area where quality cost measurements are to be applied.
- Several bases are often necessary to get a complete picture of the relative magnitude of quality costs.
- Some commonly used bases are (Campanella, 1990):
  - A labor base (such as total labor, direct labor, or applied labor)
  - A cost base (such as shop cost, operating cost, or total material and labor)
  - A sales base (such as net sales billed, or sales value of finished goods)
  - A unit base (such as the number of units produced, or the volume of output)

While actual dollars spent are usually the best indicator for determining where quality improvement projects will have the greatest impact on profits and where corrective action should be taken, unless the production rate is relatively constant, it will not provide a clear indication of quality cost improvement trends. Since the goal of the cost of quality program is improvement over time, it is necessary to adjust the data for other time-related changes such as production rate, inflation, etc. Total quality cost compared to an applicable base result in an index, which may be plotted and analyzed using control charts as described in Chap. 8.

For long-range analyses and planning, net sales is the base most often used for presentations to top management (Campanella, 1990). If sales are relatively constant over time, the quality cost analysis can be performed for relatively short spans of time. In other industries this figure must be computed over a longer time interval to smooth out large swings in the sales base. For example, in industries such as ship building or satellite manufacturing, some periods may have no deliveries, while others have large dollar amounts. It is important that the quality costs incurred be related to the sales for the same period. Consider the sales as the "opportunity" for the quality costs to happen.

Some examples of cost of quality bases are (Campanella, 1990):

- Internal failure costs as a percent of total production costs
- External failure costs as an average percent of net sales
- Procurement appraisal costs as a percent of total purchased material cost
- Operations appraisal costs as a percent of total production costs
- Total quality costs as a percent of production costs

Strategy Deployment Plan

Unlike traditional measurement systems, which tend to have a control bias, balanced scorecards are based on strategy. The idea is to realize the leadership vision using a set of linked strategies. Metrics operationalize these strategies and create a bond between the activities of the organization and the vision of the leadership.

Figure 3.5 illustrates these principles for a hypothetical organization, where the metrics are shown in rectangles on the left side of the figure. The strategy deployment plan clearly displays the linkage between the metrics and the larger issues of interest. These unobserved or latent constructs are shown in ellipses and are inferred from the metrics. This perspective helps leadership understand the limitations of metrics, as well as their value. If, for example, all of the metrics leading to shareholder perceived value are strongly positive, but surveys of the shareholders (Voice of Shareholder) indicate shareholder dissatisfaction, then the dashboard metrics are obviously inadequate and need to be revised.

The organization is pursuing a particular strategy and emphasizing certain dashboard metrics, which are shown in boldface type. Goals for these metrics will be set very high in an attempt to differentiate this organization from its competition. Goals for other metrics (key requirements) will be set to achieve competitiveness, usually by maintaining the historical levels for these metrics.

The organization's leaders believe their core competencies are in the areas of technology and customer service. They want their customers to think of them as the company to go to for the very best products completely customized to meet extremely demanding needs.

However, note that the organization's differentiators are:

1. Cost per unit
2. Revenues from new sources
3. (Customer) service relationship
4. Product introductions, (new product) revenues
5. Research deployment time

It appears that item 1 is inconsistent with the leadership vision: achieving benchmark status for items 2 to 5 would conflict with a similar goal for item 1. The plan indicates that the productivity strategy for this organization should be reevaluated. Unless the company is losing its market due to uncompetitive prices, or losing its investors due to low profits, item 1 should probably be a key requirement maintained at historical levels. If costs are extremely out of line, cost per unit might be the focus of a greater than normal amount of attention to bring it down to reasonable levels. However, it should not be shown as a *differentiator* on the strategic dashboard. The company has no desire to become a cost leader in the eyes of customers or shareholders.

Six Sigma plays a vital role in achieving the leadership vision by providing the resources needed to facilitate change where it is needed. Six Sigma projects are linked to dashboard metrics through the project selection process discussed in Chap. 4. The process involves calculating the expected impact of the project on a dashboard metric. The metrics used for Six Sigma projects are typically on a lower-level dashboard, but since the lower-level dashboard metrics flow down from the top level, the linkage is explicit. The process begins by identifying the gap between the current state and the goal for each top-level dashboard metric; Master Black Belts commonly assist with this activity. Six Sigma projects impacting differentiator dashboard metrics which show large gaps are prime candidates. This determination is usually done by Master Black Belts. This information is also very useful in selecting Black Belt candidates.

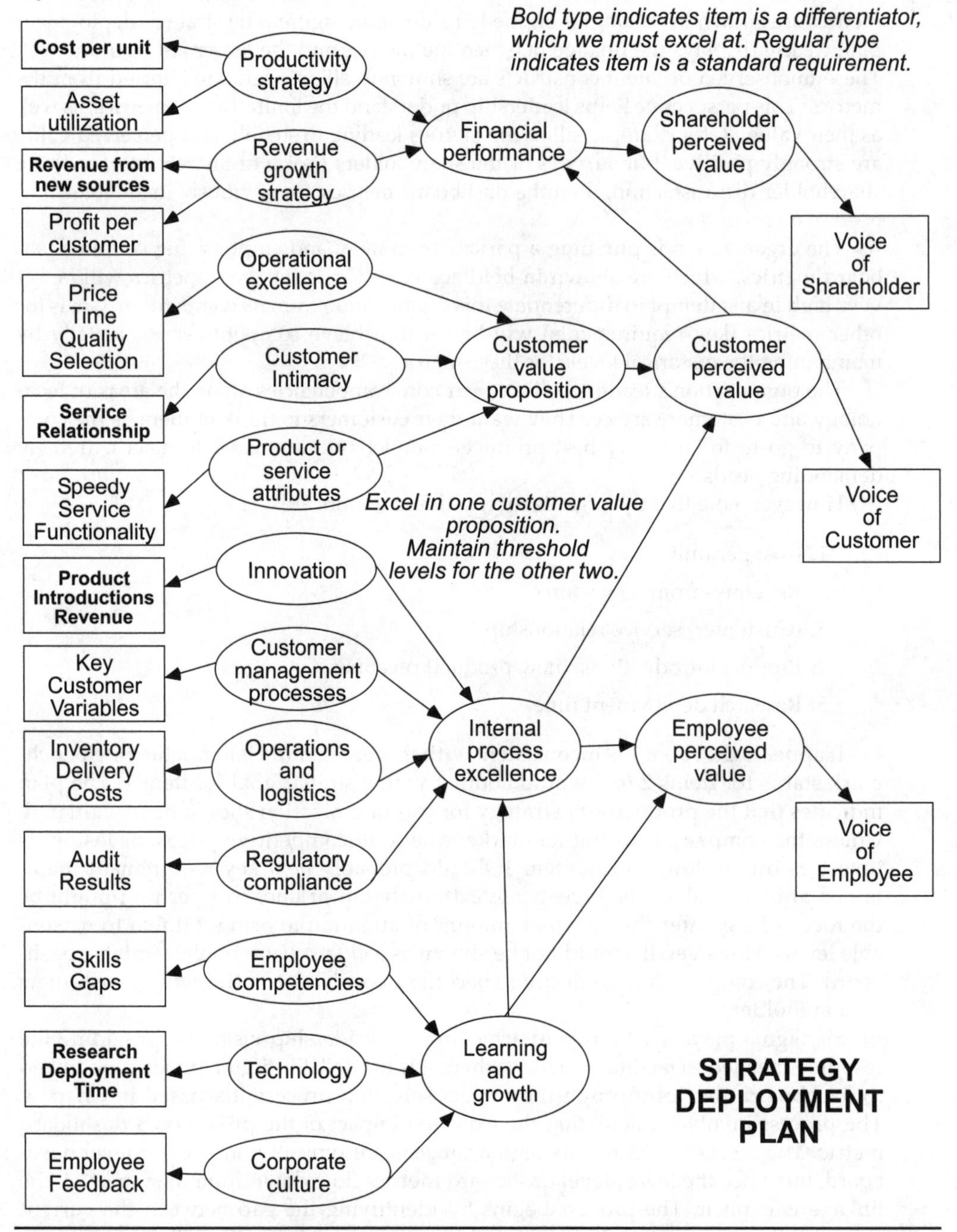

FIGURE 3.5 Strategy deployment plan for a hypothetical organization.

Candidates with backgrounds in areas where high-impact projects will be pursued may be given preference over equally qualified candidates from elsewhere in the organization.

Dashboard Design

Strategies are operationalized by metrics displayed on dashboards. Dashboard displays are designed to standardize information throughout the organization. A process owner at any level of the organization should be able to quickly recognize the meaning of the dashboard data and accelerate the learning cycle. The strategy deployment plan is merely a hypothesis. Science-based management requires testing of the hypothesis to determine if it is in reasonable agreement with the facts, and take action; or revise the strategy deployment plan accordingly.

Dashboard metrics should embody the principles of good metrics discussed earlier in the chapter:

1. Display performance over time.
2. Include statistical guidelines to help separate signal (variation from an identifiable cause) from noise (variation similar to random fluctuations).
3. Show causes of variation when known.
4. Identify acceptable and unacceptable performance (defects).
5. Be linked to higher-level dashboards (goals and strategies) or lower-level dashboards (drivers) to guide strategic activity within the organization.

The statistical process control charts described in Chaps. 7 and 8 address the first two concerns: they are the proper tool to analyze the metric over time and include guidelines (i.e. statistical control limits) to separate real process changes from the random variation expected within a stable process. SPC software can include more advanced features that allow for drill-down between upper and lower level metrics (principle 5), as well as drill-down to display and quantify causes of excessive variation (principles 3 and 4).

An example of a business-level customer dashboard based on the example strategy deployment plan described earlier is shown in Fig. 3.6. The tabular dashboard (top left in the right-hand pane of the software application window) provides a quick view of top-level metrics, with status (either predictably in control or subject to unpredictable out of control signals) and sigma level over a given period of time. For example, the *Service Time* metric has a relatively low Sigma level of 1.115, corresponding to a defect rate of approximately 13%. Its *In Control* status further indicates the error rate is inherent to the process and not due to unpredictable fluctuations: the process must be fundamentally changed to reduce its error rate from 13%. (The interpretation of control charts are further discussed in Chaps. 7 and 8).

Each of the key dashboard metrics may be drilled-down to display its behavior over time, as shown in Fig. 3.6 for the *Service Time* metric. The control chart in this example displays the observed process behavior for a 30-day window. The absence of data beyond the statistical Upper Control Limit (UCL) of 18.375 and Lower Control Limit (LCL) of 0.234 indicate the process is stable during this time period, as initially indicated in the dashboard. An interesting period of reduced variation is noted towards the middle of

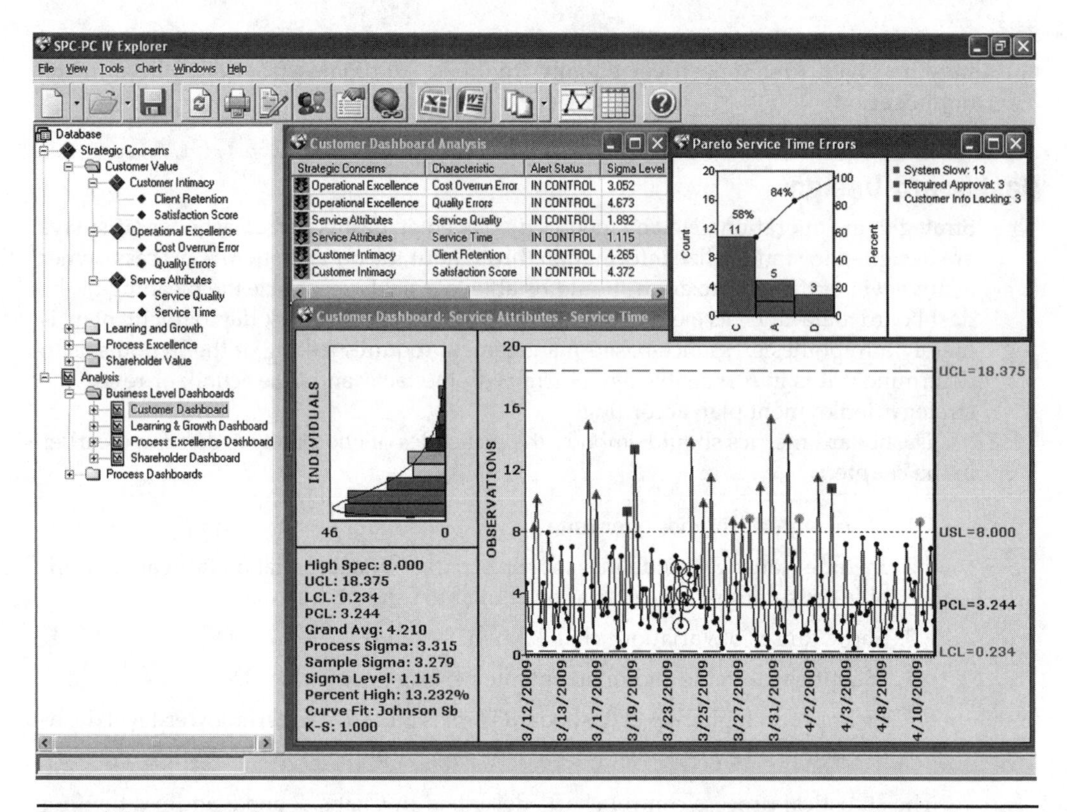

FIGURE 3.6 Example of a business level dashboard with drill down. (*Courtesy www.qualityamerica.com by permission.*)

the chart. The large circles around the data points indicate this period of relatively low variability is statistically significant; further investigation of this period may translate to understanding of process design that could lead to reduced overall variation.

The process goal (i.e. the Upper Specification Limit) is conveniently shown on the chart as 8 minutes. The visual display makes it easy to see there are 19 observations exceeding this requirement: 13 are plotted as triangles; 3 as boxes; and 3 as filled circles. Each of these symbol types has been linked to a potential cause identified during data entry by the customer support personnel: boxes associated with "System Slow;" triangles with "Required Approval"; and filled circles with "Customer Information Lacking." The time-ordered nature of the control chart shows these non-conformances to be somewhat equally distributed over time, rather than clumped in a specific time period.

Further drill-down provides one of several possible Pareto diagrams associated with the data. The Pareto diagram shown in Fig. 3.6 provides the product associated with each of the excessive service times (i.e. service time greater than 8 minutes), further partitioned by stacked boxes within each product type based on the customer service error code. This particular drill-down indicates Product C has the highest number of errors (58%), followed by Product A (an additional 26%), with the remaining errors associated with Product D. Another Pareto easily created through drill-down (and not shown) indicates these percentages roughly correspond to the percent of total products processed. In other words, all things being equal, we might expect Product C to have about 60% of the errors, since it represents that percentage of the total product processed.

Interestingly, all of the Product D errors are associated with the "Customer Information Lacking" error; a majority of the Product A errors are associated with the "Required Approval" error; and neither of these errors occurred for Product C. While not statistically significant among themselves, these are potential issues useful for project directed improvements to the process. For example, a project may be designed to reduce errors due to "Customer Information Lacking," and further analysis of similar data over a longer time period may be used to isolate the products for which this error occurs.

In this way, the dashboard provides at first glance the means to monitor the status of the organizational transformation towards the goals of the strategic plan. The dashboard's drill-down capabilities provide a much richer source of information to further understand the process complexities and potentials. Clearly, it is beneficial to design an information management system capable of collecting and disseminating information reflective of the ever-changing organizational needs.

Information Systems Requirements

Balanced scorecards begin with the highest level metrics. At any given level, dashboards will display a relatively small number of metrics. While this allows the user of the dashboard to focus on key items, it also potentially hampers their ability to understand the larger issues associated with a given process change. Information systems (IS) help address this issue through the "drill down" capability shown in the earlier section. Drill down involves disaggregating dashboard metrics into their component parts. For example, a cost-per-unit metric can be decomposed by division, plant, department, shift, worker, week, etc. These components of the higher-level metric are sometimes already on dashboards at lower levels of the organization, in which case the answer is provided in advance. However, if the established lower-level dashboard metrics can't explain the situation, other exploratory drill downs may be required. Online analytic processing (OLAP) cubes often ease the demands on the IS caused by drill down requests.

This raises an important point: in Six Sigma organizations the IS must be accessed by many more people. The attitude of many IS departments is "The data systems belong to us. If you want some data, submit a formal request." In a Six Sigma organization, this attitude is hopelessly outmoded. The demands on IS increase dramatically when Six Sigma is deployed. In addition to the creation of numerous dashboards, and the associated drill downs and problem investigations, the Black Belts and Green Belts make frequent use of IS in their projects. Six Sigma "show me the data" emphasis places more demands on the IS. In planning for Six Sigma success, companies need to assign a high-level champion to oversee the adaptation of IS to the new realities of Six Sigma. A goal is to make access as easy as possible while maintaining data security and integrity.

Although it's important to be timely, most Six Sigma data analyses don't require real-time data access. Data that are a day or a few days old will often suffice. The IS department may want to provide facilities for off-line data analysis by Six Sigma team members and Belts. A few high-end workstations capable of handling large data sets or intensive calculations are also very useful at times, especially for data mining analyses such as clustering, neural networks, or classification and decision trees.

Six Sigma technical leaders work to extract actionable knowledge from an organization's information warehouse. To assure access to the needed information, Six Sigma activities should be closely integrated with the information systems (IS) of the organization. Obviously, the skills and training of Six Sigma technical leaders must be supplemented by an investment in software and hardware.

Integrating Six Sigma with Other Information Systems Technologies

There are three information systems topics that are closely related to Six Sigma activities:

- Data warehousing
- Online analytic processing (OLAP)
- Data mining

Data warehousing relates to the data retained by the organization, and therefore available for use in Six Sigma activities. It impacts the data storage, which impacts ease of access for Six Sigma analyses. OLAP enables the analysis of large databases by persons who may not have the technical background of a Six Sigma technical leader. Data mining involves retrospective analysis of data using advanced tools and techniques. Each of these subjects will be discussed in turn.

While larger organizations clearly require an advanced approach to data warehousing, OLAP and data mining, there is much to be gained by smaller organizations in implementing appropriately sized systems. Stated simply, an Excel spreadsheet, or more likely a proliferation of these spreadsheets, is ill-suited for the longer-term data management and analysis required by Six Sigma teams. Relatively simple database designs using MS Access or SQL Server back-ends provide immense benefits over spreadsheet data storage, particularly when leveraged with user interfaces to easily drill-down the database, or query associated databases, such as Laboratory Information Systems (LIMS) or Manufacturing Resource Planning (MRP) systems.

Data Warehousing

Data warehousing has progressed rapidly. Virtually nonexistent in 1990, now every large corporation has at least one data warehouse; some have several. Hundreds of vendors offer data warehousing solutions, from software to hardware to complete systems. Few standards exist and there are as many data warehousing implementations as there are data warehouses. However, the multitiered approach to data warehousing is a model that appears to be gaining favor and recent advances in technology and decreases in prices have made this option more appealing to corporate users.

Multitiered data warehousing architecture focuses on how the data are used in the organization. While access and storage considerations may require summarization of data into multiple departmental warehouses, it is better for Six Sigma analysis if the warehouse keeps all of the detail in the data for historical analysis. The major components of this architecture are (Berry and Linoff, 1997):

- *Source systems* are where the data come from.
- *Data transport and cleansing* move data between different data stores.
- The *central repository* is the main store for the data warehouse.
- The *metadata* describes what is available and where.

- *Data marts* provide fast, specialized access for end users and applications.
- *Operational feedback* integrates decision support back into the operational systems.
- *End users* are the reason for developing the warehouse in the first place.

Figure 3.7 illustrates the multitiered approach.

Every data warehouse includes at least one of these building blocks. The data originates in the source systems and flows to the end users through the various components. The components can be characterized as hardware, software, and networks. The purpose is to deliver information, which is in turn used to create new knowledge, which is then acted on to improve business performance. In other words, the data warehouse is ultimately a component in a decision-support system.

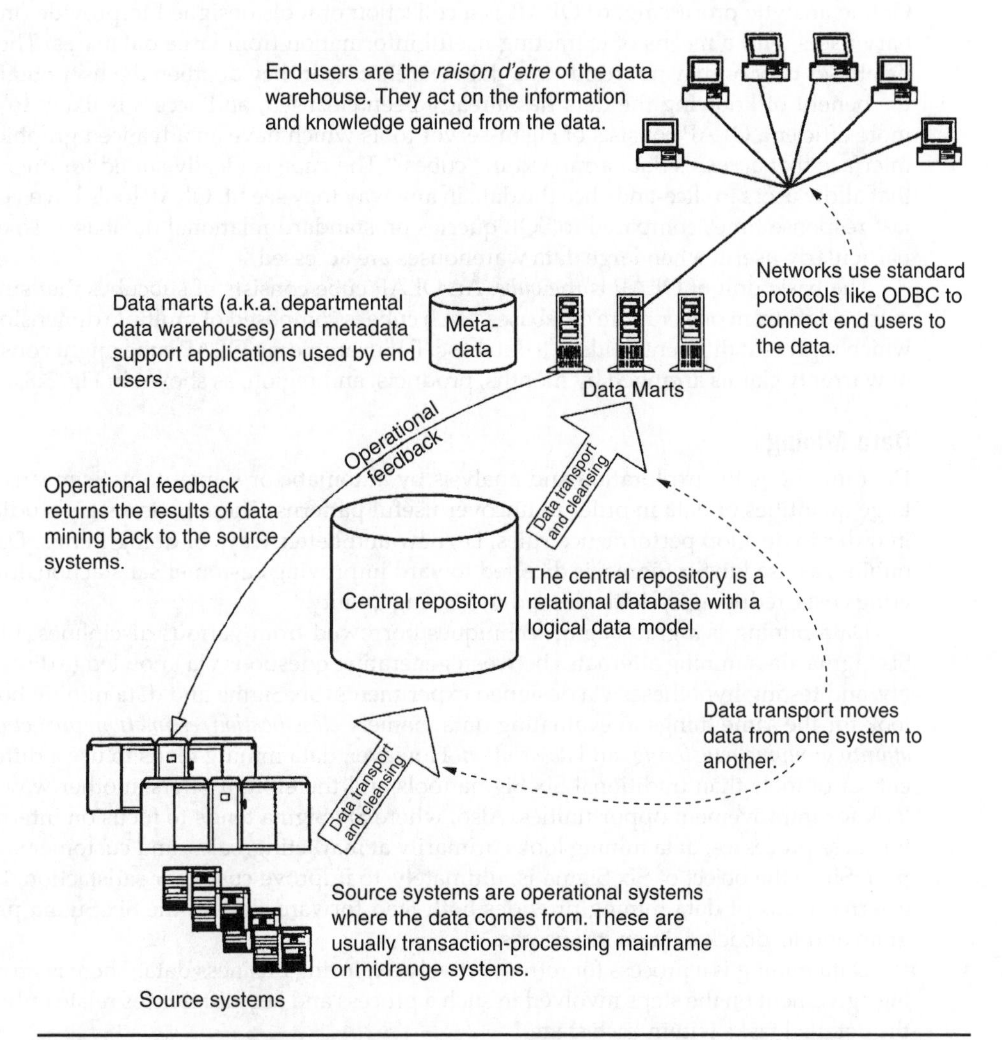
End users are the *raison d'etre* of the data warehouse. They act on the information and knowledge gained from the data.
Networks use standard protocols like ODBC to connect end users to the data.
Data marts (a.k.a. departmental data warehouses) and metadata support applications used by end users.
Meta-data
Data Marts
Operational feedback
Data transport and cleansing
Operational feedback returns the results of data mining back to the source systems.
Central repository
The central repository is a relational database with a logical data model.
Data transport moves data from one system to another.
Data transport and cleansing
Source systems are operational systems where the data comes from. These are usually transaction-processing mainframe or midrange systems.
Source systems

FIGURE 3.7 The multitiered approach to data warehousing. [*Berry and Linoff (1997). Used by permission of the publisher.*]

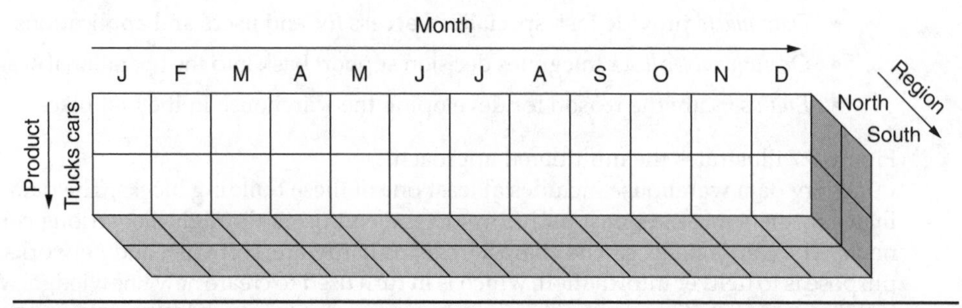

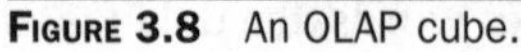

Figure 3.8 An OLAP cube.

OLAP

Online analytic processing, or OLAP, is a collection of tools designed to provide ordinary users with a means of extracting useful information from large databases. These databases may or may not reside in a data warehouse. If they do, then the user obtains the benefit of knowing the data has already been cleansed, and access is likely to be more efficient. OLAP consists of client-server tools which have an advanced graphical interface that accesses data arranged in "cubes." The cube is ideally suited for queries that allow users to slice-and-dice the data in any way they see fit. OLAP tools have very fast response times compared to SQL queries on standard relational databases, so are particularly useful when large data warehouses are accessed.

The basic unit of OLAP is the *cube*. An OLAP cube consists of subcubes that summarize data from one or more databases. Each cube is composed of multiple dimensions which represent different fields in a database. For example, an OLAP cube might consist of warranty claims arranged by months, products, and region, as shown in Fig. 3.8.

Data Mining

Data mining is the exploration and analysis by automatic or semiautomatic means of large quantities of data in order to uncover useful patterns. These patterns are studied in order to develop performance rules, i.e, new and better ways of doing things. Data mining, as used in Six Sigma, is directed toward improving customer satisfaction, lowering costs, reducing cycle times, and increasing quality.

Data mining is a grab-bag of techniques borrowed from various disciplines. Like Six Sigma, data mining alternates between generating questions via knowledge discovery, and testing hypotheses via designed experiments. Six Sigma and data mining both look for the same things in evaluating data, namely *classification, estimation, prediction, affinity grouping, clustering,* and *description*. However, data mining tends to use a different set of tools than traditional Six Sigma tools and therefore it offers another way to look for improvement opportunities. Also, where Six Sigma tends to focus on internal business processes, data mining looks primarily at marketing, sales, and customer support. Since the object of Six Sigma is, ultimately, to improve customer satisfaction, the external focus of data mining provides both feed forward data to the Six Sigma program and feedback data on its success.

Data mining is a process for retrospectively exploring business data. There is growing agreement on the steps involved in such a process and any differences relate only to the detailed tasks within each stage.*

*http://www.attar.com/tutor/deploy.htm.

- **Goal definition**—This involves defining the goal or objective for the data mining project. This should be a business goal or objective which normally relates to a business event such as arrears in mortgage repayment, customer attrition (churn), energy consumption in a process, etc. This stage also involves the design of how the discovered patterns will result in action that leads to business improvement.
- **Data selection**—This is the process of identifying the data needed for the data mining project and the sources of these data.
- **Data preparation**—This involves cleansing the data, joining/merging data sources and the derivation of new columns (fields) in the data through aggregation, calculations or text manipulation of existing data fields. The end result is normally a flat table ready for the application of the data mining itself (i.e. the discovery algorithms to generate patterns). Such a table is normally split into two data sets; one set for pattern discovery and other set for pattern verification.
- **Data exploration**—This involves the exploration of the prepared data to get a better feel prior to pattern discovery and also to validate the results of the data preparation. Typically, this involves examining descriptive statistics (minimum, maximum, average, etc.) and the frequency distribution of individual data fields. It also involves field versus field scatter plots to understand the dependency between fields.
- **Pattern discovery**—This is the stage of applying the pattern discovery algorithm to generate patterns. The process of pattern discovery is most effective when applied as an exploration process assisted by the discovery algorithm. This allows business users to interact with and to impart their business knowledge to the discovery process. For example, if creating a classification tree, users can at any point in the tree construction examine/explore the data filtering to that path, examine the recommendation of the algorithm regarding the next data field to use for the next branch then use their business judgment to decide on the data field for branching. The pattern discovery stage also involves analyzing the ability to predict occurrences of the event in data other than those used to build the model.
- **Pattern deployment**—This stage involves the application of the discovered patterns to solve the business goal of the data mining project. This can take many forms:
- **Pattern presentation**—The description of the patterns (or the graphical tree display) and their associated data statistics are included in a document or presentation.
- **Business intelligence**—The discovered patterns are used as queries against a database to derive business intelligence reports.
- **Data scoring and labeling**—The discovered patterns are used to score and/or label each data record in the database with the propensity and the label of the pattern it belongs to.
- **Decision support systems**—The discovered patterns are used to make components of a decision support system.
- **Alarm monitoring**—The discovered patterns are used as norms for a business process. Monitoring these patterns will enable deviations from normal conditions to be detected at the earliest possible time. This can be achieved by embedding the data mining tool as a monitoring component, or through the use of a classical approach, such as control charts.

- **Pattern validity monitoring**—As a business process changes over time, the validity of patterns discovered from historic data will deteriorate. It is therefore important to detect these changes at the earliest possible time by monitoring patterns with new data. Significant changes to the patterns will point to the need to discover new patterns from more recent data.

OLAP, Data Mining, and Six Sigma

OLAP is not a substitute for data mining. OLAP tools are a powerful means for reporting on data, while data mining focuses on finding hidden patterns in data. OLAP helps users explore existing theories by quickly presenting data to confirm or deny ad hoc hypotheses. It is, essentially, a semiautomated means of analysis. OLAP and data mining are complementary: both approaches complement the standard arsenal of tools and techniques used in Six Sigma. Both OLAP and data mining are used for *retrospective studies*: they are used to generate hypotheses by examining past data. Designed experiments, on the other hand, help users design *prospective studies:* they test the hypotheses generated by OLAP and data mining. Used together, Six Sigma, data mining and OLAP comprise a powerful collection of business improvement tools.

Benchmarking

Benchmarking is a topic of general interest in Six Sigma. Thus, the discussion here goes beyond the use of benchmarking in project management alone.

Benchmarking is a popular method for developing requirements and setting goals. In more conventional terms, benchmarking can be defined as measuring your performance against that of best-in-class companies, determining how the best-in-class achieve those performance levels, and using the information as the basis for your own company's targets, strategies, and implementation.

Benchmarking involves research into the best practices at the industry, firm, or process level. Benchmarking goes beyond a determination of the "industry standard" it breaks the firm's activities down to process operations and looks for the best-in-class for a particular operation. For example, to achieve improvement in their parts distribution process Xerox Corporation studied the retailer L.L. Bean.

Benchmarking goes beyond the mere setting of goals. It focuses on practices that produce superior performance. Benchmarking involves setting up partnerships that allow both parties to learn from one another. Competitors can also engage in benchmarking, providing they avoid proprietary issues.

Benchmarking projects are like any other major project. Benchmarking must have a structured methodology to ensure successful completion of thorough and accurate investigations. However, it must be flexible to incorporate new and innovative ways of assembling difficult-to-obtain information. It is a discovery process and a learning experience. It forces the organization to take an external view, to look beyond itself.

The Benchmarking Process

Camp (1989) lists the following steps for the benchmarking process:

1. Planning
    1.1. Identify what is to be benchmarked
    1.2. Identify comparative companies
    1.3. Determine data collection method and collect data
2. Analysis
    2.1. Determine current performance "gap"
    2.2. Project future performance levels
3. Integration
    3.1. Communicate benchmark findings and gain acceptance
    3.2. Establish functional goals
4. Action
    4.1. Develop action plans
    4.2. Implement specific actions and monitor progress
    4.3. Recalibrate benchmarks
5. Maturity
    5.1. Leadership position attained
    5.2. Practices fully integrated into process

The first step in benchmarking is determining what to benchmark. To focus the benchmarking initiative on critical issues, begin by identifying the process outputs most important to the customers of that process (i.e., the key quality characteristics). This step applies to every organizational function, since each one has outputs and customers. The QFD/customer needs assessment is a natural precursor to benchmarking activities.

Getting Started with Benchmarking

The essence of benchmarking is the acquisition of information. The process begins with the identification of the process that is to be benchmarked. The process chosen should be one that will have a major impact on the success of the business.

Once the process has been identified, contact a business library and request a search for the information relating to your area of interest. The library will identify material from a variety of external sources, such as magazines, journals, special reports, etc. You should also conduct research using the Internet and other electronic networking resources. However, be prepared to pare down what will probably be an extremely large list of candidates (e.g., an Internet search on the word "benchmarking" produced 20,000 hits). Don't forget your organization's internal resources. If your company has an "Intranet" use it to conduct an internal search. Set up a meeting with people in key departments, such as R&D. Tap the expertise of those in your company who routinely work with customers, competitors, suppliers, and other "outside" organizations. Often your company's board of directors will have an extensive network of contacts.

The search is, of course, not random. Look for the best of the best, not the average firm. There are many possible sources for identifying the elites. One approach is to build a compendium of business awards and citations of merit that organizations have received in business process improvement. Sources to consider are *Industry Week*'s Best

Plant's Award, National Institute of Standards and Technology's Malcolm Baldrige Award, *USA Today* and Rochester Institute of Technology's Quality Cup Award, European Foundation for Quality Management Award, Occupational Safety and Health Administration (OSHA), Federal Quality Institute, Deming Prize, Competitiveness Forum, *Fortune* magazine, United States Navy's Best Manufacturing Practices, to name just a few. You may wish to subscribe to an "exchange service" that collects benchmarking information and makes it available for a fee. Once enrolled, you will have access to the names of other subscribers—a great source for contacts.

Don't overlook your own suppliers as a source for information. If your company has a program for recognizing top suppliers, contact these suppliers and see if they are willing to share their "secrets" with you. Suppliers are predisposed to cooperate with their customers; it's an automatic door-opener. Also contact your customers. Customers have a vested interest in helping you do a better job. If your quality, cost, and delivery performance improve, your customers will benefit. Customers may be willing to share some of their insights as to how their other suppliers compare with you. Again, it isn't necessary that you get information about direct competitors. Which of your customer's suppliers are best at billing? Order fulfillment? Customer service? Keep your focus at the process level and there will seldom be any issues of confidentiality. An advantage to identifying potential benchmarking partners through your customers is that you will have a referral that will make it easier for you to start the partnership.

Another source for detailed information on companies is academic research. Companies often allow universities access to detailed information for research purposes. While the published research usually omits reference to the specific companies involved, it often provides comparisons and detailed analysis of what separates the best from the others. Such information, provided by experts whose work is subject to rigorous peer review, will often save you thousands of hours of work.

After a list of potential candidates is compiled, the next step is to choose the best three to five targets. A candidate that looked promising early in the process might be eliminated later based on the following criteria (Vaziri, 1992):

- Not the best performer
- Unwilling to share information and practices (i.e., doesn't view the benchmarking process as a mutually beneficial learning opportunity)
- Low availability and questionable reliability of information on the candidate

As the benchmarking process evolves, the characteristics of the most desirable candidates will be continually refined. This occurs as a result of a clearer understanding of your organization's key quality characteristics and critical success factors and an improved knowledge of the marketplace and other players.

This knowledge and the resulting actions tremendously strengthen an organization.

Why Benchmarking Efforts Fail

The causes of failed benchmarking projects are the same as those for other failed projects (DeToro, 1995):

- **Lack of sponsorship**—A team should submit to management a one- to four-page benchmarking project proposal that describes the project, its objectives, and potential costs. If the team can't gain approval for the project or get a sponsor, it makes little sense to proceed with a project that's not understood or appreciated or that is unlikely to lead to corrective action when completed.

- **Wrong people on team**—Who are the right people for a benchmarking team? Individuals involved in benchmarking should be the same ones who own or work in the process. It's useless for a team to address problems in business areas that are unfamiliar or where the team has no control or influence.
- **Teams don't understand their work completely**—If the benchmarking team didn't map, flowchart, or document its work process, and if it didn't benchmark with organizations that also documented their processes, there can't be an effective transfer of techniques. The intent in every benchmarking project is for a team to understand how its process works and compare it to another company's process at a detailed level. The exchange of process steps is essential for improved performance.
- **Teams take on too much**—The task a team undertakes is often so broad that it becomes unmanageable. This broad area must be broken into smaller, more manageable projects that can be approached logically. A suggested approach is to create a functional flowchart of an entire area, such as production or marketing, and identify its processes. Criteria can then be used to select a process to be benchmarked that would best contribute to the organization's objectives.
- **Lack of long-term management commitment**—Since managers aren't as familiar with specific work issues as their employees, they tend to underestimate the time, cost, and effort required to successfully complete a benchmarking project. Managers should be informed that while it's impossible to know the exact time it will take for a typical benchmarking project, there is a rule of thumb that a team of four or five individuals requires a third of their time for 5 months to complete a project.
- **Focus on metrics rather than processes**—Some firms focus their benchmarking efforts on performance targets (metrics) rather than processes. Knowing that a competitor has a higher return on assets doesn't mean that its performance alone should become the new target (unless an understanding exists about how the competitor differs in the use of its assets and an evaluation of its process reveals that it can be emulated or surpassed).
- **Not positioning benchmarking within a larger strategy**—Benchmarking is one of many Six Sigma tools—such as problem solving, process improvement, and process reengineering—used to shorten cycle time, reduce costs, and minimize variation. Benchmarking is compatible with and complementary to these tools, and they should be used together for maximum value.
- **Misunderstanding the organization's mission, goals, and objectives**—All benchmarking activity should be launched by management as part of an overall strategy to fulfill the organization's mission and vision by first attaining the short-term objectives and then the long-term goals.
- **Assuming every project requires a site visit**—Sufficient information is often available from the public domain, making a site visit unnecessary. This speeds the benchmarking process and lowers the cost considerably.
- **Failure to monitor progress**—Once benchmarking has been completed for a specific area or process benchmarks have been established and process changes implemented, managers should review progress in implementation and results.

The issues described here are discussed in other parts of this chapter and in other parts of this book. The best way of dealing with them is to prevent their occurrence by carefully planning and managing the project from the outset.

This list can be used as a checklist to evaluate project plans; if the plans don't clearly preclude these problems, then the plans are not complete.

The Benefits of Benchmarking

The benefits of competitive benchmarking include:

- Creating a culture that values continuous improvement to achieve excellence
- Enhancing creativity by devaluing the not-invented-here syndrome
- Increasing sensitivity to changes in the external environment
- Shifting the corporate mind-set from relative complacency to a strong sense of urgency for ongoing improvement
- Focusing resources through performance targets set with employee input
- Prioritizing the areas that need improvement
- Sharing the best practices between benchmarking partners

Some Dangers of Benchmarking

Benchmarking is based on learning from others, rather than developing new and improved approaches. Since the process being studied is there for all to see, benchmarking cannot give a firm a sustained competitive advantage. Although helpful, benchmarking should never be the primary strategy for improvement.

Competitive analysis is an approach to goal setting used by many firms. This approach is essentially benchmarking confined to one's own industry. Although common, competitive analysis virtually guarantees second-rate quality because the firm will always be following their competition. If the entire industry employs the approach it will lead to stagnation for the entire industry, setting them up for eventual replacement by outside innovators.

CHAPTER 4

Maximizing Resources

> The best Six Sigma projects begin not inside the business but outside it, focused on answering the question: How can we make the customer more competitive? What is critical to the customer's success? Learning the answer to that question and learning how to provide the solution is the only focus we need.
>
> Jack Welch, CEO, General Electric

This chapter addresses Six Sigma project selection and the management support activities related to project success. Projects are the core activity driving change in the Six Sigma organization. Although change also takes place due to other efforts, such as *Kaizen*, project-based change is the force that drives breakthrough and cultural transformation. In a typical Six Sigma organization about 1 percent of the workforce is engaged full time in change activities, and each of these change agents will complete between three and seven projects in a year. In addition there are another 5 percent or so part-time change agents, each of whom will complete about two smaller projects per year. The mathematics translates to about 500 major projects and 1,000 smaller projects in an organization with 10,000 employees in any given year. Clearly, learning how to effectively deal with projects is critical to Six Sigma success.

Choosing the Right Projects

Projects must be focused on the *right* goals. This is the responsibility of the senior leadership, for example, the project sponsor, Executive Six Sigma Council or equivalent group. Senior leadership is the only group with the necessary authority to designate cross-functional responsibilities and allow access to interdepartmental resources. Six Sigma projects will impact one of the major stakeholder groups: customers, shareholders, or employees. Although it is possible to calculate the impact of any given project on all three groups, I recommend that initially projects be evaluated separately for each group. This keeps the analysis relatively simple and ensures that a good stakeholder mix is represented in the project portfolio.

Types of Projects

Customer Value Projects

Many, if not most Six Sigma projects are selected because they have a positive impact on customers. To evaluate such projects one must be able to determine the linkage between business processes and customer-perceived value. Chapter 2 discussed how to create organizations that are customer-driven, which is essential. Customer-driven organizations, especially process enterprises, focus on customer value as a matter of routine. This focus will generate many Six Sigma customer value projects in the course of strategy deployment. However, in addition to the strategy-based linkage of Six Sigma projects described in Chap. 2, there is also a need to use customer demands directly to generate focused Six Sigma projects. The techniques for obtaining this linkage are the same as those used in Chap. 2. The difference is that the focus here is not on strategy deployment or budgeting, but on Six Sigma improvement projects focused on specific customer demands.

Learning what customers value is primarily determined by firsthand contact with customers through customer focus groups, interviews, surveys, etc. The connection between customer-perceived value and business processes, or "customer value streams," is established through business process mapping (see Chap. 6 and 7) and quality function deployment (QFD). The Executive Six Sigma Council and project sponsors should carefully review the results of these efforts to locate the "lever points" where Six Sigma projects will have the greatest impact on customer value.

Shareholder Value Projects

Six Sigma provides a "double-whammy" by addressing both efficiency and revenues. Revenue is impacted by improving the customer value proposition, which allows organizations to charge premium prices for superior quality, or to keep prices competitive and increase sales volume and market share due to superior quality. Improved efficiency is achieved by reducing the cost of poor quality, reducing cycle time, or eliminating waste in business processes. To determine which Six Sigma projects address the issue of business process efficiency evaluate the high-level business process maps (including SIPOC) and flow charts.

Other Six Sigma Projects

Some Six Sigma projects address intangibles, such as employee morale, regulatory concerns, or environmental issues. These projects can be just as important as those which address customer or shareholder value.

Analyzing Project Candidates

You now have a list of candidate Six Sigma projects. Assuming that the organization has limited resources, the next task is to select a subset of these projects to fund and staff.

Projects cost money, take time, and disrupt normal operations and standard routines. For these reasons projects designed to improve processes should be limited to processes that are important to the enterprise. Furthermore, projects should be undertaken only when success is highly likely. *Feasibility* is determined by considering the scope and cost of a project and the support it receives from the process owner. In this section a number of techniques and approaches are presented to help identify those projects that will be chosen for Six Sigma.

Benefit-Cost Analysis

Benefit-cost analysis can be as elaborate or as simple as the magnitude of the project expenditures demands. The Six Sigma manager is advised that most such analyses are easier to "sell" to senior management if done by (or reviewed and approved by) experts in the finance and accounting department. The plain fact is that the finance department has credibility in estimating cost and benefit that the Six Sigma department, and any other department, lacks. The best approach is to get the finance department to conduct the benefit-cost analysis with support from the other departments involved in the project. We will provide an overview of some principles and techniques that are useful in benefit-cost analysis.

A fundamental problem with performing benefit-cost analysis is that, in general, it is easier to accurately estimate costs than benefits. Costs can usually be quantified in fairly precise terms in a budget. Costs are claims on resources the firm already has. In contrast, benefits are merely predictions of future events, which may or may not actually occur. Also, benefits are often stated in units other than dollars, making the comparison of cost and benefit problematic. The problem is especially acute where quality improvement projects are concerned. For example, a proposed project may involve placing additional staff on a customer "hot line." The cost is easy to compute: X employees at a salary of $Y each, equipment, office space, supervision, etc. The benefit is much more difficult to determine. Perhaps data indicate that average time on hold will be improved, but the amount of the improvement and the probability that it will occur are speculations. Even if the time-on-hold improvement were precise, the impact on customer satisfaction would be an estimate. The association between customer satisfaction and revenues is yet another estimate. Despite these difficulties, reasonable cause-and-effect linkages can be established to form the basis for benefit-cost analysis. To compensate for the uncertainties of estimating benefits, it makes sense to demand a relatively high ratio of benefit to cost. For example, it is not unusual to have senior leadership demand a ROI of 100% in the first year on a Six Sigma project. Rather than becoming distressed at this "injustice," the Black Belt should realize that such demands are a response to the inherent difficulties in quantifying benefits.

Types of Savings

The accounting or finance department should formally define the different categories of savings. Savings are typically placed in categories such as:

Hard savings are actual reductions in dollars now being spent, such as reduced budgets, fewer employees, reduction of prices paid on purchasing contracts, and so on. Hard savings can be used to lower prices, change bid models, increase profits, or for other purposes where a high degree of confidence in the benefit is required.

Soft savings are projected reductions that should result from the project. For example, savings from less inventory, reduced testing, lower cycle times, improved yields, lower rework rates, and reduced scrap.

It is important that savings be integrated into the business systems of the organization. If the institutional framework doesn't change, the savings could eventually be lost. For example, if a Six Sigma project improves a process yield, be sure the MRP system's calculations reflect the new yields.

A System for Assessing Six Sigma Projects

Assessing Six Sigma projects is an art as well as a science. It is also critical to the success of Six Sigma, and to the individual Black Belt. Far too many Black Belts fail because they are not discriminating enough in their selection of projects. If project selection is systematically sloppy, the entire Six Sigma effort can fail.

The approach offered here is quantitative in the sense that numbers are determined and an overall project score calculated. It is subjective to the degree it requires interpretation of the situation, estimated probabilities, costs, and commitments. However, the rigor of completing this assessment process allows for better judgments regarding projects. The numbers (weights, scores, acceptable length of projects, dollar cutoffs, etc.) are recommendations which can and should be replaced by the organization's leadership to reflect organizational priorities. The scale ranges from 0 to 9 for each criterion, and the weights sum to 1.00, so the highest possible weighted score for a project is 9.

The Six Sigma department or Process Excellence function can compile summary listings of project candidates from these assessments. Sorting the list in descending order provides a guide to the final decision as to which projects to pursue. Each Black Belt or Green Belt will probably have their own list, which can also be sorted and used to guide their choices.

Worksheet 1. Six Sigma Project Evaluation

Project Name:	Date of Assessment:
Black Belt:	Master Black Belt:
Weighted Overall Project Score:	Project Number:

Criteria	Score	Weight	Weighted Score*
1. Sponsorship		0.23	
2. Benefits (specify main beneficiary) ☐ 2.1 External customer: ☐ 2.2 Shareholder: ☐ 2.3 Employee or internal customer: ☐ 2.4 Other (e.g., supplier, environment):	Overall Benefit Score ☐	0.19	
3. Availability of resources other than team		0.16	
4. Scope in terms of Black Belt effort		0.12	
5. Deliverable		0.09	
6. Time to complete		0.09	
7. Team membership		0.07	
8. Project Charter		0.03	
9. Value of Six Sigma approach		0.02	
TOTAL (sum of weighted score column)		1.00	

Note: Any criterion scores of zero must be addressed before project is approved.

*Weighted score = project's score for each criterion times the weight.

Worksheet 2. Six Sigma Project Evaluation Guidelines

1.0 Sponsorship

Score	Interpretation
9	Director-level sponsor identified, duties specified and sufficient time committed and scheduled
3	Director-level sponsor identified, duties specified and sufficient time committed but not scheduled
1	Willing Director-level sponsor who has accepted charter statement
0	Director-level sponsor not identified, or sponsor has not accepted the charter

2.0 Stakeholder Benefits*

"Tangible and verifiable benefits for a major stakeholder"

2.1 Stakeholder: External Customer

2.1.1 Customer Satisfaction

Score	Interpretation
9	Substantial and statistically significant increase in *overall* customer satisfaction or loyalty
3	Substantial and statistically significant increase in a *major subcategory* of customer satisfaction
1	Substantial and statistically significant increase in a *focused area* of customer satisfaction
0	Unclear or nonexistent customer satisfaction impact

2.1.2 Quality Improvement (CTQ)

Score	Interpretation
9	10 × or greater improvement in critical to quality (CTQ) metric
5	5 × to 10 × improvement in CTQ metric
3	2 × to 5 × improvement in CTQ metric
1	Statistically significant improvement in CTQ metric, but less than 2 × magnitude
0	Project's impact on CTQ metrics undefined or unclear

*Note: Several stakeholder benefit categories are shown in section 2. At least one stakeholder category is required. Show benefit scores for each category, then use your judgment to determine an overall benefit score for the project

2.2 Stakeholder: Shareholder

2.2.1 Financial Benefits

Score	Interpretation
9	Hard net savings (budget or bid model change) greater than $500K. Excellent ROI
5	Hard net savings between $150K and $500K. Excellent ROI
3	Hard net savings between $50K and $150K, or cost avoidance greater than $500K. Good ROI
1	Hard savings of at least $50K, or cost avoidance of between $150K and $500K. Acceptable ROI
0	Project claims a financial benefit but has hard savings less than $50K, cost avoidance less than $150K, or unclear financial benefit

2.2.2 Cycle Time Reduction

Score	Interpretation
9	Cycle time reduction that improves revenue, bid model or budget by more than $500K. Excellent ROI
5	Cycle time reduction that improves revenue, bid model or budget by $150K to $500K. Excellent ROI
3	Cycle time reduction that improves revenue, bid model or budget by $50K to $150K, or creates a cost avoidance of more than $500K. Good ROI
1	Cycle time reduction that results in cost avoidance between $150K and $500K. Acceptable ROI
0	Project claims a cycle time improvement but has hard savings less than $50K, cost avoidance less than $150K, or unclear financial benefit from the improvement in cycle time

2.2.3 Revenue Enhancement

Score	Interpretation
9	Significant increase in revenues, excellent ROI
3	Moderate increase in revenues, good ROI
1	Increase in revenues with acceptable ROI
0	Unclear or nonexistent revenue impact

2.3 Stakeholder: Employee or Internal Customer

2.3.1 Employee Satisfaction

Score	Interpretation
9	Substantial and statistically significant increase in *overall* employee satisfaction
3	Substantial and statistically significant increase in *a major element* of employee satisfaction
1	Substantial and statistically significant increase in *a focused area* of employee satisfaction
0	Unclear or nonexistent employee satisfaction impact

2.4 Stakeholder: Other

2.4.1 Specify Stakeholder: ______________________________

Benefits

Score	Interpretation
9	
5	
3	
1	
0	Unclear or nonexistent benefit

3.0 Availability of Resources Other Than Team

Score	Interpretation
9	Needed resources available when needed
3	Limited or low priority access to needed to resources
0	Resources not available, or excessive restrictions on access to resources

4.0 Scope in Terms of Black Belt Effort

Score	Interpretation
9	Projected return substantially exceeds required return
3	Projected return exceeds required return
1	Projected return approximately equals required return
0	Projected return not commensurate with required return

Required return can be calculated as follows:*

(1) Length of project (months) = ________________

(2) Proportion of Black Belt's time required (between 0 and 1) = __________

(3) Probability of success (between 0 and 1) = ____________

Required return** = \$83,333 × (1) × (2) ÷ (3) = \$ __________

Projected return: \$__________

5.0 Deliverable (Scope)

Score	Interpretation
9	New or improved process, product or service to be created is clearly and completely defined
3	New or improved process, product or service to be created is defined
0	Deliverable is poorly or incorrectly defined. For example, a "deliverable" that is really a tool such as a process map

6.0 Time to Complete

Score	Interpretation
9	Results realized in less than 3 months
3	Results realized in between 3 and 6 months
1	Results realized in 7 to 12 months
0	Results will take more than 12 months to be realized

7.0 Team Membership

Score	Interpretation
9	Correct team members recruited and time commitments scheduled
3	Correct team members recruited, time committed but not scheduled
1	Correct team members recruited
0	Team members not recruited or not available

8.0 Project Charter

Score	Interpretation
9	All elements of the project charter are complete and acceptable. Linkage between project activities and deliverable is clear
3	Project charter acceptable with minor modifications
0	Project charter requires major revisions

*Thanks to Tony Lin of Boeing Satellite Systems for this algorithm.

**Based on expected Black Belt results of \$1million/year.

9.0 Value of Six Sigma Approach (DMAIC or equivalent)

Score	Interpretation
9	Six Sigma approach essential to the success of the project. Black Belt/ Green Belt skill set required for success
3	Six Sigma approach helpful but not essential. Black Belt/Green Belt skill set can be applied
0	Usefulness of Six Sigma approach not apparent. Specific Black Belt or Green Belt skills are not necessary

Other Methods of Identifying Promising Projects

Projects should be selected to support the organization's overall strategy and mission. Because of this global perspective most projects involve the efforts of several different functional areas. Not only do individual projects tend to cut across organizational boundaries, different projects are often related to one another. To effectively manage this complexity it is necessary to integrate the planning and execution of projects across the entire enterprise. One way to accomplish this is QFD, which is discussed in detail in Chap.2. In addition to QFD and the scoring method described above, a number of other procedures are presented here to help identify a project's potential worth.

Using Pareto Analysis to Identify Six Sigma Project Candidates

Pareto principle refers to the fact that a small percentage of processes cause a large percentage of the problems. The Pareto principle is useful in narrowing a list of choices to those few projects that offer the greatest potential. When using Pareto analysis keep in mind that there may be hidden "pain signals." Initially problems create pain signals such as schedule disruptions and customer complaints. Often these *symptoms* are treated rather than their underlying "diseases"; for example, if quality problems cause schedule slippages which lead to customer complaints, the "solution" might be to keep a large inventory and sort the good from the bad. The result is that the schedule is met and customers stop complaining, but at huge cost. These opportunities are often greater than those currently causing "pain," but they are now built into business systems and therefore very difficult to see. One solution to the hidden problem phenomenon is to focus on processes rather than symptoms. Some guidelines for identifying dysfunctional processes for potential improvement are shown in Table 4.1.

The "symptom" column is useful in identifying problems and setting priorities. The "disease" column focuses attention on the underlying causes of the problem, and the "cure" column is helpful in chartering quality improvement project teams and preparing mission statements.

Prioritizing Projects with the Pareto Priority Index

After a serious search for improvement opportunities the organization's leaders will probably find themselves with more projects to pursue than they have resources. The Pareto Priority Index (PPI) is a simple way of prioritizing these opportunities. The PPI is calculated as follows (Juran and Gryna , 1993):

$$\text{PPI} = \frac{\text{Saving} \times \text{probability of success}}{\text{Cost} \times \text{time to completion (years)}} \tag{4.1}$$

Symptom	Disease	Cure
Extensive information exchange, data redundancy, rekeying	Arbitrary fragmentation of a natural process	Discover why people need to communicate with each other so often; integrate the process
Inventory, buffers, and other assets stock piled	System slack to cope with uncertainty	Remove the uncertainty
High ratio of checking and control to value-added work (excessive test and inspection, internal controls, audits, etc.)	Fragmentation	Eliminate the fragmentation, integrate processes
Rework and iteration	Inadequate feedback in a long work process	Process control
Complexity, exceptions and special causes	Accretion onto a simple base	Uncover original "clean" process and create new process(es) for special situations; eliminate excessive standardization of processes

TABLE 4.1 Dysfunctional Process Symptoms and Underlying Diseases

A close examination of the PPI equation shows that it is related to return on investment adjusted for probability of success. The inputs are, of course, estimates and the result is totally dependent on the accuracy of the inputs. The resulting number is an index value for a given project. The PPI values allow comparison of various projects. If there are clear standouts the PPI can make it easier to select a project. Table 4.2 shows the PPIs for several hypothetical projects.

Project	Savings, $ Thousands	Probability	Cost, $ Thousands	Time, Years	PPI
Reduce wave solder defects 50%	$70	0.7	$25	0.75	2.61
NC machine capability improvement	$50	0.9	$20	1.00	2.25
ISO 9001 certification	$150	0.9	$75	2.00	0.90
Eliminate customer delivery complaints	$250	0.5	$75	1.50	1.11
Reduce assembly defects 50%	$90	0.7	$30	1.50	1.40

TABLE 4.2 Illustration of the Pareto Priority Index (PPI)

The PPI indicates that resources be allocated first to reducing wave solder defects, then to improving NC machine capability, and so on. The PPI may not always give such a clear ordering of priorities. When two or more projects have similar PPIs a judgment must be made on other criteria.

Throughput-Based Project Selection

While careful planning and management of projects is undeniably important, they matter little if the projects being pursued have no impact on the bottom line (throughput). As you will see in the following section, if you choose the wrong projects it is possible to make big "improvements" in quality and productivity that have no impact whatever on the organization's net profit. Selecting which projects to pursue is of critical importance. In this section we will use the theory of constraints (TOC) to determine which project(s) to pursue.

Theory of Constraints

Every organization has constraints. Constraints come in many forms. When a production or service process has a resource constraint (i.e., it lacks a sufficient quantity of some resource to meet the market demand), then the sequence of improvement projects should be identified using very specific rules. According to Goldratt (1990), the rules are:

1. Identify the system's constraint(s). Consider a fictitious company that produces only two products, P and Q (Fig. 4.1). The market demand for P is 100 units per week and P sells for $90 per unit. The market demand for Q is 50 units per week and Q sells for $100 per unit. Assume that A, B, C, and D are workers who have different noninterchangeable skills and that each worker is available for only 2,400 minutes per week (8 hours per day, 5 days per week). For simplicity,

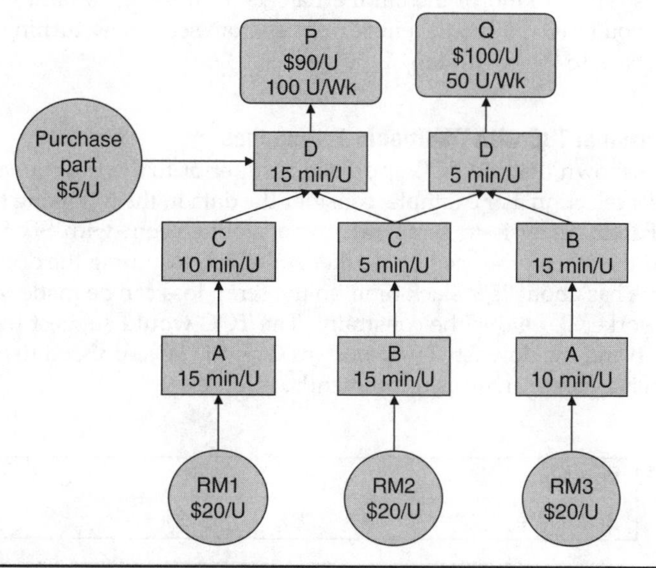

FIGURE 4.1 A simple process with a constraint.

assume that there is no variation, waste, etc. in the process. Assume this process has a constraint, worker B. This fact has profound implications for selecting Six Sigma projects.

2. Decide how to exploit the system's constraint(s). Look for Six Sigma projects that minimize waste of the constraint. For example, if the constraint is (feeding) the market demand (i.e. a capacity constraint), then we look for Six Sigma projects that provide 100% on time delivery. Let's not waste anything! If the constraint is a machine, or process step, as in this example, focus on reducing setup time, eliminating errors or scrap, and keeping the process step running as much as possible.
3. Subordinate everything else to the above decision. Choose Six Sigma projects to maximize throughput of the constraint. After completing step 2, choose projects to eliminate waste from downstream processes; once the constraint has been utilized to create something we don't want to lose it due to some downstream blunder. Then choose projects to ensure that the constraint is always supplied with adequate nondefective resources from upstream processes. We pursue upstream processes last because by definition they have slack resources, so small amounts of waste upstream that are detected before reaching the constraint are less damaging to throughput.
4. Elevate the system's constraint(s). Elevate means "Lift the restriction." This is step 4, not step 2. Often the projects pursued in steps 2 and 3 will eliminate the constraint. If the constraint continues to exist after performing steps 2 and 3, look for Six Sigma projects that provide additional resources to the constraint. These might involve, for example, purchasing additional equipment or hiring additional workers with a particular skill.
5. If, in the previous steps, a constraint has been broken, go back to step 1. There is a tendency for thinking to become conditioned to the existence of the constraint. A kind of mental inertia sets in. If the constraint has been lifted, then you must rethink the entire process from scratch. Returning to step 1 takes you back to the beginning of the cycle.

Comparison of TOC with Traditional Approaches

It can be shown that the TOC approach is superior to the traditional TQM approaches to project selection. For example, consider the data in the following table. If you were to apply Pareto analysis to scrap rates you would begin with Six Sigma projects that reduced the scrap produced by worker A. In fact, assuming the optimum product mix, worker A has about 25% slack time, so the scrap loss can be made up without shutting down worker B, who is the constraint. The TOC would suggest that the scrap loss of worker B and the downstream processes C and D be addressed first, the precise opposite of what Pareto analysis recommends.

Process		A	B	C	D
Scrap rate		8%	3%	5%	7%

Process Scrap Rates

Of course, before making a decision as to which projects to finance cost/benefit analyses are still necessary, and the probability of the project succeeding must be estimated. But by using the TOC you will at least know where to look first for opportunities.

Using Constraint Information to Focus Six Sigma Projects

Applying the TOC strategy described earlier tells us *where* in the process to focus. Adding CTX information (see Table 4.3) can help tell us which type of project to focus on, that is, should we focus on quality, cost or schedule projects? Assume that you have three Six Sigma candidate projects, all focusing on process step B, the constraint. The area addressed is correct, but which project should you pursue first? Let's assume that we learn that one project will primarily improve quality, another cost, and another schedule. Does this new information help? Definitely! Take a look at Table 4.3 to see how this information can be used. Projects in the same priority group are ranked according to their impact on throughput.

The same thought process can be applied to process steps before and after the constraint. The results are shown in Table 4.4.

Note that Table 4.4 assumes that projects *before* the constraint do not result in problems *at* the constraint. Remember, impact should always be measured in terms of throughput. If a process upstream from the constraint has an adverse impact on throughput, then it can be considered a constraint. If an upstream process *average* yield is enough to feed the constraint on the average, it may still present a problem. For example, an upstream process producing 20 units per day with an average yield of 90% will produce, on average, 18 good units. If the constraint requires 18 units, things will be okay about 50% of the time, but the other 50% of the time things won't be okay. One solution to this problem is to place a work-in-process (WIP) inventory between the process and the constraint as a safety buffer. Then on those days when the process yield is below 18 units, the inventory can be used to keep the constraint running. However,

Project Type	Discussion
CTQ	Any unit produced by the constraint is especially valuable because if it is lost as scrap additional constraint time must be used to replace it or rework it. Since constraint time determines throughput (net profit of the entire system), the loss far exceeds what appears on scrap and rework reports. CTQ projects at the constraint are the highest priority.
CTS	CTS projects can reduce the time it takes the constraint to produce a unit, which means that the constraint can produce more units. This directly impacts throughput. CTS projects at the constraint are the highest priority.
CTC	Since the constraint determines throughput, the unavailability of the constraint causes lost throughput *of the entire system*. This makes the cost of constraint down time extremely high. The cost of *operating* the constraint is usually miniscule by comparison. Also, CTC projects can have an adverse impact on quality or schedule. Thus, CTC projects at the constraint are low priority.

Table 4.3 Throughput priority of CTX Projects that Affect the Constraint

Focus of Six Sigma Project				
		Before the constraint	At the constraint	After the constraint
CTX: Characteristic addressed is critical to ...	Quality (CTQ)	△	⊙	⊙
	Cost (CTC)	O	△	O
	Schedule (CTS)	△	⊙	O

△ Low throughput priority.
O Moderate throughput priority.
⊙ High throughput priority.

Table 4.4 Project Throughput Priority versus Project Focus

there is a cost associated with carrying a WIP inventory. A Six Sigma project that can improve the yield will reduce or eliminate the need for the inventory and should be considered even if it doesn't impact the constraint directly, assuming the benefit-cost analysis justifies the project. On the other hand, if an upstream process can easily make up any deficit before the constraint needs it, then a project for the process will have a low priority.

Knowing the project's throughput priority will help you make better project selection decisions. Of course, the throughput priority is just one input into the project selection process, other factors may lead to a different decision. For example, impact on other projects, a regulatory requirement, a better payoff in the long-term, etc.

Multitasking and Project Scheduling

A Six Sigma enterprise will always have more projects to pursue than it has resources to do them. The fact that resources (usually Black Belts or Green Belts) are scarce means that projects must be scheduled, that is, some projects must be undertaken earlier than others. In such situations it is tempting to use *multitasking* of the scarce resource. Multitasking is defined as the assignment of a resource to several priorities during the same period of time. The logic is that by working on several projects or assignments simultaneously, the entire portfolio of work will be done more quickly. However, while this is true for independent resources working independent projects or subprojects in parallel, it is *not* true when applied to a single resource assigned to multiple projects or interdependent tasks within a project.

Consider the following situation. You have three Six Sigma projects, A, B, and C. A single-tasking solution is to first do A, then B, and then C. Here's the single-activity project schedule.

A	B	C
(Complete in Wk 10)	(Complete in Wk 20)	(Complete in Wk 30)

If each project takes 10 weeks to complete, then A will be completed in 10 weeks, B in 20 weeks, and C in 30 weeks. The average time to complete the three projects is $(10 + 20 + 30)/3 = 60/3 = 20$ weeks. The average doesn't tell the whole story, either. The benefits will begin as soon as the project is completed and by the end of the 30-week period project A will have been completed for 20-weeks, and project B for 10 weeks.

Now let's consider a multitasking strategy. Here we split our time equally between the three projects in a given 10-week period. That way the sponsor of projects B and C will see activity on their projects much sooner than if we used a single-task approach to scheduling. The new schedule looks like this:

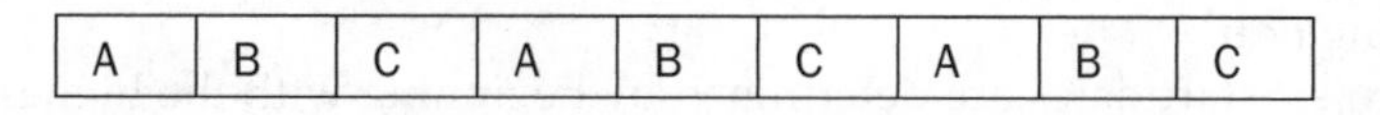

With this multitasking schedule project A will be completed in 23.3 weeks, project B in 26.7 weeks, and project C will still take 30 weeks. The completion time for project A went from 10 weeks to 23.3 weeks, for project B it went from 20 weeks to 26.7 weeks, and for project C it remained the same, 30 weeks. The overall average completion time went from 20 weeks to 26.67 weeks, a 33% deterioration in average time to complete. And this is a best-case scenario. In real life there is always some lost time when making the transition from one project to another. The Black Belt has to clear her head of what she was doing, review the next project, get the proper files ready, reawaken sponsors and team members, and so on. This can often take a considerable amount of time, which is added to the time needed to complete the projects.

Critical Chain Project Portfolio Management

Critical chain project management avoids the multitasking problem by changing the way the organization manages groups of projects, and the way the individual projects are managed.

Managing the Organization's Projects

First, at the organizational level, multitasking of key resources is stopped. People and other resources are allowed to focus on projects one at a time. This means that management must accept responsibility for prioritizing projects, and policies must be developed which mandate single-project focus and discourage multitasking. To be successful the organization must determine its capacity to complete projects. Every organization finds itself with more opportunities than it can successfully pursue with finite resources. This means that only a select portfolio of projects should be undertaken in any time interval. The constraining resource is usually a key position in the organization, say the time available by project sponsors, engineers, programmers, etc. This information can be used to determine organizational capacity and to schedule project start dates according to the availability of the key resource. This is called *project launch synchronization* and the scarce resource that paces the project system is called a *synchronizer resource*.

Synchronizer Resource Usage

Critical chain project management does not permit multitasking of scarce resources. People and equipment that are fully utilized on projects, *synchronizer resources*, are assigned to a sequence of single projects. The sequence of projects is based on enterprise priorities. If a project requires one or more synchronizer resources it is vital that your project start dates integrate the schedules of these resources. In particular, this will require that those activities that require time from a synchronizer resource (and the project as a whole) stipulate "Start no earlier than" dates. Although synchronizer resources are protected by capacity buffers and might hypothetically start at a date earlier than specified, the usual practice is to utilize any unplanned excess capacity to allow the organization to pursue additional opportunities, thereby increasing the organization's capacity to complete projects. Note that human resources are defined in terms of the skills required for the activity, not in terms of individual people. In fact, the resource manager should refrain from assigning an activity to an individual until all predecessors have been completed and the activity is ready to begin. This precludes the temptation to multitask as the individual looks ahead and sees the activity start date drawing near.

Project start dates are determined by beginning with the highest priority project and calculating the end date for the synchronizing resource based on the estimated duration of all activities that require the synchronizing resource. The second highest priority project's start date will be calculated by adding a capacity buffer to the expected end date of the first project. The third highest priority project's start date is based on the completion date of the second, and so on. If, by chance, the synchronizing resource is available before the scheduled start date, the time can be used to increase the organization's capacity to complete more projects. Figure 4.2 illustrates this strategy.

Summary and Preliminary Project Selection

At this point you have evaluated project candidates using a number of different criteria. You must now rank the projects, and make your preliminary selections. You may use Worksheet 3 to assist you with this. The reason your selections are preliminary is that you lack complete data. As they work on the project, Six Sigma project teams will continuously reevaluate it and they may uncover data which will lower or raise the

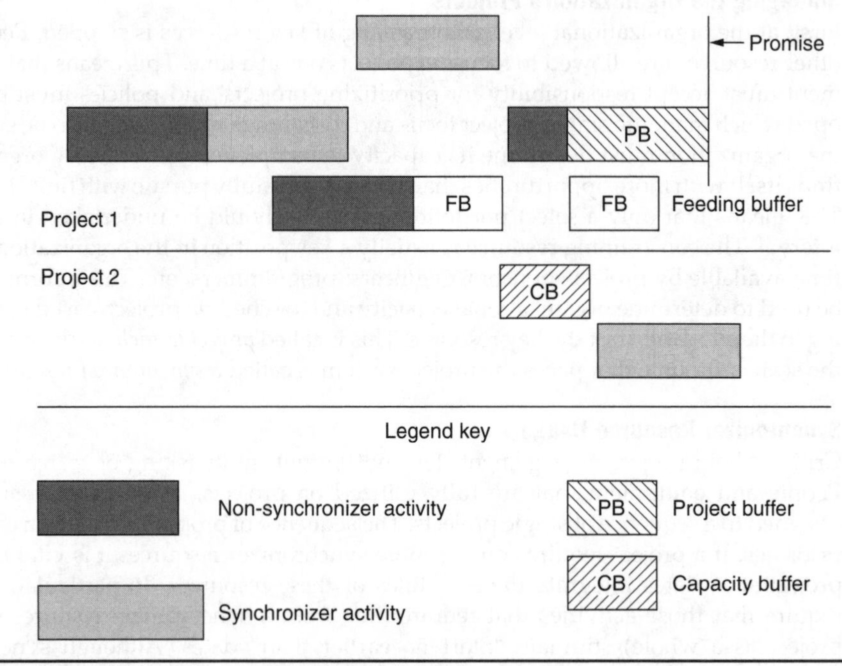

FIGURE 4.2 Critical chain scheduling illustration.

project's priority. The project sponsor is responsible for coordinating changes in priority with the process owners.

Ongoing Management Support

Once projects have been selected, management support continues through the life of the project in a number of ways. Chapter 5 discusses the project reporting requirements expected of the project team and its leader. Management sponsors provide the management interface necessary to ensure the project remains on course relative to its objectives, or to change objectives if necessary given new information discovered by the project team. At times, it will be necessary for management to reiterate its project support to clear roadblocks, as discussed below. Management must also evaluate the project results, as well as the team performance, to provide feedback to the management systems for identifying improvement opportunities. This is further discussed later in this chapter.

Internal Roadblocks

Most organizations still have a hierarchical, command-and-control organizational structure, sometimes called "smoke stacks" or "silos." The functional specialists in charge of each smoke stack tend to focus on optimizing their own functional area, often to the detriment of the organization as a whole. In addition, the hierarchy gives these managers a monopoly on the authority to act on matters related to their functional specialty. The combined effect is both a desire to resist change and the authority to resist change, which often creates insurmountable roadblocks to quality improvement projects.

It is important to realize that organizational rules are, by their nature, a barrier to change. The formal rules take the form of written standard operating procedures (SOPs). The very purpose of SOPs is to standardize behavior. Unfortunately, the quality profession has historically overemphasized formal documentation. Approaches such as ISO 9000 and ISO 14000. Risk indoctrinating formal rules that are merely responses to problems that no longer exist after the reason for their existence has passed. In an organization that is serious about its written rules even senior leaders find themselves helpless to act without submitting to a sometimes burdensome rule-changing process. In those cases, the true power in such an organization is the bureaucracy that controls the procedures. If the organization falls into the trap of creating written rules for too many things, it can find itself moribund in a fast-changing external environment. This is a recipe for disaster. While electronic document control systems can remove some of these issues, it's critical to manage them as a means toward control only when it meets the overall needs of the system, rather than as an inefficient solution to localized problems.

Restrictive rules need not take the form of management limitations on itself, procedures that define hourly work in great detail also produce barriers, for example, union work rules. Projects almost always require that work be done differently and such procedures prohibit such change. Organizations that tend to be excessive in SOPs also tend to be heavy on work rules. The combination is often deadly to quality improvement efforts.

Organization structures preserve the status quo in other ways besides formal, written restrictions in the form of procedures and rules. Another effective method of limiting change is to require permission from various departments, committees, councils, boards, experts, and so on. Even though the organization may not have a formal requirement, that "permission" be obtained, the effect may be the same, for example, "You should run that past accounting" or "Ms. Reimer and Mr. Evans should be informed about this project." When permission for vehicles for change (e.g., project budgets, plan approvals) is required from a group that meets infrequently it creates problems for project planners. Plans may be rushed so they can be presented at the next meeting, lest the project be delayed for months. Plans that need modifications may be put on hold until the next meeting, months away. Or, projects may miss the deadline and be put off indefinitely.

External Roadblocks

Modern organizations do not exist as islands. Powerful external forces take an active interest in what happens within the organization. Government bodies have created a labyrinth of rules and regulations that the organization must negotiate to utilize its human resources without incurring penalties or sanctions. The restrictions placed on modern businesses by outside regulators are challenging to say the least. When research involves people, ethical and legal concerns sometimes require that external approvals be obtained. The approvals are contingent on such issues as informed consent, safety, cost and so on.

Many industries have "dedicated" agencies to deal with. For example, the pharmaceutical industry must deal with the Food and Drug Administration (FDA). These agencies must often be consulted before undertaking projects. For example, a new treatment protocol involving a new process for treatment of pregnant women prior to labor may involve using a drug in a new way (e.g., administered on an outpatient basis instead of on an inpatient basis).

Many professionals face liability risks that are part of every decision. Often these fears create a "play it safe" mentality that acts as a barrier to change. The fear is even greater when the project involves new and untried practices and technology.

Individual Barriers to Change

Perhaps the most significant change, and therefore the most difficult, is to change ourselves. It seems to be a part of human nature to resist changing oneself. By and large, we worked hard to get where we are, and our first impulse is to resist anything that threatens our current position. Forsha (1992) provides the process for personal change shown in Fig. 4.3.

The adjustment path results in preservation of the status quo. The action path results in change. The well-known PDCA cycle can be used once a commitment to action has been made by the individual. The goal of such change is continuous *self*-improvement.

Within an organizational context, the individual's reference group plays a part in personal resistance to change. A reference group is the aggregation of people a person thinks of when they use the word "we." If "we" refers to the company, then the company is the individual's reference group and he or she feels connected to the company's success or failure. However, "we" might refer to the individual's profession or trade group, for example, "We doctors," "We engineers," "We union members." In this case the leaders shown on the formal organization chart will have little influence on the individual's attitude toward the success or failure of the project. When a project involves external reference groups with competing agendas, the task of building buy-in and consensus is daunting indeed.

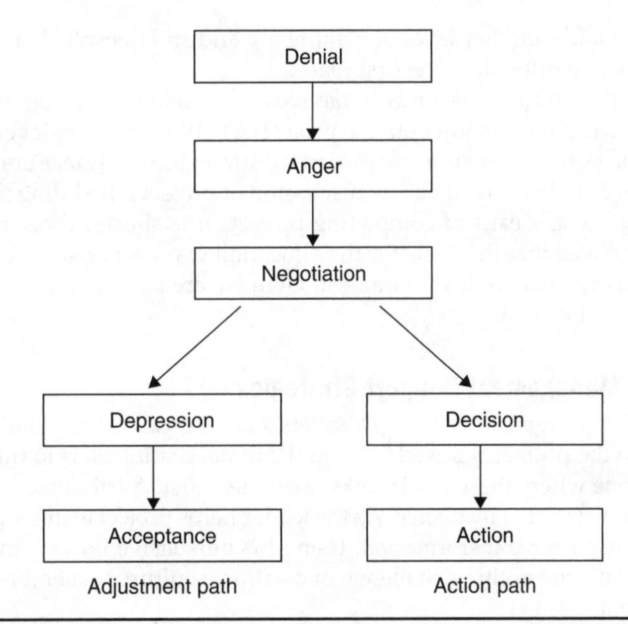

Figure 4.3 The process of personal change. From Forsha (1992). Copyright © 1992 by ASQ Quality Press, Milwaukee, WI. Used by permission.

Ineffective Management Support Strategies

Strategy 1: Command people to act as you wish—With this approach the senior leadership simply commands people to act as the leaders wish. The implication is that those who do not comply will be subjected to disciplinary action. People in less senior levels of an organization often have an inflated view of the value of raw power. The truth is that even senior leaders have limited power to rule by decree. Human beings by their nature tend to act according to their own best judgment. Thankfully, commanding that they do otherwise usually has little effect. The result of invoking authority is that the decision-maker must constantly try to divine what the leader wants them to do in a particular situation. This leads to stagnation and confusion as everyone waits on the leader. Another problem with commanding as a form of "leadership" is the simple communication problem. Under the best of circumstances people will often simply misinterpret the leadership's commands.

Strategy 2: Change the rules by decree—When rules are changed by decree the result is again confusion. What are the rules today? What will they be tomorrow? This leads again to stagnation because people don't have the ability to plan for the future. Although rules make it difficult to change, they also provide stability and structure that may serve some useful purpose. Arbitrarily changing the rules based on force (which is what "authority" comes down to) instead of a set of guiding principles does more harm than good.

Strategy 3: Authorize circumventing of the rules—Here the rules are allowed to stand, but exceptions are made for the leader's "pet projects." The result is general disrespect for and disregard of the rules, and resentment of the people who are allowed to violate rules that bind everyone else. An improvement is to develop a formal method for circumventing the rules, for example, deviation request procedures. While this is less arbitrary, it adds another layer of complexity and still doesn't change the rules that are making change difficult in the first place.

Strategy 4: Redirect resources to the project—Leaders may also use their command authority to redirect resources to the project. A better way is to develop a fair and easily understood system to ensure that projects of strategic importance are adequately funded as a matter of policy. In our earlier discussion of project scheduling we discussed "crash scheduling" as a means of completing projects in a shorter time frame. However, the assumption was that the basis for the allocation was cost or some other objective measure of the organization's best interest. Here we are talking about political clout as the basis of the allocation.

Effective Management Support Strategies

Strategy 1: Transform the formal organization and the organization's culture—By far the best solution to the problems posed by organizational roadblock is to transform the organization to one where these roadblocks no longer exist. As discussed earlier, this process can't be implemented by decree. As the leader helps project teams succeed, he will learn about the need for transformation. Using his persuasive powers the leader-champion can undertake the exciting challenge of creating a culture that embraces change instead of fighting it.

Strategy 2: *Mentoring*—In Greek mythology, Mentor was an elderly man, the trusted counselor of Odysseus, and the guardian and teacher of his son Telemachus. Today the term, "mentor" is still used to describe a wise and trusted counselor or teacher. When this person occupies an important position in the organization's hierarchy, he or she can be a powerful force for eliminating roadblocks. Modern organizations are complex and confusing. It is often difficult to determine just where one must go to solve a problem or obtain a needed resource. The mentor can help guide the project manager through this maze by clarifying lines of authority. At the same time, the mentor's senior position enables him to see the implications of complexity and to work to eliminate unnecessary rules and procedures.

Strategy 3: Identify informal leaders and enlist their support—Because of their experience, mentors often know that the person whose support the project really needs is not the one occupying the relevant box on the organization chart. The mentor can direct the project leader to the person whose opinion really has influence. For example, a project may need the approval of, say, the vice president of engineering. The engineering VP may be balking because his senior metallurgist hasn't endorsed the project.

Strategy 4: Find legitimate ways around people, procedures, resource constraints and other roadblocks—It may be possible to get approvals or resources through means not known to the project manager. Perhaps a minor change in the project plan can bypass a cumbersome procedure entirely. For example, adding an engineer to the team might automatically place the authority to approve process experiments within the team rather than in the hands of the engineering department.

Cross-Functional Collaboration

This section will address the impact of organizational structures on management of Six Sigma projects.

Six Sigma projects are process-oriented and most processes that have significant impact on quality cut across several different departments. Modern organizations, however, are hierarchical, that is, they are defined by superior/subordinate relationships. These organizations tend to focus on specialized functions (e.g., accounting, engineering). But adding value for the customer requires that several different functions work together. The ideal solution is the transformation of the organization into a structure designed to produce value without the need for a hierarchical structure. However, until that is accomplished, Six Sigma project managers will need to deal with the conflicts inherent in doing cross-functional projects in a hierarchical environment.

Project managers "borrow" people from many departments for their projects, which create matrix organizational structures. The essential feature of a matrix organization is that some people have two or more bosses or project customers. These people effectively report to multiple bosses, for example, the project manager and their own boss. Ruskin and Estes refer to people with more than one boss as *multi-bossed individuals*, and their bosses and customers *as multiple bosses*. Somewhere in the organization is a *common boss*, who resolves conflicts between multiple bosses when they are unable to do so on their own. Of course, multiple bosses can prevent conflicts by cooperation and collaboration before problems arise.

Often multi-bossed individuals are involved with several projects, further complicating the situation. Collaborative planning between the multiple bosses is necessary to determine how the time of multi-bossed individuals, and other resources, will be shared. For additional discussion of more complex matrix structures (see Ruskin and Estes, 1995)

Good communication is helpful in preventing problems. Perhaps the most important communication is frequent, informal updates of all interested parties by the project manager. More formal status reports should also be specified in the project plan and sent to people with an interest in the project. The project manager should determine who gets what information, which is often tricky due to the multi-boss status of the project manager. Some managers may not want "too much" information about their department's "business" shared with their peers from other areas. Other managers may be offended if they receive less information than everyone else. The project manager's best diplomatic skills may be required to find the right balance.

Status reports invariably indicate that the project plan is less than perfect. The process by which the plans will be adjusted should be understood in advance. The process should specify who will be permitted to make adjustments, when the adjustments will be allowed and how much authority the bosses and project manager have in making adjustments.

Negotiated agreements should be documented, while generating the minimum possible amount of additional red tape and paperwork. The documentation will save the project manager a great deal of time in resolving disputes down the road regarding who agreed to what.

Tracking Six Sigma Project Results

It is vital that information regarding results be accumulated and reported. This is useful for a variety of purposes:

- ☐ Evaluating the effectiveness of the Six Sigma project selection system
- ☐ Determining the overall return on investment
- ☐ Setting budgets

WORKSHEET 3. Project Assessment Summary

Project Description or ID Number	Project Score	PPI Priority	ROI Priority	Throughput Priority	Comments

- ☐ Appraising individual and group performance
- ☐ Setting goals and targets
- ☐ Identifying areas where more (or less) emphasis on Six Sigma is indicated
- ☐ Helping educate newcomers on the value of Six Sigma
- ☐ Answering skeptics
- ☐ Quieting cynics

A major difference between Six Sigma and failed programs of the past is the emphasis on tangible, measurable results. Six Sigma advocates a strong point of the fact that projects are selected to provide a mixture of short- and long-term paybacks that justify the investment and the effort. Unless proof is provided, any statements regarding paybacks are nothing more than empty assertions.

Data storage is becoming so inexpensive that the typical organization can afford to keep fairly massive amounts of data in databases. The limiting factor is the effort needed to enter the data into the system. This is especially important if highly trained change agents such as Master Black Belts, Black Belts, or Green Belts are needed to perform the data entry (Table 4.5).

Usually viewing access is restricted to the project data according to role played in the project, position in the organization, etc. Change access is usually restricted to the project sponsor, leader, or Black Belt. However, to the extent possible, it should be easy to "slice-and-dice" this information in a variety of ways. Periodic reports might be created summarizing results according to department, sponsor, Black Belt, etc. The system should also allow ad hoc views to be easily created, such as the simple list shown in Table 4.6.

Financial Results Validation

Six Sigma financial benefits claimed for every project must be confirmed by experts in accounting or finance. Initial savings estimates may be calculated by Black Belts or sponsors, but final results require at least the concurrence of the finance department. This should be built in from the start. The finance person assigned to work with the team should be listed in the project charter. Without this involvement the claimed savings are simply not credible. Aside from the built-in bias involved in calculating the benefit created from one's own project, there is the issue of qualifications. The best qualified people to calculate financial benefits are generally those who do such calculations for a living.

This is not to imply that the finance expert's numbers should go unchallenged. If the results appear to be unreasonable, either high or low, then they should be clearly

☐ Charter information (title, sponsor, membership, deadline etc.)
☐ Description of project in ordinary language
☐ Project status
☐ Savings type (hard, soft, cost avoidance, CTQ, etc.)
☐ Process or unit owner
☐ Key accounting information (charge numbers, etc.)
☐ Project originator
☐ Top-level strategy addressed by project
☐ Comments, issues
☐ Lessons learned
☐ Keywords (for future searches)
☐ Related documents and links
☐ Audit trail of changes
☐ Project task and schedule information

TABLE 4.5 Possible Information to Be Captured

explained in terms the sponsor understands. The Six Sigma Leader also has an interest in ensuring that the numbers are valid. Invalid results pose a threat to the viability of the Six Sigma effort itself. For example, on one project the Black Belt claimed savings of several hundred thousand dollars for "unpaid overtime." A finance person concurred. However, the Six Sigma Leader would not accept the savings, arguing quite reasonably that the company hadn't saved anything if it had never paid the overtime. This isn't to say that the project didn't have a benefit. Perhaps morale improved or turnover declined due to the shorter working hours. Care must be taken to show the benefits properly.

Team Performance Evaluation

Evaluating team performance involves the same principles as evaluating performance in general. Before one can determine how well the team's task has been done, a baseline must be established and goals must be identified. Setting goals using benchmarking and other means is discussed elsewhere in this book (see Chap. 3). Records of progress should be kept as the team pursues its goals.

Performance measures generally focus on group tasks, rather than on internal group issues. Typically, financial performance measures show a payback ratio of between 2:1 and 8:1 on team projects. Some examples of tangible performance measures are:

- Productivity
- Quality
- Cycle time
- Grievances
- Medical usage (e.g., sick days)

Project ID	Project Title	Status	Black Belt	Sponsor	Due	Savings Type	Total Savings	Costs
76	Cup Dipole Antenna	Pending approval	J Jones	Jane Doe	3/1/04	Hard	$508,000	$5,900
33	Tank assembly	Define	B Olson	Sam Smith	9/30/03	Hard	$250,000	$25,000
35	SSPA	Completed	N Hepburn	Sal Davis	10/31/03	Cost avoidance	$1.3 Million	$13,000
37	FCC RFI compliance	Control	M Littleton	Henry Little	9/30/03	Other	NA	$1,500
•	•	•	•	•	•	•	•	•
•	•	•	•	•	•	•	•	•
•	•	•	•	•	•	•	•	•

Table 4.6 A Typical View of Six Sigma Projects

- Absenteeism
- Service
- Turnover
- Dismissals
- Counseling usage

Many intangibles can also be measured. Some examples of intangibles affected by teams are:

- Employee attitudes
- Customer attitudes
- Customer compliments
- Customer complaints

The performance of the team process should also be measured. Project failure rates should be carefully monitored. A p chart can be used to evaluate the causes of variation in the proportion of team projects that succeed. Failure analysis should be rigorously conducted.

Aubrey and Felkins (1988) list the effectiveness measures shown below:

- Leaders trained
- Number of potential volunteers
- Number of actual volunteers
- Percent volunteering
- Projects started
- Projects dropped
- Projects completed/approved
- Projects completed/rejected
- Improved productivity
- Improved work environment
- Number of teams
- Inactive teams
- Improved work quality
- Improved service
- Net annual savings

Team Recognition and Reward

Recognition is a form of employee motivation in which the company identifies and thanks employees who have made positive contributions to the company's success. In an ideal company, motivation flows from the employees' pride of workmanship. When employees are enabled by management to do their jobs and produce a product or service of excellent quality, they will be motivated.

The reason recognition systems are important is not that they improve work by providing incentives for achievement. Rather, they make a statement about what is important to the company. Analyzing a company's employee recognition system provides a powerful insight into the company's values in action. These are the values that are actually driving employee behavior. They are not necessarily the same as management's stated values. For example, a company that claims to value customer satisfaction but recognizes only sales achievements probably does not have customer satisfaction as one of its values in action.

Public recognition is often better for two reasons:

1. Some (but not all) people enjoy being recognized in front of their colleagues.
2. Public recognition communicates a message to all employees about the priorities and function of the organization.

The form of recognition can range from a pat on the back to a small gift to a substantial amount of cash. When substantial cash awards become an established pattern, however, it signals two potential problems:

1. It suggests that several top priorities are competing for the employee's attention, so that a large cash award is required to control the employee's choice.
2. Regular, large cash awards tend to be viewed by the recipients as part of the compensation structure, rather than as a mechanism for recognizing support of key corporate values.

Carder and Clark (1992) list the following guidelines and observations regarding recognition:

Recognition is not a method by which management can manipulate employees. If workers are not performing certain kinds of tasks, establishing a recognition program to raise the priority of those tasks might be inappropriate. Recognition should not be used to get workers to do something they are not currently doing because of conflicting messages from management. A more effective approach is for management to first examine the current system of priorities. Only by working on the system can management help resolve the conflict.

Recognition is not compensation. In this case, the award must represent a significant portion of the employee's regular compensation to have significant impact. Recognition and compensation differ in a variety of ways:

- Compensation levels should be based on long-term considerations such as the employee's tenure of service, education, skills, and level of responsibility. Recognition is based on the specific accomplishments of individuals or groups.
- Recognition is flexible. It is virtually impossible to reduce pay levels once they are set, and it is difficult and expensive to change compensation plans.
- Recognition is more immediate. It can be given in timely fashion and therefore relate to specific accomplishments.
- Recognition is personal. It represents a direct and personal contact between employee and manager. Recognition should not be carried out in such a manner that implies that people of more importance (managers) are giving something to people of less importance (workers).

Positive reinforcement is not always a good model for recognition. Just because the manager is using a certain behavioral criterion for providing recognition, it doesn't mean that the recipient will perceive the same relationship between behavior and recognition.

Employees should not believe that recognition is based primarily on luck. An early sign of this is cynicism. Employees will tell you that management says one thing but does another.

Recognition meets a basic human need. Recognition, especially public recognition, meets the needs for belonging and self-esteem. In this way, recognition can play an important function in the workplace. According to Abraham Maslow's theory, until these needs for belonging and self-esteem are satisfied, self-actualizing needs such as pride in work, feelings of accomplishment, personal growth, and learning new skills will not come into play.

Recognition programs should not create winners and losers. Recognition programs should not recognize one group of individuals time after time while never recognizing another group. This creates a static ranking system, with all of the problems discussed earlier.

Recognition should be given for efforts, not just for goal attainment. According to Imai (1986), a manager who understands that a wide variety of behaviors are essential to the company will be interested in criteria of discipline, time management, skill development, participation, morale, and communication, as well as direct revenue production. To be able to effectively use recognition to achieve business goals, managers must develop the ability to measure and recognize such process accomplishments.

Employee involvement is essential in planning and executing a recognition program. It is essential to engage in extensive planning before instituting a recognition program or before changing a bad one. The perceptions and expectations of employees must be surveyed.

Lessons-Learned Capture and Replication

It is often possible to apply the lessons learned from a project to other processes, either internally or externally. Most companies have more than one person or organizational unit performing similar or identical tasks. Many also have suppliers and outsourcers who do work similar to that being done internally. By replicating the changes done during a project the benefits of Six Sigma can be multiplied manyfold, often at very minimal cost. Think of it as a form of benchmarking. Instead of looking for the best-in-class process for you to learn from, the Six Sigma team *created* a best-in-class process and you want to teach the new approach to others.

Unlike benchmarking, where the seeker of knowledge is predisposed to change what they are doing, the process owners who might benefit from the knowledge gained during a Six Sigma project may not even be aware that they can benefit from a change. This needs to be accounted for when planning the program for sharing lessons learned. The process is a combination of motivation, education and selling the target audience on the new approach. Chances are that those who worked the project are *not* the best ones to sell others on the new approach. They can serve as technical advisers to those who will carry the message to other areas. The Six Sigma function (Process Excellence) usually takes the lead in developing a system for replication and sharing of lessons learned.

In addition to the lessons learned about business processes, a great deal will be learned about how to conduct successful projects. In a few years even a moderately sized Six Sigma effort will complete hundreds or thousands of projects. These project lessons learned should be captured and used to help other project teams. The information is usually best expressed in simple narratives by the project Black Belt. The narratives can be indexed by search engines and used by other Black Belts in the organization. The lessons learned database is an extremely valuable asset to the Six Sigma organization.

PART II

Six Sigma Tools and Techniques

CHAPTER 5

Project Management Using DMAIC and DMADV

Part II addresses the tools and techniques commonly used in Six Sigma. Many of these tools have been used by the quality professional and applied statistician for decades. Six Sigma formalizes the use of the tools within the DMAIC and DMADV project deployment methodologies, where they are applied to real-world projects designed to deliver tangible results for identified stakeholders.

DMAIC and DMADV Deployment Models

When applied for performance improvement of an existing product, process, or service, the Define-Measure-Analyze-Improve-Control, or DMAIC model is used. DMAIC is summarized in Fig. 5.1.

The DMAIC structure provides a useful framework for creating a "gated process" for project control, as shown in Fig. 5.2. Criteria for completing a particular phase are defined and projects reviewed to determine if all of the criteria have been met before the next phase is begun. If all criteria have been satisfied, the gate (e.g., define) is "closed."

Table 5.1 shows a partial listing of tools often found to be useful in a given stage of a project. There is considerable overlap in practice.

When the project goal is the development of a new or radically redesigned product, process or service, the Define-Measure-Analyze-Design-Verify, or DMADV, model is used (Fig. 5.3). DMADV is a part of the design for Six Sigma (DFSS) toolkit. Note the similarities between the tools used, as well as the objectives.

Figure 5.4 illustrates the relationship between DMAIC and DMADV.

Projects are the means through which processes and products are systematically changed; the bridge between the planning and the doing.

Frank Gryna makes the following observations about projects (Juran and Gryna, 1988, pp. 22.18–22.19):

- An agreed-upon project is also a legitimate project. This legitimacy puts the project on the official priority list. It helps to secure the needed budgets, facilities, and personnel. It also helps those guiding the project to secure attendance at scheduled meetings, to acquire requested data, to secure permission to conduct experiments, etc.

D	Define the goals of the improvement activity, and incorporate into a Project Charter. Obtain sponsorship and assemble team.
M	Measure the existing system. Establish valid and reliable metrics to help monitor progress toward the goal(s) defined at the previous step. Establish current process baseline performance using metric.
A	Analyze the system to identify ways to eliminate the gap between the current performance of the system or process and the desired goal. Use exploratory and descriptive data analysis to help you understand the data. Use statistical tools to guide the analysis.
I	Improve the system. Be creative in finding new ways to do things better, cheaper, or faster. Use project management and other planning and management tools to implement the new approach. Use statistical methods to validate the improvement.
C	Control the new system. Institutionalize the improved system by modifying compensation and incentive systems, policies, procedures, MRP, budgets, operating instructions and other management systems. You may wish to utilize standardization such as ISO 9000 to ensure that documentation is correct. Use statistical tools to monitor stability of the new systems.

Figure 5.1 Overview of DMAIC.

- The project provides a forum of converting an atmosphere of defensiveness or blame into one of constructive actions.
- Participation in a project increases the likelihood that the participant will act on the findings.
- All breakthrough is achieved *project by project*, and in no other way.
- Effective project management will prevent a number of problems that result in its absence.
- Projects have little or no impact on the organization's success, even if successful, no one will really care.
- Missions overlap the missions of other teams. For example, Team A's mission is to reduce solder rejects, Team B's mission is to reduce wave solder rejects, Team C's mission is to reduce circuit board assembly problems.
- Projects improve processes that are scheduled for extensive redesign, relocation or discontinuation.
- Studying a huge system ("patient admitting"), rather than a manageable process ("outpatient surgery preadmission").
- Studying symptoms ("touch-up of defective solder joints") rather than root causes ("wave solder defects")
- Project deliverables are undefined. For example, "study TQM" rather than "reduce waiting time in urgent care."

There are several reasons why one should plan carefully before starting a project (Ruskin and Estes, 1995, p. 44):

1. The plan is a simulation of prospective project work, which allows flaws to be identified in time to be corrected.

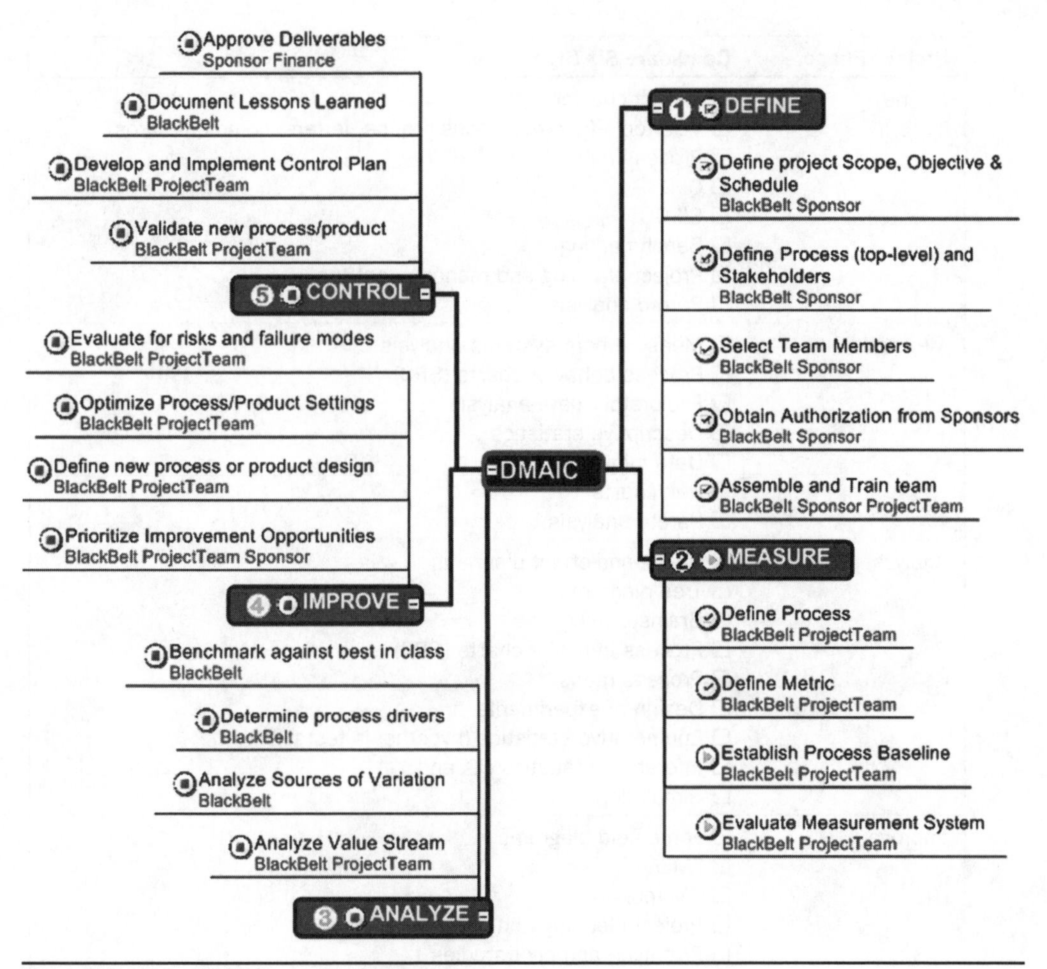

FIGURE 5.2 Using DMAIC on a Six Sigma project.

2. The plan is a vehicle for discussing each person's role and responsibilities, thereby helping direct and control the work of the project.
3. The plan shows how the parts fit together, which is essential for coordinating related activities.
4. The plan is a point of reference for any changes of scope, thereby helping project managers deal with their customers.
5. The plan helps everyone know when the objectives have been reached and therefore when to stop.

The official plan and authorization for the project is summarized in the Six Sigma Project Charter, which is outlined in Chap. 6.

Project Phase	Candidate Six Sigma Tools
Define	☐ Project charter ☐ VOC tools (surveys, focus groups, letters, comment cards) ☐ Process map ☐ QFD ☐ SIPOC ☐ Benchmarking ☐ Project planning and management tools ☐ Pareto analysis
Measure	☐ Measurement systems analysis ☐ Process behavior charts (SPC) ☐ Exploratory data analysis ☐ Descriptive statistics ☐ Data mining ☐ Run charts ☐ Pareto analysis
Analyze	☐ Cause-and-effect diagrams ☐ Tree diagrams ☐ Brainstorming ☐ Process behavior charts (SPC) ☐ Process maps ☐ Design of experiments ☐ Enumerative statistics (hypothesis tests) ☐ Inferential statistics (*X*s and *Y*s) ☐ Simulation
Improve	☐ Force field diagrams ☐ FMEA ☐ 7M tools ☐ Project planning and management tools ☐ Prototype and pilot studies ☐ Simulations
Control	☐ SPC ☐ FMEA ☐ ISO 900× ☐ Change budgets, bid models, cost estimating models ☐ Reporting system

TABLE 5.1 Six Sigma Tools Commonly Used in Each Phase of a Project

Define	**Define** the goals of the design activity.
Measure	**Measure** customer input to determine what is critical to quality from the customers' perspective. Use special methods when a completely new product or service is being designed (see the Kano Model discussions in Chap. 2). Translate customer requirements into project goals.
Analyze	**Analyze** innovative concepts for products and services to create value for the customer. Determine performance of similar best-in-class designs.
Design	**Design** new processes, products and services to deliver customer value. Use predictive models, simulation, prototypes, pilot runs, etc. to validate the design concept's effectiveness in meeting goals.
Verify	**Verify** that new systems perform as expected. Create mechanisms to ensure continued optimal performance.

FIGURE 5.3 Overview of DMADV.

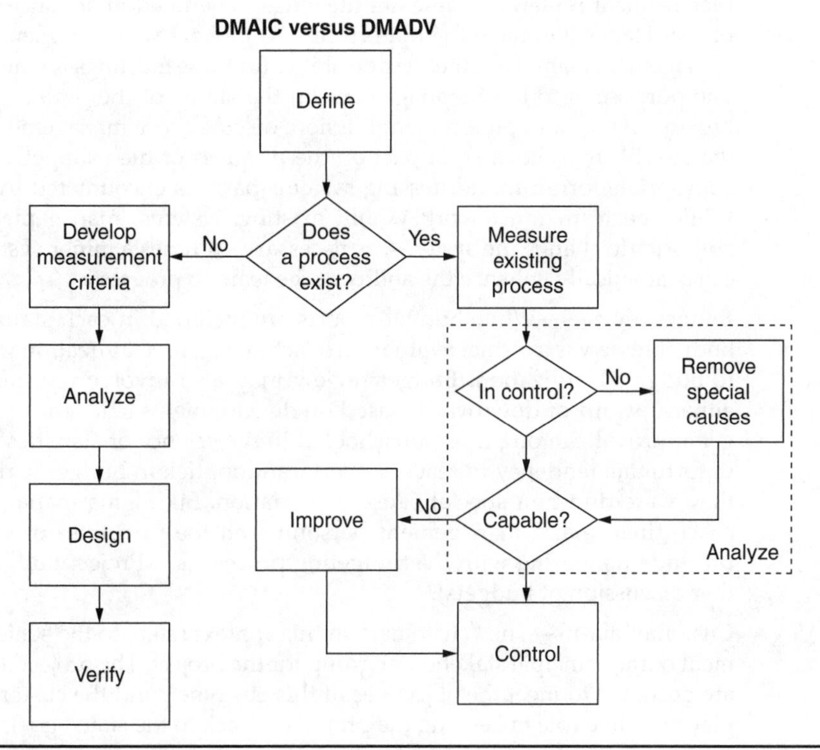

FIGURE 5.4 DMAIC and DMADV.

Project Reporting

Project information should be collected on an ongoing basis as the project progresses. Information obtained should be communicated in a timely fashion to stakeholders and sponsors, who can often help the project manager to maintain or recover the schedule. The Project Charter (further defined in Chap. 6) provides an important feedback tool in the form of a formal, written report. It provides the background for periodic reporting to various stakeholder groups:

- Status reports—Formal, periodic written reports, often with a standardized format, indicating the current project status, relative to the plan outlined in the latest distributed revision of the Charter. Where project performance does not match planned performance, the reports include information as to the cause of the problem and corrective actions to align with the plan. Remedial action may, at times, involve revising the plan, subject to the approval or the sponsor. When the project is not meeting the plan due to obstacles which the project team cannot overcome, the status report will request senior management intervention.
- Management reviews—These are meetings, scheduled in advance, where the project leader will have the opportunity to interact with key members of the management team. The chief responsibility for these meetings is management's. The purpose is to brief management on the status of the project, review the Project Charter and project team mission, discuss those management activities that are likely to have an impact on the progress of the team, etc. This is the appropriate forum for addressing systems barriers encountered by the team: While the team must work within existing systems, management has the authority to change the systems, as necessary. At times a minor system change can dramatically enhance the ability of the team to progress.
- Budget reviews—While budget reports are included in each status report, a budget review is a formal evaluation of actual resource utilization with respect to budgeted utilization. Budget review may also involve revising budgets, either upward or downward, based on developments since the original budget approval. Among those unschooled in the science of statistics there is an unfortunate tendency to react to every random tick in budget variances as if they were due to a special cause of variation. Six Sigma managers should coach finance and management personnel on the principles of variation to preclude tampering with the budgeting process (see Project Budgets for further discussion of budgets).
- Customer audits—The "customer" in this context refers to the senior management of the principal stakeholder group for the project. The project deliverables are designed to meet the objectives of this customer, and the customer should play an active role in keeping the project on track to the stated goals.
- Updating plans and timetables—The purpose of feedback is to provide information to form a basis for modifying future behavior. Since that behavior is documented in the project plans and schedules, these documents must be modified to ensure that the appropriate action is taken.
- Resource redirection—The modifications made to the plans and timetables will result in increasing or decreasing resource allocation to the project, or

accelerating or decelerating the timetable for resource utilization. The impact of these resource redirections on other projects should be evaluated by management in view of the organization's overall objectives.

Project Budgets

The process of allocating resources to be expended in the future is called *budgeting*. A listing of project expenditures, broken out into specific categories, is referred to as called the project budget.

Ruskin and Estes (1995) list the following types of project-related budgets:

Direct labor budgets are usually prepared for each work element in the project plan, then aggregated for the project as a whole. Control is usually maintained at the work element level to ensure the aggregate budget allowance is not exceeded. Budgets may be in terms of dollars or some other measure of value, such as direct labor hours expended.

Support services budgets need to be prepared because, without budgets, support services tend to charge based on actual costs, without allowances for errors, rework, etc. The discipline imposed by making budget estimates and being held to them often leads to improved efficiency and higher quality.

Purchased items budgets cover purchased materials, equipment, and services. The budgets can be based on negotiated or market prices. The issues mentioned for support services also apply here.

Budgets allocate resources to be used in the future. No one can predict the future with certainty. Thus, an important element in the budgeting process is tracking actual expenditures after the budgets have been prepared. Project managers, sometimes in conjunction with their contacts in finance, are responsible for evaluating expenditures on a periodic basis. Often, variance reports are compiled to compare the actual and budgeted expenditures. (The term "variance" is used here in the accounting sense, not the statistical sense. In accounting, a variance is simply the difference between the budgeted amount and the actual amount. An accounting variance may or may not indicate a special cause of variation; statistical techniques are required to make this determination.)

Variance reports can appear in a variety of formats. Most common are simple tables that show the actual/budgeted/variances by budget item, overall for the current period, and cumulatively for the project. Since it is unlikely that variances will be zero, an allowance is usually made, for example, 5% over or under is allowed without the need for explanations. A better approach is to plot historical data on control charts to set allowances and/or spot patterns or trends.

The project manager should review the variance data for patterns which contain useful information. Ideally, the pattern will be a mixture of positive and negative but minor variances. Assuming that this pattern is accompanied by an on-schedule project, this indicates a reasonably good budget, that is, an accurate forecasting of expenditures. Variances should be evaluated separately for each type of budget (direct labor, materials, etc.). However, the variance report for the entire project is the primary source of information concerning the status of the project in terms of resource utilization.

Since budgeted resources are generally scarce, overspending represents a serious threat to the project and, perhaps, to the organization itself. When a project overspends its budget, it depletes the resources available for other activities and projects. The project team and team leader and sponsors should design monitoring systems to detect and correct overspending before it threatens the project or the organization. Overspending is often a symptom of other problems with the project, for example, paying extra in an attempt to "catch up" after falling behind schedule, additional expenses for rework, etc.

Underspending is potentially as serious as overspending. If the project budget was prepared properly then the expenses reflect a given schedule and quality level. Underspending may reflect "cutting corners" or allowing suppliers an allowance for slower delivery. The reasons for any significant departure from the plan should be explained.

Project Records

Project records provide information that is useful both while the project is underway, and afterwards. Project records serve three basic purposes: cost accounting requirements; legal requirements; and learning.

Project records should be organized and maintained as if they were part of a single database, even if it isn't possible to keep all of the records in a single location. There should be one "official" copy of the documentation, and a person designated to act as the caretaker of this information while the project is active. Upon completion of the project, the documentation should be sent to the organization's archives. Large quality improvement projects are expensive and time-consuming undertakings. The process involved is complex and often confusing. However, a great deal can be learned from studying the "project process" where data exists. Archives of a number of projects can be used to identify common problems and patterns between the various projects. For example, project schedules may be consistently too optimistic or too pessimistic.

The following records should be kept:

- Statement of work
- Plans and schedules for projects and subprojects
- Correspondence (written and electronic)
- Written agreements
- Meeting minutes
- Action items and responsibilities
- Budgets and financial reports
- Cost-benefit analyses
- Status reports
- Presentation materials
- Documentation of changes made to plans and budgets
- Procedures followed or developed
- Notes of significant lessons learned

It is good practice for the project team to have a final meeting to perform a "post mortem" of the project. The meeting should be conducted soon after the project's completion, while memories are still fresh. The meeting will cover the lessons learned from conducting the project, and recommendations for improving the process. The minutes from these meetings should be used to educate project managers.

Low-cost Document Control and Project Management software can automatically catalog the information and provide the ability to search the entire database quickly and easily. There seems to be little reason not to store complete project information indefinitely.

Six Sigma Teams

Six Sigma teams working on projects are the primary means of deploying Six Sigma and accomplishing the goals of the enterprise. Six Sigma teams are sometimes lead by the Black Belt, but the team leader may also be a properly trained Green Belt or a Six Sigma champion who has a passion for the project. In these latter cases, the team must include a Black Belt to oversee the analysis of data, which is not a part of the Green Belt and Champion training.

Six Sigma teams are composed of groups of individuals who bring authority, knowledge, skills, abilities, and personal attributes to the project. There is nothing particularly special about Six Sigma teams compared with other work teams. They are people with different backgrounds and talents pursuing a common short-term goal. Like all groups of people, there are dynamics involved that must be understood if the mission of the team is to be accomplished. This section addresses the practices that Black Belts, Green Belts, sponsors, champions, facilitators, and leaders can employ to ensure that Six Sigma teams are successful. It focuses on:

- Stages in learning to work as a team
- The difference between group maintenance roles and group task roles
- Identifying and encouraging productive roles essential to team success
- Identifying and discouraging counterproductive behavior on teams
- Facilitating team meetings
- Dealing constructively with conflicts
- Evaluating, recognizing, and rewarding teams

Team Membership

The structure of modern organizations is based on the principle of division of labor. Most organizations today consist of a number of departments, each devoted to their own specialty. A fundamental problem is that the separate functional departments tend to optimize their own operations, often to the detriment of the organization as a whole.

Traditional organizations, in effect, create barriers between departments. Departmental managers are often forced to compete for shares of limited budgets; in other words, they are playing a "zero sum game" where another manager's gain is viewed as their department's loss. Behavioral research has shown that people engaged in zero sum games think in terms of win-lose. This leads to self-destructive and cutthroat behavior. Overcoming this tendency requires improved communication and cooperation between departments.

Interdepartmental teams are groups of people with the skills needed to deliver the value desired. Processes are designed by the team to create the value in an effective and efficient manner. Management must see to it that the needed skills exist in the organization. It is also management's job to see that they remove barriers to cooperation, as discussed in Chap. 2.

Team Dynamics Management, Including Conflict Resolution

Conflict management is usually the responsibility of the Six Sigma project team leader. If the team also includes a facilitator, then the facilitator can assist the leader by assuring that creative conflict is not repressed, but encouraged. Explore the underlying reasons for the conflict. If "personality disputes" are involved that threaten to disrupt the team meeting, arrange one-on-one meetings between the parties and attend the meetings to help mediate.

The first step in establishing an effective group is to create a consensus decision rule for the group, namely:

No judgment may be incorporated into the group decision until it meets at least tacit approval of every member of the group.

This minimum condition for group movement can be facilitated by adopting the following behaviors:

- *Avoid arguing for your own position*—Present it as lucidly and logically as possible, but be sensitive to and consider seriously the reactions of the group in any subsequent presentations of the same point.
- *Avoid "win-lose" stalemates in the discussion of opinions*—Discard the notion that someone must win and someone must lose in the discussion; when impasses occur, look for the next most acceptable alternative for all the parties involved.
- *Avoid changing your mind only to avoid conflict and to reach agreement and harmony*—Withstand pressures to yield which have no objective or logically sound foundation. Strive for enlightened flexibility; but avoid outright capitulation.
- *Avoid conflict-reducing techniques such as the majority vote, averaging, bargaining, coin-flipping, trading out, and the like*—Treat differences of opinion as indicative of an incomplete sharing of relevant information on someone's part, either about task issues, emotional data, or gut level intuitions.
- *View differences of opinion as both natural and helpful rather than as a hindrance in decision-making*—Generally, the more the ideas expressed, the greater the likelihood of conflict will be; but the richer the array of resources will be as well.
- *View initial agreement as suspect*—Explore the reasons underlying apparent agreements; make sure people have arrived at the same conclusions for either the same basic reasons or for complementary reasons before incorporating such opinions into the group decision.
- *Avoid subtle forms of influence and decision modification*—For example, when a dissenting member finally agrees, don't feel that he must be rewarded by having his own way on some subsequent point.
- *Be willing to entertain the possibility that your group can achieve all the foregoing and actually excel at its task*—Avoid doomsaying and negative predictions for group potential.

Collectively, the above steps are sometimes known as the "consensus technique." In tests it was found that 75% of the groups who were instructed in this approach significantly outperformed their best individual resources.

Stages in Group Development

Groups of many different types tend to evolve in similar ways. It often helps to know that the process of building an effective group is proceeding normally. Tuckman (1965) identified four stages in the development of a group: forming, storming, norming, and performing.

During the *forming* stage a group tends to emphasize procedural matters. Group interaction is very tentative and polite. The leader dominates the decision-making process and plays a very important role in moving the group forward.

The *storming* stage follows forming. Conflict between members, and between members and the leader, are characteristic of this stage. Members question authority as it relates to the group objectives, structure, or procedures. It is common for the group to resist the attempts of its leader to move them toward independence. Members are trying to define their role in the group.

It is important that the leader deal with the conflict constructively. There are several ways in which this may be done:

- Do not tighten control or try to force members to conform to the procedures or rules established during the forming stage. If disputes over procedures arise, guide the group toward new procedures based on a group consensus.
- Probe for the true reasons behind the conflict and negotiate a more acceptable solution.
- Serve as a mediator between group members.
- Directly confront counterproductive behavior.
- Continue moving the group toward independence from its leader.

During the *norming* stage the group begins taking responsibility, or ownership, of its goals, procedures, and behavior. The focus is on working together efficiently. Group norms are enforced on the group by the group itself.

The final stage is *performing*. Members have developed a sense of pride in the group, its accomplishments, and their role in the group. Members are confident in their ability to contribute to the group and feel free to ask for or give assistance.

Table 5.2 lists some common problems with teams, along with recommended remedial action (Scholtes, 1988).

Member Roles and Responsibilities

Productive Group Roles

There are two basic types of roles assumed by members of a group: task roles and group maintenance roles. Group task roles are those functions concerned with facilitating and coordinating the group's efforts to select, define, and solve a particular problem. The group task roles shown in Table 5.3 are generally recognized.

Another type of role played in small groups is the group maintenance roles. Group maintenance roles are aimed at building group cohesiveness and group-centered behavior. They include those behaviors shown in Table 5.4.

The development of task and maintenance roles is a vital part of the team-building process. Team building is defined as the process by which a group learns to function as a unit, rather than as a collection of individuals.

Problem	Action
Floundering	• Review the plan • Develop a plan for movement
The expert	• Talk to offending party in private • Let the data do the talking • Insist on consensus decisions
Dominating participants	• Structure participation • Balance participation • Act as gatekeeper
Reluctant participants	• Structure participation • Balance participation • Act as gatekeeper
Using opinions instead of facts	• Insist on data • Use scientific method
Rushing things	• Provide constructive feedback • Insist on data • Use scientific method
Attribution (i.e., attributing motives to people with whom we disagree)	• Don't guess at motives • Use scientific method • Provide constructive feedback
Ignoring some comments	• Listen actively • Train team in listening techniques • Speak to offending party in private
Wanderlust	• Follow a written agenda • Restate the topic being discussed
Feuds	• Talk to offending parties in private • Develop or restate ground rules

TABLE 5.2 Common Team Problems and Remedial Action

Counterproductive Group Roles

In addition to developing productive group-oriented behavior, it is also important to recognize and deal with individual roles which may block the building of a cohesive and effective team. These roles are shown in Table 5.5.

The leader's role includes that of process observer. In this capacity, the leader monitors the atmosphere during group meetings and the behavior of individuals. The purpose is to identify counterproductive behavior. Of course, once identified, the leader must tactfully and diplomatically provide feedback to the group and its members. The success of Six Sigma is, to a great extent, dependent on the performance of groups.

Management's Role

Perhaps the most important thing management can do for a group is to give it time to become effective. This requires, among other things, that management work to

Role I.D.	Description
Initiator	Proposes new ideas, tasks, or goals; suggests procedures or ideas for solving a problem or for organizing the group.
Information seeker	Asks for relevant facts related to the problem being discussed.
Opinion seeker	Seeks clarification of values related to problem or suggestion.
Information giver	Provides useful information about subject under discussion.
Opinion giver	Offers his/her opinion of suggestions made. Emphasis is on values rather than facts.
Elaborator	Gives examples.
Coordinator	Shows relationship among suggestions; points out issues and alternatives.
Orientor	Relates direction of group to agreed-upon goals.
Evaluator	Questions logic behind ideas, usefulness of ideas, or suggestions.
Energizer	Attempts to keep the group moving toward an action.
Procedure technician	Keeps group from becoming distracted by performing such tasks as distributing materials, checking seating, etc.
Recorder	Serves as the group memory.

Table 5.3 Group Task Roles

Role I.D.	Description
Encourager	Offers praise to other members; accepts the contributions of others.
Harmonizer	Reduces tension by providing humor or by promoting reconciliation; gets people to explore their differences in a manner that benefits the entire group.
Compromiser	This role may be assumed when a group member's idea is challenged; admits errors, and offers to modify his/her position.
Gatekeeper	Encourages participation, suggests procedures for keeping communication channels open.
Standard setter	Expresses standards for group to achieve, evaluates group progress in terms of these standards.
Observer/commentator	Records aspects of group process; helps group evaluate its functioning.
Follower	Passively accepts ideas of others; serves as audience in group discussions.

Table 5.4 Group Maintenance Roles

Role I.D.	Description
Aggressor	Expresses disapproval by attacking the values, ideas, or feelings of others. Shows jealousy or envy.
Blocker	Prevents progress by persisting on issues that have been resolved; resists attempts at consensus; opposes without reason.
Recognition seeker	Calls attention to himself/herself by boasting, relating personal achievements, etc.
Confessor	Uses group setting as a forum to air personal ideologies that have little to do with group values or goals.
Playboy	Displays lack of commitment to group's work by cynicism, horseplay, etc.
Dominator	Asserts authority by interrupting others, using flattery to manipulate, and claiming superior status.
Help seeker	Attempts to evoke sympathy and/or assistance from other members through "poor me" attitude.
Special interest pleader	Asserts the interests of a particular group. This group's interest matches his/her self-interest.

TABLE 5.5 Counterproductive Group Roles

maintain consistent group membership. Group members must not be moved out of the group without very good reason. Nor should there be a constant stream of new people temporarily assigned to the group. If a group is to progress through the four stages described earlier in this chapter, to the crucial performing stage, it will require a great deal of discipline from both the group and management.

Another area where management must help is creating an atmosphere within the company where groups can be effective.

Facilitation Techniques

When to Use an Outside Facilitator

It is not always necessary to have an outside party facilitate a group or team. While facilitators can often be of benefit, they may also add cost and the use of facilitators should, therefore, be carefully considered. The following guidelines can be used to determine if outside facilitation is needed (Schuman, 1996):

- Distrust or bias—In situations where distrust or bias is apparent or suspected, groups should make use of an unbiased outsider to facilitate (and perhaps convene) the group.
- Intimidation—The presence of an outside facilitator can encourage the participation of individuals who might otherwise feel intimidated.
- Rivalry—Rivalries between individuals and organizations can be mitigated by the presence of an outside facilitator.

- Problem definition—If the problem is poorly defined, or is defined differently by multiple parties, an unbiased listener and analyst can help construct an integrated, shared understanding of the problem.
- Human limits—Bringing in a facilitator to lead the group process lets members focus on the problem at hand, which can lead to better results.
- Complexity or novelty—In a complex or novel situation, a process expert can help the group do a better job of working together intellectually to solve the problem.
- Timelines—If a timely decision is required, as in a crisis situation, the use of a facilitator can speed the group's work.
- Cost—A facilitator can help the group reduce the cost of meeting—a significant barrier to collaboration.

Selecting a Facilitator

Facilitators should possess four basic capabilities (Schuman, 1996):

1. He or she should be able to anticipate the complete problem-solving and decision-making processes.
2. He or she should use procedures that support both the group's social and cognitive process.
3. He or she should remain neutral regarding content issues and values.
4. He or she should respect the group's need to understand and learn from the problem-solving process.

Facilitation works best when the facilitator:

- Takes a strategic and comprehensive view of the problem-solving and decision-making processes and selects, from a broad array, the specific methods that match the group's needs and the tasks at hand.
- Supports the group's social and cognitive processes, freeing the group members to focus their attention on substantive issues.
- Is trusted by all group members as a neutral party who has no biases or vested interest in the outcome.
- Helps the group understand the techniques being used and helps the group improve its own problem-solving processes.

Principles of Team Leadership and Facilitation

Human beings are social by nature. People tend to seek out the company of other people. This is a great strength of our species, one that enabled us to rise above and dominate beasts much larger and stronger than ourselves. It is this ability that allowed men to control herds of livestock to hunt swift antelope, and to protect themselves against predators. However, as natural as it is to belong to a group, there are certain behaviors that can make the group function more (or less) effectively than their members acting as individuals.

We will define a group as a collection of individuals who share one or more common characteristics. The characteristic shared may be simple geography, that is, the individuals are gathered together in the same place at the same time. Perhaps the group shares a common ancestry, like a family. Modern society consists of many different types of groups. The first group we join is, of course, our family. We also belong to groups of friends, sporting teams, churches, PTAs, and so on. The groups differ in many ways. They have different purposes, different time frames, and involves varying numbers of people. However, all effective groups share certain common features. In their work, *Joining Together*, Johnson and Johnson (1999) list the following characteristics of an effective group:

- Group goals must be clearly understood, be relevant to the needs of group members, and evoke from every member a high level of commitment to their accomplishment.
- Group members must communicate their ideas and feelings accurately and clearly. Effective, two-way communication is the basis of all group functioning and interaction among group members.
- Participation and leadership must be distributed among members. All should participate, and all should be listened to. As leadership needs arise, members should all feel responsibility for meeting them. The equalization of participation and leadership makes certain that all members will be involved in the group's work, committed to implementing the group's decisions, and satisfied with their membership. It also ensures that the resources of every member will be fully utilized, and increases the cohesiveness of the group.
- Appropriate decision-making procedures must be used flexibly if they are to be matched with the needs of the situation. There must be a balance between the availability of time and resources (such as member's skills) and the method of decision-making used for making the decision. The most effective way of making a decision is usually by consensus (see below). Consensus promotes distributed participation, the equalization of power, productive controversy, cohesion, involvement, and commitment.
- Power and influence need to be approximately equal throughout the group. They should be based on expertise, ability, and access to information, not on authority. Coalitions that help fulfill personal goals should be formed among group members on the basis of mutual influence and interdependence.
- Conflicts arising from opposing ideas and opinions (controversy) are to be *encouraged*. Controversies promote involvement in the group's work, quality, creativity in decision-making, and commitment to implementing the group's decisions. Minority opinions should be accepted and used. Conflicts prompted by incompatible needs or goals, by the scarcity of a resource (money, power), and by competitiveness must be negotiated in a manner that is mutually satisfying and does not weaken the cooperative interdependence of group members.
- Group cohesion needs to be high. Cohesion is based on members liking each other, each member's desire to continue as part of the group, the satisfaction of members with their group membership, and the level of acceptance, support, and trust among the members. Group norms supporting psychological safety,

individuality, creativeness, conflicts of ideas, growth, and change need to be encouraged.

- Problem-solving adequacy should be high. Problems must be resolved with minimal energy and in a way that eliminates them permanently. Procedures should exist for sensing the existence of problems, inventing and implementing solutions, and evaluating the effectiveness of the solutions. When problems are dealt with adequately, the problem-solving ability of the group is increased, innovation is encouraged, and group effectiveness is improved.
- The interpersonal effectiveness of members needs to be high. Interpersonal effectiveness is a measure of how well the consequences of your behavior match intentions.

These attributes of effective groups apply regardless of the activity in which the group is engaged. It really doesn't matter if the group is involved in a study of air defense, or planning a prom dance. The common element is that there is a group of human beings engaged in pursuit of group goals.

Facilitating the Group Task Process

Team activities can be divided into two subjects: task-related and maintenance-related. Task activities involve the reason the team was formed, its charter, and its explicit goals.

The facilitator should be selected before the team is formed and he or she should assist in identifying potential team members and leaders, and in developing the team's charter.

The facilitator also plays an important role in helping the team develop specific goals based on their charter. Goal-setting is an art and it is not unusual to find that team goals bear little relationship to what management actually had in mind when the team was formed. Common problems are goals that are too ambitious, goals that are too limited and goals that assume a cause and effect relationship without proof. An example of the latter would be a team chartered to reduce scrap assuming that Part X had the highest scrap loss (perhaps based on a week's worth of data) and setting as its goal the reduction of scrap for that part. The facilitator can provide a channel of communication between the team and management.

Facilitators can assist the team leader in creating a realistic schedule for the team to accomplish its goals. The issue of scheduling projects is covered in Chap. 6.

Facilitators should ensure that adequate records are kept on the team's projects. Records should provide information on the current status of the project. Records should be designed to make it easy to prepare periodic status reports for management. The facilitator should arrange for clerical support with such tasks as designing forms, scheduling meetings, obtaining meeting rooms, securing audio visual equipment and office supplies.

Other activities where the facilitator's assistance is needed include:

- *Meeting management*—Schedule the meeting well ahead of time. Be sure that key people are invited and that they plan to attend. Prepare an agenda and stick to it! Start on time. State the purpose of the meeting clearly at the outset. Take minutes. Summarize from time-to-time. Actively solicit input from those less talkative. Curtail the overly talkative members. Manage conflicts. Make assignments and responsibilities explicit and specific. End on time.

- *Communication*—The idea that "the quality department" can "ensure" or "control" quality is now recognized as an impossibility. To achieve quality the facilitator must enlist the support and cooperation of a large number of people outside of the team. The facilitator can relay written and verbal communication between the team and others in the organization. Verbal communication is valuable even in the era of instantaneous electronic communication. A five minute phone call can provide an opportunity to ask questions and receive answers that would take a week exchanging email and faxes. Also, the team meeting is just one communication forum, the facilitator can assist team members in communicating with one another between meetings by arranging one-on-one meetings, acting as a go-between, etc.

Facilitating the Group Maintenance Process

Study the group process. The facilitator is in a unique position to stand back and observe the group at work. Are some members dominating the group? Do facial expressions and body language suggest unspoken disagreement with the team's direction? Are quiet members being excluded from the discussion?

When these problems are observed, the facilitator should provide feedback and guidance to the team. Ask the quiet members for their ideas and input. Ask if anyone has a problem with the team's direction. Play devil's advocate to draw out those with unspoken concerns.

CHAPTER 6
The Define Phase

The key objectives within the Define phase are:

- Develop the Project Charter
  - Define scope, objectives, and schedule
  - Define the process (top-level) and its stakeholders
  - Select team members
  - Obtain authorization from sponsor
- Assemble and train the team

Project Charters

The official plan and authorization for the project is summarized in the Six Sigma Project Charter, as shown in Fig. 6.1. The Project Charter is a contract between the project team and its sponsor. As such, any changes in the critical elements of scope, objectives, or schedule require approval from the sponsor and consensus of the team.

The charter documents the *why*, *how*, *who*, and *when* of a project, include the following elements:

- Problem statement
- Project objective or purpose, including the business need addressed
- Scope
- Deliverables (i.e., objective measures of success that will be used to evaluate the effectiveness of the proposed changes, as discussed below)
- Sponsor and stakeholder groups
- Team members
- Project schedule (using Gantt or PERT as an attachment)
- Other resources required

These items are largely interrelated: as the scope increases, the timetable and the deliverables also expand. Whether initiated by management or proposed by operational personnel, many projects initially have too broad a scope. As the project cycle time increases, the tangible cost of the project deployment, such as cost due to labor and material usage, will increase. The intangible costs of the project will also increase: frustration due to lack of progress, diversion of manpower away from other activities, and

Project Charter

Project Name/Title:	**Order Processing Efficiency**	***Start Date: 9/17/07***	
Problem/Project Description: Current capacity in Sales/Customer Support area is constrained, while there are untapped opportunities for increased sales. We should limit, wherever possible, Sales involvement in order processing to free up resource for active lead follow-up and sales generation. In addition, errors and/or gaps in information acquired during Order Processing procedure have a negative impact on time required to generate, and/or receipt rate of, email marketing and software renewals to existing clients. This has an especially large potential impact, since it requires correction by senior sales staff, who might otherwise have more time to engage with clients, develop marketing efforts, or work with product development staff.			
Project Scope (Process, Product, functional areas): Limited to software products.			
Project Objectives & Goals:	*Metric*	*Baseline*	*Goal*
To decrease cycle time & costs of specific Sale Department activities: ➤ Order Processing by 50%+ ➤ Marketing to existing clients by 80+% ➤ Software renewals by 80+%	Cost/Order Time/campaign Time/update	$32 download $40 shipped 2-4 hours 2-4 hours	$16 download $20 shipped 20 minutes 20 minutes
Business Need — *Customer Impact:* Improved notification rate for renewals & upgrades; reduction in total cycle time as procedure more streamlined.			
Business Need — *Shareholder Impact:* Increased sales potential, immediately on upgrades, but also for future sales with availability of sales staff; Reduced cost for order processing. Reduced costs for marketing & renewal campaigns.			
Business Need — *Employee Impact:* Clearer responsibilities; Less interruption in process flow.			

Project Sponsor:	*Stakeholder Group:*	*Signature / Date*
Peter Keene, VP	Sales & Operations	
Team Black Belt: Patrick Killihan		
Team Members: Don Debuski	Customer Support	
Helen Winkleham	Shipping & Packaging	
Anne Sheppard	Accounting	

DEFINE	MEASURE	ANALYZE	IMPROVE	CONTROL
Objective DateComplete	*Objective DateComplete*	*Objective DateComplete*	*Objective DateComplete*	*Objective Date Complete*
➤ Project Def. 9/17/07 ➤ Top level Process Def. 9/19/07 ➤ Team Formation 9/19/07	➤ Process Definition ____ ➤ Metric Def. ____ ➤ Estimate Baseline ____	➤ Value Stream Analysis ____ ➤ Analyze Variation ____ ➤ Determine Drivers ____	➤ Implement Process ____ ➤ Assess Benefits ____ ➤ Evaluate Failure Mode ____	➤ Standardize Methods ____ ➤ Control Plan ____ ➤ Lessons Learned ____

FIGURE 6.1 Example Project Charter.

delay in realization of project benefits, to name just a few. When the project cycle time exceeds 6 months or so, these intangible costs may result in the loss of critical team members, causing additional delays in the project completion. These "world peace" projects, with laudable but unrealistic goals, generally serve to frustrate teams and undermine the credibility of the Six Sigma program.

Project Decomposition

Large projects must be broken down into smaller projects and, in turn, into specific work elements and tasks. The process of going from project objectives to tasks is called decomposition. The result of decomposition is the *project scope*: the particular area of interest and focus for the project, in process terms.

Work Breakdown Structures

Ruskin and Estes (1995) offer work breakdown structures (WBS) as a process for defining the final and intermediate products of a project and their relationships. Defining project tasks is typically complex and accomplished by a series of decompositions followed by a series of aggregations. For example, a software project to develop an SPC software application would disaggregate the customer requirements into very specific analytic requirements (e.g., the customer's requirement that the product create X-bar charts would be decomposed into analytic requirements such as subroutines for computing subgroup means and ranges, plotting data points, drawing lines, etc.). Aggregation would involve linking the various modules to produce an X-bar chart displayed on the screen.

The WBS can be represented in a tree diagram, as shown in Fig. 6.2. Tree diagrams are used to break down or stratify ideas in progressively greater detail. The objective is to partition a big idea or problem into its smaller components, to reach a level where projects are "tiny." By doing this, you will make the idea easier to understand, or the problem easier to solve. The basic idea is that, at some level, a problem's solution becomes relatively easy to find, or is isolated enough in scope as to be simpler than other solutions necessary to satisfy all possible conditions. This is the tiny level. Work takes place on the smallest elements in the tree diagram.

For example, an order processing project team for a software sales and training firm decomposed the order processing tasks based on Software or Professional Service; then by product family (e.g. Product Family A, Product Family B, Product Family C), then by the type of sale within that family (Download, Shipment, Support Renewal). The breakdown allows the team to then further consider which of these somewhat unique elements should be considered for improvement.

While the WBS provides a graphical view of the potential project elements, it provides no insight into the relative benefits to be obtained in improving a given element. Here, a Pareto analysis is often helpful. Pareto diagrams are useful in the Define, Measure and Analyze stages to focus project resources on the areas, defects, or causes yielding the highest return.

Pareto Analysis

A Pareto diagram is a vertical bar graph showing problems (or perhaps more directly *opportunities*) in a prioritized order, so it can be determined which problems or opportunities should be tackled first. The categories for the vertical bars represent mutually exclusive categories of interest. The categories are sorted in decreasing order from left to right in the Pareto diagram by their count or cost, whichever is being displayed.

In the Define phase, the Pareto diagram is used in conjunction with the WBS to quantify the opportunities inherent in each of the *tiny* elements resulting from the WBS.

For example, the data in Table 6.1 have been recorded for orders received in the Order Processing function considered in the WBS.

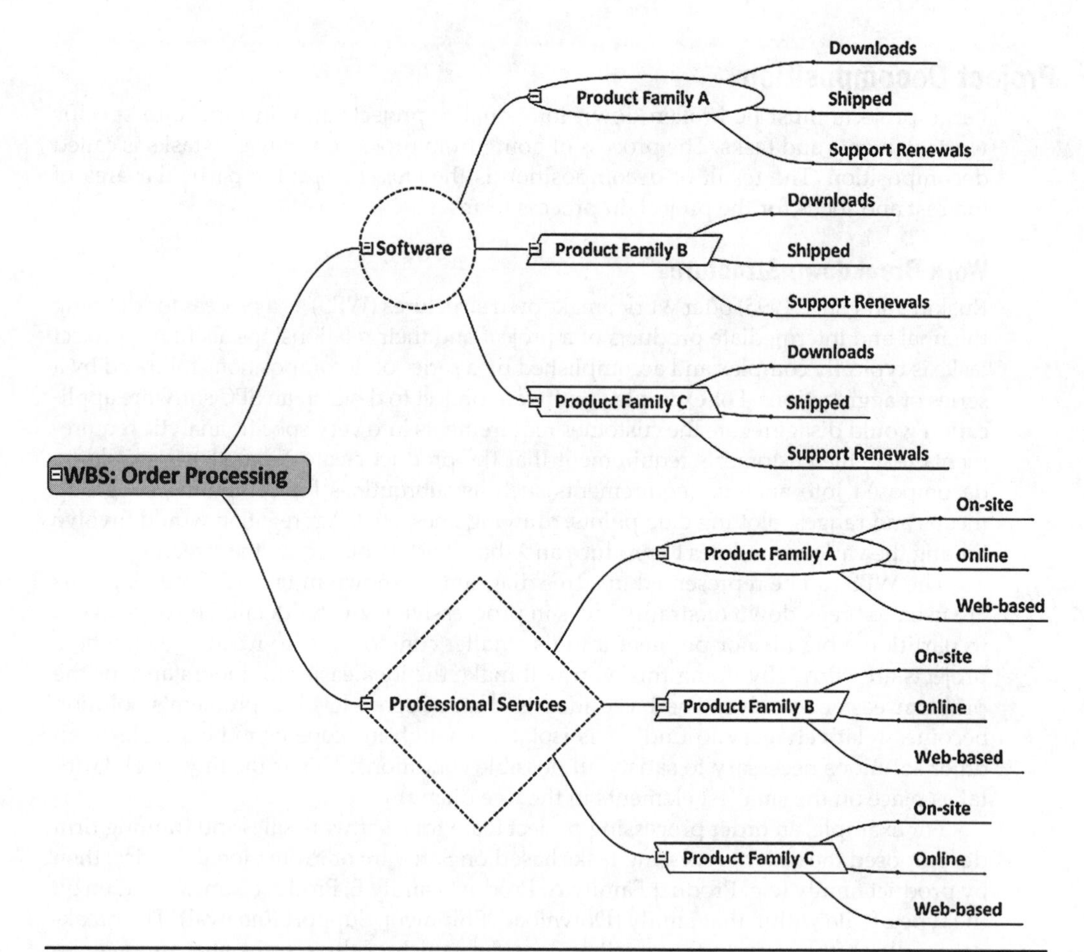

FIGURE 6.2 Example WBS for Order Processing project; constructed using *Mind Genius* software (www.qualityamerica.com by permission).

Product Family	Download	Shipment	Support
A	27	3	30
B	33	5	8
C	13	5	0

TABLE 6.1 Raw Data for Pareto Analysis

The data in Table 6.1 is shown in a massaged format in Table 6.2, to demonstrate how it is analyzed for the Pareto diagram of Fig. 6.3. In this case, a "stacked" Pareto diagram is used to display the type of order that is processed: the bottom section of each bar represents *Downloads*; the middle section *Shipments*; and the top section *Support Renewals*.

Rank	Product Family	Count	Percentage	Cum %
1	A	60	48.39	48.39
2	C	46	37.10	85.49
3	B	18	14.52	100.01

TABLE 6.2 Data Organized for Pareto Analysis

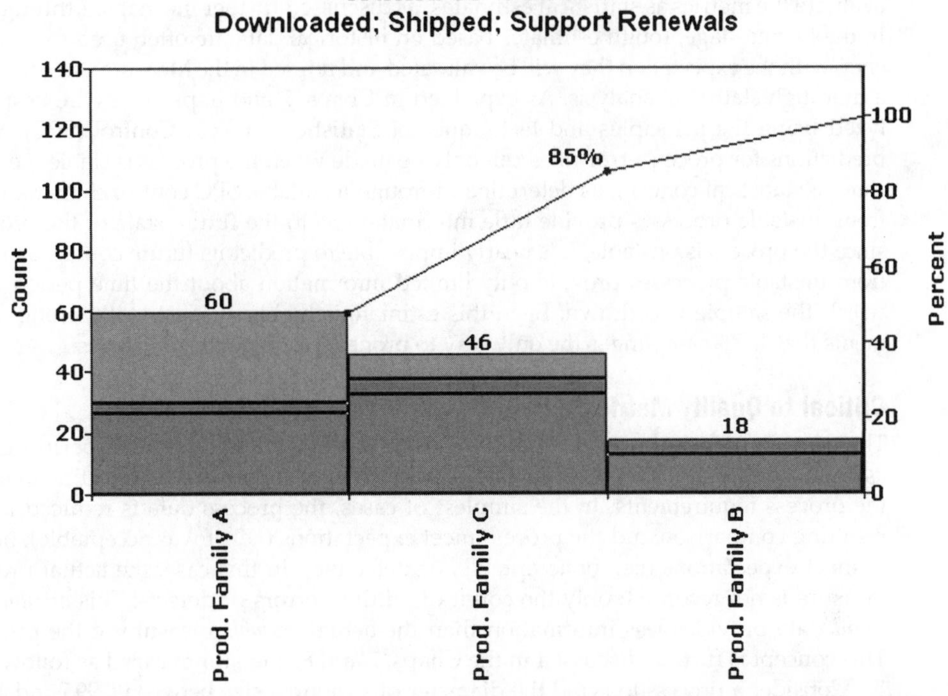

FIGURE 6.3 Example Pareto diagram constructed using *Green Belt XL* software (www.qualityamerica.com by permission).

Note that, as often happens, the final percentage is slightly different than 100%. This is due to round-off error and is nothing to worry about.

Deliverables

Six Sigma projects focus on one or more of the following three critical deliverables: cost, quality, and schedule. Factors Critical to Cost (CTC) include parameters that impact work in progress, finished goods inventory, overhead, delivery, material and labor, even when the costs can be passed on to the customer. Critical to Quality (CTQ) factors are perhaps most familiar to operational personnel since they directly impact the functional requirements specified by the internal and external customers. Critical to Schedule (CTS) factors impact the delivery time of the product or service.

While there are other CTx factors that can be considered, such as Critical to Safety, these factors are often best expressed in terms of Quality, Schedule, and/or Cost. These three prime factors are most meaningful to organizational objectives, so remain the focus of the vast majority of projects.

Most estimates of quality, schedule, and often cost are based on statistical samples. That is, it is often cost prohibitive to sample each and every potential unit (i.e., 100% sampling) of product or service for all critical characteristics, so a sub-sample of the process is taken to estimate its quality, cost or schedule metric. In that regard, it is critical to properly evaluate the metrics as statistical estimates, as discussed further in Chaps. 7 through 10. In the Define stage, rough estimates based on historical data are often used for expediency, with the expectation they will be validated and refined in the Measure stage through a thorough statistical analysis. As explained in Chaps. 7 and 8, processes are best analyzed using the principles and techniques of Statistical Process Control (SPC). Valid predictions for process error rates can only be made when the process is stable (i.e., in a state of statistical control), as determined through a suitable SPC control chart. Samples from unstable processes provide little information as to the future state of the process; since the process is unstable, it is nearly impossible to predict its future course. Samples from unstable processes provide only limited information about the time period from which the sample was drawn. Even this estimate is highly suspect, with strong arguments that 100% sampling is the only way to properly estimate its properties.

Critical to Quality Metrics

There are a number of ways to quantify Critical to Quality (CTQ) process performance, as discussed below. CTQ metrics are derived by comparing process observations against the process requirements. In the simplest of cases, the process data is reduced to the resulting comparison: did the process meet expectations (i.e., it was acceptable), or fail to meet expectations (i.e., unacceptable, or defective). In this case, the actual process measure is not recorded; only the count of failures, errors or defects. This *attribute* or *count* data provides less information than the actual *variable* measure of the process. This concept is further discussed in the Chaps. 7 and 8, and summarized as follows.

Consider a process to grind the diameter of a pin to a size between 0.995 and 1.005 inches. A pin is sampled from the process, and measured at 1.001 inches. Since its size is between the maximum and minimum diameter allowed, it meets the requirements and is not counted as a defect. Likewise, a pin measuring 1.00495 meets requirements, since the diameter is less than the 1.005 inch maximum. Yet, even if these pins are equally acceptable in their application (which is questionable), the information content of the measured value is much higher than that of the count of defects. If, for example, a vast majority of the sampled pieces have diameters approaching the maximum size, then it is possible, if not likely, that some pieces have diameters exceeding the maximum since all pieces are not sampled. That is, even though the number of pins exceeding the maximum in the sample is zero, we could use the measured variables data to reliably predict that additional samples would yield a certain percentage larger than the maximum size. Using variables (i.e., measurement) data, we can use the statistical properties of the samples to predict errors even when there is no direct evidence of errors, demonstrating the higher informational content of the variables measurements over the attribute counts.

The CTQ metrics below use an estimate of the proportion of defects in the sample to estimate the metric. The percent defective may be estimated using the actual data (e.g., the measured diameters), as well as the count of items that fail to meet the requirements. In light of the arguments provided in the preceding paragraph, it should be clear that the estimates using the counts are less reliable than those using the variables data. These concepts are discussed in further detail in Chap. 8.

Also note that the terms defects and errors apply universally to all processes with customers, regardless of industry. A service process, such as the filling of a prescription at the local pharmacy, may have a customer expectation of delivery within 30 minutes of drop-off. In this case, the delivery time might be considered a critical quality characteristic, since it relates to the quality of the service as perceived by the customer. Of course, the content of the prescription refill (i.e., whether the correct medicine of the correct potency with the correct dosage instructions) could also be considered a critical quality characteristic. It should be clear from Chaps. 2 and 3 that quality metrics may be established for a wide variety of process parameters, including those related to schedule and cost, to achieve a customer requirement, expectation or excitement (as in the Kano model). These limits may be defined by external customers, internal management, contract specifications, or regulation.

Defects-Per-Million Opportunities When considering the defects-per-million opportunities (DPMO) for a single process characteristic, DPMO is directly calculated using the estimated defect rate (i.e., percent defective; defects per hundred) by multiplying by 10^6. For example, a process with an estimated error rate of 1.2% (.012) will have a DPMO of 12,000.

In some cases, analysts seek to estimate an overall DPMO for a process. Consider the pharmacy example cited above, where there are four critical characteristics (delivery time, medicine formula, formula potency, and dosage). If a sample of 1000 prescriptions are checked for all of the four critical characteristics, and 12 prescriptions are identified which have an error with at least one of the critical characteristics, the overall number of defects is 12,000 per million, as in the case cited in the above paragraph. Yet, the DPMO is calculated as 3,000 defects-per-million *opportunities*, since there are 4 opportunities for failure in each of the 12,000 samples. Observant readers will question the validity of this metric, since the DPMO could be arbitrarily lessened by increasing the number of critical characteristics (e.g. six critical characteristics using same data results in 2,000 DPMO). For these reasons, DPMO estimates for anything beyond a single characteristic should be treated with caution at the least.

Process defect rates are best estimated using SPC control charts (detailed in Chap. 8). For attribute (count) data, the error rate for a given process characteristic is best estimated as the centerline of the *p* control chart (*p*-bar: the average error percentage). The centerline on the attribute chart is the long-term expected quality level of the process, for example, the average proportion defective for the case of the *p* chart. This is the level created by the common causes of variation inherent to the process, as discussed in further detail in Chaps. 7 and 8.

For variables (measurement) data, the Defect Rate for a given process characteristic is estimated using the variable control chart (e.g. the X-bar control chart) and assumptions regarding the distribution of the process observations.

$$Z_U = \frac{\text{upper specification} - \overline{\overline{X}}}{\hat{\sigma}} \tag{6.1}$$

$$Z_L = \frac{\overline{\overline{X}} - \text{lower specification}}{\hat{\sigma}} \tag{6.2}$$

In these equations, the upper specification limit (USL) and/or lower specification limit (LSL) represent respectively the upper and lower requirements for the process; X-DoubleBar is the process centerline and sigma is the standard deviation of the process, as estimated using the statistical control chart.

The formulas for estimating the process standard deviation are:

Range chart method:

$$\hat{\sigma} = \frac{\bar{R}}{d_2} \tag{6.3}$$

Sigma chart method:

$$\hat{\sigma} = \frac{\bar{s}}{c_4} \tag{6.4}$$

The values d_2 and c_4 are constants from Table 9 in the Appendix.

The Z_U and Z_L indices measure the process location (central tendency) relative to its standard deviation and the upper and lower requirements (respectively). If the distribution is normal, the values of Z_U (and Z_L) can be used to determine the percentage above the upper requirement (and below the lower requirement) by using Table 2 in the Appendix. The method is the same as described in the Normal Distribution section of Chap. 8 (using Z_U instead of using Z).

In general, a larger Z value is better. A value of at least +3 is required to ensure that 0.1% or less defective will be produced. A value of +4 is generally desired to allow some room for process drift. For a Six Sigma process Z_U would be +6.

For example, assume that a process is in statistical control based on an X-bar and R chart with subgroups of 5. The grand average (or centerline of the X-Bar chart) is calculated as 0.99832, and the average range (or centerline of the R chart) is calculated as 0.02205. From the table of d_2 values (Appendix Table 9), we find d_2 is 2.326 for subgroups of 5. Thus, using the equation above for calculating the process standard deviation using the Range chart method:

$$\hat{\sigma} = \frac{0.02205}{2.326} = 0.00948 \tag{6.5}$$

If the process requirements are a lower specification of 0.980 and an upper specification of 1.020 (i.e., 1:000 ± 0:020), the Z values are calculated as:

$$Z_U = \frac{\text{upper specification} - \bar{\bar{X}}}{\hat{\sigma}} = \frac{1.020 - 0.99832}{0.00948} = 2.3 \tag{6.6}$$

$$Z_L = \frac{\bar{\bar{X}} - \text{lower specification}}{\hat{\sigma}} = \frac{0.99832 - 0.980}{0.00948} = 1.9 \tag{6.7}$$

Referring to Table 2 in the Appendix we find that approximately 1.1% will be oversized (based on the Z_u value of 2.3) and approximately 2.9% will be undersized (based on the Z_L value of 1.9). Adding the percents finds a total reject rate of 4.0%, which can be equivalently expressed as a DPMO of 40,000 or a yield of 96.0%.

$$C_p = \frac{\text{engineering tolerance}}{6\hat{\sigma}}$$ where engineering tolerance = upper specification limit – lower specification limit	(6.8)
$$C_R = 100 \times \frac{6\hat{\sigma}}{\text{engineering tolerance}}$$ where engineering tolerance = upper specification limit – lower specification limit	(6.9)
$$C_M = \frac{\text{engineering tolerance}}{8\hat{\sigma}}$$ where engineering tolerance = upper specification limit – lower specification limit	(6.10)
$$Z_{\text{MIN}} = \text{Minimum}\{Z_L, Z_U\}$$	(6.11)
$$C_{\text{PK}} = \frac{Z_{\text{MIN}}}{3}$$	(6.12)
$$C_{pm} = \frac{C_p}{\sqrt{1 + \frac{(\mu - T)^2}{\hat{\sigma}^2}}}$$	(6.13)

Table 6.3 Process Capability Indices

Process Capability Indices

A process capability index is another metric used to indicate the performance of the process relative to requirements, as indicated in Table 6.3.

Perhaps the biggest drawback of using process capability indexes is that they take the analysis a step away from the data. The danger is that the analyst will lose sight of the purpose of the capability analysis, which is to improve quality. To the extent that capability indexes help accomplish this goal, they are worthwhile. To the extent that they distract from the goal, they are harmful. The analyst should continually refer to this principle when interpreting capability indexes.

C_p—This is one of the first capability indexes used. The "natural tolerance" of the process is computed as 6σ. The index simply makes a direct comparison of the process natural tolerance to the engineering requirements. Assuming the process distribution is normal and the process average is exactly centered between the engineering requirements, a C_p index of 1 would give a "capable process." However, to allow a bit of room for process drift, the generally accepted minimum value for C_p is 1.33. In general, the larger C_p is, the better. For a Six Sigma process, that is, a process that produces 3.4 defects-per-million-opportunities including a 1.5 sigma shift, the value of C_p would be 2.

The C_p index has two major shortcomings. First, it can't be used unless there are both upper and lower specifications. Second, it does not account for process centering. If the process average is not exactly centered relative to the engineering requirements, the C_p index will give misleading results. In recent years, the C_p index has largely been replaced by C_{PK} (see below).

C_R—The C_R index is equivalent to the C_p index. The index simply makes a direct comparison of the process to the engineering requirements. Assuming the process distribution is normal and the process average is exactly centered between the engineering requirements, a C_R index of 100% would give a "capable process." However, to allow a bit of room for process drift, the generally accepted maximum value for C_R is 75%. In general, the smaller C_R is the better. The C_R index suffers from the same shortcomings as the C_p index. For a Six Sigma process, that is, a process that produces 3.4 defects-per-million opportunities including a 1.5 sigma shift, the value of C_R would be 50%.

C_M—The C_M index is generally used to evaluate machine capability studies, rather than full-blown process capability studies. Since variation will increase when other sources of process variation are added (e.g., tooling, fixtures, materials, etc.), C_M uses an 8 sigma spread rather than a 6 sigma spread to represent the natural tolerance of the process. For a machine to be used on a Six Sigma process, a 10 sigma spread would be used.

Z_{MIN}—The value of Z_{MIN} is simply the smaller of the Z_L or the Z_U values whose calculation is shown in the DPMO section above. It is used in computing C_{PK}. For a Six Sigma process Z_{MIN} would be +6.

C_{PK}—The value of C_{PK} is simply Z_{MIN} divided by 3. Since the smallest value represents the nearest specification, the value of C_{PK} tells you if the process is truly capable of meeting requirements. A C_{PK} of at least +1 is required, and +1.33 is preferred. Note that C_{PK} is closely related to C_p, the difference between C_{PK} and C_p represents the potential gain to be had from centering the process. For a Six Sigma process C_{PK} would be 2.

Extending the example shown in the DPMO section above:

$$C_P = \frac{\text{engineering tolerance}}{6\hat{\sigma}} = \frac{1.020 - 0.9800}{6 \times 0.00948} = 0.703 \tag{6.14}$$

$$C_R = 100 \times \frac{6\hat{\sigma}}{\text{engineering tolerance}} = 100 \times \frac{6 \times 0.00948}{0.04} = 142.2\% \tag{6.15}$$

$$C_M = \frac{\text{engineering tolerance}}{8\hat{\sigma}} = \frac{0.04}{8 \times 0.00948} = 0.527 \tag{6.16}$$

$$Z_{MIN} = \text{Minimum}\{1.9, 2.3\} = 1.9 \tag{6.17}$$

$$C_{PK} = \frac{Z_{MIN}}{3} = \frac{1.9}{3} = 0.63 \tag{6.18}$$

Assuming that the target is precisely 1.000, we compute:

$$C_{pm} = \frac{C_p}{\sqrt{1 + \frac{\left(\bar{\bar{X}} - T\right)^2}{\hat{\sigma}^2}}} = \frac{0.703}{\sqrt{1 + \frac{(0.99832 - 1.000)^2}{0.00948^2}}} = 0.692 \tag{6.19}$$

Since the minimum acceptable value for C_p is 1, the 0.703 result indicates that this process cannot meet the requirements. Furthermore, since the C_p index doesn't consider the centering process, we know that the process can't be made acceptable by merely adjusting the process closer to the center of the requirements. Thus, we would expect the Z_L, Z_U, and Z_{MIN} values to be unacceptable as well.

The C_R value always provides the same conclusions as the C_P index. The number implies that the "natural tolerance" of the process uses 142.2% of the engineering requirement, which is, of course, unacceptable.

The C_M index should be 1.33 or greater. Obviously it is not. If this were a machine capability study the value of the C_M index would indicate that the machine was incapable of meeting the requirement.

The value of C_{PK} is only slightly smaller than that of C_P. This indicates that we will not gain much by centering the process. The actual amount we would gain can be calculated by assuming the process is exactly centered at 1.000 and recalculating Z_{MIN}. This gives a predicted total reject rate of 3.6% instead of 4.0%.

Minitab has a built-in capability analysis feature, called a "Six Pack," providing a compact display of the most important statistics and analysis. The control charts indicate if the process is in statistical control. If it's out of control, stop and find out why. The histogram and normal probability plot indicate if the normality assumption is justified. If not, you can't trust the capability indices, and should consider using Minitab's non-normal capability analysis. (An example is provided at the end of Chap. 8. The "within" capability indices C_p and C_{PK} are based on within-subgroup variation only, called short-term variability. The "overall" capability indices P_p and P_{PK} are based on total variation, sometimes called long-term variability, which includes variation within subgroups and variation between subgroups. A Capability Plot graphically compares within variability (short-term) and overall variability (long-term) to the specifications. Ideally, for a Six Sigma process, the process variability (process tolerance) will be about half of the specifications. However, the capability plot for the example shows that the process tolerance is actually wider than the specifications.

What is missing in the six pack is an estimate of the process yield. There is an option in the six pack to have this information (and a great deal more) stored in the worksheet. Alternatively, you can run Minitab's Capability Analysis (Normal) procedure and get the information along with a larger histogram. The PPM levels confirm the capability and performance indices calculations.

Rolled Throughput Yield and Sigma Level

The rolled throughput yield (RTY) summarizes defects-per-million opportunities (DPMO) data for a process or product. DPMO is the same as the parts-per-million calculated by Minitab. RTY is a measure of the overall process quality level or, as its name suggests, throughput. For a process, throughput is a measure of what comes out of a process as a function of what goes into it. For a product, throughput is a measure of the quality of the entire product as a function of the quality of its various features. Throughput combines the results of the capability analyses into a measure of overall performance.

To compute the rolled throughput yield for an N-step process (or N-characteristic product), use the following equation:

$$\text{Rolled Throughput Yield} = \left(1-\frac{\text{DPMO}_1}{1{,}000{,}000}\right)\times\left(1-\frac{\text{DPMO}_2}{1{,}000{,}000}\right)\cdots\left(1-\frac{\text{DPMO}_N}{1{,}000{,}000}\right)^{*} \quad (6.20)$$

*When calculating RTY, use the approach shown here for DPU values than 0.1. Otherwise use the DPU defective rates instead of DPMO, and calculate RTY using the exact formula RTY = exp-($dpu_1 + dpu_2 + \cdots dpu_N$).

Process Step	DPMO	dpu=DPMO/1,000,000	1 – dpu
1	5,000	0.005000	0.9950
2	15,000	0.015000	0.9850
3	1,000	0.001000	0.9990
4	50	0.000050	0.99995
Rolled throughput yield = 0.995 × 0.985 × 0.999 × 0.99995 = 0.979			

TABLE 6.4 Calculations Used to Find RTY

DPMO is the defects-per-million opportunities for step x in the process. For example, consider a four-step process with the following DPMO levels at each step (Table 6.4) (dpu is defects-per-unit).

Figure 6.4 shows the Excel spreadsheet and formula for this example. The meaning of the RTY is simple: if you started 1,000 units through this four-step process you would only get 979 units out the other end. Or, equivalently, to get 1,000 units out of this process you should start with (1,000/0.979) + 1 = 1,022 units of input. Note that the RTY is worse than the worst yield of any process or step. It is also worse than the average yield of 0.995. Many a process owner is lulled into complacency by reports showing high average process yields. They are confused by the fact that, despite high average yields, their ratio of end-of-the-line output to starting input is abysmal. Calculating RTY may help open their eyes to what is really going on. The effect of declining RTYs grows exponentially as more process steps are involved.

The sigma level equivalent for this four-step process RTY is 3.5 (see Appendix, Table 14: **Process σ levels and equivalent PPM quality levels**). This would be the estimated "process" sigma level, based on observed defects. A more precise measure of process capability will be shown in the Measure Stage, which should validate these initial estimates.

Normalized Yield and Sigma Level

To compute the normalized yield, which is a kind of average, for an N-process or N-product department or organization, use Eq. (6.21):

B11 =D6*D7*D8*D9 ← RTY equation

	A	B	C	D
5	Process Step	DPMO	DPMO/1,000,000	1-(DPMO/1,000,000)
6	1	5,000	0.005	0.995
7	2	15,000	0.015	0.985
8	3	1,000	0.001	0.999
9	4	50	0.00005	0.99995
10				
11	RTY	0.979046		

FIGURE 6.4 Excel spreadsheet for RTY.

Normalized Yield

$$= \sqrt[N]{\left(1-\frac{\text{DPMO}_1}{1{,}000{,}000}\right)\times\left(1-\frac{\text{DPMO}_2}{1{,}000{,}000}\right)\cdots\left(1-\frac{\text{DPMO}_N}{1{,}000{,}000}\right)} \tag{6.21}$$

For example, consider a four-process organization with the following DPMO levels for each process:

Process	DPMO	DPMO/1,000,000	1-(DPMO/1,000,000)
Billing	5,000	0.005000	0.9950000
Shipping	15,000	0.015000	0.9850000
Manufacturing	1,000	0.001000	0.9990000
Receiving	50	0.000050	0.9999500

$$\text{Normalized yield} = \sqrt[4]{0.995 \times 0.985 \times 0.999 \times 0.99995} = 0.99472 \tag{6.22}$$

Figure 6.5 shows the Excel spreadsheet for this example.

The sigma level equivalent of this four-process organization's normalized yield is 4.1 (see Appendix, Table 14: **Process σ levels and equivalent PPM quality levels**). This would be the estimated "organization" sigma level. Normalized yield should be considered a handy accounting device for measuring overall system quality. Because it is a type of average it is not necessarily indicative of any particular product or process yield or of how the organization's products will perform in the field. To calculate these refer to "Rolled throughput yield and sigma level" above.

Assuming every step has an equal yield, it is possible to "backsolve" to find the normalized yield required in order to get a desired RTY for the entire process, see Eq. 6.23.

$$Y_n = \sqrt[N]{\text{RTY}} = \text{RTY}^{1/N} \tag{6.23}$$

where Y_n is the yield for an individual process step and N is the total number of steps.

If the process yields are not equal, then Y_n is the required yield of the worst step in the process. For example, for a ten-step process with a desired RTY of 0.999 the worst acceptable yield for any process step $Y_n = \text{RTY}^{1/10} = (0.999)^{1/10} = 0.9999$. If all other yields are not 100% then the worst-step yield must be even higher.

B6 =(D2*D3*D4*D5)^0.25 ← Normalized yield

	A	B	C	D
1	Process	DPMO	DPMO/1,000,000	1-(DPMO/1,000,000)
2	Billing	5,000	0.005	0.995
3	Shipping	15,000	0.015	0.985
4	Manufacturing	1,000	0.001	0.999
5	Receiving	50	0.000	1.000
6	NY	0.99472		

Figure 6.5 Excel spreadsheet for calculating normalized yield.

Unfortunately, finding the RTY isn't always as straightforward as described above. In the real world you seldom find a series of process steps all neatly feeding into one another in a nice, linear fashion. Instead, you have different supplier streams, each with different volumes and different yields. There are steps that are sometimes taken and sometimes not. There are test and inspection stations, with imperfect results. There is rework and repair. The list goes on and on. In such cases it is sometimes possible to trace a particular batch of inputs through the process, monitoring the results after each step. However, this is often exceedingly difficult to control. The production and information systems are not designed to provide the kind of tracking needed to get accurate results. The usual outcome of such attempts is questionable data and disappointment.

High-end simulation software offers an alternative. With simulation you can model the individual steps, then combine the steps into a process using the software. The software will monitor the results as it "runs" the process as often as necessary to obtain the accuracy needed. Figure 6.6 shows an example. Note that the Properties dialog box is for step 12 in the process ("Right Med?"). The model is programmed to keep track of the errors encountered as a Med Order works its way through the process. Statistics are defined to calculate dpu and RTY for the process as a whole (see the Custom Statistics box in the lower right section of Fig. 6.6). Since the process is nonlinear (i.e., it includes

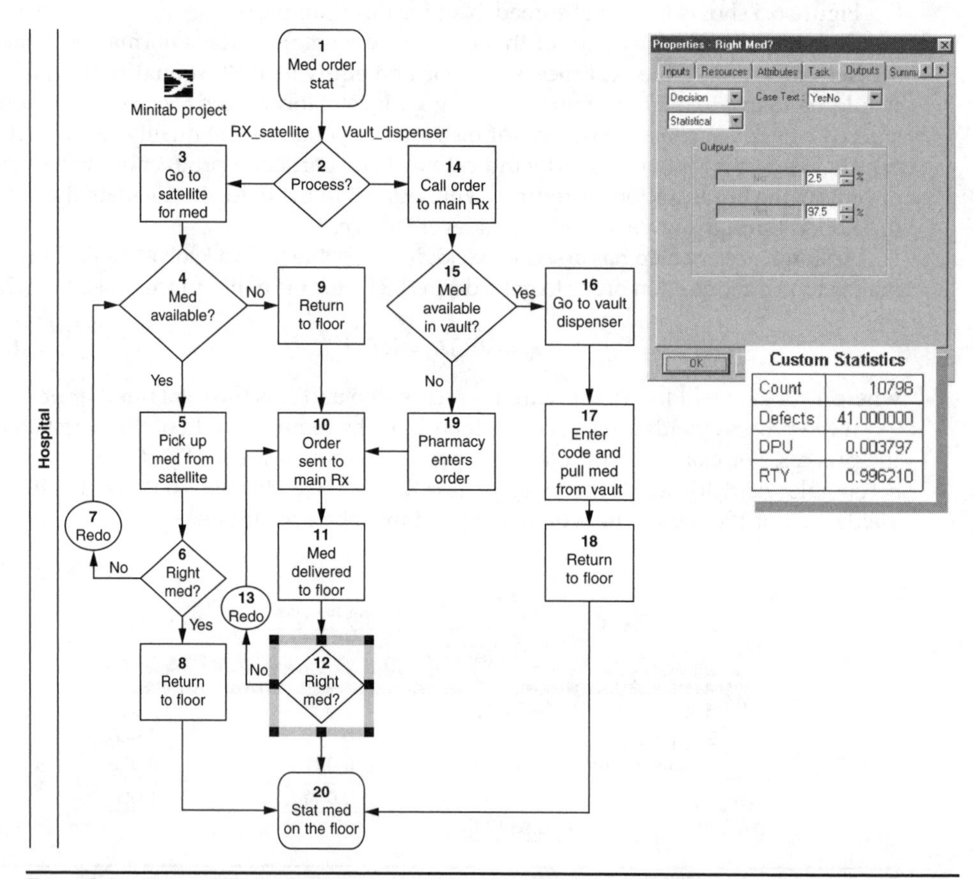

FIGURE 6.6 Finding RTY using simulation software (iGrafx Process for Six Sigma, Corel Corporation).

feedback loops) it isn't a simple matter to determine which steps would have the greatest impact on RTY. However, the software lets the Black Belt test multiple what-if scenarios to determine this. It can also link to Minitab or Excel to allow detailed data capture and analysis.

Critical to Schedule Metrics

Critical to Schedule (CTS) metrics are related to Cycle Time and scheduling efficiencies, including: Cycle Time; Process Cycle Efficiency; Process Velocity; and Overall Equipment Effectiveness.

Cycle Time

The Cycle Time for a process usually refers to the total elapsed time for the process (from process start to process completion). Note that this may not equal the total task time for the process, since there may be additional time where no task is being performed (e.g. queue time between tasks). This total elapsed time more accurately reflects the process from a customer's perspective, even though it may not reflect the internal costs of providing the service. In practice, it is suggested to clearly define the Cycle Time so that your audience is clear on its usage in your discussions.

Process Cycle Efficiency

Process Cycle Efficiency is a metric useful for prioritizing improvement opportunities. Process Cycle Efficiency is calculated by dividing the value-added time associated with a process by the total lead time of the process (George, 2002). If the process consists of only value-added activities, then the Process Cycle Efficiency would reach a theoretical maximum of 100%. In practice, Process Cycle Efficiencies will exceed 25% for processes that have been improved through the use of Lean methods. Typical Process Cycle Efficiencies for various processes is shown in Fig. 6.7.

The key to improving Process Cycle Efficiency is to reduce the Lead Time, the denominator of the equation.

Process Lead Time

The Process Lead Time is calculated by dividing the number of items in process by the completions per hour (George, 2002). For example, if it takes 2 hours on average to complete each Purchase Order, then there are 0.5 completions per hour. This is the denominator of the equation. If there are ten Purchase Orders waiting in queue (the numerator), then the Process Lead Time is 20 hours (10 divided by 1/2 equals 20).

Process type	Typical efficiency	World class efficiency
Machining	1%	20%
Fabrication	10%	25%
Assembly	15%	35%
Continuous	30%	80%
Transactional	10%	50%
Creative	5%	25%

FIGURE 6.7 Typical and world class process efficiencies. (*George, 2002*)

In other words, new orders can't be processed until the 20 hour lead time has allowed the existing Work in Process to be completed.

Velocity

Once Lead Time is known, Velocity can be calculated by dividing it into the number of value added steps (George, 2002). In the preceding example, if there are five value-added process steps in the PO process, then the Velocity may be calculated as 5 divided by 20 equals 0.25 steps per hour.

The Velocity of the process represents the responsiveness or flexibility of the process to customer demand. A long lead times results in slow velocity. Lead time is reduced, and velocity increased, when Work in Progress is reduced.

The rationale is simple: new orders from customers cannot be started until work (or items) in process is completed. Thus, the activity on new items is stalled. Consider your doctor's waiting room. The patients are Work in Progress. New patients aren't serviced by the doctor until those that arrived earlier are completed.

Overall Equipment Effectiveness

The Overall Equipment Efficiency (OEE) is a Lean metric that incorporates process availability, performance and quality in a single metric. The resulting OEE value can be used to compare and prioritize processes for improvement. It is also recommended to track and prioritize based on the individual Availability (*A*), Performance (*P*), and Quality (*Q*) components of the score, since their relative values may provide varying cost benefits.

$$\text{OEE} = A \times P \times Q$$
$$\text{A} = \text{Actual operating time / Planned time}$$
$$\text{P} = \text{Ideal cycle time / Actual cycle time}$$
$$\text{Q} = \text{Acceptable output / Total output}$$

The OEE Availability is calculated as the actual time the process is producing product (or service) divided by the amount of time that is planned for production. The planned time, by definition, excludes all scheduled shutdowns when equipment is not operational, which may include lunches, breaks, plant shutdowns for holidays and so on. The remaining portion of the planned time includes the available time for production, and the downtime. Downtime is the loss in production time due to shift changeovers, part changeovers, waiting for material, equipment failures, and so on. It is any measurable time where the process is not available for production, when production had been planned. 100% Availability implies that the process is operational 100% of the time.

OEE Performance is the efficiency of the process in minimizing its Operating Time. Performance is calculated by dividing the ideal process cycle time by the actual cycle time. It accounts for process inefficiencies, such as due to poor quality materials, operator inefficiencies, equipment slowdown, etc. 100% Performance implies the process is running at maximum velocity.

OEE Quality is the percent of total output that meets the requirement without need for any repair or rework. It might otherwise be called the first pass quality. 100% Quality implies the process is producing no errors.

Critical to Cost Metrics

Metrics that are Critical to Quality and Critical to Schedule are (by definition) often Critical to Cost as well. Costs associated with the process issues must include the effect of losses due to Hidden Factory and Customer impact, such as delays in shipments or in information exchange.

Metrics used in CTC evaluations, often to quantify and compare opportunities, include the Net Present Value (NPV) and Internal Rate of Return (IRR). NPV and IRR calculations reflect the time value of money: over time, interest is earned on our initial investments, as well as on the interest that has already been accrued from the initial investment. This is known as *compounding*.

It's important to consider the time value of money, since improvements efforts will typically require an initial investment, whose benefits will be experienced at some future time. The initial investment represents a diversion of funds, either from a lender or from some other investment opportunity. The value of a particular investment, then, depends on the cost of that diversion, which varies based on market conditions and the length of time before the investment "pays off."

Financial Analysis of Benefit and Cost

In performing benefit-cost analysis it is helpful to understand some of the basic principles of financial analysis, in particular, break-even analysis and the time value of money (TVM).

Let's assume that there are two kinds of costs:

1. *Variable costs* are those costs which are expected to change at the same rate as the firm's level of sales. As more units are sold, total variable costs will rise. Examples include sales commissions, shipping costs, hourly wages and raw materials.
2. *Fixed costs* are those costs that are constant, regardless of the quantity produced, over some meaningful range of production. Total fixed cost *per unit* will decline as the number of units increases. Examples of fixed costs include rent, salaries, depreciation of equipment, etc.

These concepts are illustrated in Fig. 6.8.

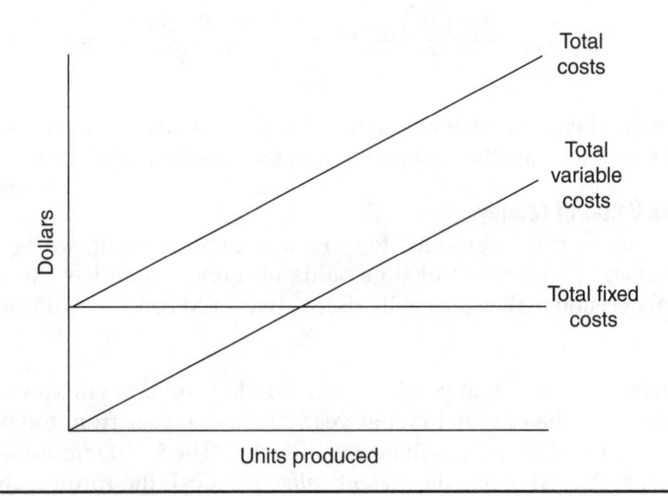

FIGURE 6.8 Fixed and variable costs.

Break-Even Points

We can define the *break-even point*, or *operating break-even point* as the level of unit sales required to make earnings before interest and taxes (EBIT) equal to zero, that is, the level of sales where profits cover both fixed and variable costs.

Let Q be the quantity sold, P the price per unit, V the variable cost per unit, and F the total fixed costs. Then the quantity $P - V$ represents the variable profit per unit and

$$Q(P-V)-F=\text{EBIT} \tag{6.24}$$

If we set EBIT equal to zero in Eq. (6.24) and solve for the break-even quantity Q^* we get:

$$Q^* = \frac{F}{P-V} \tag{6.25}$$

Example Break-Even Analysis

A publishing firm is selling books for \$30 per unit. The variable costs are \$10 per unit and fixed costs total \$100,000. The break-even point is:

$$Q^* = \frac{F}{P-V} = \frac{\$100{,}000}{\$30-\$10} = 5{,}000 \text{ units} \tag{6.26}$$

Of course, management usually wishes to earn a profit rather than to merely break even. In this case, simply set EBIT to the desired profit rather than zero in Eq. (6.24) and we get the production quantity necessary to meet management's target:

$$Q^*_{\text{TARGET}} = \frac{F+\text{EBIT}_{\text{TARGET}}}{P-V} \tag{6.27}$$

For example, if the publisher mentioned above wishes to earn a \$5,000 profit then the break-even level of sales becomes

$$Q^*_{\text{TARGET}} = \frac{F+\text{EBIT}_{\text{TARGET}}}{P-V} = \frac{\$100{,}000+\$5{,}000}{\$30-\$10} = 5{,}250 \text{ units} \tag{6.28}$$

In project benefit-cost analysis these break-even quantities are compared to the sales forecasts to determine the probability that the expected return will actually be earned.

The Time Value of Money

Because money can be invested to grow to a larger amount, we say that money has a "time value." The concept of time value of money underlies much of the theory of financial decision making. We will discuss two TVM concepts: future value and present value.

Future Value Assume that you have \$1,000 today and that you can invest this sum and earn interest at the rate of 10% per year. Then, one year from today, your \$1,000 will have grown by \$100 and it will be worth \$1,100. The \$1,100 figure is the *future value* of your \$1,000. The \$1,000 is the *present value*. Let's call the future value FV, the present value PV and the interest rate i, where i is expressed as a proportion rather than as a percentage. Then we can write this example algebraically as follows:

$$FV = PV + PV \times i = PV(1+i) \tag{6.29}$$

Now, let's say that you could invest at the 10% per year rate for two years. Then your investment would grow as follows:

Year	Starting Amount	Interest	Ending Amount
1	$1,000	$100	$1,100
2	$1,100	$110	$1,210

Observe that in year 2 you earned interest on your original $1,000 *and* on the $100 interest you earned in year 1. The result is that you earned more interest in year 2 than in year 1. This is known as *compounding*. The year time interval is known as the *compounding period*. Thus, the FV after two years is $1,210. Algebraically, here's what happened:

$$FV = [\$1{,}000(1.10)](1.10) = \$1{,}000(1.10)^2 \tag{6.30}$$

Where the value between the [ ] characters represents the value at the end of the first year. This approach can be used for any number of N compounding periods. The equation is:

$$FV = PV(1+i)^N \tag{6.31}$$

Of course, Eq. (6.31) can be solved for PV as well, which gives us the present value of some future amount of money at a given rate of interest.

$$PV = \frac{FV}{(1+i)^N} \tag{6.32}$$

Nonannual Compounding Periods Note that N can be stated in any time interval, it need not be in years. For example, if the compounding period was quarterly then N would be the number of quarters. Of course, the interest rate would also need to be stated in quarters. For example, if the $1,000 were invested for 2 years at 10% per year, compounded quarterly, then

$$FV = PV(1+i)^N = \$1{,}000\left(1+\frac{0.1}{4}\right)^{2\times4} = \$1{,}000(1+0.025)^8 = \$1{,}218.40 \tag{6.33}$$

Continuous Compounding Note that the FV is greater when a greater number of compounding periods are used. The limit is an infinite number of compounding periods, known as continuous compounding. For continuous compounding the PV and FV equations are:

$$FV = PV \times e^{i \times t} \tag{6.34}$$

$$PV = \frac{FV}{e^{i \times t}} \tag{6.35}$$

Where t is the length of time (in years) the sum is compounded, e is a constant 2.71828, and all other terms are as previously defined. For our example, we have a 2 year period which gives

$$FV = PV \times e^{i \times t} = \$1,000 \times 2.7182818^{0.1 \times 2} = \$1,221.40 \tag{6.36}$$

Net Present Value When evaluating project costs and benefits, it often happens that both costs and benefits come in *cash flow streams*, rather than in lump sums. Furthermore, the cash flow streams are uneven, that is, the amounts vary from one period to the next. The approach described above can be used for uneven cash flow streams as well. Simply compute the PV (or FV) of each cash flow separately and add the various results together. The result of applying this procedure is called *the net present value*, or NPV. The procedure, while conceptually easy to grasp, becomes tedious quite quickly. Fortunately, most spreadsheets have a built in capability to perform this analysis.

Assume that a proposed project has the projected costs and benefits shown in the table below.

Year	Cost	Benefit
1	$10,000	$0
2	$2,000	$500
3	$0	$5,000
4	$0	$10,000
5	$0	$15,000

Also assume that management wants a 12% return on their investment. What is the NPV of this project?

There are two ways to approach this question, both of which produce the same result (Fig. 6.9). One method would be to compute the net difference between the cost and benefit for each year of the project, then find the NPV of this cash flow stream. The other method is to find the NPV of the cost cash flow stream and benefit cash flow stream, then subtract.

The NPV of the cost column is $10,523; the NPV of the benefits is $18,824. The project NPV can be found by subtracting the cost NPV from the benefit NPV, or by finding the NPV of the yearly benefit minus the yearly cost. Either way, the NPV analysis indicates that this project's net present value is $8,301.

Internal Rate of Return Often in financial analysis of projects, it is necessary to determine the yield of an investment in a project given its price and cash flows. For example, this may be the way by which projects are prioritized. When faced with uneven cash flows, the solution to this type of problem is usually done by computer. For example, with Microsoft Excel, we need to make use of the internal rate of return (IRR) function. The IRR is defined as the rate of return which equates the present value of future cash flows with the cost of the investment. To find the IRR the computer uses an iterative process. In other words, the computer starts by taking an initial "guess" for the IRR, determines how close the computed PV is to the cost of the investment, and then

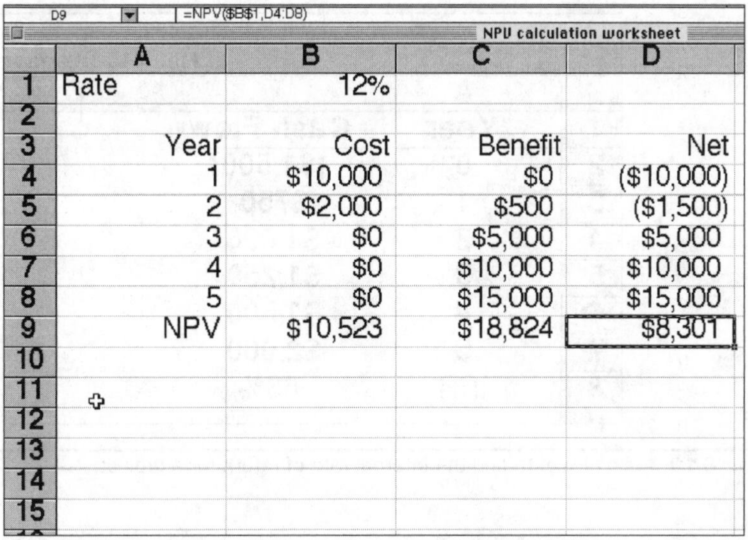

	A	B	C	D
1	Rate	12%		
2				
3	Year	Cost	Benefit	Net
4	1	$10,000	$0	($10,000)
5	2	$2,000	$500	($1,500)
6	3	$0	$5,000	$5,000
7	4	$0	$10,000	$10,000
8	5	$0	$15,000	$15,000
9	NPV	$10,523	$18,824	$8,301
10				
11				
12				
13				
14				
15				

D9 =NPV($B$1,D4:D8)

NPV calculation worksheet

FIGURE 6.9 Using Excel to find the net present value of a project.

adjusts its estimate of the IRR either upward or downward. The process is continued until the desired degree of precision has been achieved.

IRR Example A quality improvement team in a hospital has been investigating the problem of lost surgical instruments. They have determined that in the rush to get the operating room cleaned up between surgeries many instruments are accidentally thrown away with the surgical waste. A test has shown that a $1,500 metal detector can save the following amounts:

Year	Savings
1	$750
2	$1,000
3	$1,250
4	$1,500
5	$1,750

After 5 years of use the metal detector will have a scrap value of $250. To find the IRR for this cash flow stream we set up the Excel spreadsheet and solve the problem as illustrated in Fig. 6.10.

The Excel formula, shown in the window at the top of the figure, was built using the Insert Formula "wizard," with the cash flows in cells B2:B7 and an initial guess of 0.1 (10%). Note that in year 5 the $250 salvage value is added to the expected $1,750 in savings on surgical instruments. The cost is shown as a negative cash flow in year 0. Excel found the IRR to be 63%. The IRR can be one of the criteria for prioritizing projects, as an alternative to, or in addition to, using the PPI.

B8 =IRR(B2:B7,0.1)

Fig. 5.13 IRR exar

	A	B	C
1	Year	Cash Flow	
2	0	($1,500)	
3	1	$750	
4	2	$1,000	
5	3	$1,250	
6	4	$1,500	
7	5	$2,000	
8	IRR	63%	

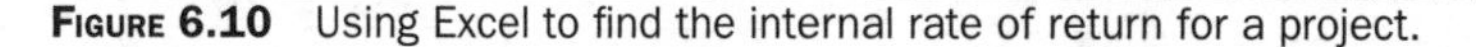

FIGURE 6.10 Using Excel to find the internal rate of return for a project.

Project Scheduling

There are a wide variety of tools and techniques available to help the project manager develop a realistic project timetable, to use the timetable to time the allocation of resources, and to track progress during the implementation of the project plan. We will review two of the most common here: Gantt charts and PERT-type systems.

Gantt Charts

A Gantt chart shows the relationships among the project tasks, along with time constraints. The horizontal axis of a Gantt chart shows the units of time (days, weeks, months, etc.). The vertical axis shows the activities to be completed. Bars show the estimated start time and duration of the various activities. Figure 6.11 illustrates a Gantt chart, modified to reflect the DMAIC milestones, as developed using MS Project software.

The *milestone* symbol represents an event rather than an activity; it does not consume time or resources. When Gantt charts are modified in this way they are sometimes called "milestone charts."

Gantt charts and milestone charts can be modified to show additional information, such as who is responsible for a task, why a task is behind schedule, remedial action planned or already taken, etc.

Although every project is unique, Six Sigma projects which use the DMAIC or DMADV frameworks have many tasks in common, at least at a general level. Many people find it helpful if they have a generic "template" they can use to plan their project activities. This is especially true when the Black Belt or Green Belt is new and has limited project management experience. Table 6.5 can be used as a planning tool by Six Sigma teams. It shows typical tasks, responsibilities and tools for each major phase of a typical Six Sigma project.

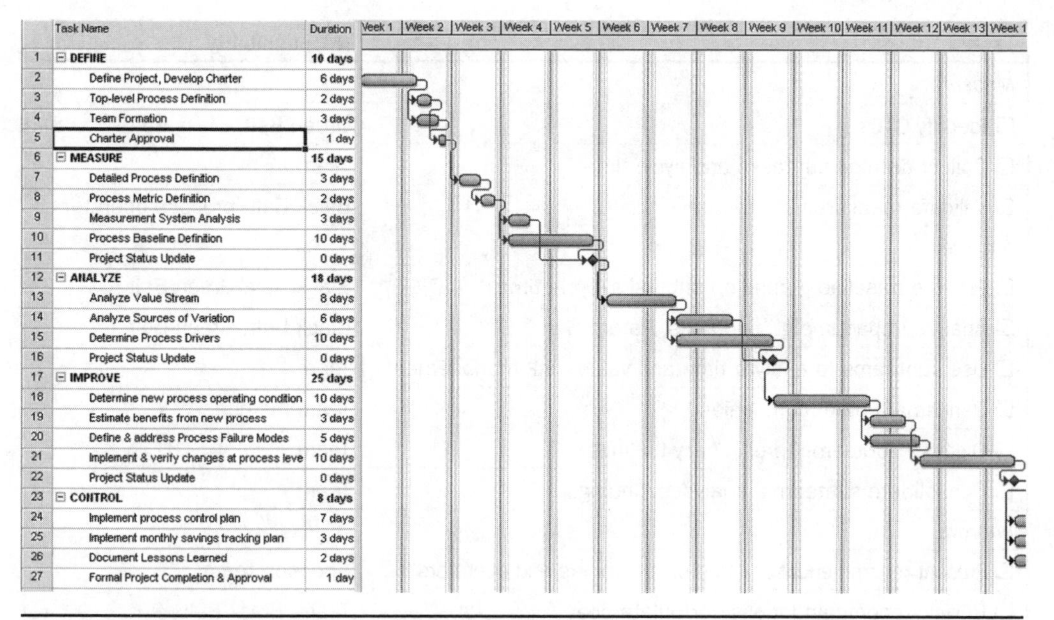

FIGURE 6.11 Enhanced Gantt chart (milestone chart).

Task	Responsibility
Charter Project	
☐ Identify opportunity for improvement	Black Belt
☐ Identify sponsor	Black Belt
☐ Estimate savings	Black Belt
☐ Draft Project Charter	Black Belt, sponsor
☐ Sponsor project review (weekly)	Sponsor, Black Belt
Define	
☐ Team selection	Sponsor, Black Belt
☐ Complete Project Charter	Black Belt
☐ Team training	Black Belt, Green Belt
☐ Review existing process documentation	Team member, process expert
☐ Define project objectives and plan	Team
☐ Present objectives and plan to management	Green Belt
☐ Define and map as-is process	Team, process expert
☐ Review and redefine problem, if necessary	Team
☐ Sponsor	

TABLE 6.5 Typical DMAIC Project Tasks and Responsibilities

Task	Responsibility
Measure	
☐ Identify CTQs	Green Belt, Black Belt
☐ Collect data on subtasks and cycle time	Team
☐ Validate measurement system	Black Belt, process operator
Analyze	
☐ Prepare baseline graphs on subtasks/cycle time	Black Belt, Green Belt
☐ Analyze impacts, e.g., subtasks, Pareto ...	Black Belt, Green Belt
☐ Use subteams to analyze time and value, risk management	Team
☐ Benchmark other companies	Team member
☐ Discuss subteams' preliminary findings	Team
☐ Consolidate subteams' analyses/findings	Team
Improve	
☐ Present recommendations to process owners and operators	Sponsor, team
☐ Review recommendations/formulate pilot	Team, Black Belt
☐ Prepare for improved process pilot	Team, process owner
☐ Test improved process (run pilot)	Process operator
☐ Analyze pilot and results	Black Belt, Green Belt
☐ Develop implementation plan	Team, process owner
☐ Prepare final presentation	Team
☐ Present final recommendations to management team	Green Belt
Control	
☐ Define control metrics	Black Belt, Green Belt, process expert
☐ Develop metrics collection tool	Black Belt
☐ Roll-out improved process	Process owner
☐ Roll-out control metrics	Process owner
☐ Monitor process monthly using control metrics	Process owner, Black Belt

TABLE 6.5 Typical DMAIC Project Tasks and Responsibilities (*Continued*)

PERT-CPM

While useful, Gantt charts and their derivatives provide limited project schedule analysis capabilities. The successful management of large-scale projects requires more rigorous planning, scheduling and coordinating of numerous interrelated activities. To aid in these tasks, formal procedures based on the use of networks and network techniques were developed beginning in the late 1950s. The most prominent of these procedures have been PERT (**P**rogram **E**valuation and **R**eview **T**echnique) and CPM

(**C**ritical **P**ath **M**ethod), both of which are also incorporated into MS Project. The two approaches are usually referred to as PERT-type project management systems. The most important difference between PERT and CPM is that originally the time estimates for the activities were assumed deterministic in CPM and were probabilistic in PERT. Today, PERT and CPM actually comprise one technique and the differences are mainly historical.

Project scheduling by PERT-CPM consists of four basic phases: planning, scheduling, improvement, and controlling. The planning phase involves breaking the project into distinct activities. The time estimates for these activities are then determined and a network (or arrow) diagram is constructed with each activity being represented by an arrow.

PERT-type systems are used to:

- Aid in planning and control of projects
- Determine the feasibility of meeting specified deadlines
- Identify the most likely bottlenecks in a project
- Evaluate the effects of changes in the project requirements or schedule
- Evaluate the effects of deviating from schedule
- Evaluate the effect of diverting resources from the project, or redirecting additional resources to the project

The ultimate objective of the scheduling phase is to construct a time chart showing the start and finish times for each activity as well as its relationship to other activities in the project. The schedule must identify activities that are "critical" in the sense that they *must* be completed on time to keep the project on schedule.

It is vital not to merely accept the schedule as a given. The information obtained in preparing the schedule can be used to improve the project schedule. Activities that the analysis indicates to be critical are candidates for improvement. Pareto analysis can be used to identify those critical elements that are most likely to lead to significant improvement in overall project completion time. Cost data can be used to supplement the time data, and the combined time/cost information analyzed using Pareto analysis.

The final phase in PERT-CPM project management is project control. This includes the use of the network diagram and Gantt chart for making periodic progress assessments

Pert Example

The activities involved, and their estimated completion times, are presented in Table 6.6. It is important that certain of these activities be done in a particular order: there is a *precedence relationship*. The network diagram graphically displays the precedence relationships involved, as shown in Fig. 6.12 (an arrow diagram). There are two time-values of interest for each event: its *earliest time of completion* and its *latest time of completion*. The earliest time for a given event is the estimated time at which the event will occur if the preceding activities are started as early as possible. The latest time for an event is the estimated time the event can occur without delaying the completion of the project beyond its earliest time. Earliest times of events are found by starting at the initial event and working forward, successively calculating the time at which each event will occur if each immediately preceding event occurs at its earliest time and each intervening activity uses only its estimated time.

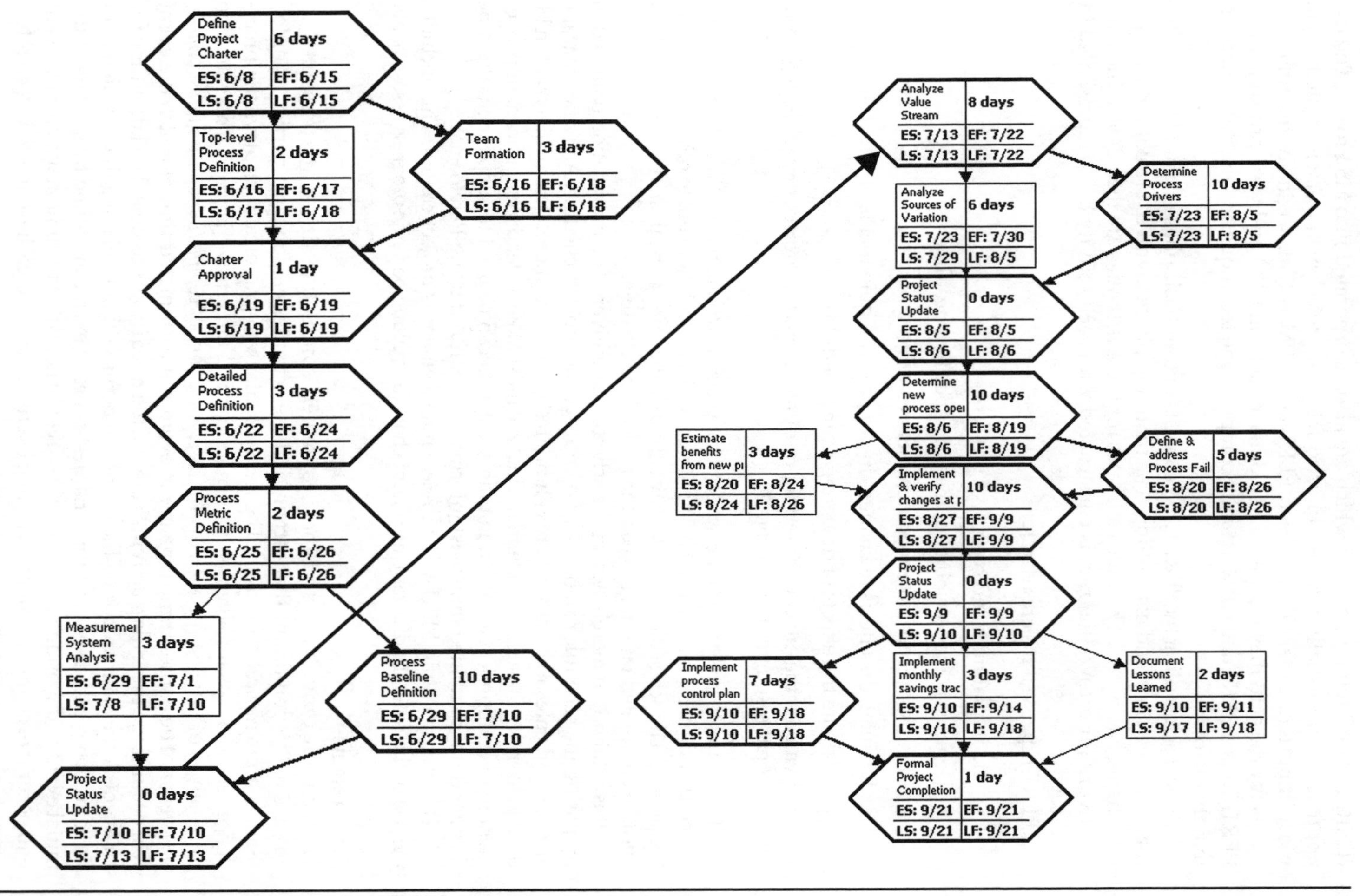
Define Project Charter
6 days
ES: 6/8
EF: 6/15
LS: 6/8
LF: 6/15
Top-level Process Definition
2 days
ES: 6/16
EF: 6/17
LS: 6/17
LF: 6/18
Team Formation
3 days
ES: 6/16
EF: 6/18
LS: 6/16
LF: 6/18
Charter Approval
1 day
ES: 6/19
EF: 6/19
LS: 6/19
LF: 6/19
Detailed Process Definition
3 days
ES: 6/22
EF: 6/24
LS: 6/22
LF: 6/24
Process Metric Definition
2 days
ES: 6/25
EF: 6/26
LS: 6/25
LF: 6/26
Measuremei System Analysis
3 days
ES: 6/29
EF: 7/1
LS: 7/8
LF: 7/10
Process Baseline Definition
10 days
ES: 6/29
EF: 7/10
LS: 6/29
LF: 7/10
Project Status Update
0 days
ES: 7/10
EF: 7/10
LS: 7/13
LF: 7/13
Analyze Value Stream
8 days
ES: 7/13
EF: 7/22
LS: 7/13
LF: 7/22
Determine Process Drivers
10 days
ES: 7/23
EF: 8/5
LS: 7/23
LF: 8/5
Analyze Sources of Variation
6 days
ES: 7/23
EF: 7/30
LS: 7/29
LF: 8/5
Project Status Update
0 days
ES: 8/5
EF: 8/5
LS: 8/6
LF: 8/6
Determine new process opei
10 days
ES: 8/6
EF: 8/19
LS: 8/6
LF: 8/19
Estimate benefits from new pi
3 days
ES: 8/20
EF: 8/24
LS: 8/24
LF: 8/26
Define & address Process Fail
5 days
ES: 8/20
EF: 8/26
LS: 8/20
LF: 8/26
Implement & verify changes at p
10 days
ES: 8/27
EF: 9/9
LS: 8/27
LF: 9/9
Project Status Update
0 days
ES: 9/9
EF: 9/9
LS: 9/10
LF: 9/10
Implement process control plan
7 days
ES: 9/10
EF: 9/18
LS: 9/10
LF: 9/18
Implement monthly savings trac
3 days
ES: 9/10
EF: 9/14
LS: 9/16
LF: 9/18
Document Lessons Learned
2 days
ES: 9/10
EF: 9/11
LS: 9/17
LF: 9/18
Formal Project Completion
1 day
ES: 9/21
EF: 9/21
LS: 9/21
LF: 9/21

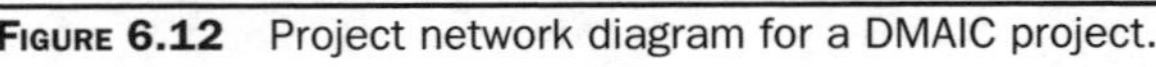
FIGURE 6.12 Project network diagram for a DMAIC project.

Slack time for an event is the difference between the latest and earliest times for a given event. Thus, assuming everything else remains on schedule, the slack for an event indicates how much delay in reaching the event can be tolerated without delaying the project completion.

Events and activities with slack times of zero are said to lie on the *critical path* for the project. A critical path for a project is defined as a path through the network such that the activities on this path have *zero slack*. All activities and events having zero slack must lie on a critical path, but no others can. Figure 6.12 shows the activities on the critical path in hexagon-shaped boxes with thick borders.

Control and Prevention of Schedule Slippage

Project managers can use the network and the information obtained from the network analysis in a variety of ways to help them manage their projects. One way is, of course, to pay close attention to the activities that lie on the critical path. Any delay in these activities will result in a delay for the project. However, the manager should also review the diagram for opportunities to modify the project plan to result in a reduction in the total project completion time. Since the network times are based on *estimates*, it is likely that the completion times will vary. When this occurs it often happens that a new critical path appears. Thus, the network should be viewed as a dynamic entity which should be revised as conditions change.

Primary causes of slippage include poor planning and poor management of the project. Outside forces beyond the control of the project manager will often play a role. However, it isn't enough to be able to simply identify "outside forces" as the cause and beg forgiveness. Astute project managers will anticipate as many such possibilities as possible and prepare contingency plans to deal with them. The process decision program chart (PDPC) is a useful tool for identifying possible events that might be encountered during the project. A portion of a PDPC, applied to one of the critical path tasks of a DMAIC project, is shown in Fig. 6.13. The emphasis of PDPC is the impact of the

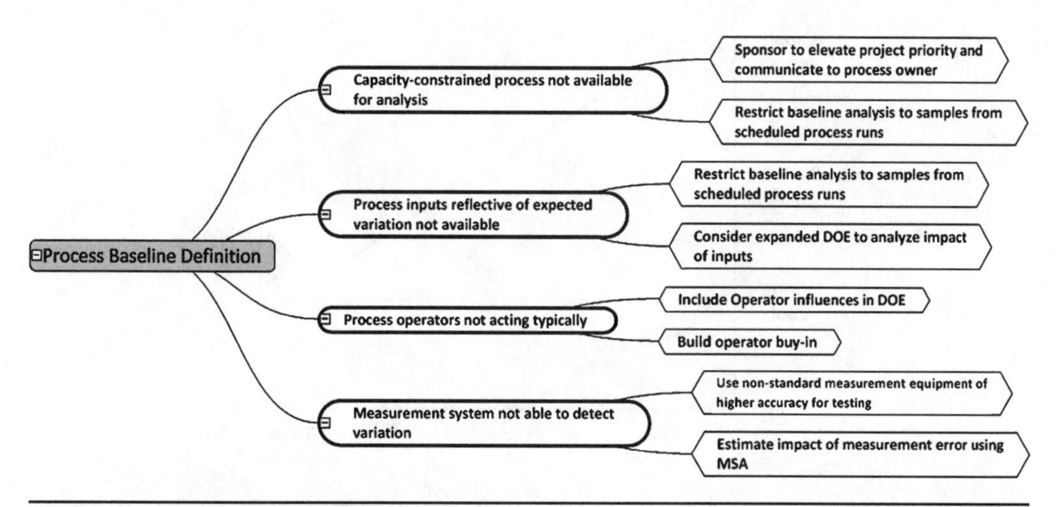

FIGURE 6.13 A portion of the PDPC for the DMAIC project example; constructed using *Mind Genius* software (www.qualityamerica.com by permission).

"failures" (problems) on project schedules. The PDPC seeks to describe specific actions to be taken to prevent the problems from occurring in the first place, and to mitigate the impact of the problems if they do occur. An enhancement to classical PDPC is the Fault Tree Analysis described in Chap. 11, which assigns subjective probabilities to the various problems and to use these to help assign priorities. The amount of detail that should go into contingency plans is a judgment call. The project manager should consider both the seriousness of the potential problem and the likelihood of its occurring.

Cost Considerations in Project Scheduling

Most project schedules can be compressed, if one is willing to pay the additional costs. For the analysis here, costs are defined to include direct elements only. Indirect costs (administration, overhead, etc.) will be considered in the final analysis. Assume that a straight-line relationship exists between the cost of performing an activity on a *normal schedule*, and the cost of performing the activity on a *crash schedule*. Also assume that there is a crash time beyond which no further time saving is possible, regardless of cost. Figure 6.14 illustrates these concepts.

For a given activity the cost-per-unit-of-time saved is found as

$$\frac{\text{crash cost} - \text{normal cost}}{\text{normal time} - \text{crash time}} \tag{6.37}$$

When deciding which activity on the critical path to improve, one should begin with the activity that has the smallest cost-per-unit-of-time saved. The project manager should be aware that once an activity time has been reduced there may be a new critical path. If so, the analysis should proceed using the updated information, that is, activities on the new critical path should be analyzed.

The data for the example are shown in Table 6.6, with additional data for costs and crash schedule times for each activity on the critical path; only critical path activities

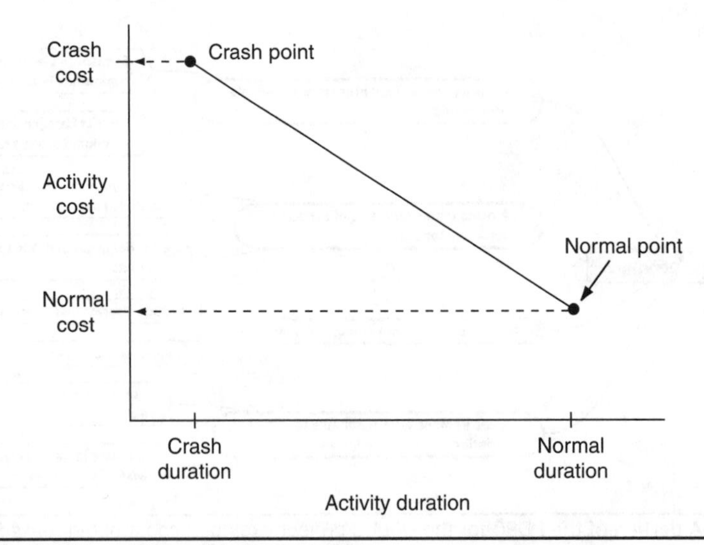

FIGURE 6.14 Cost-time relationship for an activity.

Activity	Normal Schedule		Crash Schedule		
	Time (days)	Cost	Time (days)	Cost	Slope
Define Project, Develop Charter	6	1000	5	3000	2000
Team Formation	3	1600	2	3400	1800
Charter Approval	1	7500	1	7500	–
Detailed Process Definition	3	3000	1	6000	1500
Process Metric Definition	2	4400	1	6000	1600
Process Baseline Definition	10	13750	8	17500	1875
Analyze Value Stream	8	3500	6	7000	1750
Determine Process Drivers	10	3200	8	5600	1200*
Define New Process Levels	10	3000	8	5500	1250
Define & Mitigate Failure	5	4800	1	11000	1550
Implement & Verify	10	4900	6	12000	1775
Define Control Plan	7	5600	3	12000	1600
Project Approval	1	4500	1	4500	–

TABLE 6.6 Schedule Costs for Activities Involved in DMAIC Project Example

are being considered since only they can produce an improvement in overall project duration. Thus, the first activity to consider improving would be *Determine Process Drivers*, which costs $1200 per day saved on the schedule [identified with an asterisk (*) in Table 6.6]. If additional resources could be directed towards this activity it would produce the best "bang for the buck" in terms of reducing the total time of the project. Assuming the critical path doesn't change, the next activities for cost consideration would be *Define New Process Levels*, then *Detailed Process Definition*, etc.

As activities are addressed one by one, the time it takes to complete the project will decline, while the direct costs of completing the project will increase. Figure 6.15 illustrates the cost-duration relationship graphically.

Conversely, *indirect costs* such as overhead, etc., are expected to *increase* as projects take longer to complete. When the indirect costs are added to the direct costs, total costs will generally follow a pattern similar to that shown in Fig. 6.16.

To optimize resource utilization, the project manager will seek to develop a project plan that produces the minimum cost schedule. Of course, the organization will likely have multiple projects being conducted simultaneously, which places additional constraints on resource allocation.

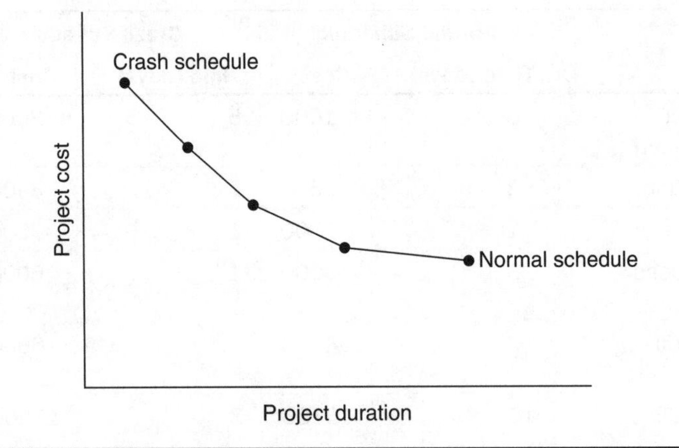

FIGURE 6.15 Direct costs as a function of project duration.

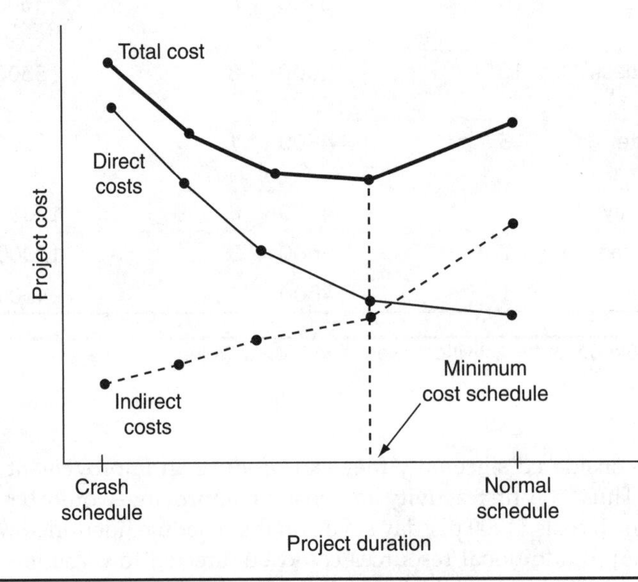

FIGURE 6.16 Total costs as a function of project duration.

Top-Level Process Definition

In the Define stage, a top-level view of the process is created to identify the broad scope of the process to be evaluated. This 30,000 foot view of the process provides critical reference for discussions on specific project objectives, calculations of project deliverables, and definition of key stakeholder groups.

Process Maps

A Process Map is used in the Define stage to document top-level process activities and their stakeholders. A *stakeholder* (also known as a *stakeholder group*) is a department, customer and/or vendor influenced by the activity or its outcome.

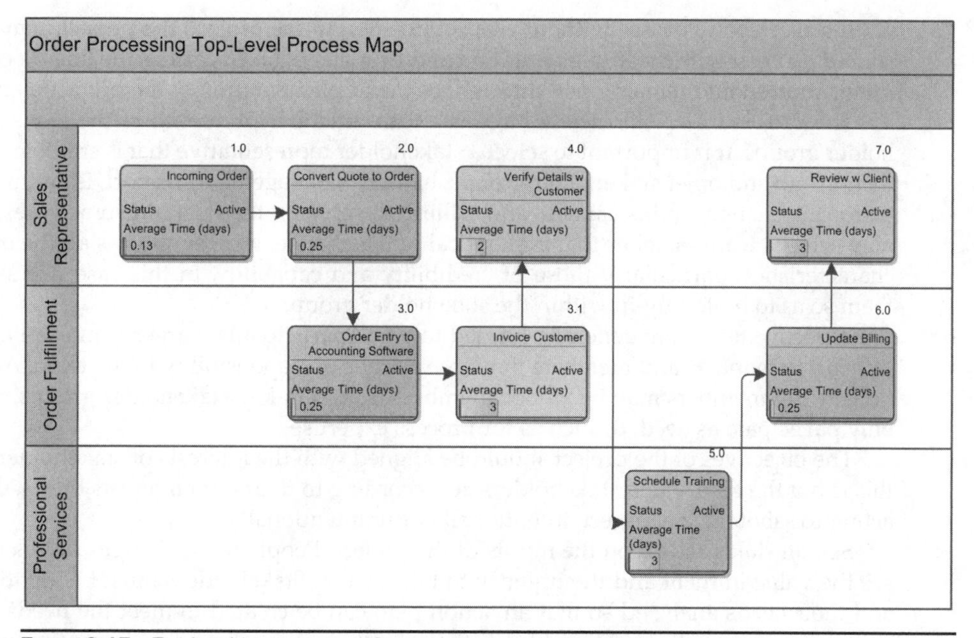

FIGURE 6.17 Top-level process map for order processing example.

In the top-level Process Map shown in Fig. 6.17, each of the broad process activities are placed in the appropriate "swim lane," where each swim lane indicates a unique stakeholder group. Note that the process activities as documented in the top-level map provide little of the detail necessary to completely define the existing process. The detailed map of process tasks and decisions is developed in the Measure stage, and referenced in the Analyze stage to uncover process complexities. The new process is fully documented on a revised Process Map in the Improve and Control stages to aid in communication with process stakeholders.

Large projects impact large number of stakeholder groups within the organization. Of greatest importance in the Define stage is to identify the key stakeholders, which are the groups that can make or break any change effort associated with the process. As discussed in Chap. 1, the key stakeholder groups must have buy-in to the process change for successful implementation of the change effort.

Assembling the Team

As soon as possible the project manager and project sponsor should arrange a short, informal meeting with managers of all stakeholder groups to inform them that a project is proposed in their functional area, with the permission and direction of the senior executives. This meeting represents an invitation for the middle managers to participate in the project by contributing a member of their team, or to challenge the scope, objective or business need addressed by the project proposal. It is important to allow the managers time to consider and act on the leadership's decision to pursue the project. If concrete information suggests that tampering or sabotage is occurring, the project manager or process owner should immediately bring it to the attention of the senior executives who approved the Project Charter. The senior leadership should resolve the issue promptly.

If a week or so passes without clear opposition to the project, the project manager should proceed with the implementation of the project plan. Of course, the lines of communication should remain open throughout the implementation of the project plan.

The project team will ideally have one team member from each of the key stakeholder groups. It is important to select a stakeholder representative that is credible with their group and other stakeholder groups, has local management support, is enthusiastic for the change, and is capable and willing to serve on the team. In some cases, we may select a team member that is skeptical of the change, if he or she has all the other characteristics, particularly those of credibility and capability. In this case, we select them so as to build buy-in within the stakeholder group.

Effective teams are generally limited to 5 to 7 participants. Larger teams are more difficult to manage, and members may lose a sense of responsibility to the team. Additional team members may be ad hoc members from non-key stakeholder groups, who only participate as needed, such as for process expertise.

The objectives of the project should be aligned with the interests of stakeholders. If this is not the case, when stakeholders act according to their own interests they will be acting to sabotage the project, intentionally or unintentionally.

Sell all stakeholders on the merits of the project. People resist changes unless they see the value in them and the urgency to take action. Stakeholders must be identified and their needs analyzed so that an action plan can be created to meet the needs and gain commitment. To avoid problems, the project team must constantly communicate with the stakeholders.

Stakeholder focus groups are a method that allows group members to evaluate the potential impact of a plan by identifying the stakeholders affected by or having influence over the project plan. The focus group approach is a highly structured method in which the project team first identifies the stakeholders and their assumptions, then brings those identified together to elicit their responses to the proposed project (see Chap. 3 for a discussion of the focus group technique). The team then rates these assumptions for importance to the stakeholders and importance to the plan. A stakeholder satisfaction plan may be developed to ensure the support of key individuals and groups.

CHAPTER 7

The Measure Phase

The objectives of the Measure stage include:

1. Process definition: to ensure the specific process under investigation is clearly defined.
2. Metric definition: to define a reliable means of measuring the process, relative to the project deliverables.
3. Establish the process baseline: to quantify the current operating results as a means of verifying previously-defined business needs, and to properly substantiate improvement results.
4. Evaluate measurement system: to validate the reliability of data for drawing meaningful conclusions.

For DFSS applications, where DMADV is used, the objectives of the Measure stage will be limited to defining the key metrics and development of a measurement system and plan for obtaining measurements once the new design becomes operational.

An argument can be made for asserting that quality begins with measurement. Only when quality is quantified, can meaningful discussion about improvement begin. Conceptually, measurement is quite simple: measurement is the assignment of numbers to observed phenomena according to certain rules. Measurement is a requirement of any science, including management science.

Process Definition

A process consists of repeatable tasks, carried out in a specific order. If processes cannot be defined as a series of repeatable tasks, then there may be multiple processes in effect, even an infinite number of processes, or simply the lack of a well-defined process.

It's not uncommon to discover that situation when interviewing process personnel. Operational workers may customize a process to address situations seen in practice, which may not get communicated to all the relevant parties. In this way, customers will experience significant variation depending on the shift or even the specific personnel processing their order. Sometimes this results in improved product or service, and sometimes not. In any event, since we seek to understand the actual process in the Measure stage, the input of the process personnel is necessary. Later, in the Improve stage, we will document a desired process after receiving input from all stakeholders.

There are several useful tools available for defining the process.

- Flowcharts are particularly useful for highlighting process complexities.
- Process maps provide an additional level of detail to indicate functional responsibilities for each process step. Process maps were previously discussed in Chap. 6.
- SIPOC is a tool for identifying the process inputs, outputs and stakeholders.

Generally, these tools will be used in conjunction with one another.

Flowcharts

A flowchart is a simple graphical tool for documenting the flow of a process. In the flowchart, each task is represented by a symbol. There is an ANSI standard which lists symbol types, primarily for computing processes, but most practitioners find rectangles appropriate for most tasks, and diamonds for decision tasks. Decisions should have only two outcomes (yes or no), so decision points must be phrased in this manner. For example, rather than having three options A, B, and C at a decision point, the decision would be constructed as a series of decisions, each of which has a simple yes or no resolution. The first decision might be Option A?, whose No path leads to a second decision of the form Option B? Since Option C is the only remaining choice, the No path from the second decision leads to Option C.

A rather simple flowchart is shown in Fig. 7.1. Notice the diamond decision points have two outcomes: one outcome continues down the main process flow, while the other diverts to a secondary path. Also note that these secondary paths may result in a jump to a later point in the process (as shown in the first decision's "yes" path) or to a prior point in the process (as shown in the second decision's "yes" path). Decision paths, as well as endpoints for processes, may also branch to other process flowcharts, as indicated by the circle in the last step of this process. In this example, the gray-toned symbols indicate external process steps.

Flowcharts are used in the Measure stage to document the current (as-is) process. In the Analyze phase, the flowchart will be reviewed to uncover complexities in the form of an excessive number of decision points that may contribute to delays or even defects. We can use symbol color or shape to indicate process delays, functional responsibility for each step (for example, oval is customer service), or points in the process where measurements are taken.

SIPOC

Virtually all Six Sigma projects address business processes that have an impact on a top-level enterprise strategy. In previous chapters, a great deal of attention was devoted to developing a list of project candidates by meticulously linking projects and strategies using dashboards, QFD, structured decision making, business process mapping, and many other tools and techniques. However, Six Sigma teams usually find that although this approach succeeds in identifying important projects, these projects tend to have too large a scope to be completed within the time and budget constraints. More work is needed to clearly define that portion of the overall business process to be improved by the project. One way to do this is to apply process flowcharting or mapping to subprocesses until reaching the part of the process that has been assigned to the team for improvement. A series of questions are asked, such as:

1. For which stakeholder does this process primarily exist?
2. What value does it create? What output is produced?

Order Processing: Measure Stage Process Definition

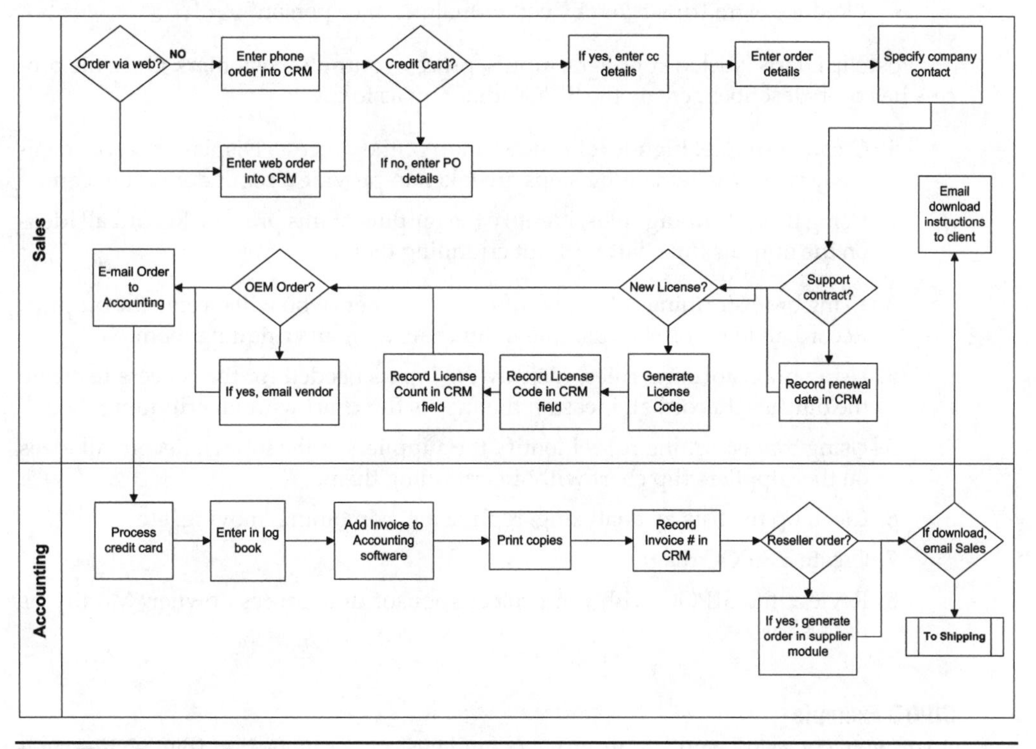

Figure 7.1 Example flowchart.

3. Who is the owner of this process?
4. Who provides inputs to this process?
5. What are the inputs?
6. What resources does this process use?
7. What steps create the value?
8. Are there subprocesses with natural start and end points?

These questions, which are common to nearly all processes addressed by Six Sigma projects, have been arranged into a standard format known as SIPOC. SIPOC stands for Suppliers-Inputs-Process-Outputs-Customers.

SIPOCs begin with people who know something about the process. This may involve people who are not full-time members of the Six Sigma team. Bring the people together in a room and conduct a "focused brainstorming" session. To begin, briefly describe the process and obtain consensus on the definition. For example:

- "Make it easy for the customer to reach technical support by phone"
- "Reduce the space needed to store tooling"
- "Reduce the downtime on the Niad CNC machine"

- "Get roofing crew to the work site on time"
- "Reduce extra trips taken by copier maintenance person"

Post flip charts labeled suppliers, inputs, process, outputs, customers. Once the process has been described, create the SIPOC diagram as follows:

1. Create a simple, high-level process map of the process. Display this conspicuously while the remaining steps are taken to provide a reminder to the team.
2. Using brainstorming rules, identify the outputs of this process. Record all ideas on the outputs flip chart without critiquing them.
3. Using brainstorming rules, identify the customers who will receive the outputs. Record all ideas on the customers flip chart without critiquing them.
4. Using brainstorming rules, identify the inputs needed for the process to create the outputs. Record all ideas on the Inputs flip chart without critiquing them.
5. Using brainstorming rules identify the suppliers of the inputs. Record all ideas on the suppliers flip chart without critiquing them.
6. Clean up the lists by analyzing, rephrasing, combining, moving, etc.
7. Create a SIPOC diagram.
8. Review the SIPOC with the project sponsor and process owner. Modify as necessary.

SIPOC Example

A software company wants to improve overall customer satisfaction (Big Y). Research has indicated that a key component of overall satisfaction is satisfaction with technical support (Little Y). Additional drill down of customer comments indicates that one important driver of technical support satisfaction is the customer's perception that it is easy to contact technical support. There are several different types of technical support available, such as self-help built into the product, the web, or the phone. The process owner commissioned Six Sigma projects for each type of contact. This team's charter is telephone support.

To begin, the team created the process map shown in Fig. 7.2.

Next the team determined that there were different billing options and created a work breakdown structure with each billing option being treated as a subproject. For this example we will follow the subproject relating to the billing-by-the-minute (BBTM) option. After completing the process described above, the team produced the SIPOC shown in Fig. 7.3.

Note that the process is mapped at a very low level. At this level the process map is usually linear, with no decision boxes shown. The typical SIPOC shows the process as it is supposed to behave. Optionally, the SIPOC can show the unintended or undesirable outcomes, as shown in Fig. 7.4.

This "bad process" SIPOC is used only for team troubleshooting. It helps the team formulate hypotheses to be tested during the Analyze phase.

SIPOC analyses focus on the Xs that drive the Ys. It helps the team understand which "dials to turn" to make the top-level dashboard's Big Y move. In the example, let's assume that the team collects information and determines that a significant

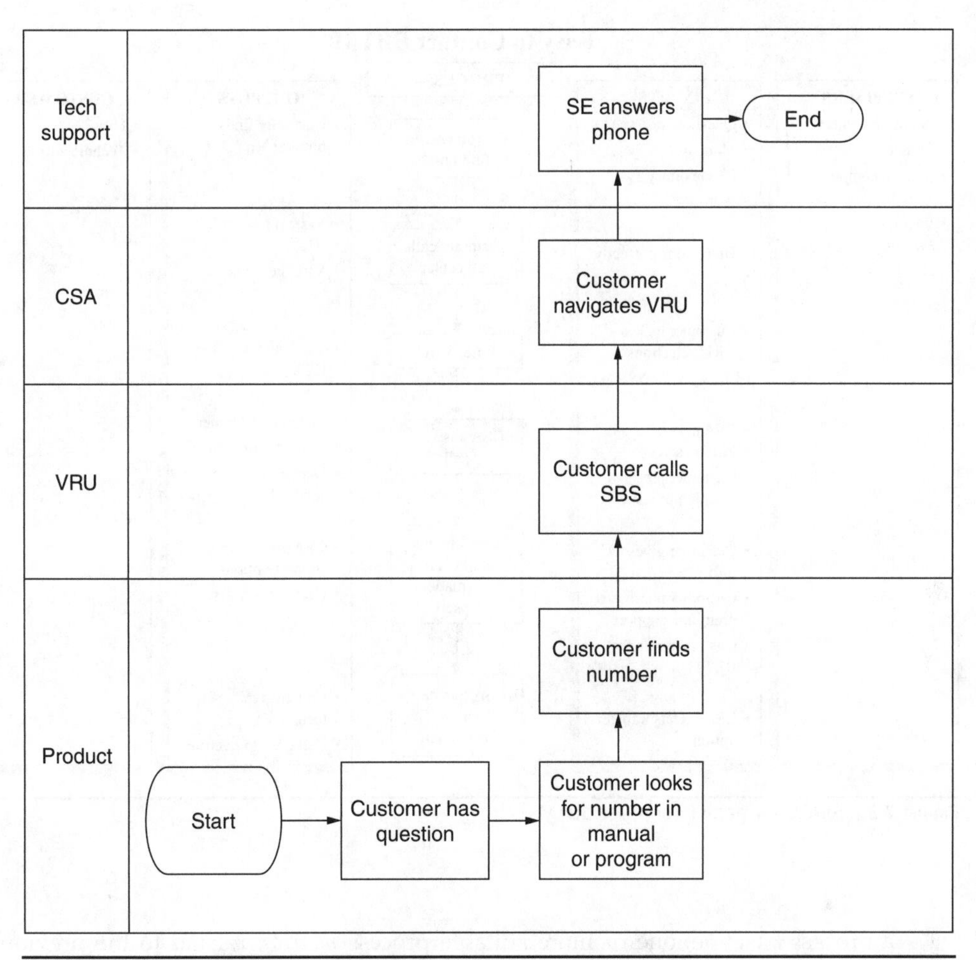

FIGURE 7.2 Process map for contacting technical support by telephone.

percentage of the customers can't find the phone number for technical support. A root cause of the problem then is the obscure location of the support center phone number. Improving overall customer satisfaction is linked to making it easier for the customer to locate the correct number, perhaps by placing a big, conspicuous sticker with the phone number on the cover of the manual. The Big Y and the root cause X are separated by several levels, but the process mapping and SIPOC analysis chain provides a methodology for making the connection.

The SIPOC will be further reviewed in the Analyze phase to remove non-value-added outputs, inputs, or tasks.

Metric Definition

A key objective of the Measure stage is to establish a process baseline, as discussed later in this chapter. The process baseline provides a quantifiable measure of the process performance before any improvement efforts have been initiated. In establishing

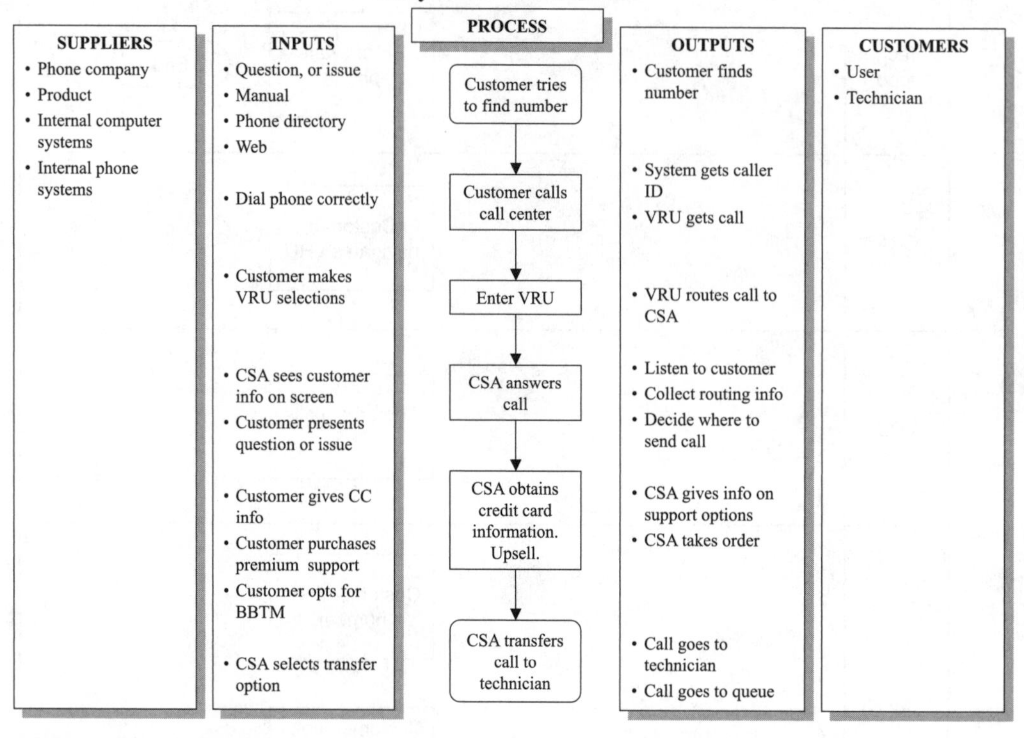

FIGURE 7.3 SIPOC for easy to contact BBTM.

a process baseline, one or more suitable process *metrics*, related to the previously-defined project deliverable, is defined and measured. These metrics are sometimes easily defined with existing process expertise. For example, if key customers are concerned with the *time to delivery* for an order, then that may be a suitable metric, assuming it meets the conditions discussed below. In some case, however, the customers' key metrics do not provide a meaningful measure of process performance, as related to the deliverables previously defined. For example, in chemical and some manufacturing processes, it's not uncommon for the customers' key characteristics to be process endpoint measurements, whose properties are defined much earlier by internal process settings unknown to the end users. In those cases, the internal metrics, or the project deliverables themselves, may provide suitable process performance estimates. In all cases, the metric must be evaluated for its validity and accuracy, as discussed later in this chapter.

A *measurement* is simply a numerical assignment to an observation. Measurements convey certain information about the relationship between the element observed and other elements. Measurement involves a theoretical domain, an area of substantive concern represented as an empirical relational system, and a domain represented by a particular selected numerical relational system. There is a mapping function that carries us from the empirical system into the numerical system. The numerical system is manipulated

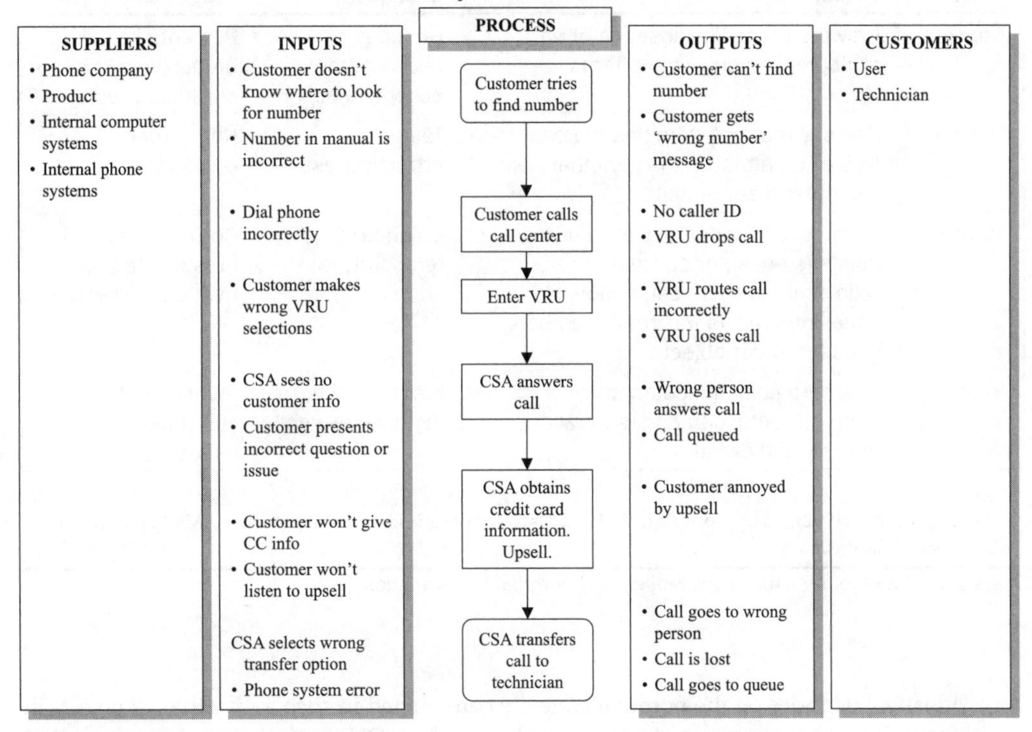

Figure 7.4 SIPOC for undesirable outcomes.

and the results of the manipulation are studied to help the analyst better understand the empirical system.

In reality, measurement is problematic: the analyst can never know the "true" value of the element being measured. The numbers provide information on a certain scale and they represent measurements of some unobservable variable of interest. Some measurements are richer than others, that is, some measurements provide more information than other measurements. The information content of a number is dependent on the scale of measurement used. This scale determines the types of statistical analyses that can be properly employed in studying the numbers. Until one has determined the scale of measurement, one cannot know if a given method of analysis is valid.

Measurement Scales

The four measurement scales are: nominal, ordinal, interval, and ratio. Harrington (1992) summarizes the properties of each scale in Table 7.1.

Numbers on a *nominal scale* aren't measurements at all; they are merely *category labels* in numerical form. Nominal measurements might indicate membership in a group (1 = male, 2 = female) or simply represent a designation (John Doe is 43 on the team). Nominal scales represent the simplest and weakest form of measurement. Nominal variables are perhaps best viewed as a form of classification rather than as a measurement scale.

Scale	Definition	Example	Statistics
Nominal	Only the presence/absence of an attribute; can only count items	Go/no go; success/fail; accept/reject	Percent; proportion; chi-square tests
Ordinal	Can say that one item has more or less of an attribute than another item; can order a set of items	Taste; attractiveness	Rank-order correlation
Interval	Difference between any two successive points is equal; often treated as a ratio scale even if assumption of equal intervals is incorrect; can add, subtract, order objects	Calendar time; temperature	Correlations; t-tests; F-tests; multiple regression
Ratio	True zero point indicates absence of an attribute; can add, subtract, multiply and divide	Elapsed time; distance; weight	t-test; F-test; correlations; multiple regression

Table 7.1 Types of Measurement Scales and Permissible Statistics

Ideally, categories on the nominal scale are constructed in such a way that all objects in the universe are members of one and only one class. Data collected on a nominal scale are called *attribute data*. The only mathematical operations permitted on nominal scales are = (equality, which shows that an object possesses the attribute of concern) or ≠ (inequality).

An ordinal variable is one that has a natural ordering of its possible values, but for which the distances between the values are undefined. An example is product preference rankings such as good, better, or best. Ordinal data can be analyzed with the mathematical operators, = (equality), ≠ (inequality), > (greater than), and < (less than). There are a wide variety of statistical techniques which can be applied to ordinal data including the Pearson correlation, discussed in Chap. 10. Other ordinal models include odds-ratio measures, log-linear models and logit models, both of which are used to analyze cross-classifications of ordinal data presented in contingency tables. In quality management, ordinal data are commonly converted into nominal data and analyzed using binomial or Poisson models. For example, if parts were classified using a poor-good-excellent ordering, the quality analyst might plot a *p chart* of the proportion of items in the poor category.

Interval scales consist of measurements where the ratios of differences are invariant. For example, 90°C = 194°F, 180°C = 356°F, 270°C = 518°F, 360°C = 680°F. Now 194°F/90°C ≠ 356°F/180°C, but

$$\frac{356°\text{F} - 194°\text{F}}{680°\text{F} - 518°\text{F}} = \frac{180°\text{C} - 90°\text{C}}{360°\text{C} - 270°\text{C}}$$

Conversion between two interval scales is accomplished by the transformation

$$y = ax + b \qquad a > 0$$

For example,

$$°F = 32 + \left(\frac{9}{5} \times °C\right)$$

where $a = 9/5$, $b = 32$. As with ratio scales, when permissible transformations are made statistical, results are unaffected by the interval scale used. Also, 0° (on either scale) is arbitrary. In this example, zero does not indicate an absence of heat.

Ratio scale measurements are so called because measurements of an object in two different metrics are related to one another by an invariant ratio. For example, if an object's mass was measured in pounds (x) and kilograms (y), then $x/y = 2.2$ for all values of x and y. This implies that a change from one ratio measurement scale to another is performed by a transformation of the form $y = ax$, $a > 0$; for example, pounds = 2.2 × kilograms. When permissible transformations are used, statistical results based on the data are identical regardless of the ratio scale used. Zero has an inherent meaning: in this example it signifies an absence of mass.

Discrete and Continuous Data

A more general classification of measurements may also be made, which is also useful in defining suitable probability distributions and analysis tools discussed later in this chapter. Data are said to be *discrete* when they take on only a finite number of points that can be represented by the nonnegative integers. An example of discrete data is the number of defects in a sample. Data are said to be *continuous* when they exist on an interval, or on several intervals. An example of continuous data is the measurement of pH.

For most purposes, nominal and ordinal data are considered discrete, while intervals and ratios possess qualities of continuous data. While discrete data may take on integer form only, a continuous data value may be defined theoretically to an infinite number of decimal places, assuming it could be measured as such. In the real world, even though length or time, for example, could theoretically be measured to an infinite number of decimal places, we are be limited by our measurement system. If the variation in length or time is smaller than the smallest unit of measure, the resulting data is essentially discrete, since the same data value is recorded for the majority of the data. Similar information content could be obtained by counting the number of items with that dimension.

Fundamentally, any item measure should meet two tests:

1. The item measures what it is intended to measure (i.e., it is *valid*).
2. A remeasurement would order individual responses in the same way (i.e., it is *reliable*).

The remainder of this chapter describes techniques and procedures designed to ensure that measurement systems produce numbers with these properties.

Process Baseline Estimates

The process baseline is best described as "what were things like before the project?" There are several reasons for obtaining this information:

- *To determine if the project should be pursued*—Although the project charter provides a business case for the project, it sometimes happens that additional, detailed information fails to support it. It may be that the situation isn't as bad as people think, or the project may be addressing an unimportant aspect of the problem.
- *To orient the project team*—The process baseline helps the team identify CTQs and other hard metrics. The information on the historic performance of these metrics may point the team to strategies. For example, if the process is erratic and unstable the team would pursue a different strategy than if it was operating at a consistently poor level.
- *To provide data that will be used to estimate savings*—Baseline information will be invaluable when the project is over and the team is trying to determine the magnitude of the savings or improvement. Many a Black Belt has discovered after the fact that the information they need is no longer available after the completion of the project, making it impossible to determine what benefit was obtained. For example, a project that streamlined a production control system was aimed at improving morale by reducing unpaid overtime worked by exempt employees. However, no measure of employee morale was obtained ahead of time; nor was the unpaid overtime documented anywhere. Consequently, the Black Belt wasn't able to substantiate his claims of improvement and his certification (and pay increase) was postponed.

Process baselines are a critical part of any process improvement effort, as they provide the reference point for assertions of benefits attained. In the absence of a proper baseline estimate, there can be no credible evidence of sustainable improvements, except perhaps in the most obvious of cases. In the vast majority of cases, when the proper measurement system is applied, process observations will vary to the extent that real changes in the process cannot be evidenced without the use of meaningful statistics.

Consider the following scenario: An improvement team uses Lean techniques to reduce the time to process an order. A random sample of 25 orders before the change had an average time to process of 3-1/2 hours. A random sample of 25 orders after the change has an average time to process of 2 hours. The team asserts they have decreased the order processing time by more than 40%. Is that credible? Would it be more credible if a confidence interval for the order processing time was calculated for the original 25 orders, and the improvement asserted only if the new average of 2 hours fell outside of that original interval?

The correct answer is *NO*, in both cases. An improvement cannot be asserted without showing that the new process is significantly different, from a statistical point of view, and the use of confidence intervals is the wrong statistical tool for analyzing the process. It makes the mistake of applying enumerative statistical concepts to an analytic statistical situation. There is no evidence that the estimate of 3-1/2 hours for the first 25 samples is a valid estimate of the process, since we have not shown that the process is stable (consistent over time). It could be that a portion of the first 25 orders had processing times quite a bit larger than 3-1/2 hours, and another portion quite a bit less that 3-1/2 hours, perhaps because of the difference between clerical staff or the type or number of line items included in the order. The confidence interval calculation assumes the data is from a single population, yet fails to prove the process is in fact stable, suitable of being represented by a single distribution.

Some appropriate analytic statistics questions might be:

- Is the process central tendency stable over time?
- Is the process dispersion stable over time?
- Is the process distribution consistent over time?

If any of the above are answered "no," then what is the cause of the instability? To help answer this question, ask "what is the nature of the variation as revealed by the patterns?" when plotted in time-sequence and stratified in various ways.

If none of the above are answered "no," then, and only then, we can ask such questions as

- Is the process meeting the requirements?
- Can the process meet the requirements?
- Can the process be improved by recentering it?
- How can we reduce variation in the process?

Enumerative and Analytic Studies

Deming (1975) defines enumerative and analytic studies as follows:

Enumerative study—a study in which action will be taken on the universe.

Analytic study—a study in which action will be taken on a process to improve performance in the future.

The term "universe" is defined in the usual way: the entire group of interest, for example, people, material, units of product, which possess certain properties of interest. An example of an enumerative study would be sampling an isolated lot to determine the quality of the lot.

In an analytic study the focus is on a *process* and how to improve it. The focus is the *future*. Thus, unlike enumerative studies which make inferences about the universe actually studied, analytic studies are interested in a universe which has yet to be produced. Table 7.2 compares analytic studies with enumerative studies (Provost, 1988).

Deming (1986) points out that "Analysis of variance, t-tests, confidence intervals, and other statistical techniques taught in the books, however interesting, are inappropriate because they provide no basis for prediction and because they bury the information contained in the order of production." These traditional statistical methods have their place, but they are widely abused in the real world. When this is the case, statistics do more to cloud the issue than to enlighten.

Analytic study methods provide information for *inductive thinking*, rather than the largely *deductive* approach of enumerative statistics. Analytic methods are primarily graphical devices such as run charts in the simplest case or statistical control charts in the more general case. Analytic statistics provide operational guidelines, rather than precise

Item	Enumerative Study	Analytic Study
Aim	Parameter estimation	Prediction
Focus	Universe	Process
Method of access	Counts, statistics	Models of the process (e.g., flow charts, cause and effects, mathematical models)
Major source of uncertainty	Sampling variation	Extrapolation into the future
Uncertainty quantifiable	Yes	No
Environment for the study	Static	Dynamic

TABLE 7.2 Important Aspects of Analytic Studies

calculations of probability. Thus, such statements as "There is a 0.13% probability of a Type I error when acting on a point outside a three-sigma control limit" are false (the author admits to having made this error in the past). The future cannot be predicted with a known level of confidence. Instead, based on knowledge obtained from every source, including analytic studies, one can state that one has a certain degree of belief (e.g., high, low) that such and such will result from such and such action on a process.

Another difference between the two types of studies is that enumerative statistics proceed from predetermined hypotheses while analytic studies try to help the analyst generate new hypotheses. In the past, this extremely worthwhile approach has been criticized by some statisticians as "fishing" or "rationalizing." However, this author believes that using data to develop plausible explanations retrospectively is a perfectly legitimate way of creating new theories to be tested. To refuse to explore possibilities suggested by data is to take a very limited view of the scope of statistics in quality improvement and control.

Although most applications of Six Sigma are analytic, there are times when enumerative statistics prove useful. These enumerative methods will be discussed in more detail in the Analyze phase, where they are useful in quantifying sources of variation. The analyst should keep in mind that analytic methods will be needed to validate the conclusions developed with the use of the enumerative methods, to ensure their relevance to the process under study.

In the Measure stage of DMAIC, statistical process control (SPC) charts (i.e., analytical statistics) are used to define the process baseline. If the process is statistically stable, as evidenced by the SPC charts, then process capability and sigma level estimates can be used to quantify the performance of the process relative to requirements.

If the process is not statistically stable, as evidenced by the SPC analysis, then clarity is sought regarding the causes of variation, as discussed in the following sections. If the special causes of variation can be easily identified and removed, and a stable baseline process established, then the requirements for the baseline objective has been met. If not, as is often the case, then our baseline estimate has provided critical information useful in our Analysis phase, where the causes of the process instability can be investigated under the controlled conditions of a designed experiment.

Of course, the baseline estimate will rely upon the validity of the measurement system, which is discussed in the final section of this chapter. While it may seem obvious that the measurement system should be evaluated before effort is expended taking baseline measurements, the measurement system verification is presented after the baseline section for two reasons:

1. A key analytical technique used for validating the measurement system is the control chart. Readers should become familiar with their use and application in process baselines before their special application toward measurement system validation is discussed.
2. When a Six Sigma project's objectives include improvement to CTQ or CTS metrics, the Measure stage baseline provides validation of the initial conditions estimated in the Define stage. Significant improvements or changes to the measurement system before the baseline will bias the baseline relative to the initial conditions. Instead, the baseline should be estimated first using the existing methods, then repeated (after improvements to the measurement system) if the measurement system error is significant. Adherence to this recommendation will lend further credibility to the improvements, since the project team can demonstrate replication of the error as well as the resulting improvement. In some cases, the measurement error is found to be the most significant cause of process variation. In those cases, the focus of the Six Sigma project changes at this stage to concentrate on improvement to the measurement system, rather than the process itself.

The statistical control chart will provide estimates of the process location (i.e., its mean or median) and its variation (i.e., process standard deviation). Of key importance, however, is how these current conditions compare with existing requirements or project objectives. These considerations are answered with a proper understanding of the statistical distribution, as will be discussed.

Principles of Statistical Process Control

A central concept in statistical process control (SPC) is that every measurable phenomenon is a statistical distribution. In other words, an observed set of data constitutes a sample of the effects of unknown common causes. It follows that, after we have done everything to eliminate special causes of variations, there will still remain a certain amount of variability exhibiting the state of control. Figure 7.5 illustrates the relationships between common causes, special causes, and distributions.

There are three basic properties of a distribution: location, spread, and shape. The *location* refers to the typical value of the distribution, such as the mean. The *spread* of the distribution is the amount by which smaller values differ from larger ones. The standard deviation and variance are measures of distribution spread. The *shape* of a distribution is its pattern—peakedness, symmetry, etc. A given phenomenon may have any one of a number of distribution shapes, for example, the distribution may be bell-shaped, rectangular-shaped, etc.

Central Limit Theorem

The central limit theorem can be stated as follows:

Irrespective of the shape of the distribution of the population or universe, the distribution of average values of samples drawn from that universe will tend toward a normal distribution as the sample size grows without bound.

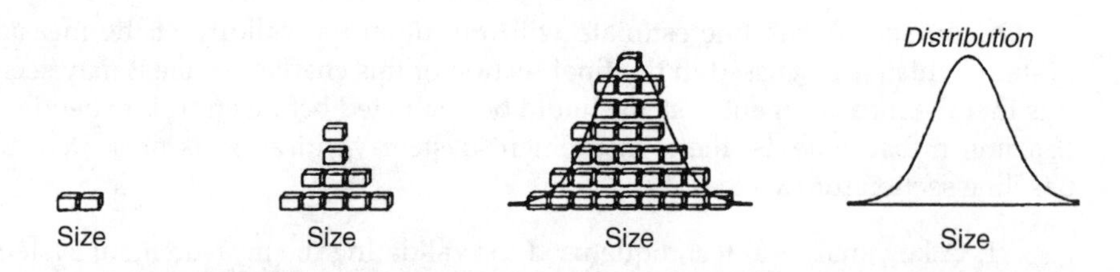

Figure 7.5 Distributions. (*From Continuing Process Control and Process Capability Improvement, p. 4a. Copyright © 1983 by Ford Motor Company. Used by permission of the publisher.*)

It can also be shown that the average of sample averages will equal the average of the universe and that the standard deviation of the averages equals the standard deviation of the universe divided by the square root of the sample size. Shewhart performed experiments that showed that small sample sizes were needed to get approximately normal distributions from even wildly non-normal universes. Figure 7.6 was created by Shewhart using samples of four measurements.

The practical implications of the central limit theorem are immense. Consider that without the central limit theorem effects, we would have to develop a separate statistical model for every non-normal distribution encountered in practice. This would be the only way to determine if the system were exhibiting chance variation. Because of the central limit theorem we can use *averages* of small samples to evaluate *any* process using the normal distribution. The central limit theorem is the basis for the most powerful of statistical process control tools, Shewhart control charts.

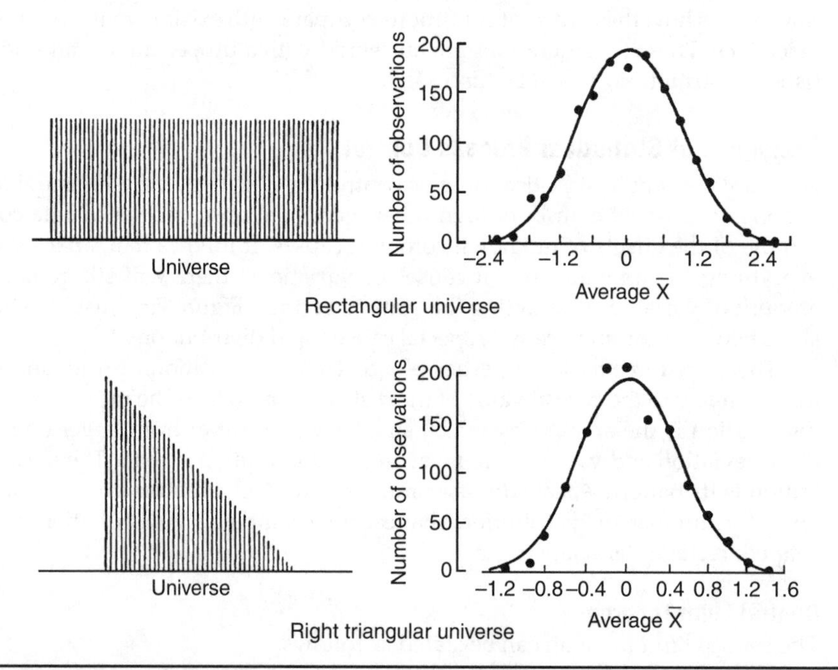

Figure 7.6 Illustration of the central limit theorem. (*From Shewhart (1931, 1980). Figure 59. Copyright © 1931, 1980 by ASQC Quality Press. Used by permission of the publisher.*)

Common and Special Causes of Variation

Shewhart (1931, 1980) defined *control* as follows:

A phenomenon will be said to be controlled when, through the use of past experience, we can predict, at least within limits, how the phenomenon may be expected to vary in the future. Here it is understood that prediction within limits means that we can state, at least approximately, the probability that the observed phenomenon will fall within the given limits.

The critical point in this definition is that control is not defined as the complete absence of variation. Control is simply a state where all variation is *predictable*. A controlled process isn't necessarily a sign of good management, nor is an out-of-control process necessarily producing nonconforming product.

In all forms of prediction there is an element of risk. For our purposes, we will call any unknown random cause of variation a *chance cause* or a *common cause*, the terms are synonymous and will be used as such. If the influence of any particular chance cause is very small, and if the number of chance causes of variation is very large and relatively constant, we have a situation where the variation is predictable within limits. You can see from the definition above, that a system such as this qualifies as a controlled system. Where Dr. Shewhart used the term chance cause, Dr. W. Edwards Deming coined the term *common cause* to describe the same phenomenon. Both terms are encountered in practice.

Needless to say, not all phenomena arise from constant systems of common causes. At times, the variation is caused by a source of variation that is not part of the constant system. These sources of variation were called *assignable causes* by Shewhart, *special causes* of variation by Deming. Experience indicates that special causes of variation can usually be found without undue difficulty, leading to a process that is less variable.

Statistical tools are needed to help us effectively separate the effects of special causes of variation from chance cause variation. This leads us to another definition: Statistical process control—the use of valid analytical statistical methods to identify the existence of special causes of variation in a process.

The basic rule of statistical process control is variation from common-cause systems should be left to chance, but special causes of variation should be identified and eliminated.

This is Shewhart's original rule. However, the rule should not be misinterpreted as meaning that variation from common causes should be ignored. Rather, common-cause variation is explored "off-line." That is, we look for long-term process improvements to address common-cause variation.

Figure 7.7 illustrates the need for statistical methods to determine the category of variation.

The answer to the question "should these variations be left to chance?" can only be obtained through the use of statistical methods. Figure 7.8 illustrates the basic concept.

In short, variation between the two "control limits" designated by the dashed lines will be deemed as variation from the common-cause system. Any variability beyond these fixed limits will be assumed to have come from special causes of variation. We will call any system exhibiting only common-cause variation, "statistically controlled." It must be noted that the control limits are not simply pulled out of the air, they are calculated from actual process data using valid statistical methods. Figure 7.7 is shown as Fig. 7.9, only with the control limits drawn on it; notice that process (a) is exhibiting variations from special causes, while process (b) is not. This implies that the type of

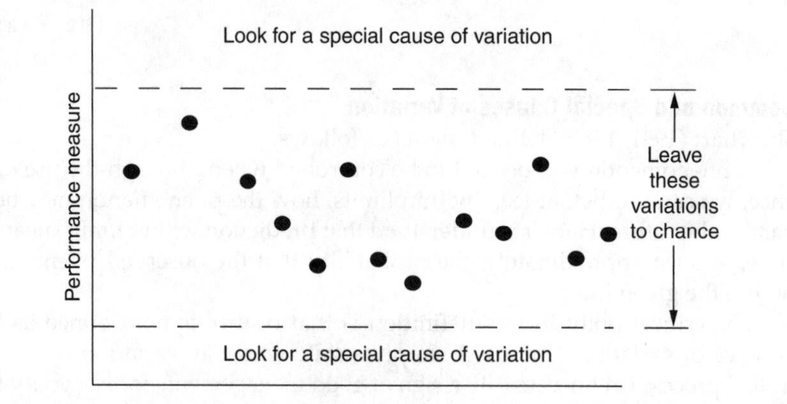

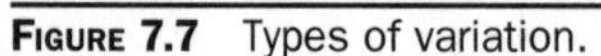
FIGURE 7.7 Types of variation.

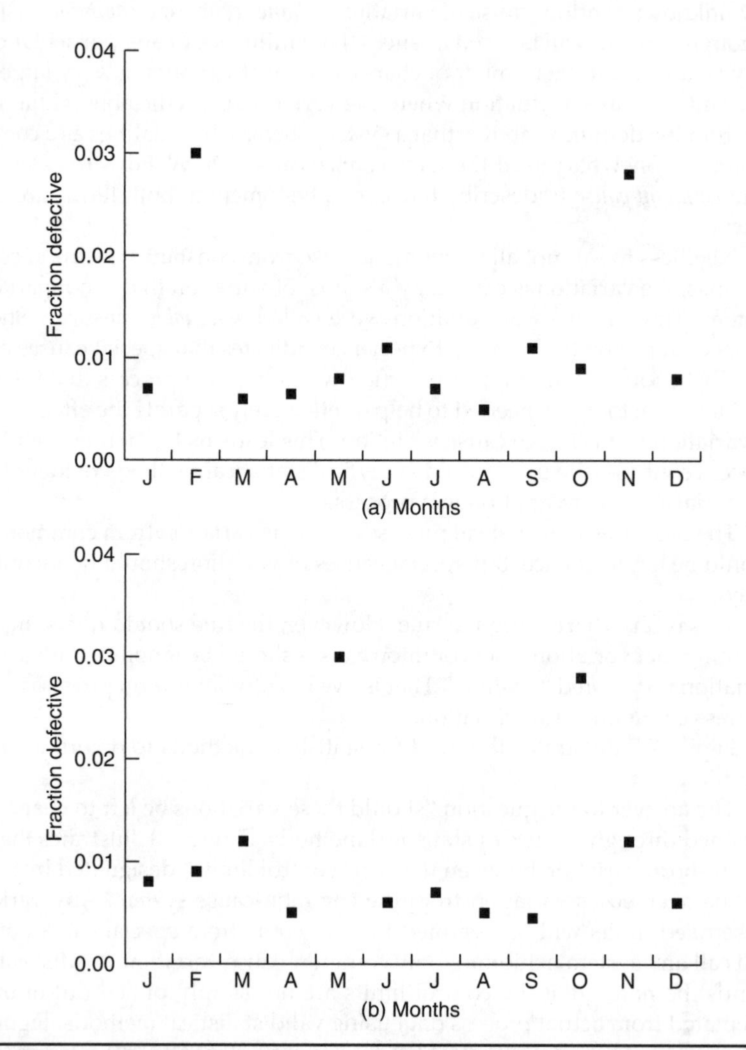

FIGURE 7.8 Should these variations be left to chance? (*From Shewhart (1931, 1980). P 13. Copyright © 1931, 1980 by ASQC Quality Press. Used by permission of the publisher.*)

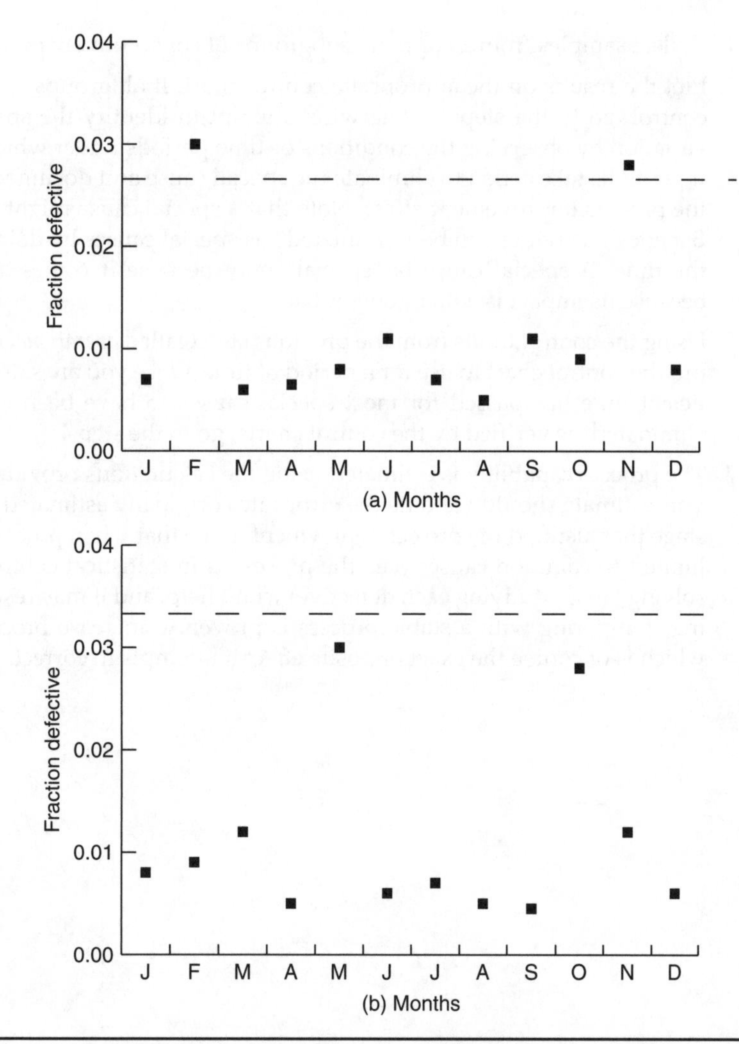

FIGURE 7.9 Charts from Fig. 7.7 with control limits shown. (*From Shewhart (1931, 1980). P 13. Copyright © 1931, 1980 by ASQC Quality Press. Used by permission of the publisher.*)

action needed to reduce the variability in each case is of a different nature. Without statistical guidance there could be endless debate over whether special or common causes were to blame for variability.

Estimating Process Baselines Using Process Capability Analysis

This section presents several methods of analyzing the data using a statistical control chart to determine the extent to which the process meets requirements.

1. Collect samples from 25 or more subgroups of consecutively produced units.
2. Plot the results on the appropriate control chart. If all groups are in statistical control, go to the step 3. Otherwise, attempt to identify the special cause of variation by observing the conditions or time periods under which they occur. If possible, take action to eliminate the special cause and document it as part of the process improvement effort. Note that a special cause might be beneficial. Beneficial activities can be "eliminated" as special causes by doing them all of the time. A special cause is "special" only because it comes and goes, not because its impact is either good or bad.
3. Using the control limits from the previous step (called *operational control limits*), put the control chart to use for a period of time. Once you are satisfied that sufficient time has passed for most special causes to have been identified and eliminated, as verified by the control charts, go to the step 4.
4. The process capability is estimated using the calculations provided in Chap. 6. This estimate should validate the error rates originally estimated in the Define stage that justified the project deployment. Note that when process variation is limited to common causes (i.e., the process is in statistical control), "problem solving" (e.g., studying each defective) won't help, and it may result in tampering. Tampering with a stable process is proven to increase process variation, which is of course the exact opposite effect it attempts to correct.

CHAPTER 8

Process Behavior Charts

Control Charts for Variables Data

In statistical process control (SPC), the mean, range, and standard deviation are the statistics most often used for analyzing measurement data. Control charts are used to monitor these statistics. An out-of-control point for any of these statistics is an indication that a special cause of variation is present and that an immediate investigation should be made to identify the special cause.

Averages and Ranges Control Charts

Averages charts are statistical tools used to evaluate the central tendency of a process over time. Ranges charts are statistical tools used to evaluate the dispersion or spread of a process over time.

Averages charts answer the question: "Has a special cause of variation caused the central tendency of this process to change over the time period observed?" *Ranges charts* answer the question: "Has a special cause of variation caused the process distribution to become more or less consistent?" Averages and ranges charts can be applied to many continuous variables such as weight, size, response time, etc.

The basis of the control chart is the *rational subgroup*. Rational subgroups (see "Rational Subgroup Sampling") are composed of items which were produced under essentially the same conditions. The average and range are computed for each subgroup separately, then plotted on the control chart. Each subgroup's statistics are compared to the control limits, and patterns of variation between subgroups are analyzed.

Subgroup Equations for Averages and Ranges Charts

$$\overline{X} = \frac{\text{sum of subgroup measurements}}{\text{subgroup size}} \tag{8.1}$$

$$R = \text{largest in subgroup} - \text{smallest in subgroup} \tag{8.2}$$

Control Limit Equations for Averages and Ranges Charts

Control limits for both the averages and the ranges charts are computed such that it is highly unlikely that a subgroup average or range from a stable process would fall outside of the limits. All control limits are set at plus and minus three standard deviations from the center line of the chart. Thus, the control limits for subgroup averages are plus and minus three standard deviations of the mean from the grand average; the control limits for the subgroup ranges are plus and minus three standard deviations of the range from the average range. These control limits are quite robust with respect to non-normality in the process distribution.

To facilitate calculations, constants are used in the control limit equations. Appendix 9 provides control chart constants for subgroups of 25 or less. The derivation of the various control chart constants is shown in Burr (1976, pp. 97–105).

Control Limit Equations for Ranges Charts

$$\bar{R} = \frac{\text{sum of subgroup ranges}}{\text{number of subgroups}} \tag{8.3}$$

$$\text{LCL} = D_3\bar{R} \tag{8.4}$$

$$\text{UCL} = D_4\bar{R} \tag{8.5}$$

Control Limit Equations for Averages Charts Using $\bar{R}$

$$\bar{\bar{X}} = \frac{\text{sum of subgroup averages}}{\text{number of subgroups}} \tag{8.6}$$

$$\text{LCL} = \bar{\bar{X}} - A_2\bar{R} \tag{8.7}$$

$$\text{UCL} = \bar{\bar{X}} + A_2\bar{R} \tag{8.8}$$

Example of Averages and Ranges Control Charts

Table 8.1 contains 25 subgroups of five observations each.

The control limits are calculated from these data as follows:

Ranges control chart example

$$\bar{R} = \frac{\text{sum of subgroup ranges}}{\text{number of subgroups}} = \frac{369}{25} = 14.76$$

$$\text{LCL}_R = D_3\bar{R} = 0 \times 14.76 = 0$$

$$\text{UCL}_R = D_4\bar{R} = 2.115 \times 14.76 = 31.22$$

Since it is not possible to have a subgroup range less than zero, the LCL is not shown on the control chart for ranges.

Averages control chart example

$$\bar{\bar{X}} = \frac{\text{sum of subgroup averages}}{\text{number of subgroups}} = \frac{2{,}487.5}{25} = 99.5$$

$$\text{LCL}_{\bar{X}} = \bar{\bar{X}} - A_2\bar{R} = 99.5 - 0.577 \times 14.76 = 90.97$$

$$\text{UCL}_{\bar{X}} = \bar{\bar{X}} + A_2\bar{R} = 99.5 + 0.577 \times 14.76 = 108.00$$

The completed averages and ranges control charts are shown in Fig. 8.1.

The charts shown in Fig. 8.3 show a process in statistical control. This merely means that we can predict the limits of variability for this process. To determine the capability

Sample 1	Sample 2	Sample 3	Sample 4	Sample 5	X-Bar	Range	Sigma
110	93	99	98	109	101.8	17	7.396
103	95	109	95	98	100.0	14	6.000
97	110	90	97	100	98.8	20	7.259
96	102	105	90	96	97.8	15	5.848
105	110	109	93	98	103.0	17	7.314
110	91	104	91	101	99.4	19	8.325
100	96	104	93	96	97.8	11	4.266
93	90	110	109	105	101.4	20	9.290
90	105	109	90	108	100.4	19	9.607
103	93	93	99	96	96.8	10	4.266
97	97	104	103	92	98.6	12	4.930
103	100	91	103	105	100.4	14	5.550
90	101	96	104	108	99.8	18	7.014
97	106	97	105	96	100.2	10	4.868
99	94	96	98	90	95.4	9	3.578
106	93	104	93	99	99.0	13	6.042
90	95	98	109	110	100.4	20	8.792
96	96	108	97	103	100.0	12	5.339
109	96	91	98	109	100.6	18	8.081
90	95	94	107	99	97.0	17	6.442
91	101	96	96	109	98.6	18	6.804
108	97	101	103	94	100.6	14	5.413
96	97	106	96	98	98.6	10	4.219
101	107	104	109	104	105.0	8	3.082
96	91	96	91	105	95.8	14	5.718

Table 8.1 Data for X-Bar, Ranges and Sigma Control Charts

of the process with respect to requirements one must use the methods described later in the process capability analysis section.

Averages and Standard Deviation (Sigma) Control Charts

Averages and standard deviation control charts are conceptually identical to averages and ranges control charts. The difference is that the subgroup standard deviation is used to measure dispersion rather than the subgroup range. The subgroup standard deviation is statistically more efficient than the subgroup range for subgroup sizes greater than 2. This efficiency advantage increases as the subgroup size increases. The inefficiency of the range statistic becomes significant if the subgroup size is 10 or larger,

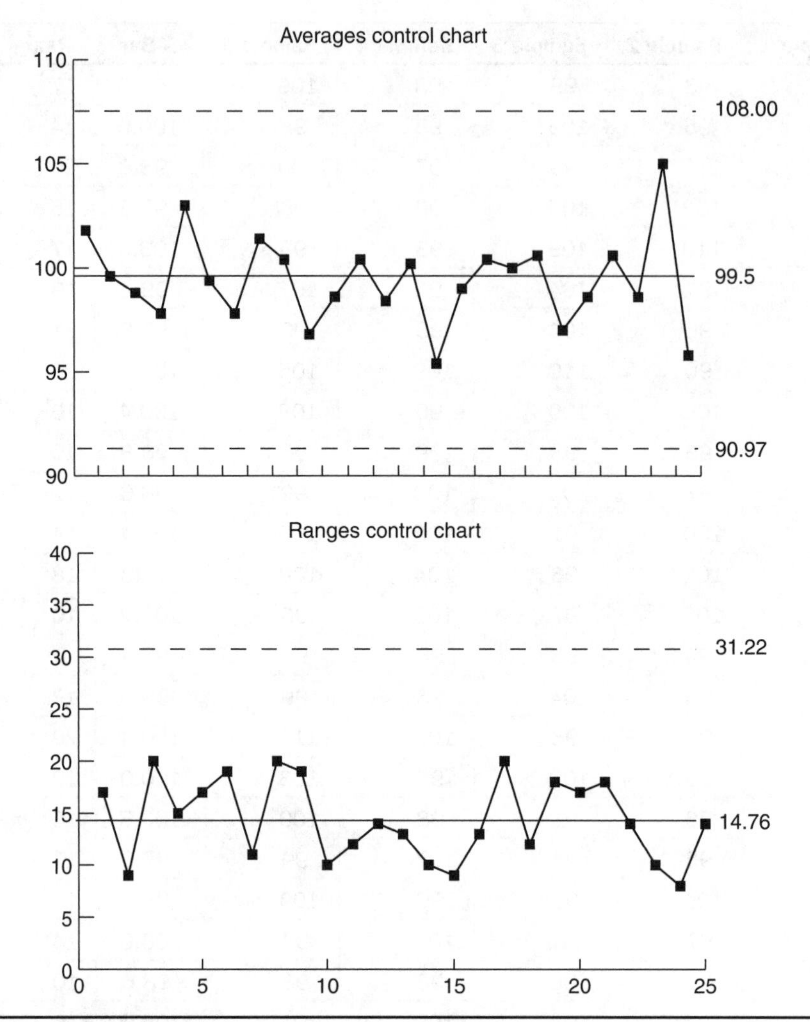

FIGURE 8.1 Completed averages and ranges control charts.

so range charts are not recommended for these large subgroup sizes. However, since Six Sigma analysts will invariably use computer software in their analyses, the standard deviation chart is recommended for all subgroup sizes.

Subgroup Equations for Averages and Sigma Charts

$$\bar{X} = \frac{\text{sum of subgroup measurements}}{\text{subgroup size}} \tag{8.9}$$

$$s = \sqrt{\frac{\sum_{i=1}^{n}(X_i - \bar{X})^2}{n-1}} \tag{8.10}$$

The standard deviation, s, is computed separately for each subgroup, using the subgroup average rather than the grand average. This is an important point; using the grand average would introduce special cause variation if the process were out of control, thereby underestimating the process capability, perhaps significantly.

Control Limit Equations for Averages and Sigma Charts

Control limits for both the averages and the sigma charts are computed such that it is highly unlikely that a subgroup average or sigma from a stable process would fall outside of the limits. All control limits are set at plus and minus three standard deviations from the center line of the chart. Thus, the control limits for subgroup averages are plus and minus three standard deviations of the mean from the grand average. The control limits for the subgroup sigmas are plus and minus three standard deviations of sigma from the average sigma. These control limits are quite robust with respect to non-normality in the process distribution.

To facilitate calculations, constants are used in the control limit equations. Appendix 9 provides control chart constants for subgroups of 25 or less.

Control Limit Equations for Sigma Charts Based Ons $\bar{s}$

$$\bar{s} = \frac{\text{sum of subgroup sigmas}}{\text{number of subgroups}} \tag{8.11}$$

$$\text{LCL} = B_3\bar{s} \tag{8.12}$$

$$\text{UCL} = B_4\bar{s} \tag{8.13}$$

Control Limit Equations for Averages Charts Based on $\bar{s}$

$$\bar{\bar{X}} = \frac{\text{sum of subgroup averages}}{\text{number of subgroups}} \tag{8.14}$$

$$\text{LCL} = \bar{\bar{X}} - A_3\bar{s} \tag{8.15}$$

$$\text{UCL} = \bar{\bar{X}} + A_3\bar{s} \tag{8.16}$$

Example of Averages and Standard Deviation Control Charts

To illustrate the calculations and to compare the range method to the standard deviation results, the data used in the previous example will be reanalyzed using the subgroup standard deviation rather than the subgroup range.

The control limits are calculated from this data as follows:

Sigma control chart

$$s = \frac{\text{sum of subgroup sigmas}}{\text{number of subgroups}} = \frac{155.45}{25} = 6.218$$

$$\text{LCL}_s = B_3\bar{s} = 0 \times 6.218 = 0$$

$$\text{UCL}_s = B_4\bar{s} = 2.089 \times 6.218 = 12.989$$

Since it is not possible to have a subgroup sigma less than zero, the LCL is not shown on the control chart for sigma for this example.

Averages control chart

$$\overline{\overline{X}} = \frac{\text{sum of subgroup averages}}{\text{number of subgroups}} = \frac{2{,}487.5}{25} = 99.5$$

$$\text{LCL}_{\overline{X}} = \overline{\overline{X}} - A_3\overline{s} = 99.5 - 1.427 \times 6.218 = 90.63$$

$$\text{UCL}_{\overline{X}} = \overline{\overline{X}} + A_3\overline{s} = 99.5 + 1.427 \times 6.218 = 108.37$$

The completed averages and sigma control charts are shown in Fig. 8.2. Note that the control limits for the averages chart are only slightly different than the limits calculated using ranges.

Note that the conclusions reached are the same as when ranges were used.

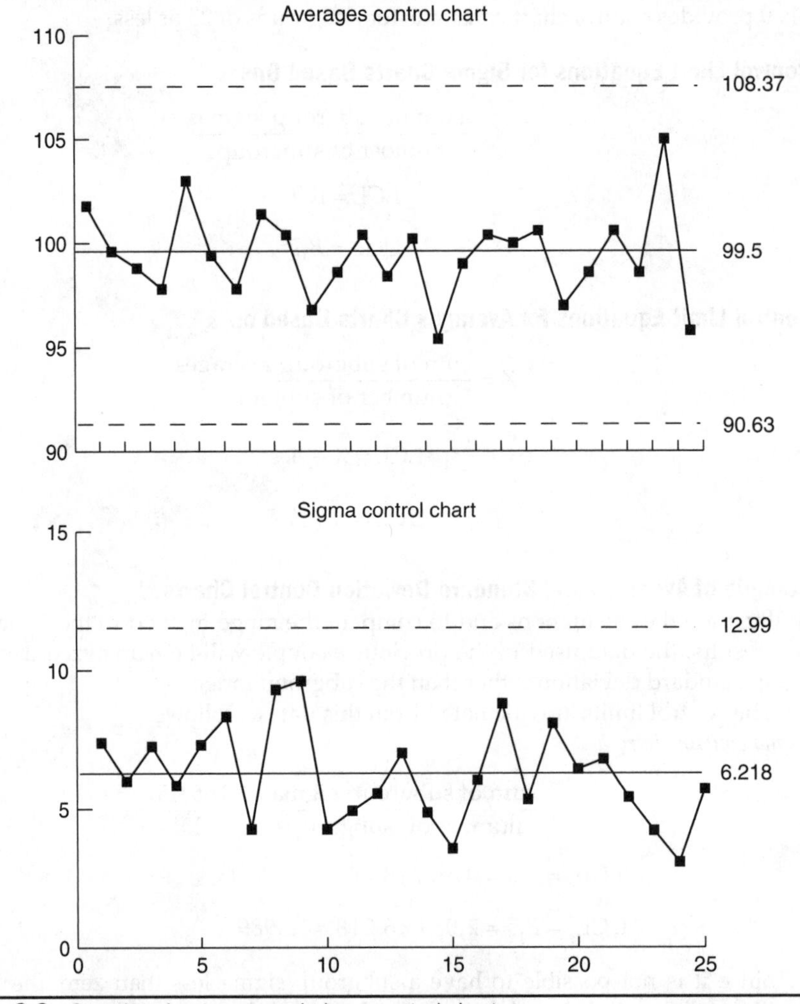

FIGURE 8.2 Completed averages and sigma control charts.

Control Charts for Individual Measurements (*X* Charts)

Individuals control charts are statistical tools used to evaluate the central tendency of a process over time. They are also called *X charts or moving range charts*. Individuals control charts are used when it is not feasible to use averages for process control. There are many possible reasons why averages control charts may not be desirable: observations may be expensive to get (e.g., destructive testing), output may be too homogeneous over short time intervals (e.g., pH of a solution), the production rate may be slow and the interval between successive observations long, etc. Control charts for individuals are often used to monitor batch process, such as chemical processes, where the within-batch variation is so small relative to between-batch variation that the control limits on a standard $\bar{X}$ chart would be too close together. Range charts (sometimes called moving range charts in this application) are used to monitor dispersion between the successive individual observations.*

Calculations for Moving Ranges Charts

As with averages and ranges charts, the range is computed as shown in previous section,

$$R = \text{largest in subgroup} - \text{smallest in subgroup}$$

Where the subgroup is a consecutive pair of process measurements. The range control limit is computed as was described for averages and ranges charts, using the D_4 constant for subgroups of 2, which is 3.267. That is,

$$\text{LCL} = 0 \text{ (for } n = 2)$$

$$\text{UCL} = 3.267 \times R\text{-bar}$$

Control Limit Equations for Individuals Charts

$$\bar{X} = \frac{\text{sum of measurements}}{\text{number of measurements}} \tag{8.17}$$

$$\text{LCL} = \bar{X} - E_2\bar{R} = \bar{X} - 2.66 \times \bar{R} \tag{8.18}$$

$$\text{UCL} = \bar{X} + E_2\bar{R} = \bar{X} + 2.66 \times \bar{R} \tag{8.19}$$

Where $E_2 = 2.66$ is the constant used when individual measurements are plotted, and $\bar{R}$ is based on subgroups of $n = 2$.

Example of Individuals and Moving Ranges Control Charts

Table 8.2 contains 25 measurements. To facilitate comparison, the measurements are the first observations in each subgroup used in the previous average/ranges and average/standard deviation control chart examples.

---

*There is some debate over the value of moving *R* charts. Academic researchers have failed to show statistical value in their usage. However, many practitioners contend that moving *R* charts provide valuable additional information useful in troubleshooting.

Sample 1	Range
110	None
103	7
97	6
96	1
105	9
110	5
100	10
93	7
90	3
103	13
97	6
103	6
90	13
97	7
99	2
106	7
90	16
96	6
109	13
90	19
91	1
108	17
96	12
101	5
96	5

TABLE 8.2 Data for Individuals and Moving Ranges Control Charts

The control limits are calculated from this data as follows:

Moving ranges control chart control limits

$$\bar{R} = \frac{\text{sum of ranges}}{\text{number of ranges}} = \frac{196}{24} = 8.17$$

$$\text{LCL}_R = D_3\bar{R} = 0 \times 8.17 = 0$$

$$\text{UCL}_R = D_4\bar{R} = 3.267 \times 8.17 = 26.69$$

Since it is not possible to have a subgroup range less than zero, the LCL is not shown on the control chart for ranges.

Individuals control chart control limits

$$\bar{X} = \frac{\text{sum of measurements}}{\text{number of measurements}} = \frac{2{,}475}{25} = 99.0$$

$$\text{LCL}_X = \bar{X} - E_2\bar{R} = 99.0 - 2.66 \times 8.17 = 77.27$$

$$\text{UCL}_X = \bar{X} + E_2\bar{R} = 99.0 + 2.66 \times 8.17 = 120.73$$

The completed individuals and moving ranges control charts are shown in Fig. 8.3. In this case, the conclusions are the same as with averages charts. However, averages charts always provide tighter control than *X* charts. In some cases, the additional

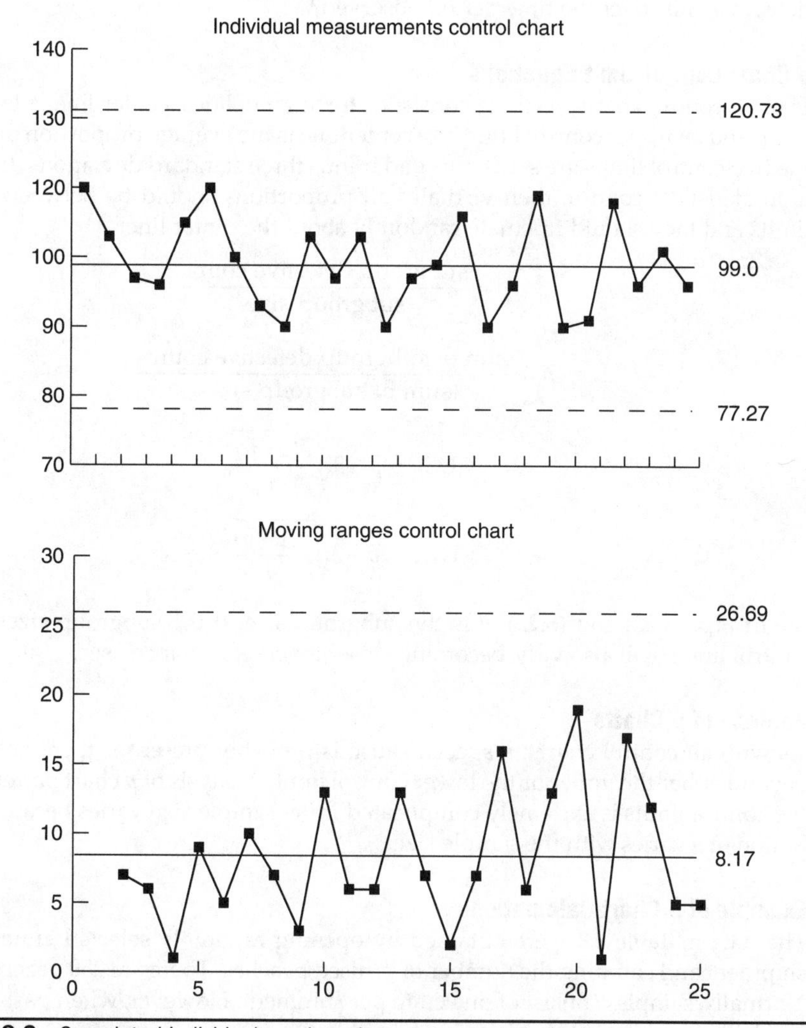

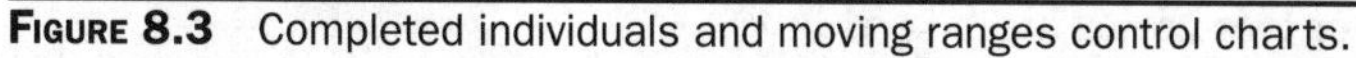
FIGURE 8.3 Completed individuals and moving ranges control charts.

sensitivity provided by averages charts may not be justified on either an economic or an engineering basis. When this happens, the use of averages charts will merely lead to wasting money by investigating special causes that are of minor importance.

Control Charts for Attributes Data

Control Charts for Proportion Defective (*p* Charts)

p charts are statistical tools used to evaluate the proportion defective, or proportion nonconforming, produced by a process.

p charts can be applied to any variable where the appropriate performance measure is a unit count. *p* charts answer the question: "Has a special cause of variation caused the central tendency of this process to produce an abnormally large or small number of defective units over the time period observed?"

p Chart Control Limit Equations

Like all control charts, *p* charts consist of three guidelines: center line, a lower control limit, and an upper control limit. The center line is the average proportion defective and the two control limits are set at plus and minus three standard deviations. If the process is in statistical control, then virtually all proportions should be between the control limits and they should fluctuate randomly about the center line.

$$p = \frac{\text{subgroup defective count}}{\text{subgroup size}} \tag{8.20}$$

$$\bar{p} = \frac{\text{sum of subgroup defective counts}}{\text{sum of subgroup sizes}} \tag{8.21}$$

$$\text{LCL} = \bar{p} - 3\sqrt{\frac{\bar{p}(1-\bar{p})}{n}} \tag{8.22}$$

$$\text{UCL} = \bar{p} + 3\sqrt{\frac{\bar{p}(1-\bar{p})}{n}} \tag{8.23}$$

In Eqs. (8.22) and (8.23), n is the subgroup size. If the subgroup sizes varies, the control limits will also vary, becoming closer together as n increases.

Analysis of *p* Charts

As with all control charts, a special cause is probably present if there are any points beyond either the upper or the lower control limit. Analysis of *p* chart patterns between the control limits is extremely complicated if the sample size varies because the distribution of *p* varies with the sample size.

Example of *p* Chart Calculations

The data in Table 8.3 were obtained by opening randomly selected crates from each shipment and counting the number of bruised peaches. There are 250 peaches per crate. Normally, samples consist of one crate per shipment. However, when part-time help is available, samples of two crates are taken.

Shipment No.	Crates	Peaches	Bruised	P
1	1	250	47	0.188
2	1	250	42	0.168
3	1	250	55	0.220
4	1	250	51	0.204
5	1	250	46	0.184
6	1	250	61	0.244
7	1	250	39	0.156
8	1	250	44	0.176
9	1	250	41	0.164
10	1	250	51	0.204
11	2	500	88	0.176
12	2	500	101	0.202
13	2	500	101	0.202
14	1	250	40	0.160
15	1	250	48	0.192
16	1	250	47	0.188
17	1	250	50	0.200
18	1	250	48	0.192
19	1	250	57	0.228
20	1	250	45	0.180
21	1	250	43	0.172
22	2	500	105	0.210
23	2	500	98	0.196
24	2	500	100	0.200
25	2	500	96	0.192
	Totals	8,000	1,544	

TABLE 8.3 Raw Data for *p* Chart

Using the above data the center line and control limits are found as follows:

$$p = \frac{\text{subgroup defective count}}{\text{subgroup size}}$$

These values are shown in the last column of Table 8.3.

$$\bar{p} = \frac{\text{sum of subgroup defective counts}}{\text{sum of subgroup size}} = \frac{1{,}544}{8{,}000} = 0.193$$

which is constant for all subgroups.

$n = 250$ (1 crate):

$$\text{LCL} = \bar{p} - 3\sqrt{\frac{\bar{p}(1-\bar{p})}{n}} = 0.193 - 3\sqrt{\frac{0.193 \times (1-0.193)}{250}} = 0.118$$

$$\text{UCL} = \bar{p} + 3\sqrt{\frac{\bar{p}(1-\bar{p})}{n}} = 0.193 + 3\sqrt{\frac{0.193 \times (1-0.193)}{250}} = 0.268$$

$n = 500$ (2 crates):

$$\text{LCL} = 0.193 - 3\sqrt{\frac{0.193 \times (1-0.193)}{500}} = 0.140$$

$$\text{UCL} = 0.193 + 3\sqrt{\frac{0.193 \times (1-0.193)}{500}} = 0.246$$

The control limits and the subgroup proportions are shown in Fig. 8.4.

Pointers for Using *p* Charts

Determine if "moving control limits" are really necessary. It may be possible to use the average sample size (total number inspected divided by the number of subgroups) to calculate control limits. For instance, with our example the sample size doubled from 250 peaches to 500 but the control limits hardly changed at all. Table 8.4 illustrates the different control limits based on 250 peaches, 500 peaches, and the average sample size which is 8,000 ÷ 25 = 320 peaches.

Notice that the conclusions regarding process performance are the same when using the average sample size as they are using the exact sample sizes. This is usually the case if the variation in sample size isn't too great. There are many rules of thumb, but most of them are extremely conservative. The best way to evaluate limits based on the average sample size is to check it out as shown above. SPC is all about improved decision-making. In general, use the most simple method that leads to correct decisions.

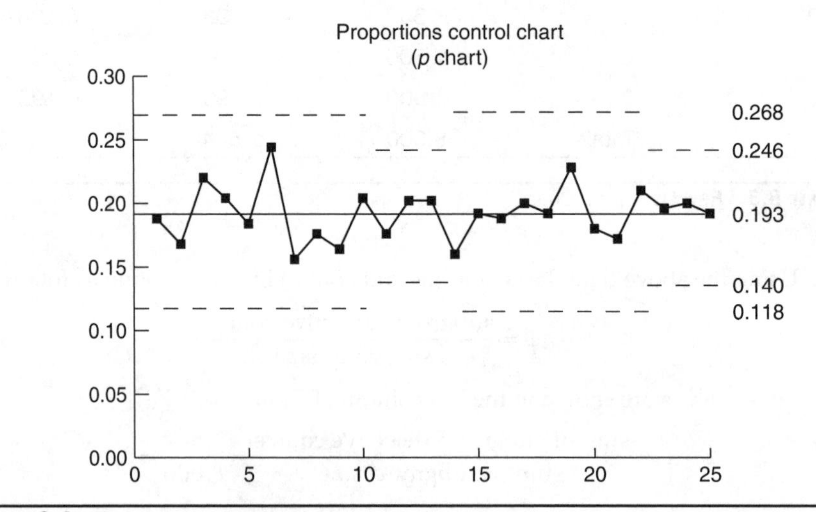

FIGURE 8.4 Completed *p* control chart.

Sample Size	Lower Control Limit	Upper Control Limit
250	0.1181	0.2679
500	0.1400	0.2460
320	0.1268	0.2592

Table 8.4 Effect of Using Average Sample Size

Control Charts for Count of Defectives (*np* Charts)

np charts are statistical tools used to evaluate the count of defectives, or count of items nonconforming, produced by a process. *np* charts can be applied to any variable where the appropriate performance measure is a unit count and the subgroup size is held constant. Note that wherever an *np* chart can be used, a *p* chart can be used too.

Control Limit Equations for *np* Charts

Like all control charts, *np* charts consist of three guidelines: center line, a lower control limit, and an upper control limit. The center line is the average count of defectives-per-subgroup and the two control limits are set at plus and minus three standard deviations. If the process is in statistical control, then virtually all subgroup counts will be between the control limits, and they will fluctuate randomly about the center line.

$$np = \text{subgroup defective count} \tag{8.24}$$

$$n\bar{p} = \frac{\text{sum of subgroup defective counts}}{\text{number of subgroups}} \tag{8.25}$$

$$\text{LCL} = n\bar{p} - 3\sqrt{n\bar{p}(1-\bar{p})} \tag{8.26}$$

$$\text{UCL} = n\bar{p} + 3\sqrt{n\bar{p}(1-\bar{p})} \tag{8.27}$$

Note that

$$\bar{p} = \frac{n\bar{p}}{n} \tag{8.28}$$

Example of *np* Chart Calculation

The data in Table 8.5 were obtained by opening randomly selected crates from each shipment and counting the number of bruised peaches. There are 250 peaches per crate (constant n is required for *np* charts).

Using the above data the center line and control limits are found as follows:

$$n\bar{p} = \frac{\text{sum of subgroup defective counts}}{\text{number of subgroups}} = \frac{838}{30} = 27.93$$

$$\text{LCL} = n\bar{p} - 3\sqrt{n\bar{p}(1-\bar{p})} = 27.93 - 3\sqrt{27.93 \times \left(1 - \frac{27.93}{250}\right)} = 12.99$$

$$\text{UCL} = n\bar{p} + 3\sqrt{n\bar{p}(1-\bar{p})} = 27.93 + 3\sqrt{27.93 \times \left(1 - \frac{27.93}{250}\right)} = 42.88$$

The control limits and the subgroup defective counts are shown in Fig. 8.5.

Shipment No.	Bruised Peaches
1	20
2	28
3	24
4	21
5	32
6	33
7	31
8	29
9	30
10	34
11	32
12	24
13	29
14	27
15	37
16	23
17	27
18	28
19	31
20	27
21	30
22	23
23	23
24	27
25	35
26	29
27	23
28	23
29	30
30	28
Total	838

TABLE 8.5 Raw Data for *np* Chart

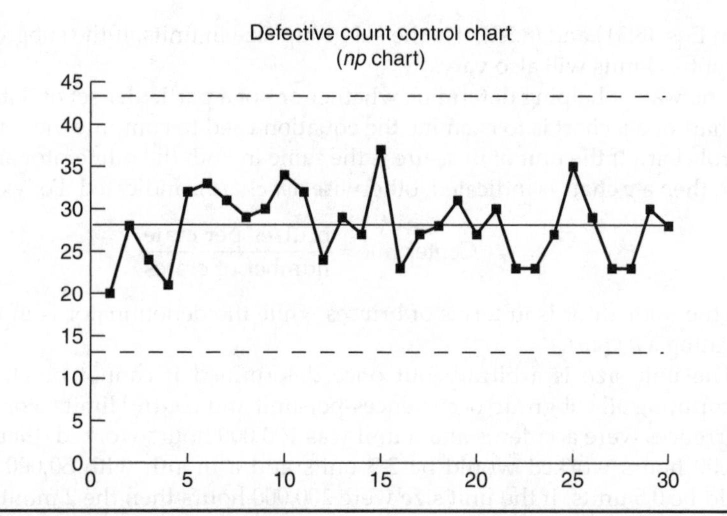

FIGURE 8.5 Completed *np* control chart.

Control Charts for Average Occurrences-Per-Unit (*u* Charts)

u charts are statistical tools used to evaluate the average number of occurrences-per-unit produced by a process. *u* charts can be applied to any variable where the appropriate performance measure is a count of how often a particular event occurs. *u* charts answer the question: "Has a special cause of variation caused the central tendency of this process to produce an abnormally large or small number of occurrences over the time period observed?" Note that, unlike *p* or *np* charts, *u* charts do not necessarily involve counting physical items. Rather, they involve counting of *events*. For example, when using a *p* chart one would count bruised peaches. When using a *u* chart one would count the *bruises*.

Control Limit Equations for *u* Charts

Like all control charts, *u* charts consist of three guidelines: center line, a lower control limit, and an upper control limit. The center line is the average number of occurrences-per-unit and the two control limits are set at plus and minus three standard deviations. If the process is in statistical control then virtually all subgroup occurrences-per-unit should be between the control limits and they should fluctuate randomly about the center line.

$$u = \frac{\text{subgroup count of occurrences}}{\text{subgroup size in units}} \tag{8.29}$$

$$\bar{u} = \frac{\text{sum of subgroup occurrences}}{\text{sum of subgroup sizes in units}} \tag{8.30}$$

$$\text{LCL} = \bar{u} - 3\sqrt{\frac{\bar{u}}{n}} \tag{8.31}$$

$$\text{UCL} = \bar{u} + 3\sqrt{\frac{\bar{u}}{n}} \tag{8.32}$$

In Eqs. (8.31) and (8.32), n is the subgroup size in units. If the subgroup size varies, the control limits will also vary.

One way of helping determine whether or not a particular set of data is suitable for a u chart or a p chart is to examine the equation used to compute the center line for the control chart. If the unit of measure is the same in both the numerator and the denominator, then a p chart is indicated, otherwise a u chart is indicated. For example, if

$$\text{Center line} = \frac{\text{bruises per crate}}{\text{number of crates}} \tag{8.33}$$

then the numerator is in terms of bruises while the denominator is in terms of crates, indicating a u chart.

The unit size is arbitrary but once determined it cannot be changed without recomputing all subgroup occurrences-per-unit and control limits. For example, if the occurrences were accidents and a unit was 100,000 hours worked, then a month with 250,000 hours worked would be 2.5 units and a month with 50,000 hours worked would be 0.5 units. If the unit size were 200,000 hours then the 2 months would have 1.25 and 0.25 units respectively. The equations for the center line and control limits would "automatically" take into account the unit size, so the control charts would give identical results regardless of which unit size is used.

Analysis of *u* Charts

As with all control charts, a special cause is probably present if there are any points beyond either the upper or the lower control limit. Analysis of u chart patterns between the control limits is extremely complicated when the sample size varies and is usually not done.

Example of *u* Chart

The data in Table 8.6 were obtained by opening randomly selected crates from each shipment and counting the number of bruises on peaches. There are 250 peaches per crate. Our unit size will be taken as one full crate, that is, we will be counting crates rather than the peaches themselves. Normally, samples consist of one crate per shipment. However, when part-time help is available, samples of two crates are taken.

Using the above data the center line and control limits are found as follows:

$$u = \frac{\text{subgroup count of occurrences}}{\text{subgroup size in units}}$$

These values are shown in the last column of Table 8.6.

$$\bar{u} = \frac{\text{sum of subgroup count of occurrences}}{\text{sum of subgroup unit sizes}} = \frac{1{,}544}{32} = 48.25$$

which is constant for all subgroups.

$n = 1$ unit:

$$\text{LCL} = \bar{u} - 3\sqrt{\frac{\bar{u}}{n}} = 48.25 - 3\sqrt{\frac{48.25}{1}} = 27.411$$

$$\text{UCL} = \bar{u} + 3\sqrt{\frac{\bar{u}}{n}} = 48.25 + 3\sqrt{\frac{48.25}{1}} = 69.089$$

Shipment No.	Units (Crates)	Flaws	Flaws-Per-Unit
1	1	47	47
2	1	42	42
3	1	55	55
4	1	51	51
5	1	46	46
6	1	61	61
7	1	39	39
8	1	44	44
9	1	41	41
10	1	51	51
11	2	88	44
12	2	101	50.5
13	2	101	50.5
14	1	40	40
15	1	48	48
16	1	47	47
17	1	50	50
18	1	48	48
19	1	57	57
20	1	45	45
21	1	43	43
22	2	105	52.5
23	2	98	49
24	2	100	50
25	2	96	48
Totals	32	1,544	

Table 8.6 Raw Data for *u* Chart

n = 2 units:

$$\text{LCL} = 48.25 - 3\sqrt{\frac{48.25}{2}} = 33.514$$

$$\text{UCL} = 48.25 + 3\sqrt{\frac{48.25}{2}} = 62.986$$

The control limits and the subgroup occurrences-per-unit are shown in Fig. 8.6.

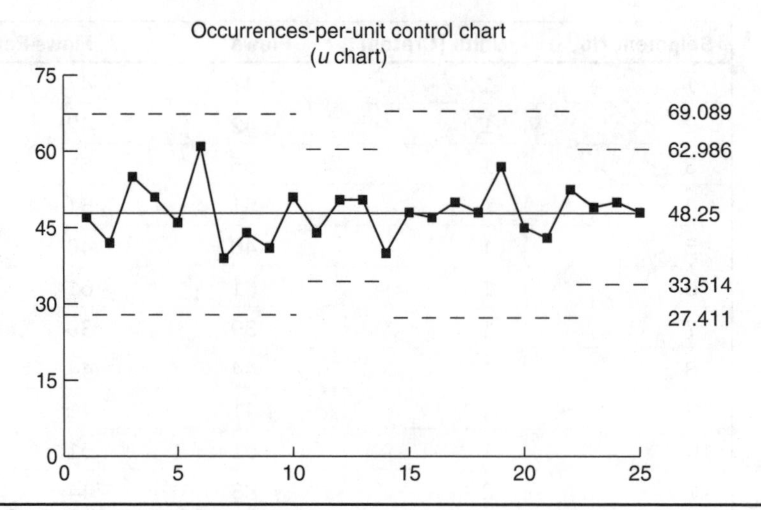

FIGURE 8.6 Completed *u* control chart.

The reader may note that the data used to construct the *u* chart were the same as those used for the *p* chart, except that we considered the counts as being counts of occurrences (bruises) instead of counts of physical items (bruised peaches). The practical implications of using a *u* chart when a *p* chart should have been used, or vice versa, are usually not serious. The decisions based on the control charts will be quite similar in most cases encountered in Six Sigma regardless of whether a *u* or a *p* chart is used.

Control Charts for Counts of Occurrences-Per-Unit (*c* Charts)

c charts are statistical tools used to evaluate the number of occurrences-per-unit produced by a process. *c* charts can be applied to any variable where the appropriate performance measure is a count of how often a particular event occurs and samples of constant size are used. *c* charts answer the question: "Has a special cause of variation caused the central tendency of this process to produce an abnormally large or small number of occurrences over the time period observed?" Note that, unlike *p* or *np* charts, *c* charts do not involve counting physical items. Rather, they involve counting of *events*. For example, when using an *np* chart one would count bruised peaches. When using a *c* chart one would count the *bruises*.

Control Limit Equations for c Charts

Like all control charts, *c* charts consist of three guidelines: center line, a lower control limit, and an upper control limit. The center line is the average number of occurrences-per-unit and the two control limits are set at plus and minus three standard deviations. If the process is in statistical control then virtually all subgroup occurrences-per-unit should be between the control limits and they should fluctuate randomly about the center line.

$$\bar{c} = \frac{\text{sum of subgroup occurrences}}{\text{number of subgroups}} \tag{8.34}$$

$$\text{LCL} = \bar{c} - 3\sqrt{\bar{c}} \tag{8.35}$$

$$\text{UCL} = \bar{c} + 3\sqrt{\bar{c}} \tag{8.36}$$

One way of helping determine whether or not a particular set of data is suitable for a *c* chart or an *np* chart is to examine the equation used to compute the center line for the control chart. If the unit of measure is the same in both the numerator and the denominator, then a *p* chart is indicated, otherwise a *c* chart is indicated. For example, if

$$\text{Center line} = \frac{\text{bruises}}{\text{number of crates}}$$

then the numerator is in terms of bruises while the denominator is in terms of crates, indicating a *c* chart.

The unit size is arbitrary but, once determined, it cannot be changed without recomputing all subgroup occurrences-per-unit and control limits.

Analysis of *c* Charts

As with all control charts, a special cause is probably present if there are any points beyond either the upper or the lower control limit. Analysis of *c* chart patterns between the control limits is shown later in this chapter.

Example of *c* Chart

The data in Table 8.7 were obtained by opening randomly selected crates from each shipment and counting the number of bruises. There are 250 peaches per crate. Our unit size will be taken as one full crate, that is, we will be counting crates rather than the peaches themselves. Every subgroup consists of one crate. If the subgroup size varied, a *u* chart would be used.

Using the above data the center line and control limits are found as follows:

$$\bar{c} = \frac{\text{sum of subgroup occurrences}}{\text{number of subgroups}} = \frac{1{,}006}{30} = 33.53$$

$$\text{LCL} = \bar{c} - 3\sqrt{\bar{c}} = 33.53 - 3\sqrt{33.53} = 16.158$$

$$\text{UCL} = \bar{c} - 3\sqrt{8\bar{c}} = 33.53 + 3\sqrt{33.53} = 50.902$$

The control limits and the occurrence counts are shown in Fig. 8.7.

Control Chart Selection

Selecting the proper control chart for a particular data set is a simple matter if approached properly. The proper approach is illustrated in Fig. 8.8.

To use the decision tree, begin at the leftmost node and determine if the data are measurements or counts. If measurements, then select the control chart based on the subgroup size. If the data are counts, then determine if the counts are of occurrences or pieces. An aid in making this determination is to examine the equation for the process average. If the numerator and denominator involve the same units, then a *p* or *np* chart is indicated. If different units of measure are involved, then a *u* or *c* chart is indicated. For example, if the average is in accidents-per-month, then a *c* or *u* chart is indicated because the numerator is in terms of accidents but the denominator is in terms of time.

Shipment No.	Flaws
1	27
2	32
3	24
4	31
5	42
6	38
7	33
8	35
9	35
10	39
11	41
12	29
13	34
14	34
15	43
16	29
17	33
18	33
19	38
20	32
21	37
22	30
23	31
24	32
25	42
26	40
27	21
28	23
29	39
30	29
Total	1,006

TABLE 8.7 Raw Data for *c* Chart

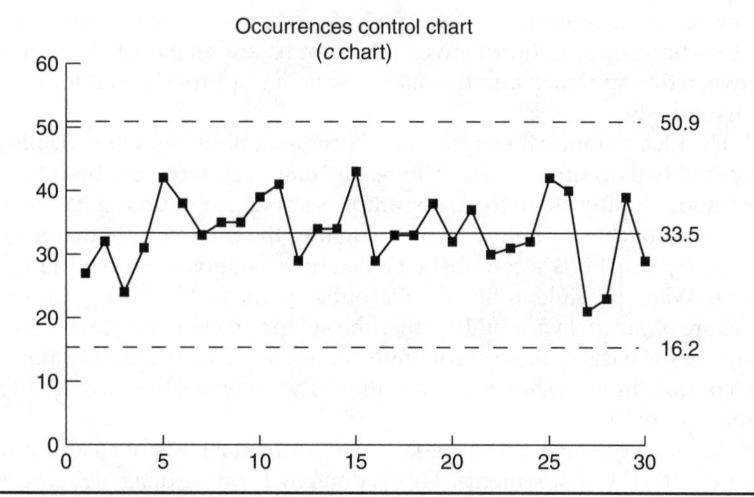

Figure 8.7 Completed *c* control chart.

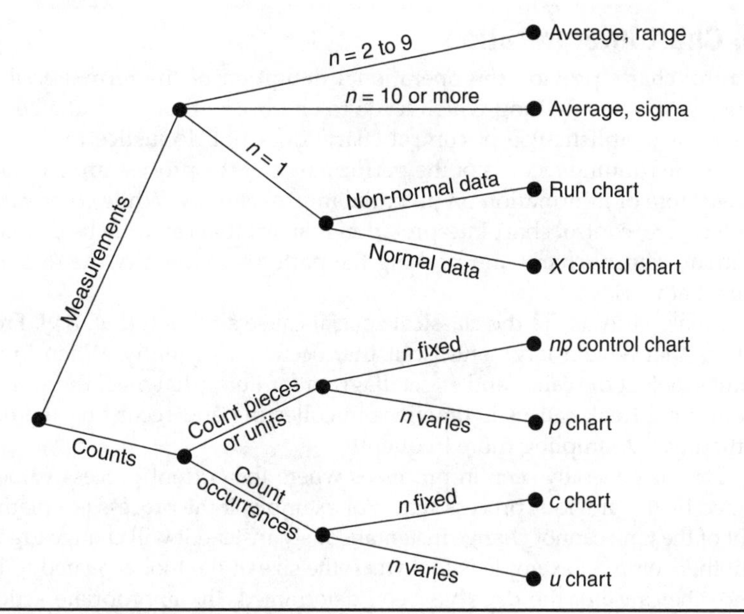

Figure 8.8 Control chart selection decision tree.

Rational Subgroup Sampling

The basis of all control charts is the *rational subgroup*. Rational subgroups are composed of items which were produced under essentially the same conditions. The statistics, for example, the average and range, are computed for each subgroup separately, then plotted on the control chart. When possible, rational subgroups are formed by using consecutive units. Each subgroup's statistics are compared to the control limits, and patterns of variation between subgroups are analyzed. Note the sharp contrast between this approach and the *random sampling* approach used for enumerative statistical methods.

The idea of rational subgrouping becomes a bit fuzzy when dealing with X charts, or individuals control charts. The reader may well wonder about the meaning of the term subgrouping when the "subgroup" is a single measurement. The basic idea underlying control charts of all types is to identify the capability of the process. The mechanism by which this is accomplished is careful formation of rational subgroups as defined above. When possible, rational subgroups are formed by using consecutive units. The measure of process variability, either the subgroup standard deviation or the subgroup range, is the basis of the control limits for averages. Conceptually, this is akin to basing the control limits on short-term variation. These control limits are used to monitor variation over time.

As far as possible, this approach also forms the basis of establishing control limits for individual measurements. This is done by forming quasi-subgroups using pairs of consecutive measurements. These "subgroups of 2" are used to compute ranges. The ranges are used to compute the control limits for the individual measurements.

Control Chart Interpretation

Control charts provide the operational definition of the term *special cause*. A special cause is simply anything which leads to an observation beyond a control limit. However, this simplistic use of control charts does not do justice to their power. Control charts are running records of the performance of the process and, as such, they contain a vast store of information on potential improvements. While some guidelines are presented here, control chart interpretation is an art that can only be developed by looking at many control charts and probing the patterns to identify the underlying system of causes at work.

Freak patterns are the classical special cause situation (Fig. 8.9). Freaks result from causes that have a large effect but that occur infrequently. When investigating freak values look at the cause and effect diagram for items that meet these criteria. The key to identifying freak causes is timelines in collecting and recording the data. If you have difficulty, try sampling more frequently.

Drift is generally seen in processes where the current process value is partly determined by the previous process state. For example, if the process is a plating bath, the content of the tank cannot change instantaneously, instead it will change gradually (Fig. 8.10). Another common example is tool wear: the size of the tool is related to its previous size. Once the cause of the drift has been determined, the appropriate action can be taken. Whenever economically feasible, the drift should be eliminated, for example, install an automatic chemical dispenser for the plating bath, or make automatic compensating adjustments to correct for tool wear. Note that the total process variability increases when drift is allowed, which adds cost. When drift elimination is not possible, the control chart can be modified in one of two ways:

1. Make the slope of the center line and control limits match the natural process drift. The control chart will then detect departures from the natural drift.
2. Plot *deviations* from the natural or expected drift.

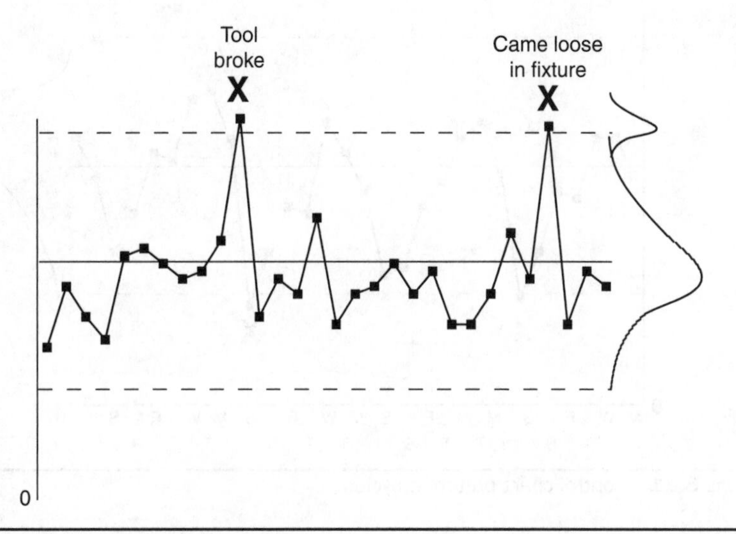

Figure 8.9 Control chart patterns: freaks.

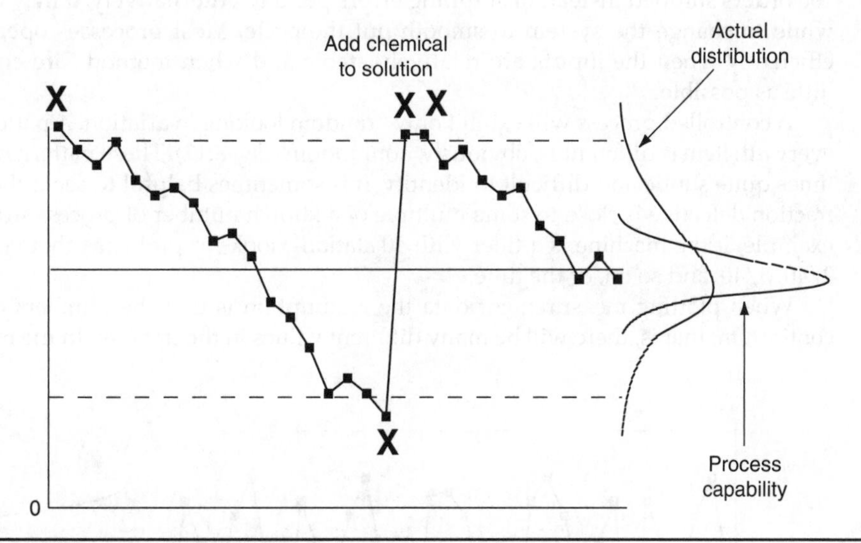

Figure 8.10 Control chart patterns: drift.

Cycles often occur due to the nature of the process. Common cycles include hour of the day, day of the week, month of the year, quarter of the year, week of the accounting cycle, and so on (Fig. 8.11). Cycles are caused by modifying the process inputs or methods according to a regular schedule. The existence of this schedule and its effect on the process may or may not be known in advance. Once the cycle has been discovered, action can be taken. The action might be to adjust the control chart by plotting

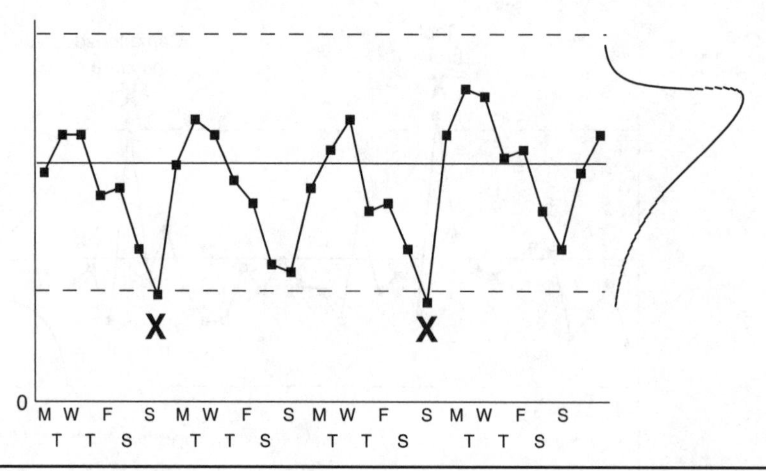

FIGURE 8.11 Control chart patterns: cycles.

the control measure against a variable base. For example, if a day-of-the-week cycle exists for shipping errors because of the workload, you might plot shipping errors per 100 orders shipped instead of shipping errors per day. Alternatively, it may be worthwhile to change the system to smooth out the cycle. Most processes operate more efficiently when the inputs are relatively stable and when methods are changed as little as possible.

A controlled process will exhibit only "random looking" variation. A pattern where every *n*th item is different is, obviously, nonrandom (Fig. 8.12). These patterns are sometimes quite subtle and difficult to identify. It is sometimes helpful to see if the average fraction defective is close to some multiple of a known number of process streams. For example, if the machine is a filler with 40 stations, look for problems that occur 1/40, 2/40, 3/40, and so on, of the time.

When plotting measurement data the assumption is that the numbers exist on a continuum, that is, there will be many different values in the data set. In the real world,

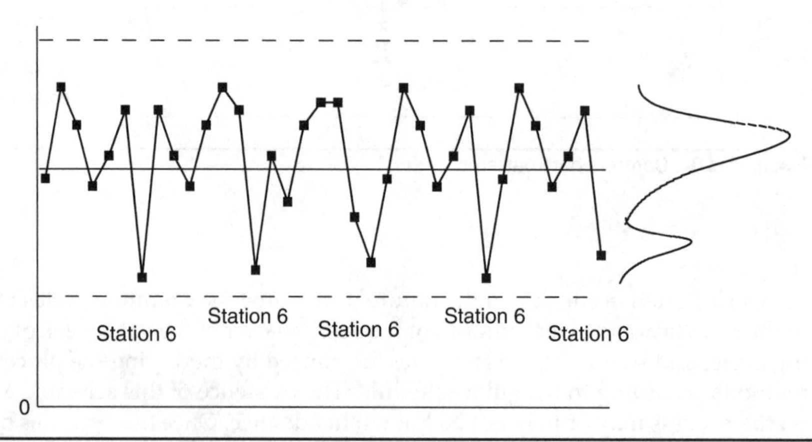

FIGURE 8.12 Control chart patterns: repeating patterns.

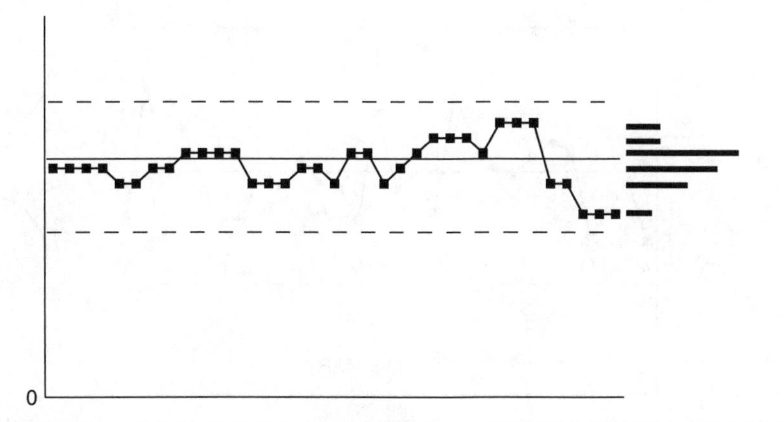

FIGURE 8.13 Control chart patterns: discrete data.

the data are never completely continuous (Fig. 8.13). It usually doesn't matter much if there are, say, 10 or more different numbers. However, when there are only a few numbers that appear over-and-over it can cause problems with the analysis. A common problem is that the *R* chart will underestimate the average range, causing the control limits on both the average and range charts to be too close together. The result will be too many "false alarms" and a general loss of confidence in SPC.

The usual cause of this situation is inadequate gage resolution. The ideal solution is to obtain a gage with greater resolution. Sometimes the problem occurs because operators, inspectors, or computers are rounding the numbers. The solution here is to record additional digits.

The reason SPC is done is to accelerate the learning process and to eventually produce an improvement. Control charts serve as historical records of the learning process and they can be used by others to improve other processes. When an improvement is realized the change should be written on the old control chart; its effect will show up as a less variable process. These charts are also useful in communicating the results to leaders, suppliers, customers, and others interested in quality improvement (Fig. 8.14).

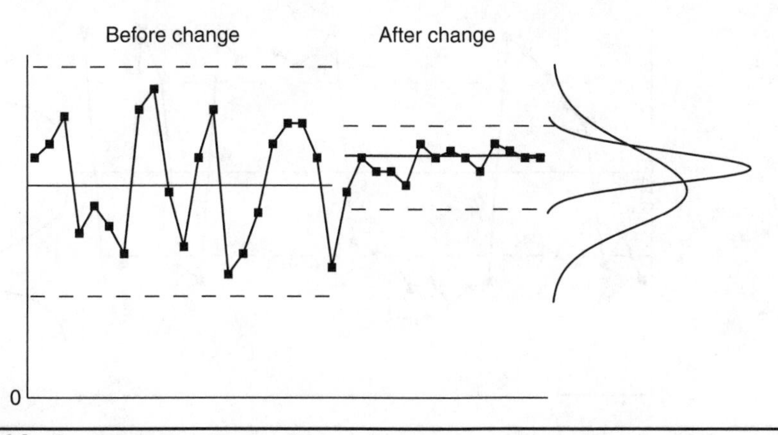

FIGURE 8.14 Control chart patterns: planned changes.

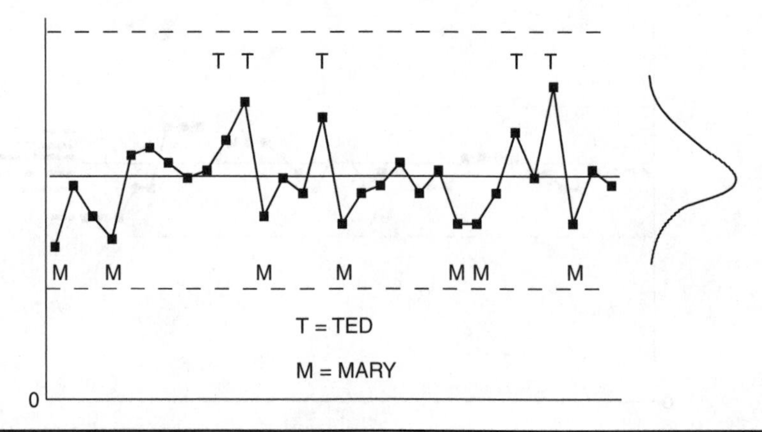

Figure 8.15 Control chart patterns: suspected differences.

Seemingly random patterns on a control chart are evidence of unknown causes of variation, which is not the same as *uncaused* variation. There should be an ongoing effort to reduce the variation from these so-called common causes. Doing so requires that the unknown causes of variation be identified. One way of doing this is a retrospective evaluation of control charts. This involves brainstorming and preparing cause and effect diagrams, then relating the control chart patterns to the causes listed on the diagram. For example, if "operator" is a suspected cause of variation, place a label on the control chart points produced by each operator (Fig. 8.15). If the labels exhibit a pattern, there is evidence to suggest a problem. Conduct an investigation into the reasons and set up controlled experiments (prospective studies) to test any theories proposed. If the experiments indicate a true cause and effect relationship, make the appropriate process improvements. Keep in mind that a statistical *association* is not the same thing as a causal *correlation*. The observed association must be backed up with solid subject-matter expertise and experimental data.

Mixture exists when the data from two different cause systems are plotted on a single control chart (Fig. 8.16). It indicates a failure in creating rational subgroups. The

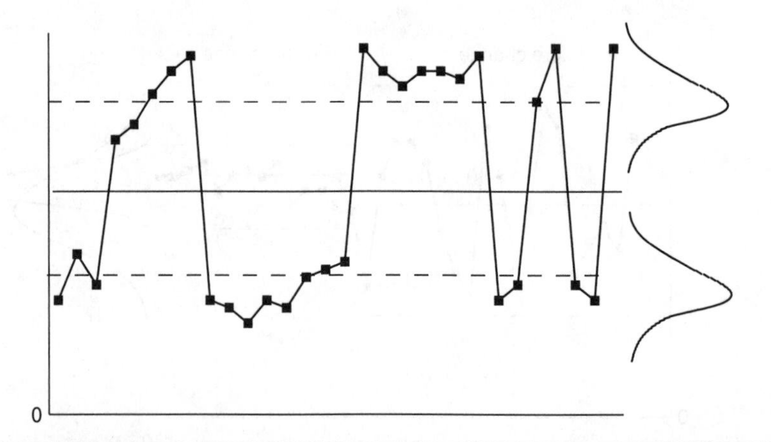

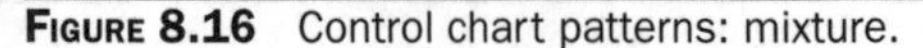

Figure 8.16 Control chart patterns: mixture.

underlying differences should be identified and corrective action taken. The nature of the corrective action will determine how the control chart should be modified.

Mixture example 1
The mixture represents two different operators who can be made more consistent. A single control chart can be used to monitor the new, consistent process.

Mixture example 2
The mixture is in the number of emergency room cases received on Saturday evening, versus the number received during a normal week. Separate control charts should be used to monitor patient-load during the two different time periods.

Run Tests

If the process is stable, then the distribution of subgroup averages will be approximately normal. With this in mind, we can also analyze the *patterns* on the control charts to see if they might be attributed to a special cause of variation. To do this, we divide a normal distribution into zones, with each zone one standard deviation wide. Figure 8.17 shows the approximate percentage we expect to find in each zone from a stable process.

Zone C is the area from the mean to the mean plus or minus one sigma, zone B is from plus or minus one sigma to plus or minus two sigma, and zone A is from plus or minus two sigma to plus or minus three sigma. Of course, any point beyond three sigma (i.e., outside of the control limit) is an indication of an out-of-control process.

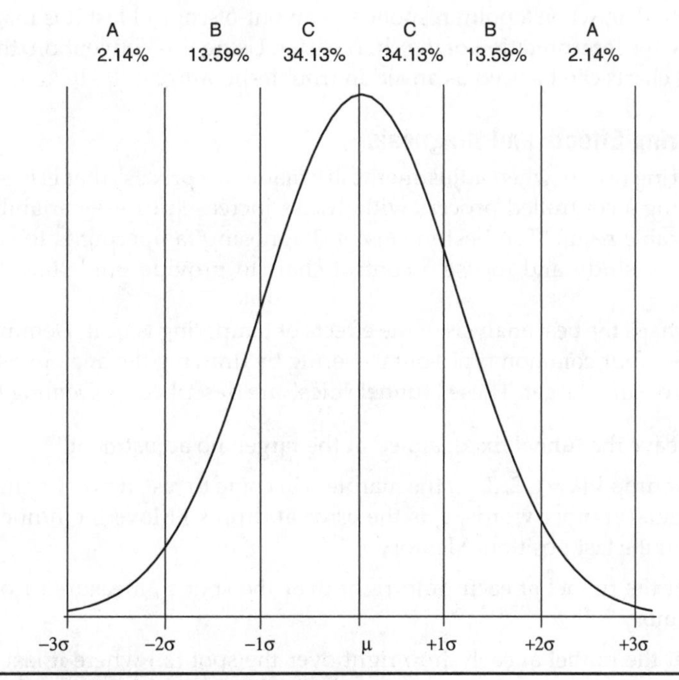

FIGURE 8.17 Percentiles for a normal distribution.

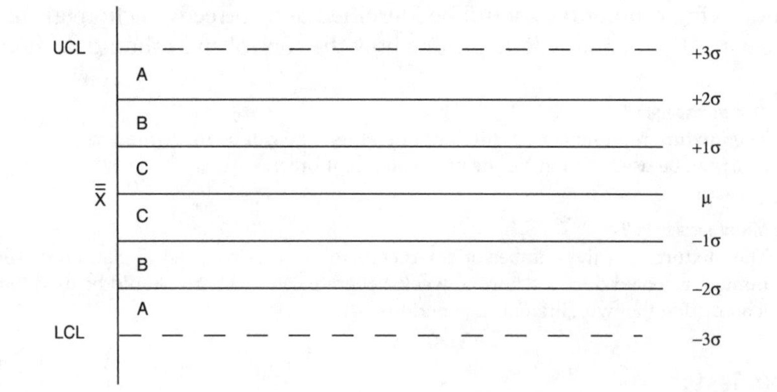

FIGURE 8.18 Zones on a control chart.

Since the control limits are at plus and minus three standard deviations, finding the one and two sigma lines on a control chart is as simple as dividing the distance between the grand average and either control limit into thirds, which can be done using a ruler. This divides each half of the control chart into three zones. The three zones are labeled A, B, and C, as shown in Fig. 8.18.

Based on the expected percentages in each zone, sensitive run tests can be developed for analyzing the patterns of variation in the various zones. Remember, the existence of a nonrandom pattern means that a special cause of variation was (or is) probably present. The averages, *np* and *c* control chart run tests are shown in Fig. 8.19.

Note that, when a point responds to an out-of-control test it is marked with an "X" to make the interpretation of the chart easier. Using this convention, the patterns on the control charts can be used as an aid in troubleshooting.

Tampering Effects and Diagnosis

Tampering occurs when adjustments are made to a process that is in statistical control. Adjusting a controlled process will always increase process variability, an obviously undesirable result. The best means of diagnosing tampering is to conduct a process capability study and to use a control chart to provide guidelines for adjusting the process.

Perhaps the best analysis of the effects of tampering is from Deming (1986). Deming describes four common types of tampering by drawing the analogy of aiming a funnel to hit a desired target. These "funnel rules" are described by Deming (1986, p. 328):

"Leave the funnel fixed, aimed at the target, no adjustment."

"At drop k ($k = 1, 2, 3, \ldots$) the marble will come to rest at point z_k, measured from the target. (In other words, z_k is the error at drop k.) Move the funnel the distance $-z_k$ from the last position. Memory 1."

"Set the funnel at each drop right over the spot z_k, measured from the target. No memory."

"Set the funnel at each drop right over the spot (z_k) where it last came to rest. No memory."

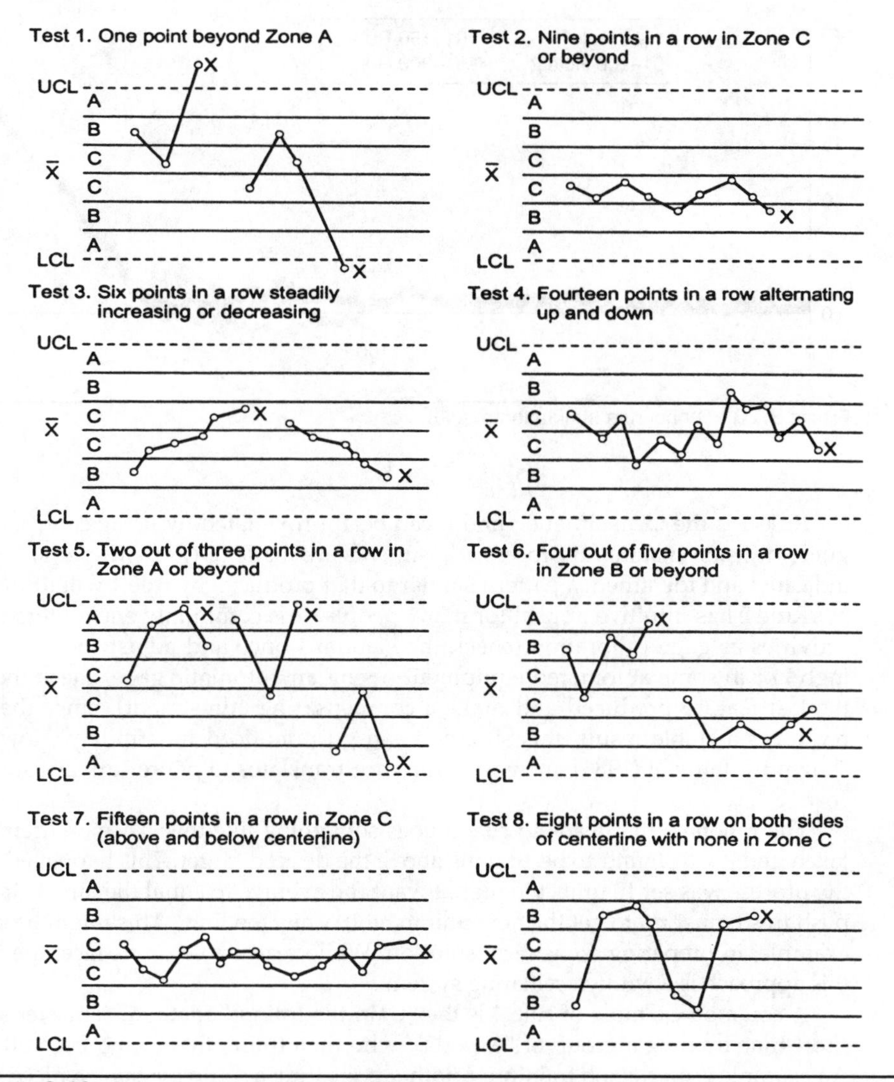

FIGURE 8.19 Tests for out-of-control patterns on control charts. From Nelson (1984) 237–239.

Rule 1 is the best rule for stable processes. By following this rule, the process average will remain stable and the variance will be minimized. Rule 2 produces a stable output but one with twice the variance of rule 1. Rule 3 results in a system that "explodes," that is, a symmetrical pattern will appear with a variance that increases without bound. Rule 4 creates a pattern that steadily moves away from the target, without limit (see Fig. 8.20).

At first glance, one might wonder about the relevance of such apparently abstract rules. However, upon more careful consideration, one finds many practical situations where these rules apply.

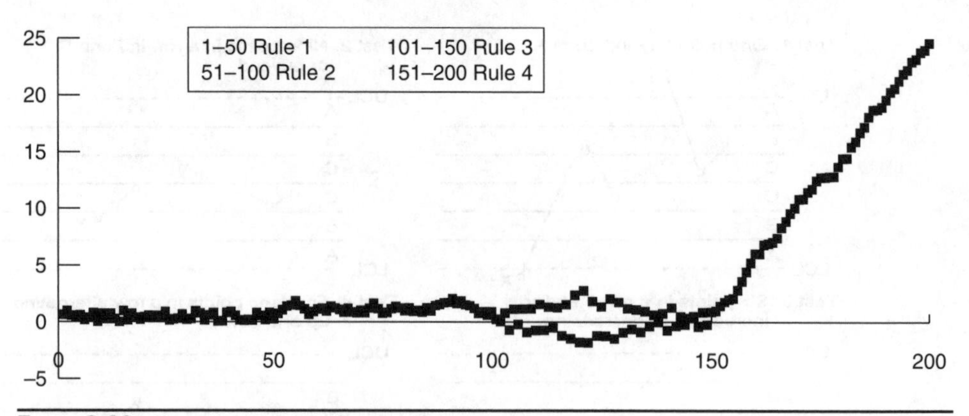

Figure 8.20 Funnel rule simulation results.

Rule 1 is the ideal situation and it can be approximated by using control charts to guide decision-making. If process adjustments are made only when special causes are indicated and identified, a pattern similar to that produced by rule 1 will result.

Rule 2 has intuitive appeal for many people. It is commonly encountered in such activities as gage calibration (check the standard once and adjust the gage accordingly) or in some automated equipment (using an automatic gage, check the size of the last feature produced and make a compensating adjustment). Since the system produces a stable result, this situation can go unnoticed indefinitely. However, as shown by Taguchi (1986), increased variance translates to poorer quality and higher cost.

The rationale that leads to rule 3 goes something like this: "A measurement was taken and it was found to be 10 units above the desired target. This happened because the process was set 10 units too high. I want the average to equal the target. To accomplish this I must try to get the next unit to be 10 units too low." This might be used, for example, in preparing a chemical solution. While reasonable on its face, the result of this approach is a wildly oscillating system.

A common example of rule 4 is the "train-the-trainer" method. A master spends a short time training a group of "experts," who then train others, who train others, etc. An example is on-the-job training. Another is creating a setup by using a piece from the last job. Yet another is a gage calibration system where standards are used to create other standards, which are used to create still others, and so on. Just how far the final result will be from the ideal depends on how many levels deep the scheme has progressed.

Short Run Statistical Process Control Techniques

Short production runs are a way of life with many manufacturing companies. In the future, this will be the case even more often. The trend in manufacturing has been toward smaller production runs with product tailored to the specific needs of individual customers. Henry Ford's days of "the customer can have any color, as long as it's black" have long since passed.

Classical SPC methods, such as $\overline{X}$ and R charts, were developed in the era of mass production of identical parts. Production runs often lasted for weeks, months, or even years. Many of the "SPC rules of thumb" currently in use were created for this situation. For example, the rule that control limits not be calculated until data are available from at least 25 subgroups of five. This may not have been a problem in 1930, but it certainly is today. In fact, many entire production runs involve fewer parts than required to start a standard control chart!

Many times the usual SPC methods can be modified slightly to work with short and small runs. For example, $\overline{X}$ and R control charts can be created using moving averages and moving ranges (Pyzdek, 1989). However, there are SPC methods that are particularly well suited to application on short or small runs.

Variables Data

Variables data, sometimes called continuous data, involve measurements such as size, weight, pH, temperature, etc. In theory data are variables data if no two values are exactly the same. In practice this is seldom the case. As a rough rule of thumb you can consider data to be variables data if at least 10 different values occur and repeat values make up no more than 20% of the data set. If this is not the case, your data may be too discrete to use standard control charts. Consider trying an attribute procedure such as the demerit charts described later in this chapter. We will discuss the following approaches to SPC for short or small runs:

1. **Exact method**. Tables of special control chart constants are used to create X, $\overline{X}$ and R charts that compensate for the fact that a limited number of subgroups are available for computing control limits. The exact method is also used to compute control limits when using a code value chart or stabilized X or $\overline{X}$ and R charts (see below). The exact method allows the calculation of control limits that are correct when only a small amount of data is available. As more data become available the exact method updates control limits until, finally, no further updates are required and standard control chart factors can be used (Pyzdek, 1992a).
2. **Code value charts**. Control charts created by subtracting nominal or other target values from actual measurements. These charts are often standardized so that measurement units are converted to whole numbers. For example, if measurements are in thousandths of an inch a reading of 0.011 inches above nominal would be recorded simply as "11." Code value charts enable the user to plot several parts from a given process on a single chart, or to plot several features from a single part on the same control chart. The exact method can be used to adjust the control limits when code value charts are created with limited data.
3. **Stabilized control charts for variables**. Statisticians have known about normalizing transformations for many years. This approach can be used to create control charts that are independent of the unit of measure and scaled in such a way that several different characteristics can be plotted on the same control chart. Since stabilized control charts are independent of the unit of measure, they can be thought of as true *process control charts*. The exact method adjusts the control limits for stabilized charts created with limited data.

Exact Method of Computing Control Limits for Short and Small Runs

This procedure, adapted from Hillier (1969) and Proschan and Savage (1960), applies to short runs or any situation where a small number of subgroups will be used to set up a control chart. It consists of three stages:

1. Finding the process (establishing statistical control)
2. Setting limits for the remainder of the initial run
3. Setting limits for future runs

The procedure correctly compensates for the uncertainties involved when computing control limits with small amounts of data.

Stage One: Find the Process

1. Collect an initial sample of subgroups (g). The factors for the recommended minimum number of subgroups are shown in Appendix 10 enclosed in a dark box. If it is not possible to get the minimum number of subgroups, use the appropriate control chart constant for the number of subgroups you actually have.
2. Using Appendix 10 compute the Range chart control limits using the equation upper control limit for ranges (UCL_R) = $D_{4F} \times \bar{R}$. Compare the subgroup ranges to the UCL_R and drop any out-of-control groups. Repeat the process until all remaining subgroup ranges are smaller than UCL_R.
3. Using the $\bar{R}$ value found in step 2, compute the control limits for the averages or individuals chart. The control limits are found by adding and subtracting $A_{2F} \times \bar{R}$ from the overall average. Drop any subgroups that have out-of-control averages and recompute. Continue until all remaining values are within the control limits. Go to stage two.

Stage Two: Set Limits for Remainder of the Initial Run

1. Using Appendix 10 compute the control limits for the remainder of the run. Use the A_{2S} factors for the $\bar{X}$ and the D_{4S} factors for the R chart; g = the number of groups used to compute stage one control limits.

Stage Three: Set Limits for a Future Run

1. After the run is complete, combine the raw data from the entire run and perform the analysis as described in stage one above. Use the results of this analysis to set limits for the next run, following the stage two procedure. If more than 25 groups are available, use a standard table of control chart constants.

Notes

1. Stage three assumes that there are no special causes of variation between runs. If there are, the process may go out of control when using the stage three control limits. In these cases, remove the special causes. If this isn't possible, apply this procedure to each run separately (i.e., start over each time).
2. This approach will lead to the use of standard control chart tables when enough data are accumulated.

3. The control chart constants for the first stage are A_{2F} and D_{4F} (the F subscript stands for first stage); for the second stage use A_{2S} and D_{4S}. These factors correspond to the A_2 and D_4 factors usually used, except that they are adjusted for the small number of subgroups actually available.

Setup Approval Procedure

The following procedure can be used to determine if a setup is acceptable using a relatively small number of sample units.

1. After the initial setup, run 3 to 10 pieces *without adjusting the process.*
2. Compute the average and the range of the sample.
3. Compute $T = \left[\dfrac{\text{average} - \text{target}}{\text{range}}\right]$

   Use absolute values (i.e., ignore any minus signs). The target value is usually the specification midpoint or nominal.
4. If T is less than the critical T in Table 8.8 accept the setup. Otherwise adjust the setup to bring it closer to the target. NOTE: there is approximately 1 chance in 20 that an on-target process will fail this test.

Example

Assume we wish to use SPC for a process that involves producing a part in lots of 30 parts each. The parts are produced approximately once each month. The control feature on the part is the depth of a groove and we will be measuring every piece. We decide to use subgroups of size three and to compute the stage one control limits after the first five groups. The measurements obtained are shown in Table 8.9.

n	3	4	5	6	7	8	9	10
Critical *T*	0.885	0.529	0.388	0.312	0.263	0.230	0.205	0.186

Table 8.8 Critical Value for Setup Acceptance

	Sample Number				
Subgroup Number	**1**	**2**	**3**	$\bar{X}$	R
1	0.0989	0.0986	0.1031	0.1002	0.0045
2	0.0986	0.0985	0.1059	0.1010	0.0074
3	0.1012	0.1004	0.1000	0.1005	0.0012
4	0.1023	0.1027	0.1000	0.1017	0.0027
5	0.0992	0.0997	0.0988	0.0992	0.0009

Table 8.9 Raw Data for Example of Exact Method

Using the data in Table 8.9 we can compute the grand average and average range as

$$\text{Grand average} = 0{:}10053$$

$$\text{Average range } (\bar{R}) = 0{:}00334$$

From Appendix 10 we obtain the first stage constant for the range chart of D_{4F} = 2.4 in the row for $g = 5$ groups and a subgroup size of 3. Thus,

$$\text{UCL}_R = D_{4F} \times \bar{R} = 2.4 \times 0.00334 = 0.0080$$

All of the ranges are below this control limit, so we can proceed to the analysis of the averages chart. If any R was above the control limit, we would try to determine why before proceeding.

For the averages chart we get

$$\text{LCL}_{\bar{X}} = \text{grand average} + A_{2F} \times \bar{R}$$

$$= 0.10053 - 1.20 \times 0.00334 = 0.09652 \text{ (rounded)}$$

$$\text{UCL}_{\bar{X}} = \text{grand average} + A_{2F} \times \bar{R}$$

$$= 0.10053 + 1.20 \times 0.00334 = 0.10454 \text{ (rounded)}$$

All of the subgroup averages are between these limits. Now setting limits for the remainder of the run we use $D_{4S} = 3.4$ and $A_{2S} = 1.47$. This gives, after rounding,

$$\text{UCL}_R = 0.01136$$

$$\text{LCL}_{\bar{X}} = 0.09562$$

$$\text{UCL}_{\bar{X}} = 0.10544$$

If desired, this procedure can be repeated when a larger number of subgroups becomes available, say 10 groups. This would provide somewhat better estimates of the control limits, but it involves considerable administrative overhead. When the entire run is finished you will have 10 subgroups of 3 per subgroup. The data from all of these subgroups should be used to compute stage one and stage two control limits. The resulting stage two control limits would then be applied to the *next run* of this part number.

By applying this method in conjunction with the code value charts or stabilized charts described below, the control limits can be applied to the next parts produced on this process (assuming the part-to-part difference can be made negligible). Note that if the standard control chart factors were used the limits for *both* stages would be (values are rounded)

$$\text{UCL}_R = 0.00860$$
$$\text{LCL}_{\bar{X}} = 0.09711$$
$$\text{UCL}_{\bar{X}} = 0.10395$$

As the number of subgroups available for computing the control limits increases, the "short run" control limits approach the standard control limits. However, if the standard control limits are used when only small amounts of data are available there is a greater chance of erroneously rejecting a process that is actually in control (Hillier, 1969).

Code Value Charts

This procedure allows the control of multiple features with a single control chart. It consists of making a simple transformation to the data, namely

$$\hat{x} = \frac{X - \text{target}}{\text{unit of measure}} \tag{8.37}$$

The resulting $\hat{x}$ values are used to compute the control limits and as plotted points on the $\overline{X}$ and R charts. This makes the target dimension irrelevant for the purposes of SPC and makes it possible to use a single control chart for several different features or part numbers.

Example

A lathe is used to produce several different sizes of gear blanks, as is indicated in Fig. 8.21.

Product engineering wants all of the gear blanks to be produced as near as possible to their nominal size. Process engineering believes that the process will have as little deviation for larger sizes as it does for smaller sizes. Quality engineering believes that the inspection system will produce approximately the same amount of measurement error for larger sizes as for smaller sizes. Process capability studies and measurement error studies confirm these assumptions. (I hope you are starting to get the idea that a number of assumptions are being made and that they must be valid before using code value charts.)

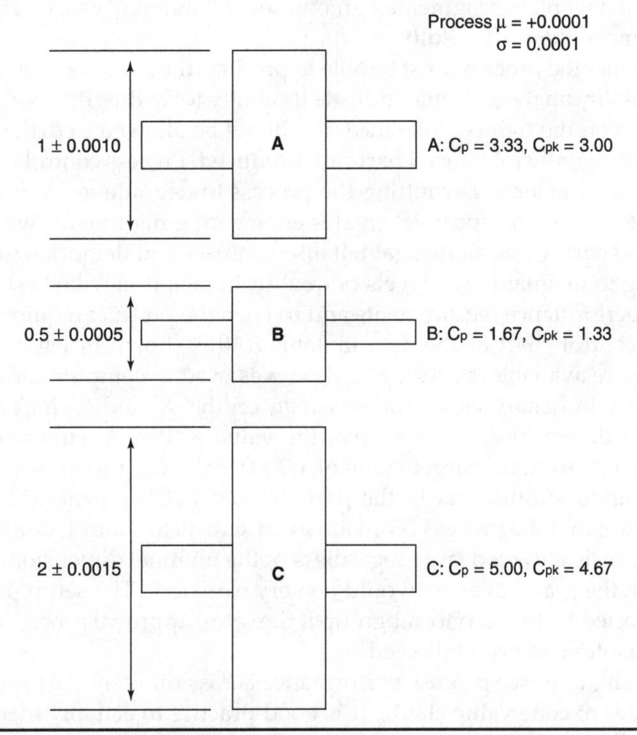

FIGURE 8.21 Some of the gear blanks to be machined.

Part	Nominal	No.	Sample Number			$\bar{X}$	R
			1	2	3		
A	1.0000	1	4	3	25	10.7	22
		2	3	3	39	15.0	36
		3	16	12	10	12.7	6
B	0.5000	4	21	24	10	18.3	14
		5	6	8	4	6.0	4
		6	19	7	21	15.7	14
C	2.0000	7	1	11	4	5.3	10
		8	1	25	8	11.3	24
		9	6	8	7	7.0	2

TABLE 8.10 Deviation from Target in Hundred-Thousandths

Based on these conclusions, the code value chart is recommended. By using the code value chart the amount of paperwork will be reduced and more data will be available for setting control limits. Also, the process history will be easier to follow since the information won't be fragmented among several different charts. The data in Table 8.10 show some of the early results.

Note that the process must be able to produce the *tightest tolerance* of ±0.0005 inches. The capability analysis should indicate its ability to do this; that is, C_{PK} should be at least 1.33 based on the tightest tolerance. It will not be allowed to drift or deteriorate when the less stringently toleranced parts are produced. Process control is independent of the product requirements. Permitting the process to degrade to its worst acceptable level (from the product perspective) creates engineering nightmares when the more tightly toleranced parts come along again. It also confuses and demoralizes operators and others trying to maintain high levels of quality. In fact, it may be best to publish only the process performance requirements and to keep the product requirements secret.

The control chart of the data in Table 8.10 is shown in Fig. 8.22. Since only nine groups were available, the exact method was used to compute the control limits. Note that the control chart shows the *deviations* on the $\bar{X}$ and *R* chart axes, not the actual measured dimensions, For example, the value of Part A, subgroup 1, sample 1 was +0.00004 inch from the target value of 1.0000 inch and it is shown as a deviation of +4 hundred-thousandths; that is, the part checked 1.00004 inches. The stage one control chart shows that the process is obviously in statistical control, but it is producing parts that are consistently too large regardless of the nominal dimension. If the process were on target, the grand average would be very close to 0. The setup problem would have been detected by the second subgroup if the setup approval procedure described earlier in this chapter had been followed.

This ability to see process performance across different part numbers is one of the advantages of code value charts. It is good practice to actually identify the changes in part numbers on the charts, as is done in Fig. 8.22.

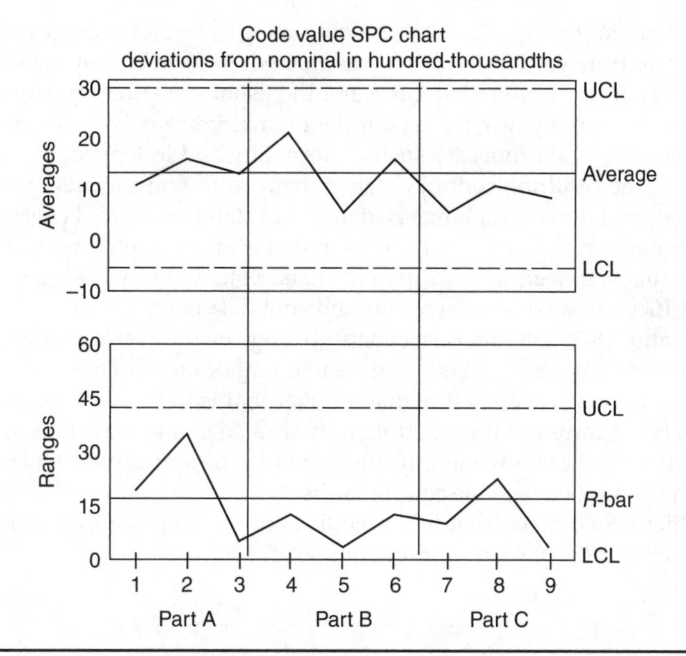

FIGURE 8.22 Code value chart of Table 8.10 data.

Stabilized Control Charts for Variables

All control limits, for standard sized runs or short and small runs, are based on methods that determine if a process statistic falls within limits that might be expected from chance variation (common causes) alone. In most cases, the statistic is based on actual measurements from the process and it is in the same unit of measure as the process measurements. As we saw with code value charts, it is sometimes useful to transform the data in some way. With code value charts we used a simple transformation that removed the effect of changing nominal and target dimensions. While useful, this approach still requires that all measurements be in the same units of measurement, for example, all inches, all grams, etc. For example, all of the variables on the control chart for the different gear blanks had to be in units of hundred-thousandths of an inch. If we had also wanted to plot, for example, the perpendicularity of two surfaces on the gear blank we would have needed a separate control chart because the units would be in degrees instead of inches.

Stabilized control charts for variables overcome the units of measure problem by converting all measurements into standard, nondimensional units. Such "standardizing transformations" are not new, they have been around for many years and they are commonly used in all types of statistical analyses. The two transformations we will be using here are shown in Eqs. (8.38) and (8.39).

$$\frac{(\bar{X} - \text{grand average})}{\bar{R}} \tag{8.38}$$

$$\frac{R}{\bar{R}} \tag{8.39}$$

As you can see, Eq. 8.38 involves subtracting the grand average from each subgroup average (or from each individual measurement if the subgroup size is one) and dividing the result by $\bar{R}$ Note that this is not the usual statistical transformation where the denominator is σ. By using $\bar{R}$ as our denominator instead of s we are sacrificing some desirable statistical properties such as normality and independence to gain simplicity. However, the resulting control charts remain valid and the false alarm risk based on points beyond the control limits is identical to standard control charts. Also, as with all transformations, this approach suffers in that it involves plotting numbers that are not in the usual engineering units people are accustomed to working with. This makes it more difficult to interpret the results and spot data entry errors.

Equation (8.39) divides each subgroup range by the average range. Since the numerator and denominator are both in the same unit of measurement, the unit of measurement cancels and we are left with a number that is in terms of the number of average ranges, $\bar{R}$'s. It turns out that control limits are also in the same units, that is, to compute standard control limits we simply multiply $\bar{R}$ by the appropriate table constant to determine the width between the control limits.

Hillier (1969) noted that this is equivalent to using the transformations shown in Eqs. (8.38) and (8.39) with control limits set at

$$-A_2 \leq \frac{(\bar{X} - \text{grand average})}{\bar{R}} \leq A_2 \tag{8.40}$$

for the individuals or averages chart. Control limits are

$$D_3 \leq \frac{R}{\bar{R}} \leq D_4 \tag{8.41}$$

for the range chart. Duncan (1974) described a similar transformation for attribute charts, p charts in particular (see below), and called the resulting chart a "stabilized p chart." We will call charts of the transformed variables data stabilized charts as well.

Stabilized charts allow you to plot multiple units of measurement on the same control chart. The procedure described in this chapter for stabilized variables charts requires that all subgroups be of the same size.* The procedure for stabilized attribute charts, described later in this chapter allows varying subgroup sizes. When using stabilized charts the control limits are always fixed. The raw data are "transformed" to match the scale determined by the control limits. When only limited amounts of data are available, the constants in Appendix Table 12 should be used for computing control limits for stabilized variables charts. As more data become available, the Appendix Table 9 constants approach the constants in standard tables of control chart factors. Table 8.11 summarizes the control limits for stabilized averages, stabilized ranges, and stabilized individuals control charts. The values for A_2, D_3, and D_4 can be found in standard control chart factor tables.

Example

A circuit board is produced on an electroplating line. Three parameters are considered important for SPC purposes: lead concentration of the solder plating bath, plating

---

*The procedure for stabilized attribute charts, described later in this chapter, allows varying subgroup sizes.

Stage	Available Groups		Chart			Appendix Table
			$\bar{X}$	R	X	
One	25 or less	LCL	$-A_{2F}$	None	$-A_{2F}$	12
		Average	0	1	0	
		UCL	$+A_{2F}$	D_{4F}	$+A_{2F}$	
Two	25 or less	LCL	$-A_{2S}$	None	$-A_{2S}$	12
		Average	0	1	0	
		UCL	$+A_{2S}$	D_{4S}	$+A_{2S}$	
One or two	More than 25	LCL	$-A_2$	D_3	−2.66	9
		Average	0	1	0	
		UCL	$+A_2$	D_4	+2.66	

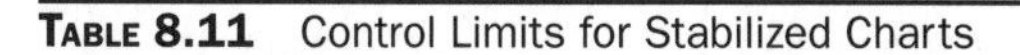

TABLE 8.11 Control Limits for Stabilized Charts

thickness, and resistance. Process capability studies have been done using more than 25 groups; thus, based on Table 8.11 the control limits are

$$-A_2 \leq \bar{X} \leq A_2$$

for the averages control chart, and

$$-D_3 \leq R \leq D_4$$

for the ranges control chart. The actual values of the constants A_2, D_3, and D_4 depend on the subgroup size; for subgroups of three $A_2 = 1.023$, $D_3 = 0$ and $D_4 = 2.574$.

The capabilities are shown in Table 8.12.

A sample of three will be taken for each feature. The three lead concentration samples are taken at three different locations in the tank. The results of one such set of sample measurements is shown in Table 8.13, along with their stabilized values.

On the control chart *only the extreme values are plotted*. Figure 8.23 shows a stabilized control chart for several subgroups. Observe that the feature responsible for the plotted point is written on the control chart. If a long series of largest or smallest values comes from the same feature it is an indication that the feature has changed. If the process is in statistical control for all features, the feature responsible for the extreme values will vary randomly.

Feature Code	Feature	GrandAvg.	Avg. Range
A	Lead %	10%	1%
B	Plating thickness	0.005 in	0.0005 in
C	Resistance	0.1 Ω	0.0005 Ω

TABLE 8.12 Process Capabilities for Example

Number	Lead % (A)	Thickness (B)	Resistance (C)
1	11%	0.0050 in	0.1000 Ω
2	11%	0.0055 in	0.1010 Ω
3	8%	0.0060 in	0.1020 Ω
$\bar{X}$	10%	0.0055 in	0.1010 Ω
R	3%	0.0010 in	0.0020 Ω
$(x-\bar{x})/\bar{R}$	0	1	2
$R/\bar{R}$	3	2	4

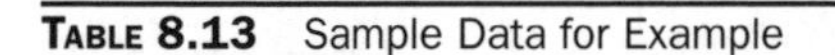

TABLE 8.13 Sample Data for Example

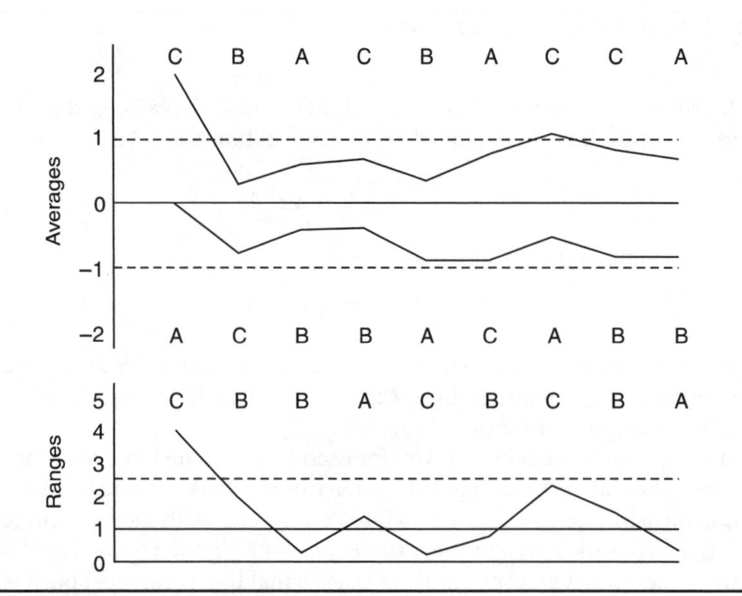

FIGURE 8.23 Stabilized control chart for variables.

When using stabilized charts it is possible to have a single control chart accompany a particular part or lot of parts through the entire production sequence. For example, the circuit boards described above could have a control chart that shows the results of process and product measurement for characteristics at all stages of production. The chart would then show the "processing history" for the part or lot. The advantage would be a coherent log of the production of a given part. Table 8.14 illustrates a process control plan that could possibly use this approach.

A caution is in order if the processing history approach is used. When small and short runs are common, the history of a given process can be lost among the charts of many different parts. This can be avoided by keeping a separate chart for each distinct

Operation	Feature	$\bar{X}$	$\bar{R}$	*n*
Clean	Bath pH	7.5	0.1	3/h
	Rinse contamination	100 ppm	5 ppm	3/h
	Cleanliness quality rating	78	4	3 pcs/h
Laminate	Riston thickness	1.5 min	0.1 mm	3 pcs/h
	Adhesion	7 in–lb	0.2 in–lb	3 pcs/h
Plating	Bath lead %	10%	1%	3/h
	Thickness	0.005 in	0.0005 in	3 pcs/h
	Resistance	0.1 Ω	0.0005 Ω	3 pcs/h

TABLE 8.14 PWB Fab Process Capabilities and SPC Plan

process; additional paperwork is involved, but it might be worth the effort. If the additional paperwork burden becomes large, computerized solutions may be worth investigating.

Attribute SPC for Small and Short Runs

When data are difficult to obtain, as is usual when small or short runs are involved, variables SPC should be used if at all possible. A variables measurement on a continuous scale contains more information than a discrete attributes classification provides. For example, a machine is cutting a piece of metal tubing to length. The specifications call for the length to be between 0.990 and 1.010 inches with the preferred length being 1.000 inch exactly. There are two methods available for checking the process. Method 1 involves measuring the length of the tube with a micrometer and recording the result to the nearest 0.001 inch. Method 2 involves placing the finished part into a "go/no-go gage." With method 2 a part that is shorter than 0.990 inch will go into the "no-go" portion of the gage, while a part that is longer than 1.010 inches will fail to go into the "go" portion of the gage. With method 1 we can determine the size of the part to within 0.001 inch. With method 2 we can only determine the size of the part to within 0.020 inch; that is, either it is within the size tolerance, it's too short, or it's too long. If the process could hold a tolerance of *less than* 0.020 inch, method 1 would provide the necessary information to hold the process to the variability it is capable of holding. Method 2 would not detect a process drift until out of tolerance parts were actually produced.

Another way of looking at the two different methods is to consider each part as belonging to a distinct category, determined by the part's length. Method 1 allows any part that is within tolerance to be placed into one of twenty categories. When out of tolerance parts are considered, method 1 is able to place parts into even more than twenty different categories. Method 1 also tells us if the part is in the best category, namely within ±0.001 inch of 1.000 inch; if not, we know how far the part is from the best category. With method 2 we can place a given part into only three categories: too short, within tolerance, or too long. A part that is far too short will be placed in the same category as a part that is only slightly short. A part that is barely within tolerance will be placed in the same category as a part that is exactly 1.000 inch long.

In spite of the disadvantages, it is sometimes necessary to use attributes data. Special methods must be used for attributes data used to control short run processes. We will describe two such methods:

- Stabilized attribute control charts
- Demerit control charts

Stabilized Attribute Control Charts

When plotting attribute data statistics from short run processes two difficulties are typically encountered:

1. Varying subgroup sizes
2. A small number of subgroups per production run

Item 1 results in messy charts with different control limits for each subgroup, distorted chart scales that mask significant variations, and chart patterns that are difficult to interpret because they are affected by both sample size changes and true process changes. Item 2 makes it difficult to track long-term process trends because the trends are broken up among many different control charts for individual parts. Because of these things, many people believe that SPC is not practical unless large and long runs are involved. This is not the case. In many cases stabilized attribute charts can be used to eliminate these problems. Although somewhat more complicated than classical control charts, stabilized attribute control charts offer a way of realizing the benefits of SPC with processes that are difficult to control any other way.

Stabilized attribute charts may be used if a process is producing part features that are essentially the same from one part number to the next. Production lot sizes and sample sizes can vary without visibly affecting the chart.

Example One A lathe is being used to machine terminals of different sizes. Samples (of different sizes) are taken periodically and inspected for burrs, nicks, tool marks, and other visual defects.

Example Two A printed circuit board hand assembly operation involves placing electrical components into a large number of different circuit boards. Although the boards differ markedly from one another, the hand assembly operation is similar for all of the different boards.

Example Three A job-shop welding operation produces small quantities of "one order only" items. However, the operation always involves joining parts of similar material and similar size. The process control statistic is weld imperfections per 100 inches of weld.

The techniques used to create stabilized attribute control charts are all based on corresponding classical attribute control chart methods. There are four basic types of control charts involved:

1. Stabilized *p* charts for proportion of defective units per sample
2. Stabilized *np* charts for the number of defective units per sample
3. Stabilized *c* charts for the number of defects per unit
4. Stabilized *u* charts for the average number of defects per unit

All of these charts are based on the transformation

$$Z = \frac{\text{sample statistic} - \text{process average}}{\text{process standard deviation}} \tag{8.42}$$

In other words, stabilized charts are plots of the number of standard deviations (plus or minus) between the sample statistic and the long-term process average. Since control limits are conventionally set at ±3 standard deviations, stabilized control charts always have the lower control limit at −3 and the upper control limit at +3. Table 8.15 summarizes the control limit equations for stabilized control charts for attributes.

When applied to long runs, stabilized attribute charts are used to compensate for varying sample sizes; process averages are assumed to be constant. However, stabilized attribute charts can be created even if the process average varies. This is often done when applying this technique to short runs of parts that vary a great deal in average quality. For example, a wave soldering process used for several missiles had boards that varied in complexity from less than 100 solder joints to over 1,500 solder joints. Tables 8.16 and 8.17 show how the situation is handled to create a stabilized u chart. The unit size is 1,000 leads, set arbitrarily. It doesn't matter what the unit size is set to, the calculations will still produce the correct result since the actual number of leads is divided by the unit size selected. $\bar{u}$ is the average number of defects per 1,000 leads.

Example Four From the process described in Table 8.16, ten TOW missile boards of type E are sampled. Three defects were observed in the sample. Using Tables 8.15 and 8.16, Z is computed for the subgroup as follows:

$$\sigma = \sqrt{\bar{u}/n}, \text{ we get } \bar{u} = 2 \text{ from Table 12.17.}$$

$$n = \frac{50 \times 10}{1{,}000} = 0.5 \text{ units}$$

$$\sigma = \sqrt{2/0.5} = \sqrt{4} = 2$$

$$u = \frac{\text{number of defects}}{\text{number of units}} = \frac{3}{0.5} = 6 \text{ defects per unit}$$

$$Z = \frac{u - \bar{u}}{\sigma} = \frac{6-2}{2} = \frac{4}{2} = 2$$

Attribute	Chart	Sample Statistic	ProcessAverage	Process σ	Z
Proportion of defective units	p chart	p	$\bar{p}$	$\sqrt{\bar{p}(1-\bar{p})/n}$	$(p-\bar{p})/\sigma$
Number of defective units	np chart	np	$\overline{np}$	$\sqrt{\overline{np}(1-\bar{p})}$	$(np-\overline{np})/\sigma$
Defects per unit	c chart	c	$\bar{c}$	$\sqrt{\bar{c}}$	$(c-\bar{c})/\sigma$
Average defects per unit	u chart	u	$\bar{u}$	$\sqrt{\bar{u}/n}$	$(u-\bar{u})/\sigma$

TABLE 8.15 Stabilized Attribute Chart Statistics

Missile	Board	Leads	Units/Board	$\bar{u}$
Phoenix	A	1,650	1.65	16
	B	800	0.80	9
	C	1,200	1.20	9
TOW	D	80	0.08	4
	E	50	0.05	2
	F	100	0.10	1

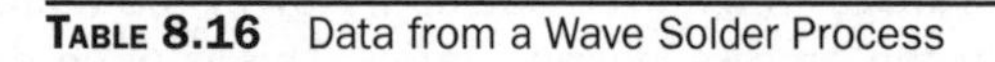
TABLE 8.16 Data from a Wave Solder Process

Since Z is between −3 and +3 we conclude that the process has not gone out of control; that is, it is not being influenced by a special cause of variation.

Table 8.17 shows the data for several samples from this process. The resulting control chart is shown in Fig. 8.24. Note that the control chart indicates that the process was better than average when it produced subgroups 2 and 3 and perhaps 4. Negative Z values mean that the defect rate is below (better than) the long-term process average. Groups 7 and 8 show an apparent deterioration in the process with group 7 being out of control. Positive Z values indicate a defect rate above (worse than) the long-term process average.

The ability to easily see process trends and changes like these in spite of changing part numbers and sample sizes is the big advantage of stabilized control charts. The disadvantages of stabilized control charts are:

1. They convert a number that is easy to understand, the number of defects or defectives, into a confusing statistic with no intuitive meaning.
2. They involve tedious calculation.

No.	Board	$\bar{u}$	Units	Sampled	n	σ	Defects	u	Z
1	E	2	0.05	10	0.50	2.00	3	6.00	2.00
2	A	16	1.65	1	1.65	3.11	8	4.85	−3.58
3	A	16	1.65	1	1.65	3.11	11	6.67	−3.00
4	B	9	0.80	1	0.80	3.35	0	0.00	−2.68
5	F	1	0.10	2	0.20	2.24	1	5.00	1.79
6	E	2	0.05	5	0.25	2.83	2	8.00	2.12
7	C	9	1.20	1	1.20	2.74	25	20.83	4.32
8	D	4	0.08	5	0.40	3.16	5	12.50	2.69
9	B	9	0.80	1	0.80	3.35	7	8.75	−0.07
10	B	9	0.80	1	0.80	3.35	7	8.75	−0.07

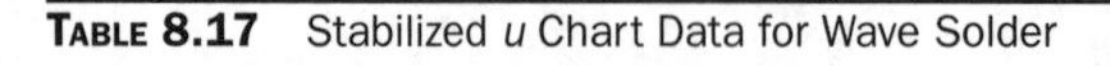
TABLE 8.17 Stabilized *u* Chart Data for Wave Solder

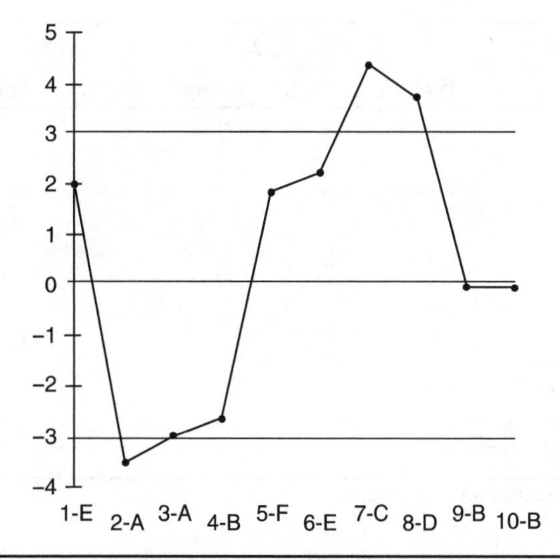

FIGURE 8.24 Control chart of Z values from Table 8.17.

Item 1 can only be corrected by training and experience applying the technique. Item 2 can be handled with computers; the calculations are simple to perform with a spreadsheet. Table 8.17 can be used as a guide to setting up the spreadsheet. Inexpensive handheld computers can be used to perform the calculations right at the process, thus making the results available immediately.

Demerit Control Charts

As described above, there are two kinds of data commonly used to perform SPC: variables data and attributes data. When short runs are involved we can seldom afford the information loss that results from using attribute data. However, the following are ways of extracting additional information from attribute data:

1. Making the attribute data "less discrete" by adding more classification categories.
2. Assigning weights to the categories to accentuate different levels of quality

Consider a process that involves fabricating a substrate for a hybrid micro-circuit. The surface characteristics of the substrate are extremely important. The "ideal part" will have a smooth surface, completely free of any visible flaws or blemishes. However, parts are sometimes produced with stains, pits, voids, cracks, and other surface defects. Although undesirable, most of the less than ideal parts are still acceptable to the customer.

If we were to apply conventional attribute SPC methods to this process the results would probably be disappointing. Since very few parts are actually rejected as unacceptable, a standard p chart or stabilized p chart would probably show a flat line at "zero defects" most of the time, even though the quality level might be less than the target ideal part. Variables SPC methods can't be used because attributes data such as "stains" are not easily measured on a variables scale. Demerit control charts offer an effective method of applying SPC in this situation.

To use demerit control charts we must determine how many imperfections of each type are found in the parts. Weights are assigned to the different categories. The quality

Subgroup Number →		1		2		3	
Attribute	**Weight**	**Freq.**	**Score**	**Freq.**	**Score**	**Freq.**	**Score**
Light stain	1	3	3				
Dark stain	5			1	5	1	5
Small blister	1			2	2	1	1
Medium blister	5	1	5				
Pit: 0.01–0.05 mm	1					3	3
Pit: 0.06–0.10 mm	5			2	10		
Pit: larger than 0.10 mm	10	1	10				
Total demerits →			18		17		9

TABLE 8.18 Demerit Scores for Substrates

score for a given sample is the sum of the weights times the frequencies of each category. Table 8.18 illustrates this approach for the substrate example.

If the subgroup size is kept constant, the average for the demerit control chart is computed as follows (Burr, 1976):

$$\text{Average} = \bar{D} = \frac{\text{sum of subgroup demerits}}{\text{number of subgroups}} \tag{8.43}$$

Control limits are computed in two steps. First compute the weighted average defect rate for each category. For example, there might be the following categories and weights:

Category	Weight
Major	10
Minor	5
Incidental	1

Three average defect rates, one each for major, minor, and incidental, could be computed using the following designations:

$\bar{c}_1$ = average number of major defects per subgroup

$\bar{c}_2$ = average number of minor defects per subgroup

$\bar{c}_3$ = average number of incidental defects per subgroup

The corresponding weights might be $W_1 = 10$, $W_2 = 5$, $W_3 = 1$. Using this notation we compute the demerit standard deviation for this three category example as

$$\sigma_D = \sqrt{W_1^2\bar{c}_1 + W_2^2\bar{c}_2 + W_3^2\bar{c}_3} \tag{8.44}$$

For the general case the standard deviation is

$$\sigma_D = \sqrt{\sum_{i=1}^{k} W_i^2\bar{c}_i} \tag{8.45}$$

The control limits are

$$LCL = \bar{D} - 3\sigma_D \quad (8.46)$$

$$UCL = \bar{D} + 3\sigma_D \quad (8.47)$$

If the lower control limit is negative, it is set to zero.

Simplified Quality Score Charts

The above procedure, while correct, may sometimes be too burdensome to implement effectively. When this is the case a simplified approach may be used. The simplified approach is summarized as follows:

1. Classify each part in the subgroup into the following classes (weights are arbitrary).

Class	Description	Points
A	Preferred quality. All product features at or very near targets.	10
B	Acceptable quality. Some product features have departed significantly from target quality levels, but they are a safe distance from the reject limits.	5
C	Marginal quality. One or more product features are in imminent danger of exceeding reject limits.	1
D	Reject quality. One or more product features fail to meet minimum acceptability requirements.	0

2. Plot the total scores for each subgroup, keeping the subgroup sizes constant.
3. Treat the total scores as if they were variables data and prepare an individuals and moving range control chart or an $\bar{X}$ and R chart. These charts are described in Pyzdek (1989) and in most texts on SPC.

Summary of Short-Run SPC

Small runs and short runs are common in modern business environments. Different strategies are needed to deal with these situations. Advance planning is essential. Special variables techniques were introduced which compensate for small sample sizes and short runs by using special tables or mathematically transforming the statistics and charts. Attribute short run SPC methods were introduced that make process patterns more evident when small runs are produced. Demerit and scoring systems were introduced that extract more information from attribute data.

SPC Techniques for Automated Manufacturing

Many people erroneously believe that statistics are not needed when automated manufacturing processes are involved. Since we have measurements from every unit produced, they reason, sampling methods are inappropriate. We will simply correct the process when the characteristic is not on target.

This attitude reflects a fundamental misunderstanding of the relationship between a process and the output of a process. It also shows a lack of appreciation for the intrinsic variability of processes and of measurements. The fact is, even if you have a "complete" data record of every feature of every part produced, you still have only a sample of the output of the process. The process is future-oriented in time, while the record of measurements is past-oriented. Unless statistical control is attained, you will be unable to use the data from past production to predict the variability from the process in the future (See "Common and Special Causes of Control" for the definition of control). And without statistical tools you have no sound basis for the belief that statistical control exists.

Another reason process control should be based on an understanding and correct use of statistical methods is the effect of making changes without this understanding. Consider, for example, the following process adjustment rule:

> Measure the diameter of the gear shaft. If the diameter is above the nominal size, adjust the process to reduce the diameter. If the diameter is below the nominal size, adjust the process to increase the diameter.

The problem with this approach is described by Deming's "funnel rules" (see above). This approach to process control will increase the variability of a statistically controlled process by 141%, certainly not what the process control analyst had in mind. The root of the problem is a failure to realize that the part measurement is a sample from the process and, although it provides information about the state of the process, the information is incomplete. Only through using proper statistical methods can the information be extracted, analyzed and understood.

Problems with Traditional SPC Techniques

A fundamental assumption underlying traditional SPC techniques is that the observed values are independent of one another. Although the SPC tools are quite insensitive to moderate violations of this assumption (Wheeler, 1991), automated manufacturing processes often breach the assumption by enough to make traditional methods fail (Alwan and Roberts, 1989). By using scatter diagrams, as described in Chap. 10, you can determine if the assumption of independence is satisfied for your data. If not, you should consider using the methods described below instead of the traditional SPC methods.

A common complaint about nonstandard SPC methods is that they are usually more complex than the traditional methods (Wheeler, 1991). This is often true. However, when dealing with automated manufacturing processes the analysis is usually handled by a computer. Since the complexity of the analysis is totally invisible to the human operator, it makes little difference. Of course, if the operator will be required to act based on the results, he or she must understand how the results are to be used. The techniques described in this chapter which require human action are interpreted in much the same way as traditional SPC techniques.

Special and Common Cause Charts

When using traditional SPC techniques the rules are always the same, namely

1. As long as the variation in the statistic being plotted remains within the control limits, leave the process alone.
2. If a plotted point exceeds a control limit, look for the cause.

This approach works fine as long as the process remains static. However, the means of many automated manufacturing processes often drift because of inherent process factors. In other words, the drift is produced by *common causes*. In spite of this, there may be known ways of intervening in the process to compensate for the drift. Traditionalists would say that the intervention should be taken in such a way that the control chart exhibits only random variation. However, this may involve additional cost. Mindlessly applying arbitrary rules to achieve some abstract result, like a stable control chart, is poor practice. All of the options should be considered.

One alternative is to allow the drift to continue until the cost of intervention equals the cost of running off-target. This alternative can be implemented through the use of a "common cause chart." This approach, described in Alwan and Roberts (1989) and Abraham and Whitney (1990), involves creating a chart of the process mean. However, unlike traditional $\bar{X}$ charts, there are no control limits. Instead, *action limits* are placed on the chart. Action limits differ from control limits in two ways:

1. They are computed based on costs rather than on statistical theory.

Since the chart shows variation from common causes, violating an action limit does not result in a search for a special cause. Instead, a prescribed action is taken to bring the process closer to the target value.

These charts are called "common cause charts" because the changing level of the process is due to built-in process characteristics. The process mean is tracked by using exponentially weighted moving averages (EWMA). While somewhat more complicated than $\bar{X}$ traditional charts, EWMA charts have a number of advantages for automated manufacturing:

- They can be used when processes have inherent drift.
- EWMA charts provide a forecast of where the next process measurement will be. This allows feed-forward control.
- EWMA models can be used to develop procedures for dynamic process control, as described later in this section.

EWMA Common Cause Charts

When dealing with a process that is essentially static, the predicted value of the average of every sample is simply the grand average. EWMA charts, on the other hand, use the actual process data to determine the predicted process value for processes that may be drifting. If the process has trend or cyclical components, the EWMA will reflect the effect of these components. Also, the EWMA chart produces a forecast of what the *next* sample mean will be; the traditional $\bar{X}$ chart merely shows what the process was doing at the time the sample was taken. Thus, the EWMA chart can be used to take preemptive action to prevent a process from going too far from the target.

If the process has inherent nonrandom components, an EWMA common cause chart should be used. This is an EWMA chart with economic action limits instead of control limits. EWMA control charts, which are described in the next section, can be used to monitor processes that vary within the action limits.

The equation for computing the EWMA is

$$\text{EWMA} = \hat{y}_t + \lambda(y_t - \hat{y}_t) \tag{8.48}$$

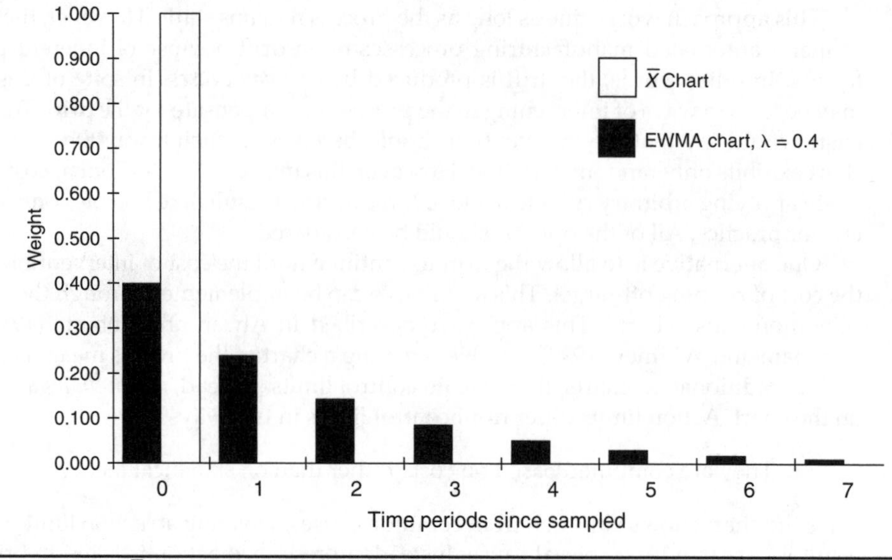

FIGURE 8.25 $\overline{X}$ versus EWMA weighting.

In this equation $\hat{y}_t$ is the predicted value of y at time period t, y_t is the actual value at time period t, and λ is a constant between 0 and 1. If λ is close to 1, Eq. (8.48) will give little weight to historic data; if λ is close to 0 then current observations will be given little weight. EWMA can also be thought of as the forecasted process value at time period $t + 1$, in other words, EWMA $= \hat{y}_{t+1}$.

Since most people already understand the traditional $\overline{X}$ chart, thinking about the relationship between $\overline{X}$ charts and EWMA charts can help you understand the EWMA chart. It is interesting to note that traditional $\overline{X}$ charts give 100% of the weight to the current sample and 0% to past data. This is roughly equivalent to setting $\lambda = 1$ on an EWMA chart. In other words, the traditional $\overline{X}$ chart can be thought of as a special type of EWMA chart where past data are considered to be unimportant (assuming run tests are not applied to the Shewhart chart). This is equivalent to saying that the data points are all independent of one another. In contrast, the EWMA chart uses the information from all previous samples. Although Eq. (8.48) may look as though it is only using the results of the most recent data point, in reality the EWMA weighting scheme applies progressively less weight to each sample result as time passes. Figure 8.25 compares the weighting schemes of EWMA and $\overline{X}$ charts.

In contrast, as λ approaches 0 the EWMA chart begins to behave like a cusum chart. With a cusum chart all previous points are given equal weight. Between the two extremes the EWMA chart weights historical data in importance somewhere between the traditional Shewhart chart and the cusum chart. By changing the value of λ the chart's behavior can be "adjusted" to the process being monitored.

In addition to the weighting, there are other differences between the EWMA chart and the $\overline{X}$ chart. The "forecast" from the $\overline{X}$ chart is always the same: the next data point will be equal to the historical grand average. In other words, the $\overline{X}$ chart treats all data points as coming from a process that doesn't change its central tendency (implied when the forecast is always the grand average).*

When using an $\bar{X}$ chart it is not essential that the sampling interval be kept constant. After all, the process is supposed to behave as if it were static. However, the EWMA chart is designed to account for process drift and, therefore, the sampling interval should be kept constant when using EWMA charts. This is usually not a problem with automated manufacturing.

Example

Krisnamoorthi (1991) describes a mold line that produces green sand molds at the rate of about one per minute. The molds are used to pour cylinder blocks for large size engines. Application of SPC to the process revealed that the process had an assignable cause that could not be eliminated from the process. The mold sand, which was partly recycled, tended to increase and decrease in temperature based on the size of the block being produced and the number of blocks in the order. Sand temperature is important because it affects the compactability percent, an important parameter. The sand temperature could not be better controlled without adding an automatic sand cooler, which was not deemed economical. However, the effect of the sand temperature on the compactability percent could be made negligible by modifying the amount of water added to the sand so feed-forward control was feasible.

Although Krishnamoorthi doesn't indicate that EWMA charts were used for this process, it is an excellent application for EWMA common cause charts. The level of the sand temperature doesn't really matter, as long as it is known. The sand temperature tends to drift in cycles because the amount of heated sand depends on the size of the casting and how many are being produced. A traditional control chart for the temperature would indicate that sand temperature is out-of-control, which we already know. What is really needed is a method to predict what the sand temperature will be the next time it is checked, then the operator can add the correct amount of water so the effect on the sand compactability percent can be minimized. This will produce an in-control control chart for compactability percent, which is what really matters.

The data in Table 8.19 show the EWMA calculations for the sand temperature data. Using a spreadsheet program, Microsoft Excel for Windows, the optimal value of λ, that is the value which provided the "best fit" in the sense that it produced the smallest sum of the squared errors, was found to be close to 0.9. Figure 8.26 shows the EWMA common cause chart for this data, and the raw temperature data as well. The EWMA is a forecast of what the sand temperature will be the next time it is checked. The operator can adjust the rate of water addition based on this forecast.

EWMA Control Limits

Although it is not always necessary to put control limits on the EWMA chart, as shown by the above example, it is possible to do so when the situation calls for it. Three sigma control limits for the EWMA chart are computed based on

$$\sigma^2_{\text{EWMA}} = \sigma^2\left[\frac{\lambda}{(2-\lambda)}\right] \tag{8.49}$$

*We aren't saying this situation actually exists, we are just saying that the $\bar{X}$ treats the process as if this were true. Studying the patterns of variation will often reveal clues to making the process more consistent, even if the process variation remains within the control limits.

Sand Temperature	EWMA	Error
125	125.00*	0.00
123	125.00	−2.00†
118	123.20‡	−5.20s
116	118.52	−2.52
108	116.25	−8.25
112	108.83	3.17
101	111.68	−10.68
100	102.07	−2.07
98	100.21	−2.21
102	98.22	3.78
111	101.62	9.38
107	110.6	−3.06
112	107.31	4.69
112	111.53	0.47
122	111.95	10.05
140	121.00	19.00
125	138.10	−13.10
130	126.31	3.69
136	129.63	6.37
130	135.36	−5.36
112	130.54	−18.54
115	113.85	1.15
100	114.89	−14.89
113	101.49	11.51
111	111.85	−0.85
128	111.08	16.92
122	126.31	−4.31
142	122.43	19.57
134	140.64	−6.04
130	134.60	−4.60
131	130.46	0.54
104	130.95	−26.95
84	106.69	−22.69
86	86.27	−0.27
99	86.03	12.97
90	97.70	−7.70
91	90.77	0.23
90	90.98	−0.98
101	90.10	10.90

*The starting EWMA is either the target, or, if there is no target, the first observation.

†Error = Actual observation − EWMA. E.g., −2 = 123 − 125.

‡Other than the first sample, all EWMAs are computed as EWMA = last EWMA + $\lambda \times$ error. E.g., 123.2 = 125 + 0.9 × (−2).

TABLE 8.19 Data for EWMA Chart of Sand Temperature

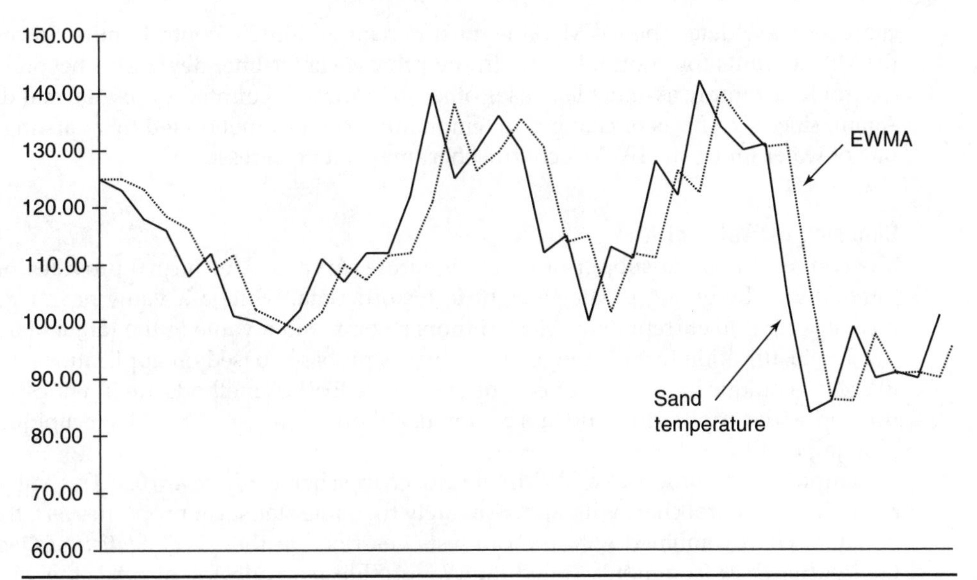

FIGURE 8.26 EWMA chart of sand temperature.

For the sand temperature example above, $\lambda = 0{:}9$ which gives

$$\sigma^2_{\text{EWMA}} = \sigma^2\left[\frac{0.9}{(2-0.9)}\right] = 0.82\sigma^2$$

σ^2 is estimated using all of the data. For the sand temperature data $\sigma = 15.37$ so EWMA $\sigma\,\text{EWMA} = 15.37 \times \sqrt{0.82} = 13.92$. The 3σ control limits for the EWMA chart are placed at the grand average plus and minus 41.75. Figure 8.27 shows the control

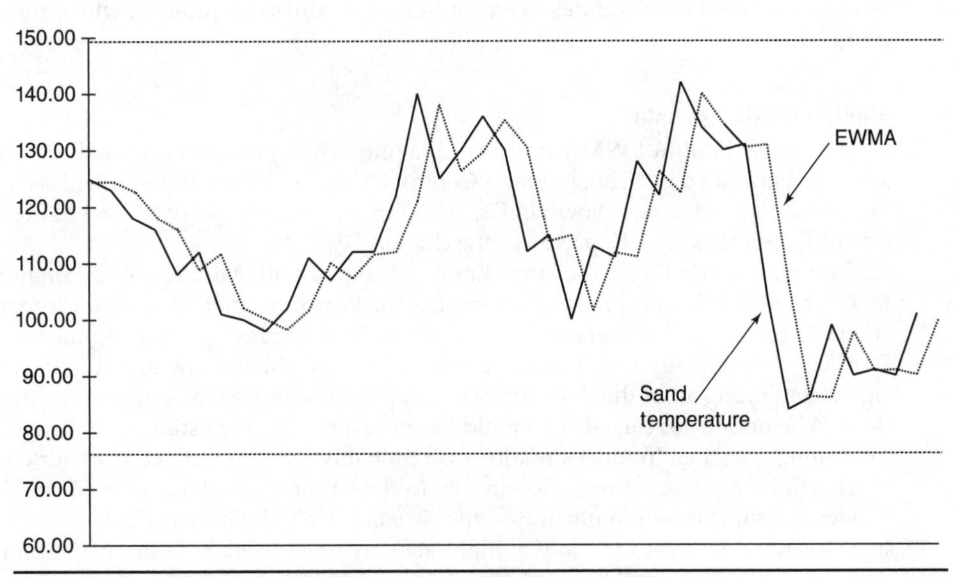

FIGURE 8.27 EWMA control chart of sand temperature.

chart for these data. The EWMA line must remain within the control limits. Since the EWMA accounts for "normal drift" in the process center line, deviations beyond the control limits imply assignable causes other than those accounted for by normal drift. Again, since the effects of changes in temperature can be ameliorated by adjusting the rate of water input, the EWMA control chart may not be necessary.

Choosing the Value of λ

The choice of λ is the subject of much literature. A value λ of near 0 provides more "smoothing" by giving greater weight to historic data, while a λ value near 1 gives greater weight to current data. Most authors recommend a value in the range of 0.2 to 0.3. The justification for this range of λ values is probably based on applications of the EWMA technique in the field of economics, where EWMA methods are in widespread use. Industrial applications are less common, although the use of EWMA techniques is growing rapidly.

Hunter (1989) proposes a EWMA control chart scheme where $\lambda = 0.4$. This value of λ provides a control chart with approximately the same statistical properties as a traditional $\bar{X}$ chart combined with the run tests described in the *AT&T Statistical Quality Control Handbook* (commonly called the Western Electric Rules). It also has the advantage of providing control limits that are exactly half as wide as the control limits on a traditional $\bar{X}$ chart. Thus, to compute the control limits for an EWMA chart when λ is 0.4 you simply compute the traditional $\bar{X}$ chart (or X chart) control limits and divide the distance between the upper and lower control limits by two. The EWMA should remain within these limits.

As mentioned above, the optimal value of λ can be found using some spreadsheet programs. The sum of the squared errors is minimized by changing the value of λ. If your spreadsheet doesn't automatically find the minimum, it can be approximated manually by changing the cell containing λ or by setting up a range of λ values and watching what happens to the cell containing the sum of the squared errors. A graph of the error sum of the squares versus different λ values can indicate where the optimum λ lies.

Minitab EWMA Example

Minitab has a built-in EWMA analysis capability. We will repeat our analysis for the sand temperature data. Choose Stat > Control Charts > EWMA and you will see a dialog box similar to the one shown in Fig. 8.28. Entering the weight of 0.9 and a subgroup size of 1, then clicking OK, produces the chart in Fig. 8.29.

You may notice that the control limits calculated with Minitab are different than those calculated in the previous example. The reason is that Minitab's estimate of sigma is based on the average moving range. This method gives a sigma value of 7.185517, substantially less than the estimate of 15.37 obtained by simply calculating sigma combining all of the data. Minitab's approach removes the effect of the process drift. Whether or not this effect should be removed from the estimate of sigma is an interesting question. In most situations we probably want to remove it so our control chart will be more sensitive, allowing us to detect more special causes for removal. However, as this example illustrates, the situation isn't always clear cut. In the situation described by the example we might actually want to include the variation from drift into the control limit calculations to prevent operator tampering.

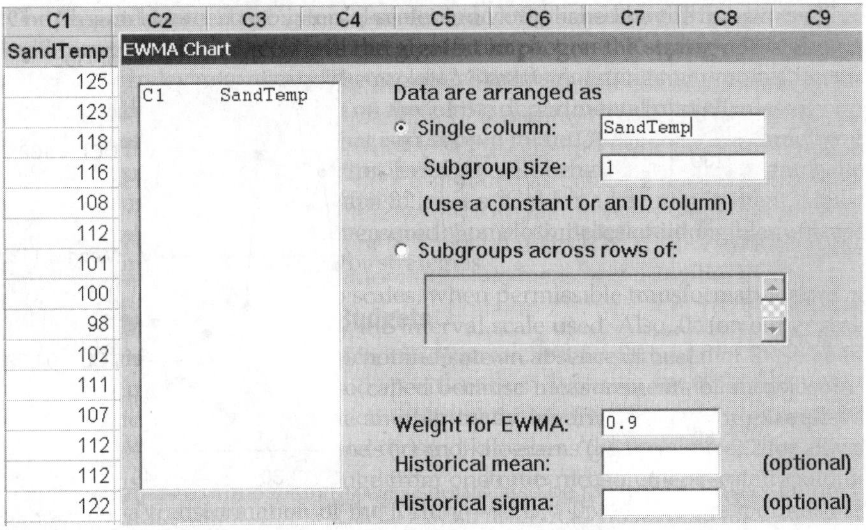

Figure 8.28 Minitab EWMA dialog box.

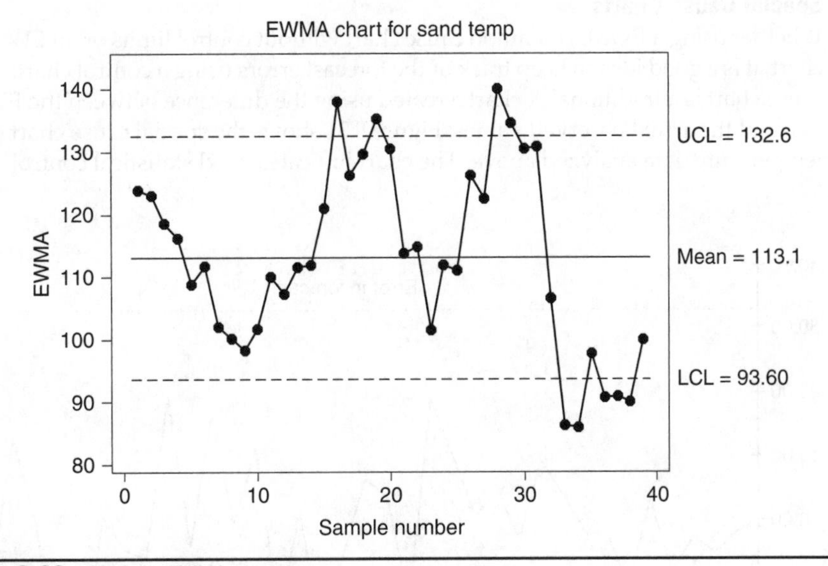

Figure 8.29 Minitab EWMA chart.

EWMA Control Charts versus Individuals Charts

In many cases an individuals control chart (*I* chart) will give results comparable to the EWMA control chart. When this is the case it is usually best to opt for the simpler *I* chart. An *I* chart is shown in Fig. 8.30 for comparison with the EWMA chart. The results are very similar to the EWMA chart from Minitab.

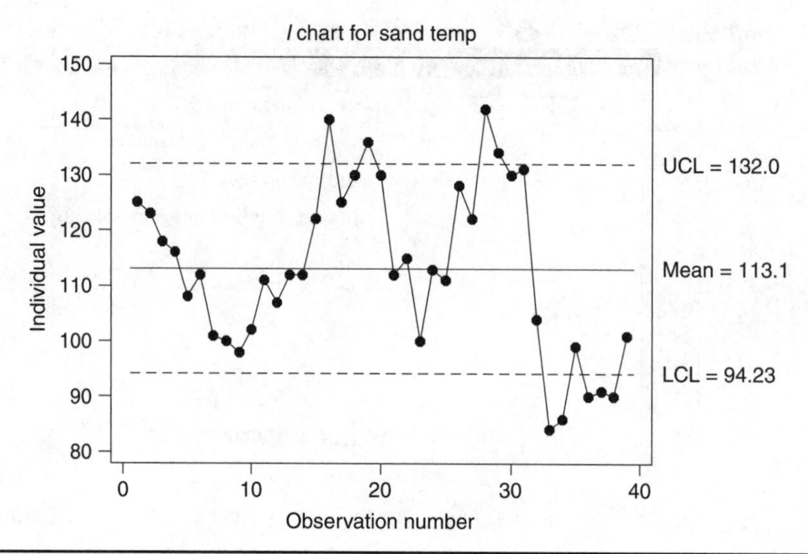

FIGURE 8.30 *I* chart for sand temperature.

Special Cause Charts

Whether using a EWMA common cause chart without control limits or an EWMA control chart, it is a good idea to keep track of the forecast errors using a control chart. The special cause chart is a traditional *X* chart, created using the difference between the EWMA forecast and the actual observed values. Figure 8.31 shows the special cause chart of the sand temperature data analyzed above. The chart indicates good statistical control.

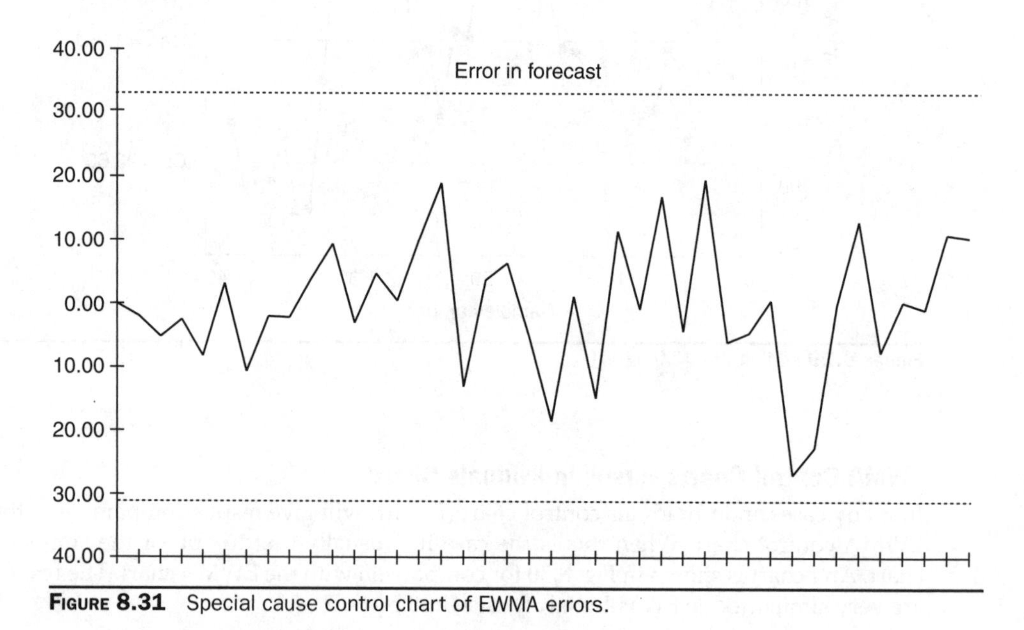

FIGURE 8.31 Special cause control chart of EWMA errors.

SPC and Automatic Process Control

As SPC has grown in popularity its use has been mandated with more and more processes. When this trend reached automated manufacturing processes there was resistance from process control analysts who were applying a different approach with considerable success (Palm, 1990). Advocates of SPC attempted to force the use of traditional SPC techniques as feedback mechanisms for process control. This inappropriate application of SPC was correctly denounced by process control analysts. SPC is designed to serve a purpose fundamentally different than automatic process control (APC). SPC advocates correctly pointed out that APC was not a cure-all and that many process controllers added variation by making adjustments based on data analysis that was statistically invalid.

Both SPC and APC have their rightful place in Six Sigma. APC attempts to dynamically control a process to minimize variation around a target value. This requires valid statistical analysis, which is the domain of the statistical sciences. SPC makes a distinction between special causes and common causes of variation. If APC responds to all variation as if it were the same it will result in missed opportunities to reduce variation by attacking it at the source. A process that operates closer to the target without correction will produce less variation overall than a process that is frequently returned to the target via APC. However, at times APC must respond to *common cause variation that can't be economically eliminated*, for example, the mold process described above. Properly used, APC can greatly reduce variability in the output.

Hunter (1986) shows that there is a statistical equivalent to the PID control equation commonly used. The PID equation is

$$u(t) = Ke(t) + \frac{K}{T_1}\int_0^1 e(s)ds + KT_D\left(\frac{d_e}{d_t}\right) \tag{8.50}$$

The "PID" label comes from the fact that the first term is a proportional term, the second an integral term and the third a derivative term. Hunter modified the basic EWMA equation by adding two additional terms. The result is the *empirical control equation*.

$$\hat{y}_{t+1} = \hat{y}_t + \lambda e_t + \lambda_2 \Sigma e_t + \lambda_3 \nabla e_t \tag{8.51}$$

The term ∇e_t means the first difference of the errors e_t, that is, $\nabla e_t = e_t - e_{t-1}$. Like the PID equation, the empirical control equation has a proportional, an integral and a differential term. It can be used by APC or the results can be plotted on a common cause chart and reacted to by human operators, as described above. A special cause chart can be created to track the errors in the forecast from the empirical control equation. Such an approach may help to bring SPC and APC together to work on process improvement.

Distributions

Methods of Enumeration

Enumeration involves counting techniques for very large numbers of possible outcomes. This occurs for even surprisingly small sample sizes. In Six Sigma, these methods are commonly used in a wide variety of statistical procedures.

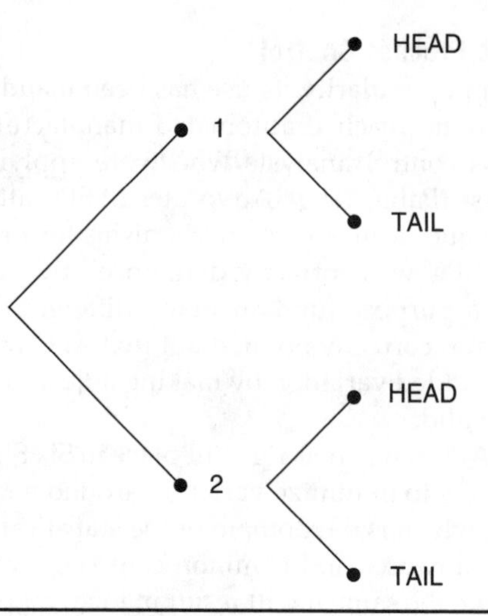

FIGURE 8.32 Multiplication principle applied to coin flips.

The basis for all of the enumerative methods described here is the multiplication principle. The multiplication principle states that the number of possible outcomes of a series of experiments is equal to the product of the number of outcomes of each experiment. For example, consider flipping a coin twice. On the first flip there are two possible outcomes (heads/tails) and on the second flip there are also two possible outcomes. Thus, the series of two flips can result in $2 \times 2 = 4$ outcomes. Figure 8.32 illustrates this example.

An ordered arrangement of elements is called a *permutation*. Suppose that you have four objects and four empty boxes, one for each object. Consider how many different ways the objects can be placed into the boxes. The first object can be placed in any of the four boxes. Once this is done there are three boxes to choose from for the second object, then two boxes for the third object and finally one box left for the last object. Using the multiplication principle you find that the total number of arrangements of the four objects into the four boxes is $4 \times 3 \times 2 \times 1 = 24$. In general, if there are n positions to be filled with n objects there are

$$n(n-1)\ldots(2)(1) = n! \tag{8.52}$$

possible arrangements. The symbol $n!$ is read n factorial. By definition, $0! = 1$.

In applying probability theory to discrete variables in quality control we frequently encounter the need for efficient methods of counting. One counting technique that is especially useful is combinations. The combination formula is shown in Eq. (8.53).

$$C_r^n = \frac{n!}{r!(n-r)!} \tag{8.53}$$

Combinations tell how many unique ways you can arrange n objects taking them in groups of r objects at a time, where r is a positive integer less than or equal to n. For example, to determine the number of combinations we can make with the letters X, Y, and Z in groups of 2 letters at a time, we note that $n = 3$ letters, $r = 2$ letters at a time and use the above formula to find

$$C_2^3 = \frac{3!}{2!(3-2)!} = \frac{3 \times 2 \times 1}{(2 \times 1)(1)} = \frac{6}{2} = 3$$

The 3 combinations are XY, XZ, and YZ. Notice that this method does not count reversing the letters as separate combinations, that is, XY and YX are considered to be the same.

Frequency and Cumulative Distributions

Distributions are a set of numbers collected from a well-defined universe of possible measurements arising from a property or relationship under study. Distributions show the way in which the probabilities are associated with the numbers being studied. Assuming a state of statistical control, by consulting the appropriate distribution one can determine the answer to such questions as:

- What is the probability that x will occur?
- What is the probability that a value less than x will occur?
- What is the probability that a value greater than x will occur?
- What is the probability that a value will occur that is between x and y?

By examining plots of the distribution shape, one can determine how rapidly or slowly probabilities change over a given range of values. In short, distributions provide a great deal of information.

A frequency distribution is an empirical presentation of a set of observations. If the frequency distribution is *ungrouped*, it simply shows the observations and the frequency of each number. If the frequency distribution is *grouped*, then the data are assembled into cells, each cell representing a subset of the total range of the data. The frequency in each cell completes the grouped frequency distribution. Frequency distributions are often graphically displayed in histograms or stem-and-leaf plots.

While histograms and stem-and-leaf plots show the frequency of specific values or groups of values, analysts often wish to examine the *cumulative frequency* of the data. The cumulative frequency refers to the total up to and including a particular value. In the case of grouped data, the cumulative frequency is computed as the total number of observations up to and including a cell boundary. Cumulative frequency distributions are often displayed on an *ogive*, as depicted in Fig. 8.33.

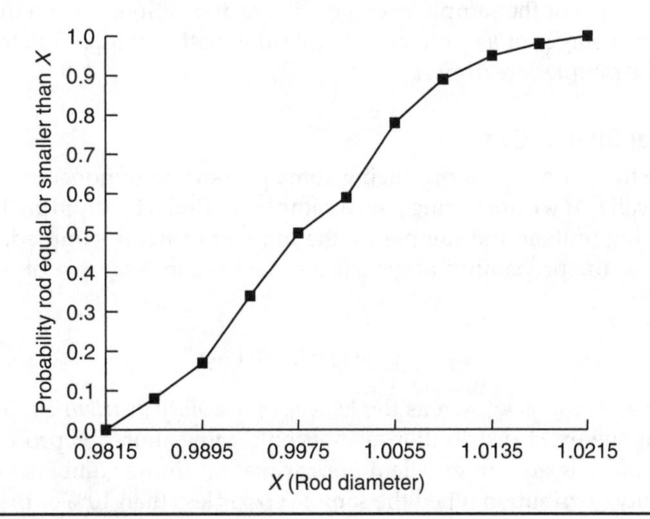

FIGURE 8.33 Ogive of rod diameter data.

Sampling Distributions

In most Six Sigma projects involving enumerative statistics, we deal with samples, not populations. We now consider the estimation of certain characteristics or parameters of the distribution from the data.

The empirical distribution assigns the probability $1/n$ to each X_i in the sample, thus the mean of this distribution is

$$\bar{X} = \sum_{i=1}^{n} X_i \frac{1}{n} \tag{8.54}$$

The symbol $\bar{X}$ is called "X bar." Since the empirical distribution is determined by a sample, $\bar{X}$ is simply called the *sample mean*.

The sample variance is given by

$$S^2 = \frac{1}{n-1} \sum_{i=1}^{n} (X_i - \bar{X})^2 \tag{8.55}$$

This equation for S^2 is commonly referred to as the *unbiased sample variance*. The *sample standard deviation* is given by

$$S = \sqrt{S^2} = \sqrt{\frac{\sum_{i=1}^{n} (X_i - \bar{X})^2}{n-1}} \tag{8.56}$$

Another sampling statistic of special interest in Six Sigma is the standard deviation of the sample average, also referred to as the *standard error of the mean* or simply the standard error. This statistic is given by

$$S_{\bar{X}} = \frac{S}{\sqrt{n}} \tag{8.57}$$

As can be seen, the standard error of the mean is inversely proportional to the square root of the sample size. That is, the larger the sample size, the smaller the standard deviation of the sample average. This relationship is shown in Fig. 8.34. It can be seen that averages of $n = 4$ have a distribution half as variable as the population from which the samples are drawn.

Binomial Distribution

Assume that a process is producing some proportion of nonconforming units, which we will call p. If we are basing p on a sample we find p by dividing the number of nonconforming units in the sample by the number of items sampled. The equation that will tell us the probability of getting x defectives in a sample of n units is shown by Eq. (8.58).

$$P(x) = C_x^n p^x (1-p)^{n-x} \tag{8.58}$$

This equation is known as the *binomial probability distribution*. In addition to being useful as the exact distribution of nonconforming units for processes in continuous production, it is also an excellent approximation to the cumbersome hypergeometric probability distribution when the sample size is less than 10% of the lot size.

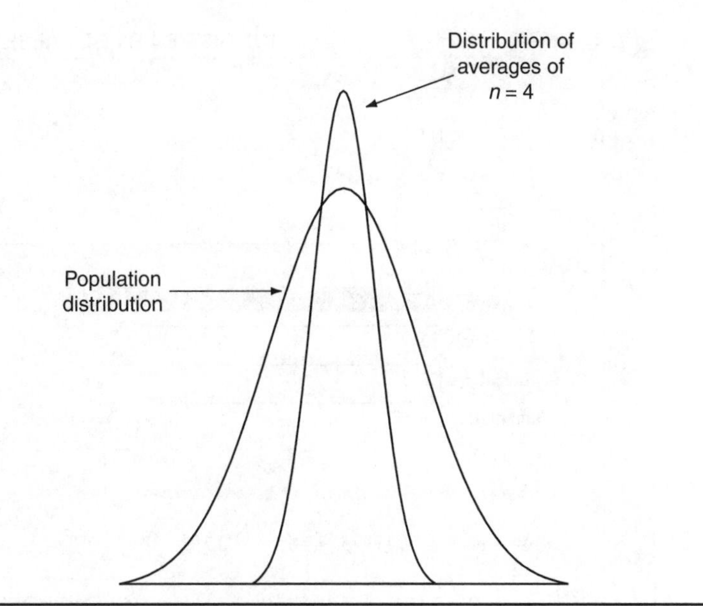

FIGURE 8.34 Effect of sample size on the standard error.

Example of Applying the Binomial Probability Distribution

A process is producing glass bottles on a continuous basis. Past history shows that 1% of the bottles have one or more flaws. If we draw a sample of 10 units from the process, what is the probability that there will be 0 nonconforming bottles?

Using the above information, $n = 10$, $p = .01$, and $x = 0$. Substituting these values into Eq. (8.58) gives us

$$p(0) = C_0^{10} 0.01^0 (1 - 0.01)^{10-0} = 1 \times 1 \times 0.99^{10} = 0.904 = 90.4\%$$

Another way of interpreting the above example is that a sampling plan "inspect 10 units, accept the process if no nonconformances are found" has a 90.4% probability of accepting a process that is averaging 1% nonconforming units.

Example of Binomial Probability Calculations Using Microsoft Excel®

Microsoft Excel has a built-in capability to analyze binomial probabilities. To solve the above problem using Excel, enter the sample size, p value, and x value as shown in Fig. 8.35. Note the formula result near the bottom of the screen.

Poisson Distribution

Another situation encountered often in quality control is that we are not just concerned with *units* that don't conform to requirements, instead we are concerned with the number of nonconformances themselves. For example, let's say we are trying to control the quality of a computer. A complete audit of the finished computer would almost certainly reveal some nonconformances, even though these nonconformances might be of minor importance (for example, a decal on the back panel might not be perfectly straight). If we tried to use the hypergeometric or binomial probability distributions to evaluate sampling plans for this situation, we would find they didn't work because our

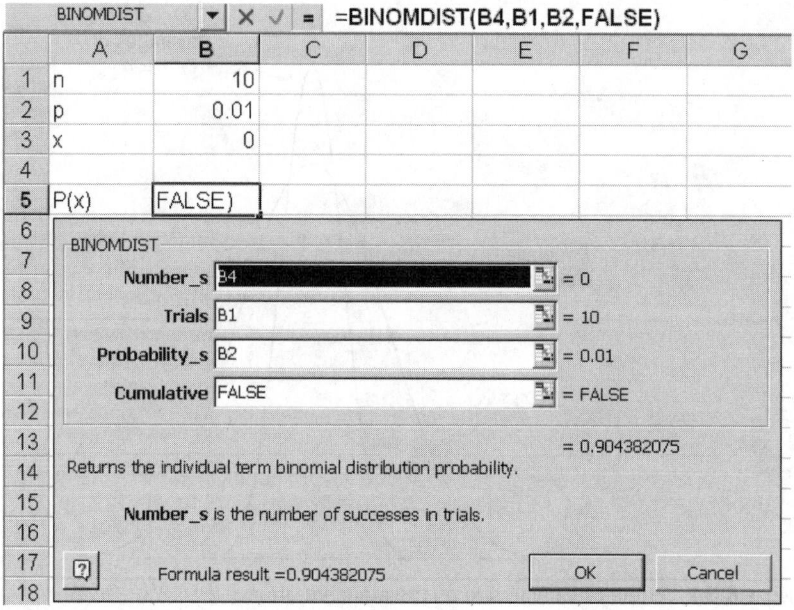

Figure 8.35 Example of finding binomial probability using Microsoft Excel.

lot or process would be composed of 100% nonconforming units. Obviously, we are interested not in the units per se, but in the non-conformances themselves. In other cases, it isn't even possible to count sample units per se. For example, the number of accidents must be counted as occurrences. The correct probability distribution for evaluating counts of non-conformances is the *Poisson distribution*. The pdf is given in Eq. (8.59).

$$p(x) = \frac{\mu^x e^{-\mu}}{x!} \tag{8.59}$$

In Eq. (8.59), μ is the average number of nonconformances per unit, x is the number of nonconformances in the sample, and e is the constant approximately equal to 2.7182818. $P(x)$ gives the probability of exactly x occurrences in the sample.

Example of Applying the Poisson Distribution

A production line is producing guided missiles. When each missile is completed, an audit is conducted by an Air Force representative and every nonconformance to requirements is noted. Even though any major nonconformance is cause for rejection, the prime contractor wants to control minor nonconformances as well. Such minor problems as blurred stencils, small burrs, etc., are recorded during the audit. Past history shows that on the average each missile has 3 minor nonconformances. What is the probability that the next missile will have 0 nonconformances?

We have $\mu = 3$, $x = 0$. Substituting these values into Eq. (8.59) gives us

$$P(0) = \frac{3^0 e^{-3}}{0!} = \frac{1 \times 0.05}{1} = 0.05 = 5\%$$

In other words, 100% − 5% = 95% of the missiles will have at least one nonconformance.

The Poisson distribution, in addition to being the exact distribution for the number of non-conformances, is also a good approximation to the binomial distribution in certain cases. To use the Poisson approximation, you simply let $\mu = np$ in Eq. (8.59). Juran (1988) recommends considering the Poisson approximation if the sample size is at least 16, the population size is at least 10 times the sample size, and the probability of occurrence p on each trial is less than 0.1. The major advantage of this approach is that it allows you to use the tables of the Poisson distribution, such as in Appendix 7. Also, the approach is useful for designing sampling plans.

Example of Poisson Probability Calculations Using Microsoft Excel

Microsoft Excel has a built-in capability to analyze Poisson probabilities. To solve the above problem using Excel, enter the average and x values as shown in Fig. 8.36. Note the formula result near the bottom of the screen.

Hypergeometric Distribution

Assume we have received a lot of 12 parts from a distributor. We need the parts badly and are willing to accept the lot if it has fewer than 3 nonconforming parts. We decide to inspect only 4 parts since we can't spare the time to check every part. Checking the sample, we find 1 part that doesn't conform to the requirements. Should we reject the remainder of the lot?

This situation involves sampling without replacement. We draw a unit from the lot, inspect it, and draw another unit from the lot. Furthermore, the lot is quite small, the sample is 25% of the entire lot. The formula needed to compute probabilities for this

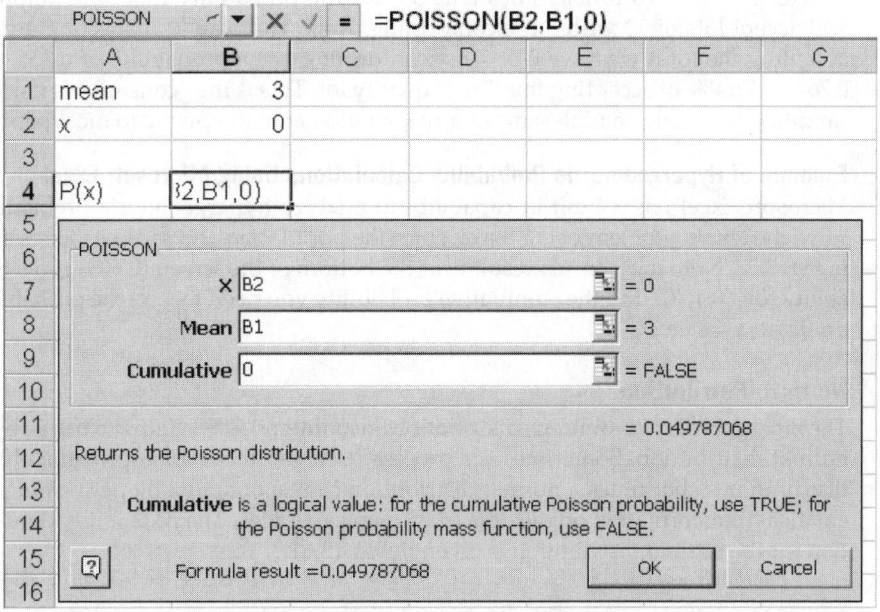

POISSON =POISSON(B2,B1,0)

	A	B	C	D	E	F	G
1	mean	3					
2	x	0					
3							
4	P(x)	32,B1,0)					

POISSON
X B2 = 0
Mean B1 = 3
Cumulative 0 = FALSE
= 0.049787068
Returns the Poisson distribution.
Cumulative is a logical value: for the cumulative Poisson probability, use TRUE; for the Poisson probability mass function, use FALSE.
Formula result =0.049787068 OK Cancel

FIGURE 8.36 Example of finding Poisson probability using Microsoft Excel.

procedure is known as the hypergeometric probability distribution, and it is shown in Eq. (8.60).

$$p(x) = \frac{C_{n-x}^{N-m} C_x^m}{C_n^N} \quad (8.60)$$

In Eq. (8.60), N is the lot size, m is the number of defectives in the lot, n is the sample size, x is the number of defectives in the sample, and $P(x)$ is the probability of getting exactly x defectives in the sample. Note that the numerator term C_{n-x}^{N-m} gives the number of combinations of non-defectives while C_x^m is the number of combinations of defectives. Thus the numerator gives the total number of arrangements of samples from lots of size N with m defectives where the sample n contains exactly x defectives. The term C_n^N the denominator is the total number of combinations of samples of size n from lots of size N, regardless of the number of defectives. Thus, the probability is a ratio of the likelihood of getting the result under the assumed conditions.

For our example, we must solve the above equation for $x = 0$ as well as $x = 1$, since we would also accept the lot if we had no defectives. The solution is shown as follows.

$$P(0) = \frac{C_{4-0}^{12-3} C_0^3}{C_4^{12}} = \frac{126 \times 1}{495} = 0.255$$

$$P(1) = \frac{C_{4-1}^{12-3} C_1^3}{C_4^{12}} = \frac{84 \times 3}{495} = \frac{252}{495} = 0.509$$

$$P(1 \text{ or less}) = P(0) + P(1)$$

Adding the two probabilities tells u0000s the probability that our sampling plan will accept lots of 12 with 3 nonconforming units. The plan of inspecting 4 parts and accepting the lot if we have 0 or 1 nonconforming has a probability of 0.255 + 0.509 = 0.764, or 76.4%, of accepting this "bad" quality lot. This is the "consumer's risk" for this sampling plan. Such a high sampling risk would be unacceptable to most people.

Example of Hypergeometric Probability Calculations Using Microsoft Excel

Microsoft Excel has a built-in capability to analyze hypergeometric probabilities. To solve the above problem using Excel, enter the population and sample values as shown in Fig. 8.37. Note the formula result near the bottom of the screen (0.509) gives the probability for $x = 1$. To find the cumulative probability you need to sum the probabilities for $x = 0$ and $x = 1$ etc.

Normal Distribution

The most common continuous distribution encountered in Six Sigma work is, by far, the normal distribution. Sometimes the process itself produces an approximately normal distribution, other times a normal distribution can be obtained by performing a mathematical transformation on the data or by using averages. The probability density function for the normal distribution is given by Eq. (8.61).

$$f(x) = \frac{1}{\sigma\sqrt{2\pi}} e^{-(x-\mu)^2/2\sigma^2} \quad (8.61)$$

	A	B	C
21	**Prob of less than or equal to sample x in a sample of n from a population N with population X successes**		
22	Population N	12	
23	Population X	3	
24	Sample n	4	
25	Answer	0.763636364	cop
26	Sample x	0	1
27	Prob	0.254545455	,$B$22)

HYPGEOMDIST

Sample_s C$26 = 1

Number_sample $B$24 = 4

Population_s $B$23 = 3

Number_pop $B$22 = 12

= 0.509090909

Returns the hypergeometric distribution.

Sample_s is the number of successes in the sample.

Formula result =0.509090909 OK Cancel

Figure 8.37 Example of finding hypergeometric probability using Microsoft Excel.

If $f(x)$ is plotted versus x, the well-known "bell curve" results. The normal distribution is also known as the Gaussian distribution. An example is shown in Fig. 8.38.

In Eq. (8.61), μ is the population average or mean and σ is the population standard deviation. These parameters have been discussed earlier in this chapter.

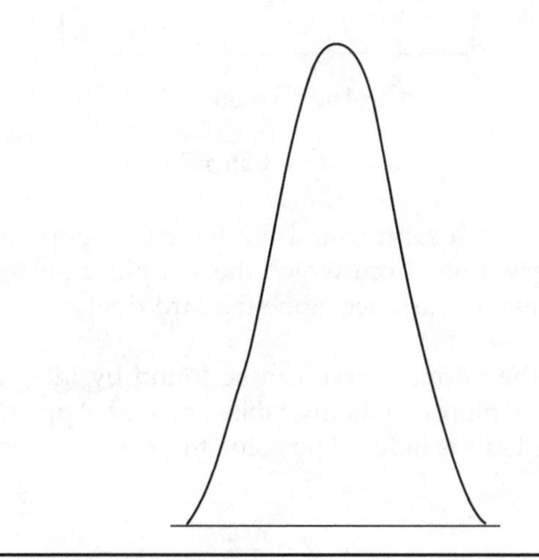

Figure 8.38 The normal/Gaussian curve.

Example of Calculating μ, σ^2, and σ

Find μ, σ^2, and σ for the following data:

Table 8.20 gives the equation for the population mean as:

i	x_i
1	17
2	23
3	5

$$\mu = \frac{1}{N}\sum_{i=1}^{N} x_i \tag{8.62}$$

To find the mean for our data we compute

$$\mu = \frac{1}{3}(17+23+5) = 15$$

The variance and standard deviation are both measures of dispersion or spread. The equations for the population variance σ^2 and standard deviation σ are given in Table 8.21.

$$\sigma^2 = \sum\nolimits_{i=i}^{N} \frac{(x_i - \mu)^2}{N}$$

$$\sigma = \sqrt{\sigma^2} \tag{8.63}$$

Referring to the data above with a mean μ of 15, we compute σ^2 and σ as follows:

i	x_i	$x_i - \mu$	$(x_i - \mu)^2$
1	17	2	4
2	23	8	64
3	5	−10	100
			Sum 168

$$\sigma^2 = 168/3 = 56$$

$$\sigma = \sqrt{\sigma^2} = \sqrt{56} \approx 7.483$$

Usually we have only a sample and not the entire population. A population is the entire set of observations from which the sample, a subset, is drawn. Calculations for the sample mean, variance, and standard deviation were shown earlier in this chapter.

The areas under the normal curve can be found by integrating Eq. (8.61) using numerical methods, but, more commonly, tables are used. Appendix 2 gives areas under the normal curve. The table is indexed by using the Z transformation, which is

$$Z = \frac{x_{i-\mu}}{\sigma} \tag{8.64}$$

for population data, or

$$Z = \frac{x_i - \overline{X}}{s} \tag{8.65}$$

for sample data.

By using the Z transformation, we can convert any normal distribution into a normal distribution with a mean of 0 and a standard deviation of 1. Thus, we can use a single normal table to find probabilities.

Example

The normal distribution is very useful in predicting long-term process yields. Assume we have checked the breaking strength of a gold wire bonding process used in microcircuit production and we have found that the process average strength is 9# and the standard deviation is 4#. The process distribution is normal. If the engineering specification is 3# minimum, what percentage of the process will be below the low specification?

Since our data are a sample, we must compute Z using Eq. (8.65).

$$Z = \frac{3-9}{4} = \frac{-6}{4} = -1.5$$

Figure 8.39 illustrates this situation.

Entering in Appendix 2 for Z = −1:5, we find that 6.68% of the area is below this Z value. Thus 6.68% of our breaking strengths will be below our low specification limit of 3. In quality control applications, we usually try to have the average at least three standard deviations away from the specification. To accomplish this, we would have to improve the process by either raising the average breaking strength or reducing the process standard deviation, or both.

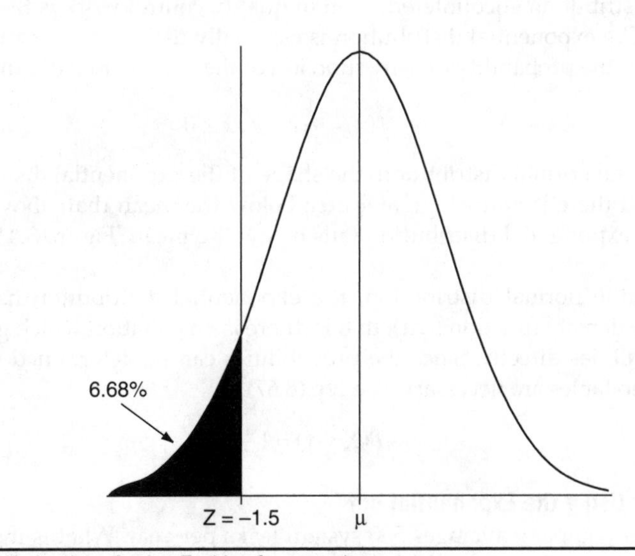

FIGURE 8.39 Illustration of using *Z* tables for normal areas.

NORMDIST =NORMDIST(B3,B1,B2,1)

	A	B	C	D	E	F
1	Average	9				
2	Sigma	4				
3	x	3				
4						
5	P(less than x)	31,B2,1)				

NORMDIST

X B3 = 3

Mean B1 = 9

Standard_dev B2 = 4

Cumulative 1 = TRUE

= 0.066807229

Returns the normal cumulative distribution for the specified mean and standard deviation.

Cumulative is a logical value: for the cumulative distribution function, use TRUE; for the probability mass function, use FALSE.

Formula result =0.066807229 OK Cancel

FIGURE 8.40 Example of finding normal probability using Microsoft Excel.

Example of Normal Probability Calculations Using Microsoft Excel

Microsoft Excel has a built-in capability to analyze normal probabilities. To solve the above problem using Excel, enter the average, sigma and x values as shown in Fig. 8.40. The formula result near the bottom of the screen gives the desired probability.

Exponential Distribution

Another distribution encountered often in quality control work is the exponential distribution. The exponential distribution is especially useful in analyzing reliability. The equation for the probability density function of the exponential distribution is

$$f(x) = \frac{1}{\mu} e^{-x/\mu}, \; x \geq 0 \tag{8.66}$$

Unlike the normal distribution, the shape of the exponential distribution is highly skewed and there is a much greater area below the mean than above it. In fact, over 63% of the exponential distribution falls below the mean. Figure 8.41 shows an exponential pdf.

Unlike the normal distribution, the exponential distribution has a closed form cumulative density function (cdf), that is, there is an equation which gives the cumulative probabilities directly. Since the probabilities can be determined directly from the equation, no tables are necessary. See Eq. (8.67).

$$P(X \leq x) = 1 - e^{-x/\mu} \tag{8.67}$$

Example of Using the Exponential cdf

A city water company averages 500 system leaks per year. What is the probability that the weekend crew, which works from 6 p.m. Friday to 6 a.m. Monday, will get no calls?

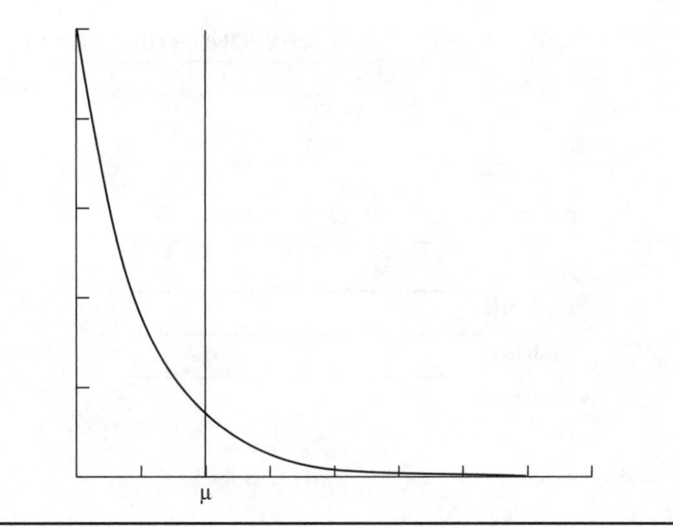

Figure 8.41 Exponential pdf curve.

We have μ = 500 leaks per year, which we must convert to leaks per hour. There are 365 days of 24 hours each in a year, or 8760 hours. Thus, mean time between failures (MTBF) is 8760/500 = 17.52 hours. There are 60 hours between 6 p.m. Friday and 6 a.m. Monday. Thus $x = 60$. Using Eq. (8.67) gives

$$P(X \leq 60) = 1 - e^{-60/17.52} = 0.967 = 96.7\%$$

Thus, the crew will get to loaf away 3.3% of the weekends.

Example of Exponential Probability Calculations Using Microsoft Excel

Microsoft Excel has a built-in capability to analyze exponential probabilities. To solve the above problem using Excel, enter the average and x values as shown in Fig. 8.42. Note that Excel uses "lambda" rather than the average in its calculations; lambda is the reciprocal of the average. The formula result near the bottom of the screen gives the desired probability.

Example of Non-Normal Capability Analysis Using Minitab

Minitab has a built-in capability to perform process capability analysis for non-normal data which will be demonstrated with an example. The process involved is technical support by telephone. A call center has recorded the total time it takes to "handle" 500 technical support calls. Handle time is a total cycle time metric which includes gathering preliminary information, addressing the customer's issues, and performing post-call tasks. It is a CTQ metric that also impacts the shareholder. It has been determined that the upper limit on handle time is 45 minutes. Once the data has been collected, it can be analyzed as follows:

Phase 1—Check for Special Causes: To begin we must determine if special causes of variation were present during our study. A special cause is operationally defined as

EXPONDIST =EXPONDIST(B2,1/B1,1)

	A	B	C	D	E	F
1	MTBF	17.52				
2	x	60				
3						
4	P(less than x)	,1/B1,1)				

EXPONDIST

X B2 = 60

Lambda 1/B1 = 0.057077626

Cumulative 1 = TRUE

= 0.96743957

Returns the exponential distribution. See Help for the equations used.

Cumulative is a logical value for the function to return: the cumulative distribution function = TRUE; the probability density function = FALSE.

Formula result =0.96743957 OK Cancel

FIGURE 8.42 Example of finding exponential probability using Microsoft Excel.

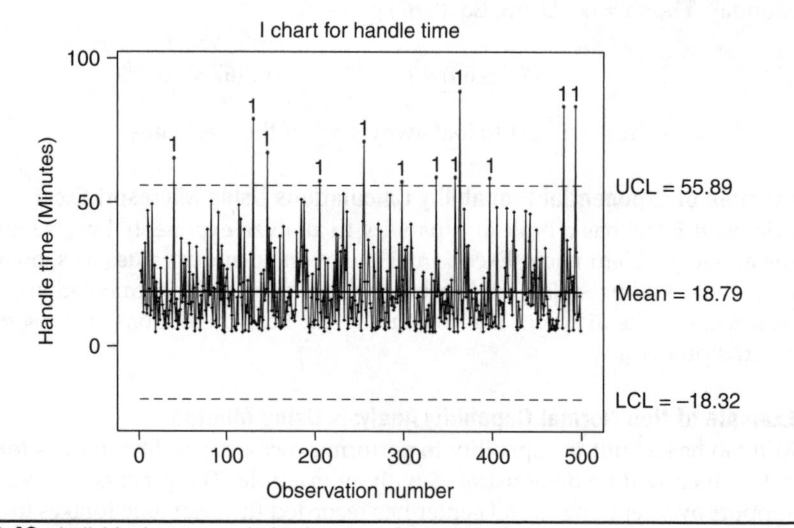

FIGURE 8.43 Individuals process behavior chart for handle time.

points beyond one of the control limits. Some authors recommend that individuals control charts be used for all analysis, so we'll try this first, see Fig. 8.43.

There are 12 out-of-control points in the chart shown in Fig. 8.43, indicating that special causes are present. However, a closer look will show that there's something odd about the chart. Note that the lower control limit (LCL) is—18:32. Since we are talking about handle time, it is impossible to obtain any result that is less than zero. A reasonable process owner might argue that if the LCL is in the wrong place (which it obviously is), then the upper control limit (UCL) may be as well. Also, the data appear to be strangely cut-off near the bottom. Apparently the individuals chart is not the best way to analyze data like these.

But what can be done? Since we don't know if special causes were present, we can't determine the proper distribution for the data. Likewise, if we don't know the distribution of the data we can't determine if special causes are present because the control limits may be in the wrong place. This may seem to be a classic case of "which came first, the chicken or the egg?" Fortunately there is a way out. The central limit theorem tells us that stable distributions produce normally distributed averages, even when the individuals data are not normally distributed. Since "stable" means no special causes, then a process with non-normal *averages* would be one that is influenced by special causes, which is precisely what we are looking for. We created subgroups of 10 in Minitab (i.e., observations 1 to 10 are in subgroup 1, observations 11 to 20 are in subgroup 2, etc.) and tested the normality of the averages. The probability plot in Fig. 8.44 indicates that the averages are normally distributed. (Note that EWMA charts and moving average charts are both valid alternatives to the use of *X*-bar charts in this example, since the central Limit theorem applies equally well to their plotted groups).

Figure 8.45 shows the control chart for the process using averages instead of individuals. The chart indicates that the process is in statistical control. The process average is stable at 18.79 minutes. The LCL is comfortably above zero at 5.9 minutes; any average below this is an indication that things are better than normal and we'd want to know why in case we can do it all of the time. Any average above 31.67 minutes indicates worse than normal behavior and we'd like to find the reason and fix it. Averages between these two limits are normal for this process.

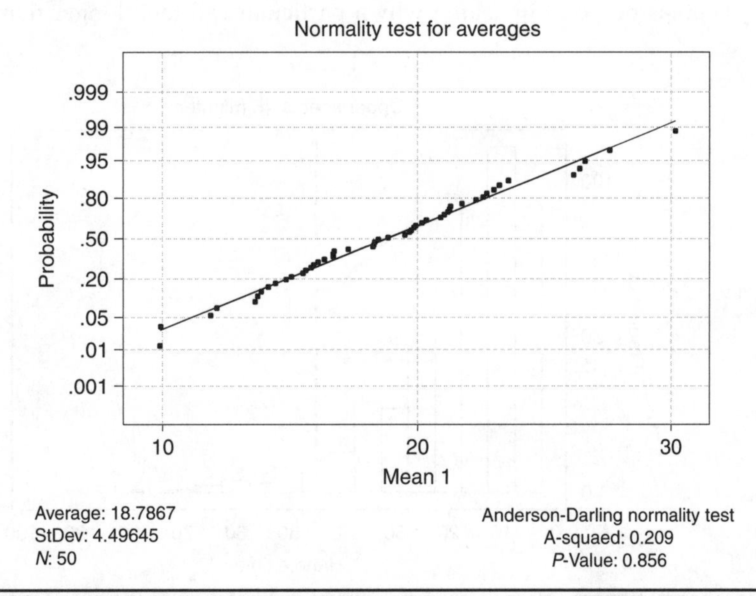

FIGURE 8.44 Normality test of subgroups of $n = 10$.

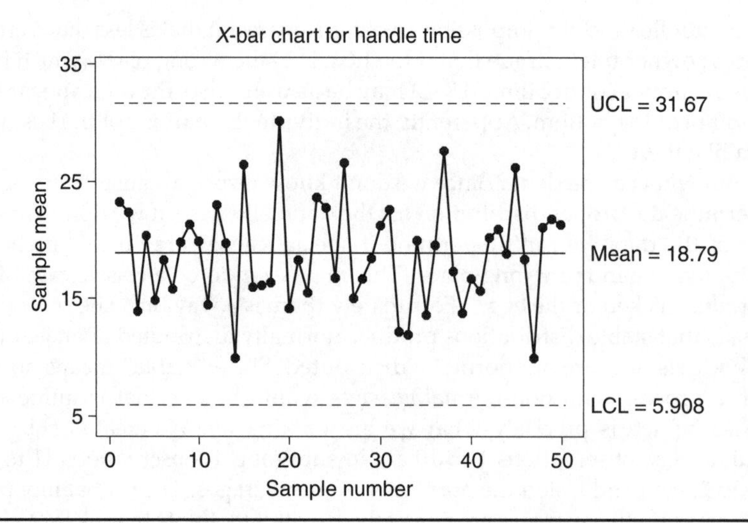

Figure 8.45 Averages of handle time ($n = 10$ per subgroup).

Phase 2—Examine the Distribution: Now that stability has been determined, we can trust the histogram of individual observations to give us an accurate display of the distribution of handle times. The histogram shows the distribution of actual handle times, which we can compare to the upper specification limit of 45 minutes. This couldn't be done with the control chart in Fig. 8.45 because it shows averages, not individual times. Figure 8.46 shows the histogram of handle time with the management upper requirement of 45 minutes drawn in. Obviously a lot of calls exceed the 45 minute requirement. Since the control chart is stable, we know that this is what we can expect from this process. There is no point in asking why a particular call took longer than 45 minutes.

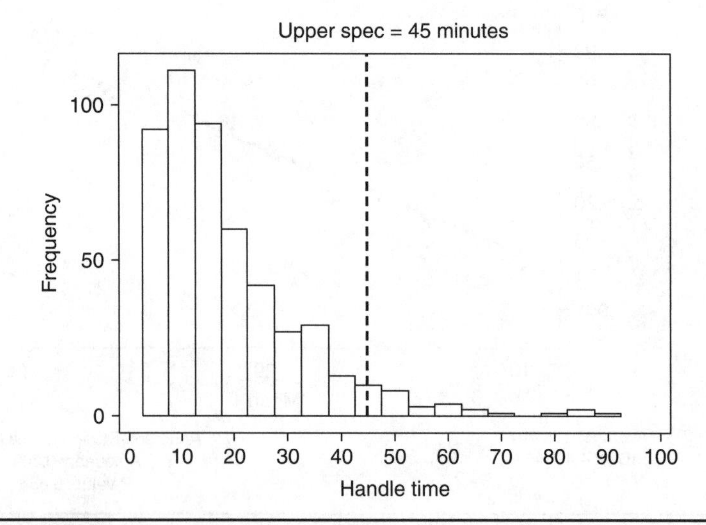

Figure 8.46 Histogram of handle time.

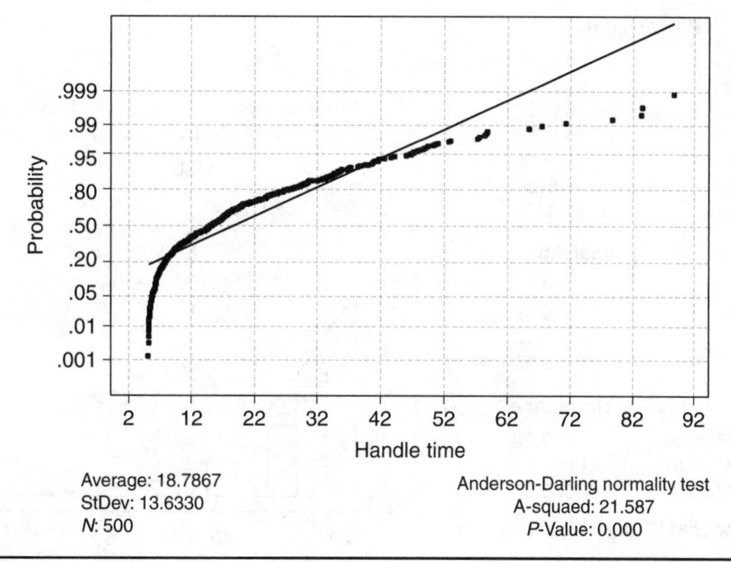

FIGURE 8.47 Normality test of handle time.

The answer is "It's to be expected for this process, unless we make a fundamental change to the process." If management doesn't like the answer they'll need to sponsor one or more Six Sigma projects to improve the process.

Phase 3—Predicting the LongTerm Defect Rate for the Process: The histogram makes it visually clear that the process distribution is non-normal. This conclusion can be tested statistically with Minitab by going to Stats > Basic Statistics > Normality test. Minitab presents the data in a chart specially scaled so that normally distributed data will plot as a straight line (Fig. 8.47). The vertical axis is scaled in cumulative probability and the horizontal in actual measurement values. The plot shows that the data are not even close to falling on the straight line, and the *P*-value of 0 confirms that the data are not normal.*

To make a prediction about the defect rate we need to find a distribution that fits the data reasonably well. Minitab offers an option that performs capability analysis using the Weibull rather than the normal distribution. Choose Stat > Quality Tools > Capability Analysis (Weibull) and enter the column name for the handle time data. The output is shown in Fig. 8.48.

Minitab calculates process performance indices rather than process capability indices (i.e., P_{PK} instead of C_{PK}). This means that the denominator for the indices is the overall standard deviation rather than the standard deviation based on the control chart's within-subgroup variability. This is called the long-term process capability, which Minitab labels as "Overall (LT) Capability." When the process is in statistical control, as this one is, there will be little difference in the estimates of the standard deviation. When

*The null hypothesis is that the data are normally distributed. The *P*-value is the probability of obtaining the observed results if the null hypothesis were true. In this case, the probability is 0.

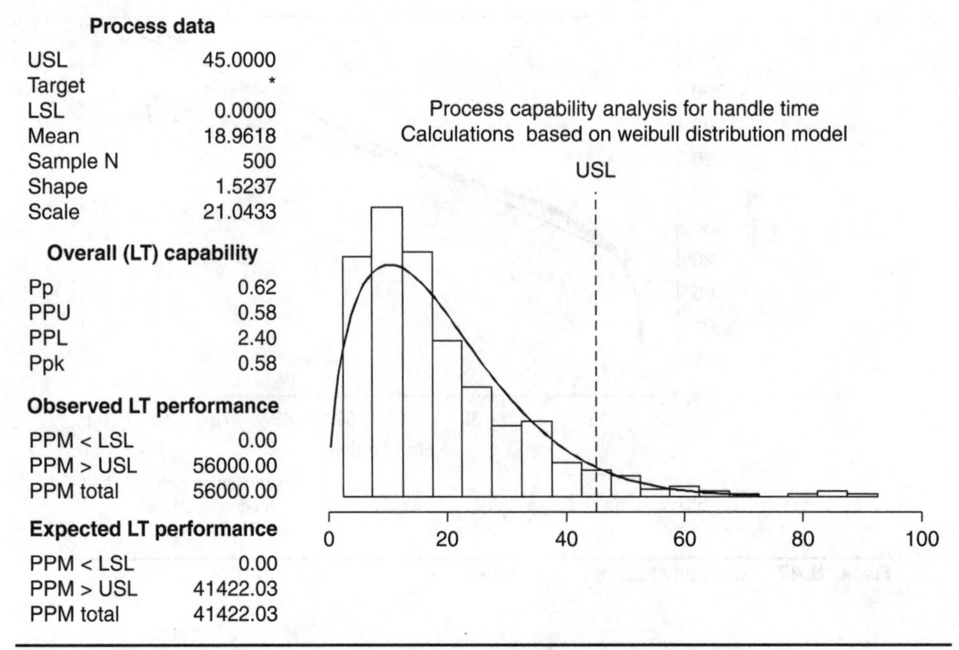

FIGURE 8.48 Capability analysis of handle times based on the Weibull distribution.

the process is not in statistical control the short-term capability estimates have no meaning, and the long-term estimates are of dubious value as well. Process performance indices are interpreted in exactly the same way as their process capability counterparts. Minitab's analysis indicates that the process is not capable ($P_{PK} < 1$). The estimated long-term performance of the process is 41,422 defects per million calls. The observed performance is even worse, 56,000 defects per million calls. The difference is a reflection of lack of fit. The part of the Weibull curve we're most interested in is the tail area above 45, and the curve appears to drop off more quickly than the actual data. When this is the case it is better to estimate the long-term performance using the actual defect count rather than Minitab's estimates.

CHAPTER 9

Measurement Systems Evaluation

A good measurement system possesses certain properties. First, it should produce a number that is "close" to the actual property being measured, that is, it should be *accurate*. Second, if the measurement system is applied repeatedly to the same object, the measurements produced should be close to one another, that is, it should be *repeatable*. Third, the measurement system should be able to produce accurate and consistent results over the entire range of concern, that is, it should be *linear*. Fourth, the measurement system should produce the same results when used by any properly trained individual, that is, the results should be *reproducible*. Finally, when applied to the same items the measurement system should produce the same results in the future as it did in the past, that is, it should be *stable*. The remainder of this section is devoted to discussing ways to ascertain these properties for particular measurement systems. In general, the methods and definitions presented here are consistent with those described by the Automotive Industry Action Group (AIAG) MSA Reference Manual (3rd ed.).

Definitions

Bias: The difference between the average measured value and a reference value is referred to as *bias*. The reference value is an agreed-upon standard, such as a standard traceable to a national standards body (see below). When applied to attribute inspection, bias refers to the ability of the attribute inspection system to produce agreement on inspection standards. Bias is controlled by *calibration*, which is the process of comparing measurements to standards. The concept of bias is illustrated in Fig. 9.1.

Repeatability: AIAG defines repeatability as the variation in measurements obtained with one measurement instrument when used several times by one appraiser, while measuring the identical characteristic on the same part. Variation obtained when the measurement system is applied repeatedly under the same conditions is usually caused by conditions inherent in the measurement system.

ASQ defines *precision* as "The closeness of agreement between randomly selected individual measurements or test results. NOTE: The standard deviation of the error of measurement is sometimes called 'imprecision'". This is similar to what we are calling repeatability. Repeatability is illustrated in Fig. 9.2.

Reproducibility: Reproducibility is the variation in the average of the measurements made by different appraisers using the same measuring instrument when measuring the identical characteristic on the same part. Reproducibility is illustrated in Fig. 9.3.

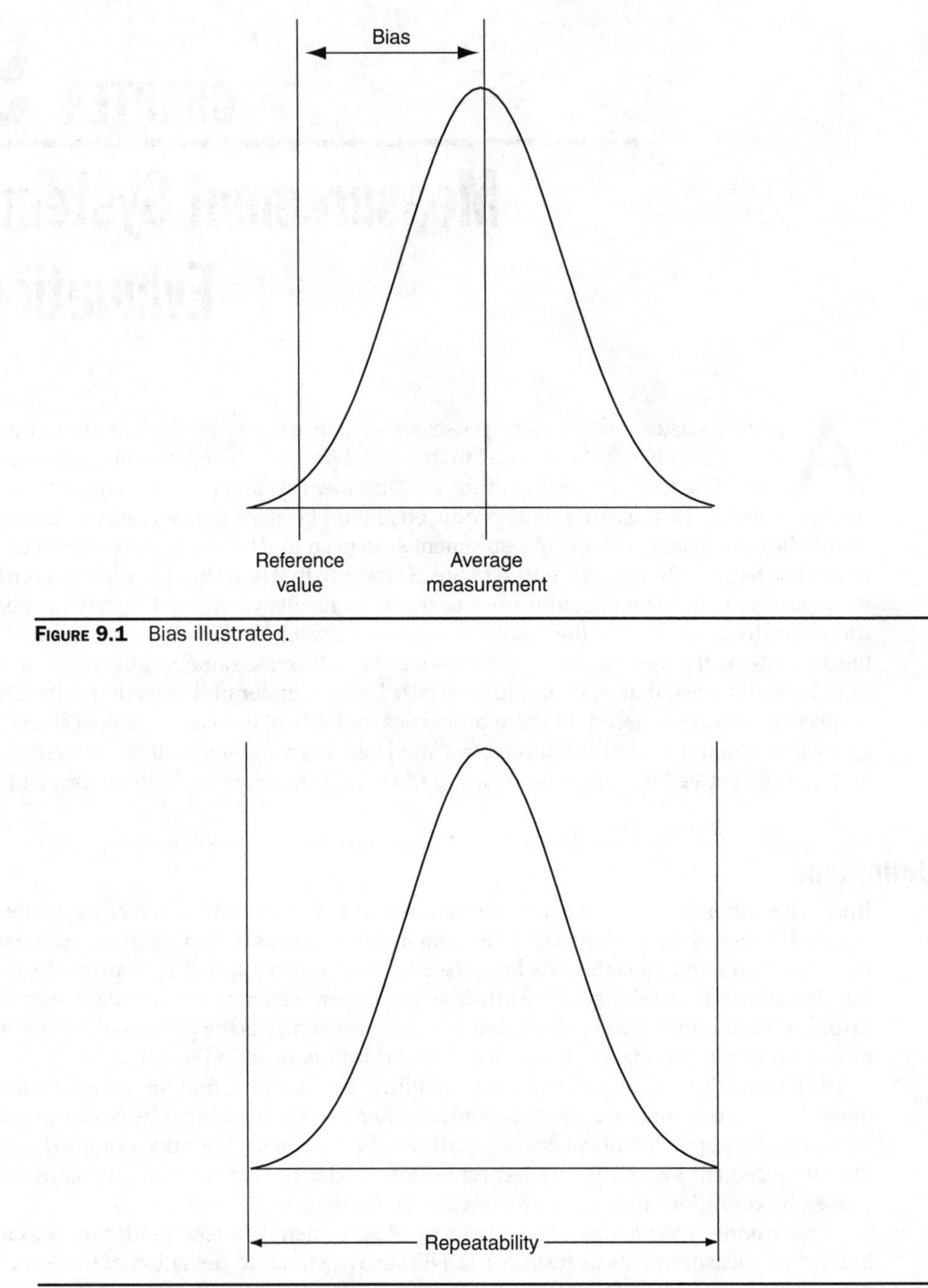

FIGURE 9.1 Bias illustrated.

FIGURE 9.2 Repeatability illustrated.

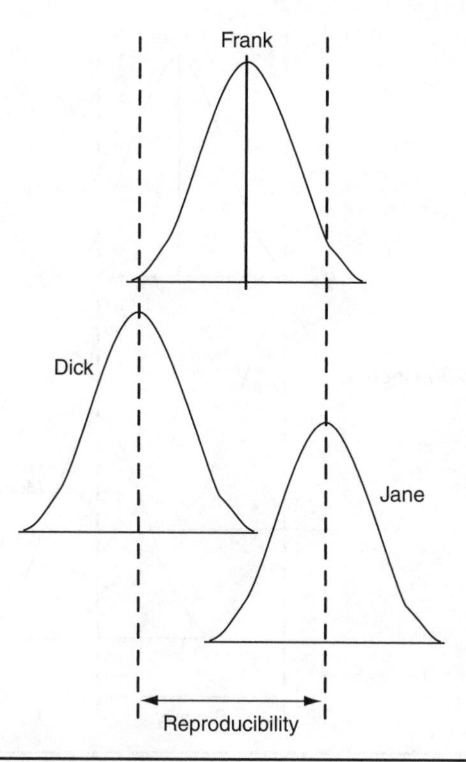

Figure 9.3 Reproducibility illustrated.

Stability: Stability is the total variation in the measurements obtained with a measurement system on the same master or parts when measuring a single characteristic over an extended time period. A system is said to be stable if the results are the same at different points in time. Stability is illustrated in Fig. 9.4.

Linearity: the difference in the bias values through the expected operating range of the gage. Linearity is illustrated in Fig. 9.5.

Historically, calibration has been the standard approach to limit the effects of bias, long considered the fundamental source of measurement error. Modern measurement system analysis goes well beyond calibration. A gage can be perfectly accurate when checking a standard and still be entirely unacceptable for measuring a product or controlling a process. This section illustrates techniques for quantifying discrimination, stability, bias, repeatability, reproducibility and variation for a measurement system. Control charts are used to provide graphical portrayals of the measurement processes, enabling the analyst to detect special causes that numerical methods alone would not detect.

Measurement System Discrimination

Discrimination, sometimes called resolution, refers to the ability of the measurement system to divide measurements into "data categories." All parts within a particular data category will measure the same. For example, if a measurement system has a

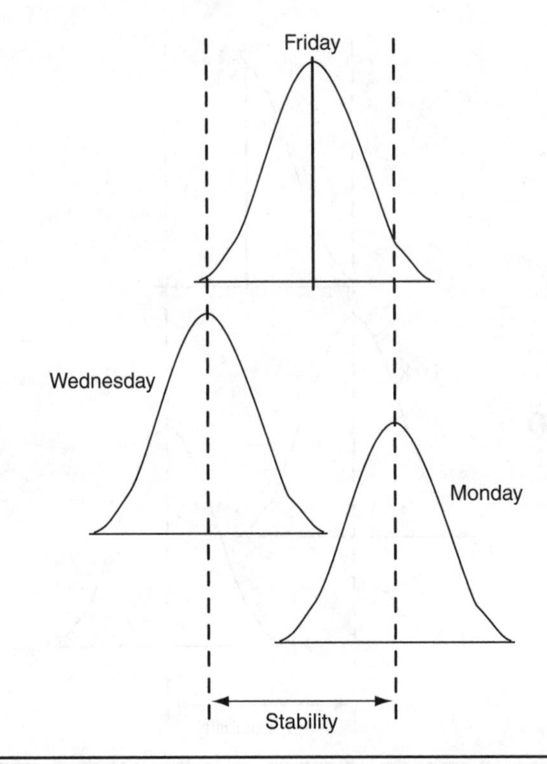

Figure 9.4 Stability illustrated.

resolution of 0.001 inch, then items measuring 1.0002, 1.0003, 0.9997 would all be placed in the data category 1.000, that is, they would all measure 1.000 inch with this particular measurement system. A measurement system's discrimination should enable it to divide the region of interest into many data categories. In Six Sigma, the region of interest is the smaller of the tolerance (the high specification minus the low specification) or six standard deviations. A measurement system should be able to divide the region of interest into at least five data categories. For example, if a process was capable (i.e., Six Sigma is less than the tolerance) and $\sigma = 0.0005$, then a gage with a discrimination of 0.0005 would be acceptable (six data categories), but one with a discrimination of 0.001 would not (three data categories). When unacceptable discrimination exists, the range chart shows discrete "jumps" or "steps." This situation is illustrated in Fig. 9.6.

Note that on the control charts shown in Fig. 9.6, the data plotted are the same, except that the data on the bottom two charts were rounded to the nearest 25. The effect is most easily seen on the *R* chart, which appears highly stratified. As sometimes happens (but not always), the result is to make the *X*-bar chart go out of control, even though the process is in control, as shown by the control charts with unrounded data. The remedy is to use a measurement system capable of additional discrimination, that is, add more significant digits. If this cannot be done, it is possible to adjust the control limits for the round-off error by using a more involved method of computing the control limits, see Pyzdek (1992a, pp. 37–42) for details.

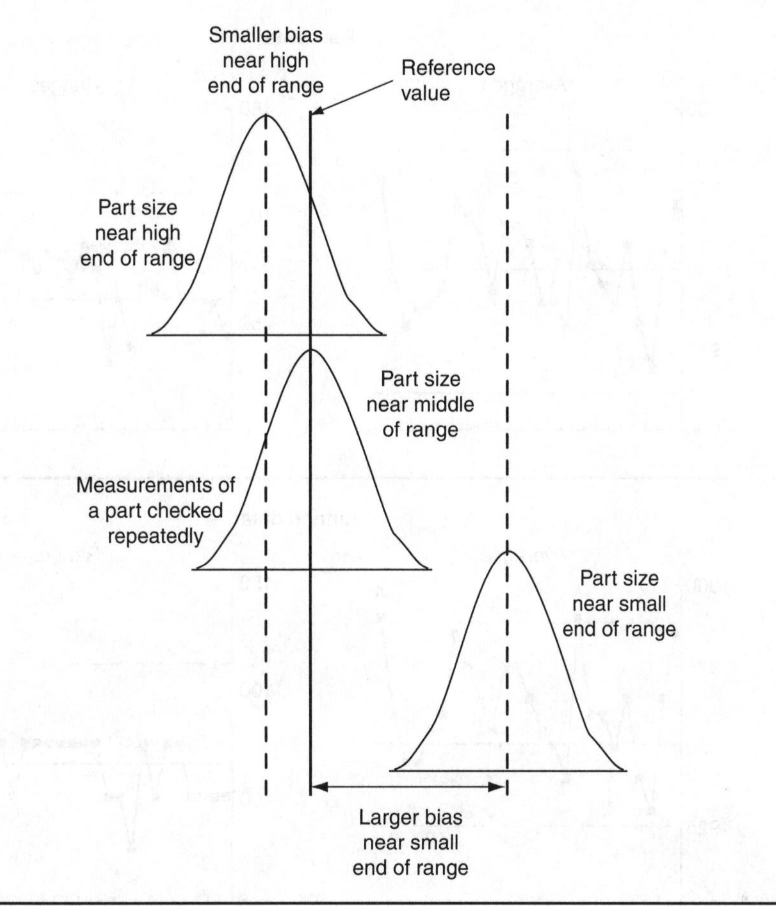

Figure 9.5 Linearity illustrated.

Stability

Measurement system stability is the change in bias over time when using a measurement system to measure a given master part or standard. *Statistical stability* is a broader term that refers to the overall consistency of measurements over time, including variation from *all causes*, including bias, repeatability, reproducibility, etc. A system's statistical stability is determined through the use of control charts. Averages and range charts are typically plotted on measurements of a standard or a master part. The standard is measured repeatedly over a short time, say an hour; then the measurements are repeated at predetermined intervals, say weekly. Subject matter expertise is needed to determine the subgroup size, sampling intervals and measurement procedures to be followed. Control charts are then constructed and evaluated. A (statistically) stable system will show no out-of-control signals on an *X*-control chart of the averages' readings. No "stability number" is calculated for statistical stability; the system either is or is not statistically stable.

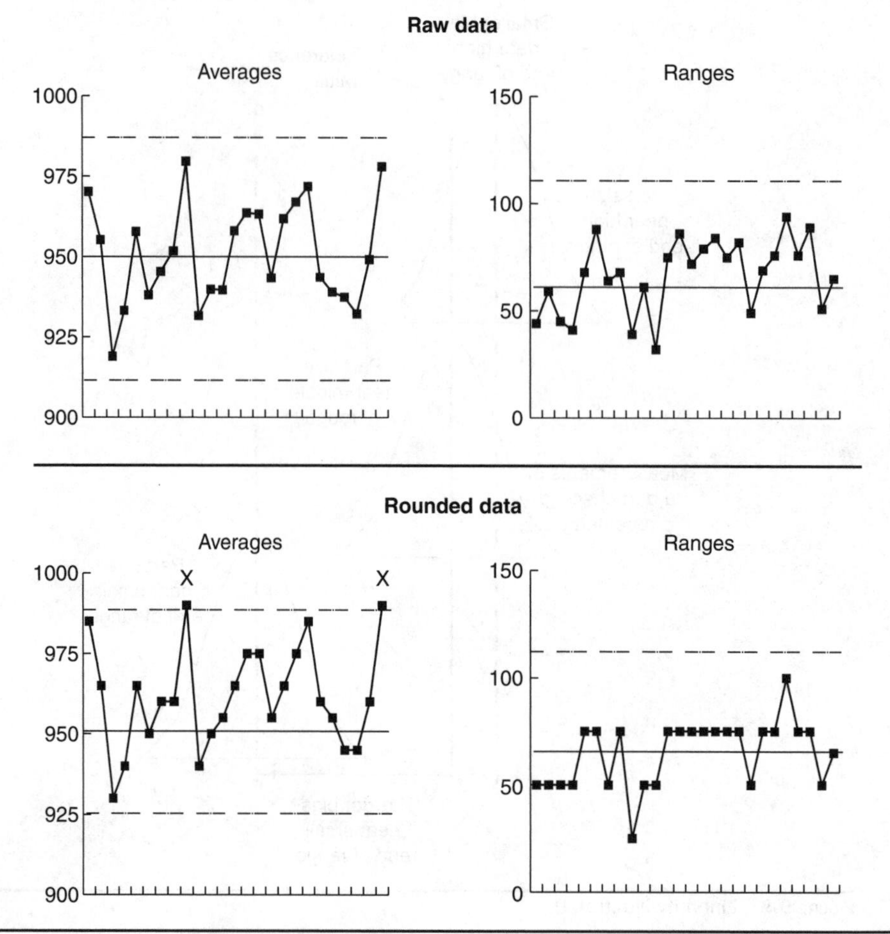

Figure 9.6 Inadequate gage discrimination on a control chart.

Once statistical stability has been achieved, but not before, measurement system stability can be determined. One measure is the process standard deviation based on the R or s chart.

R chart method:

$$\hat{\sigma} = \frac{\overline{R}}{d_2} \tag{9.1}$$

s chart method:

$$\hat{\sigma} = \frac{\overline{s}}{c_2} \tag{9.2}$$

The values d_2 and c_4 are constants from Table 9 in the Appendix.

Bias

Bias is the difference between an observed average measurement result and a reference value. Estimating bias involves identifying a standard to represent the reference value, then obtaining multiple measurements on the standard. The standard might be a master part whose value has been determined by a measurement system with much less error than the system under study, or by a standard traceable to NIST. Since parts and processes vary over a range, bias is measured at a point within the range. If the gage is nonlinear, bias will not be the same at each point in the range (see the definition of linearity defined earlier).

Bias can be determined by selecting a single appraiser and a single reference part or standard. The appraiser then obtains a number of repeated measurements on the reference part. Bias is then estimated as the difference between the average of the repeated measurement and the known value of the reference part or standard.

Example of Computing Bias

A standard with a known value of 25.4 mm is checked 10 times by one mechanical inspector using a dial caliper with a resolution of 0.025 mm. The readings obtained are:

25.425	25.425	25.400	25.400	25.375
25.400	25.425	25.400	25.425	25.375

The average is found by adding the 10 measurements together and dividing by 10,

$$\overline{X} = \frac{254.051}{10} = 25.4051 \text{ mm}$$

The bias is the average minus the reference value, that is,

$$\begin{aligned} \text{bias} &= \text{average} - \text{reference value} \\ &= 25.4051 \text{ mm} - 25.400 \text{ mm} = 0.0051 \text{ mm} \end{aligned}$$

The bias of the measurement system can be stated as a percentage of the tolerance or as a percentage of the process variation. For example, if this measurement system were to be used on a process with a tolerance of ±0.25 mm then

$$\begin{aligned} \% \text{ bias} &= 100 \times |\text{bias}| / \text{tolerance} \\ &= 100 \times 0.0051/0.5 = 1\% \end{aligned}$$

This is interpreted as follows: this measurement system will, on average, produce results that are 0.0051 mm larger than the actual value. This difference represents 1% of the allowable product variation. The situation is illustrated in Fig. 9.7.

Repeatability

A measurement system is repeatable if its variability is consistent. Consistent variability is operationalized by constructing a range or sigma chart based on repeated measurements of parts that cover a significant portion of the process variation or the tolerance,

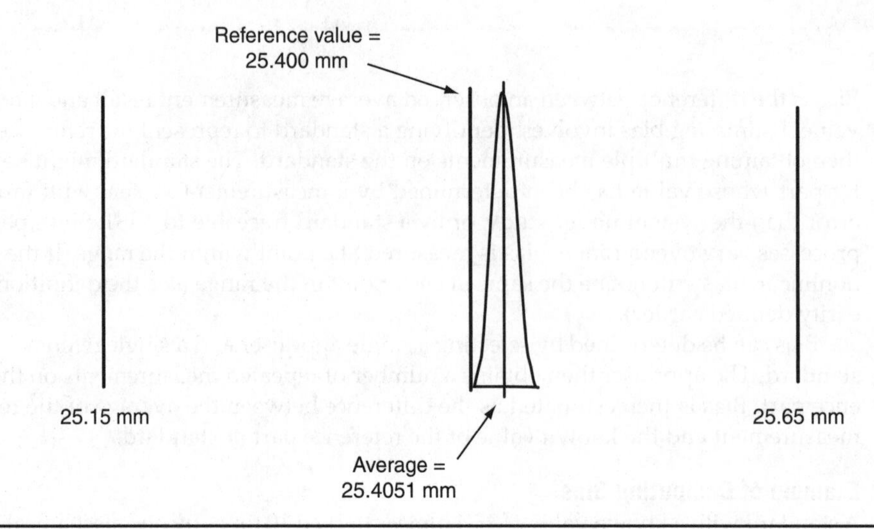

FIGURE 9.7 Bias example illustrated.

whichever is greater. If the range or sigma chart is out of control, then special causes are making the measurement system inconsistent. If the range or sigma chart is in control then repeatability can be estimated by finding the standard deviation based on either the average range or the average standard deviation. The equations used to estimate sigma are shown in the example below.

Example of Estimating Repeatability

The data in Table 9.1 are from a measurement study involving two inspectors. Each inspector checked the surface finish of five parts, each part was checked twice by each

Part	Reading 1	Reading 2	Average	Range
		Inspector 1		
1	111.9	112.3	112.10	0.4
2	108.1	108.1	108.10	0.0
3	124.9	124.6	124.75	0.3
4	118.6	118.7	118.65	0.1
5	130.0	130.7	130.35	0.7
		Inspector 2		
1	111.4	112.9	112.15	1.5
2	107.7	108.4	108.05	0.7
3	124.6	124.2	124.40	0.4
4	120.0	119.3	119.65	0.7
5	130.4	130.1	130.25	0.3

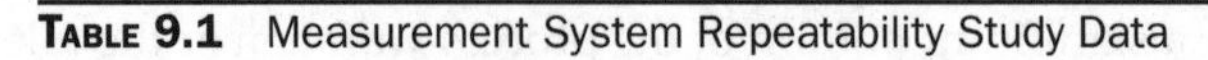

TABLE 9.1 Measurement System Repeatability Study Data

inspector. The gage records the surface roughness in μ-inches (micro-inches). The gage has a resolution of 0.1 μ-inches.

We compute:

Ranges chart

$$\bar{R} = 0.51$$

$$\text{UCL} = D_4\bar{R} = 3.267 \times 0.51 = 1.67$$

Averages chart

$$\bar{\bar{X}} = 118.85$$

$$\text{LCL} = \bar{\bar{X}} - A_2\bar{R} = 118.85 - 1.88 \times 0.51 = 118.65$$

$$\text{UCL} = \bar{\bar{X}} + A_2\bar{R} = 118.85 + 1.88 \times 0.51 = 119.05$$

The data and control limits are displayed in Fig. 9.8. The *R* chart analysis shows that all of the *R* values are less than the upper control limit. This indicates that the measurement system's variability is consistent, that is, there are no special causes of variation.

Note that many of the averages are outside of the control limits. This is the way it should be! Consider that the spread of the X-bar chart's control limits is based on the average range, which is based on the repeatability error. If the averages *were* within the control limits it would mean that the part-to-part variation was less than the variation due to gage repeatability error, an undesirable situation. Because the *R* chart is in control we can now estimate the standard deviation for repeatability or gage variation:

$$\sigma_e = \frac{\bar{R}}{d_2^*} \tag{9.3}$$

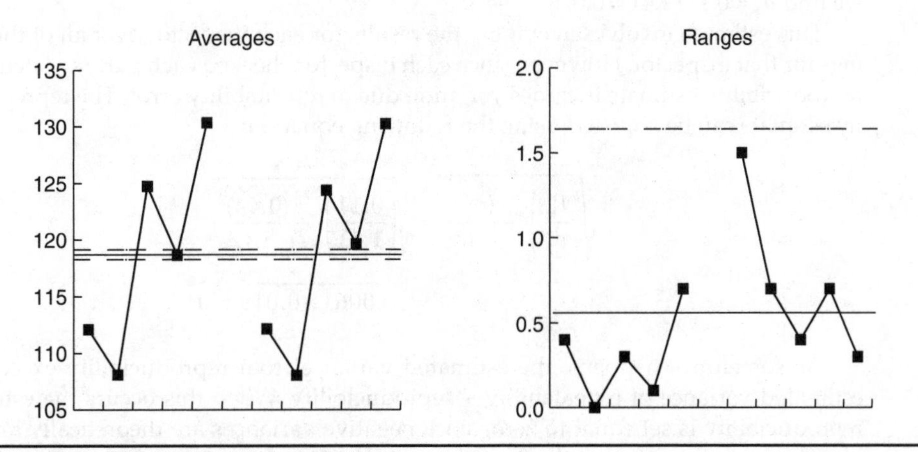

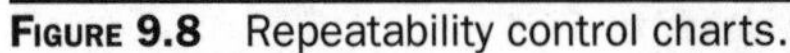
FIGURE 9.8 Repeatability control charts.

where d_2^* is obtained from Appendix 11. Note that we are using d_2^* not d_2. The d_2^* are adjusted for the small number of subgroups typically involved in gage R&R studies. Appendix 11 is indexed by two values: m is the number of repeat readings taken (m = 2 for the example), and g is the number of parts times the number of inspectors ($g = 5 \times 2 = 10$ for the example). This gives, for our example

$$\sigma_e = \frac{\overline{R}}{d_2^*} = \frac{0.51}{1.16} = 0.44$$

Report the Repeatability as percent of process sigma by dividing σ_e by Part to Part Sigma (see below).

Reproducibility

A measurement system is reproducible when different appraisers produce consistent results. Appraiser-to-appraiser variation represents a bias due to appraisers. The appraiser bias, or reproducibility, can be estimated by comparing each appraiser's average with that of the other appraisers. The standard deviation of reproducibility (σ_o) is estimated by finding the range between appraisers (R_o) and dividing by d_2^*. Percent Reproducibility is calculated by dividing σ_o by Part to Part Sigma (see below).

Reproducibility Example (AIAG Method)

Using the data shown in the previous example, each inspector's average is computed and we find:

$$\begin{aligned}\text{Inspector 1 average} &= 118.79\ \mu\text{-inches}\\ \text{Inspector 2 average} &= 118.90\ \mu\text{-inches}\\ \text{Range} = R_o &= 0.11\ \mu\text{-inches}\end{aligned}$$

Looking in Table 11 in the Appendix for one subgroup of two appraisers we find $d_2^* = 1.41$ ($m = 2, g = 1$), since there is only one range calculation $g = 1$. Using these results we find $d_2^* = 0.11/1.41 = 0.078$.

This estimate involves averaging the results for each inspector over all of the readings for that inspector. However, since each inspector checked each part repeatedly, this reproducibility estimate includes variation due to repeatability error. The reproducibility estimate can be adjusted using the following equation:

$$\sqrt{\left(\frac{R_o}{d_2^*}\right)^2 - \frac{(\sigma_e)^2}{nr}} = \sqrt{\left(\frac{0.11}{1.41}\right)^2 - \frac{(0.44)^2}{5 \times 2}}$$

$$= \sqrt{0.0061 - 0.019} = 0$$

As sometimes happens, the estimated variance from reproducibility exceeds the estimated variance of repeatability + reproducibility. When this occurs the estimated reproducibility is set equal to zero, since negative variances are theoretically impossible. Thus, we estimate that the reproducibility is zero.

The measurement system standard deviation is

$$\sigma_m = \sqrt{\sigma_e^2 + \sigma_o^2} = \sqrt{(0.44)^2 + 0} = 0.44$$

report the Measurement System Error as percent of process sigma by dividing σ_m by Part to Part Sigma (see below).

Reproducibility Example (Alternative Method)

One problem with the above method of evaluating reproducibility error is that it does not produce a control chart to assist the analyst with the evaluation. The method presented here does this. This method begins by rearranging the data in Table 9.1 so that all readings for any given part become a single row. This is shown in Table 9.2.

Observe that when the data are arranged in this way, the R value measures the combined range of repeat readings plus appraisers. For example, the smallest reading for part 3 was from inspector 2 (124.2) and the largest was from inspector 1 (124.9). Thus, R represents two sources of measurement error: repeatability and reproducibility.

The control limits are calculated as follows:

Ranges chart

$$\bar{R} = 100$$

$$\text{UCL} = D_4\bar{R} = 2.282 \times 1.00 = 2.282$$

Note that the subgroup size is 4.

Averages chart

$$\bar{\bar{X}} = 118.85$$

$$\text{LCL} = \bar{\bar{X}} - A_2\bar{R} = 118.85 - 0.729 \times 1 = 188.12$$

$$\text{UCL} = \bar{\bar{X}} + A_2\bar{R} = 118.85 + 0.729 \times 1 = 119.58$$

	Inspector 1		Inspector 2			
Part	**Reading 1**	**Reading 2**	**Reading 1**	**Reading 2**	***X* bar**	***R***
1	111.9	112.3	111.4	112.9	112.125	1.5
2	108.1	108.1	107.7	108.4	108.075	0.7
3	124.9	124.6	124.6	124.2	124.575	0.7
4	118.6	118.7	120	119.3	119.15	1.4
5	130	130.7	130.4	130.1	130.3	0.7
				Averages →	118.845	1

TABLE 9.2 Measurement Error Data for Reproducibility Evaluation

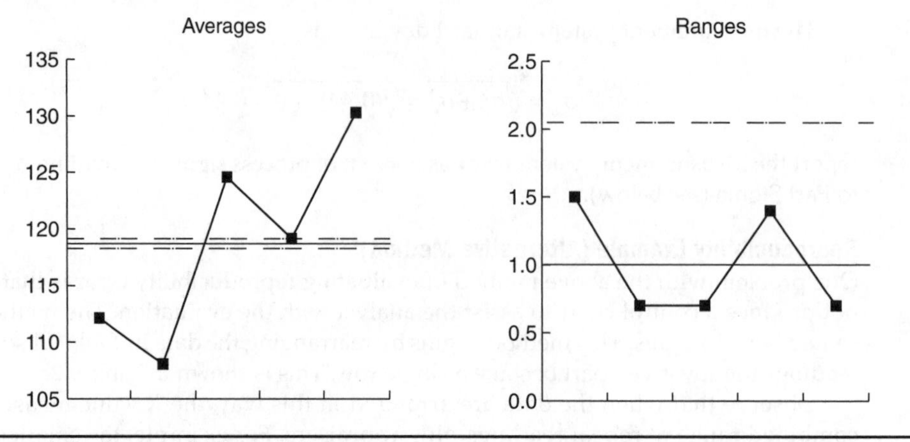

FIGURE 9.9 Reproducibility control charts.

The data and control limits are displayed in Fig. 9.9. The *R* chart analysis shows that all of the *R* values are less than the upper control limit. This indicates that the measurement system's variability due to the combination of repeatability and reproducibility is consistent, that is, there are no special causes of variation.

Using this method, we can also estimate the standard deviation of reproducibility plus repeatability, as we can find $\sigma_o = R_o/d_2^* = 1/2.08 = 0.48$. Now we know that variances are additive, so

$$\sigma^2_{\text{repeatability+reproducibility}} = \sigma^2_{\text{repeatability}} + \sigma^2_{\text{reproducibility}} \tag{9.4}$$

which implies that

$$\sigma_{\text{reproducibility}} = \sqrt{\sigma^2_{\text{repeatability+reproducibility}} - \sigma^2_{\text{repeatability}}}$$

In a previous example, we computed $\sigma_{\text{repeatability}} = 0.44$. Substituting these values gives

$$\sigma_{\text{reproducibility}} = \sqrt{\sigma^2_{\text{repeatability+reproducibility}} - \sigma^2_{\text{repeatability}}}$$

$$= \sqrt{(0.48)^2 - (0.44)^2} = 0.19$$

The resulting σ_m is estimated as 0.48.

Part-to-Part Variation

The *X*-bar charts show the part-to-part variation. To repeat, if the measurement system is adequate, *most of the parts will fall outside of the X-bar chart control limits*. If fewer than half of the parts are beyond the control limits, then the measurement system is not capable of detecting normal part-to-part variation for this process.

Part-to-part variation can be estimated once the measurement process is shown to have adequate discrimination and to be stable, accurate, linear (see below), and consistent with respect to repeatability and reproducibility. If the part-to-part standard deviation is to be estimated from the measurement system study data, the following procedures are followed:

1. Plot the average for each part (across all appraisers) on an averages control chart, as shown in the reproducibility error alternate method.
2. Confirm that at least 50% of the averages fall outside the control limits. If not, find a better measurement system for this process.
3. Find the range of the part averages, R_p.
4. Compute $\sigma_p = R_p/d_2^*$, the part-to-part standard deviation. The value of d_2^* is found in Table 11 in the Appendix using m = the number of parts and $g = 1$, since there is only one R calculation.
5. The total process standard deviation is found as $\sigma_t = \sqrt{\sigma_m^2 + \sigma_p^2}$.

Once the above calculations have been made, the overall measurement system can be evaluated.

1. The %EV = $100 \times (\sigma_e/\sigma_T)\%$
2. The %AV = $100 \times (\sigma_o/\sigma_T)\%$
3. The percent repeatability and reproducibility (R&R) is $100 \times (\sigma_m/\sigma_t)\%$.
4. The number of distinct data categories that can be created with this measurement system is $1.41 \times (\mathrm{PV}/\mathrm{R\&R})$.

Example of Measurement System Analysis Summary

1. Plot the average for each part (across all appraisers) on an averages control chart, as shown in the reproducibility error alternate method. Done earlier, see Fig. 9.8.
2. Confirm that at least 50% of the averages fall outside the control limits. If not, find a better measurement system for this process. 4 of the 5 part averages, or 80%, are outside of the control limits. Thus, the measurement system error is acceptable.
3. Find the range of the part averages, R_p.

$$R_p = 130.33 - 108.075 = 22.23$$

4. Compute $\sigma_p = R_p/d_2^*$, the part-to-part standard deviation. The value of d_2^* is found in Table 11 in the Appendix using m = the number of parts and $g = 1$, since there is only one R calculation.

$$m = 5, g = 1, d_2^* = 2.48, \sigma_p = 22.23/2{:}48 = 8.96$$

5. The total process standard deviation is found as $\sigma_t = \sqrt{\sigma_m^2 + \sigma_p^2}$

$$\sigma_t = \sqrt{\sigma_m^2 + \sigma_p^2} = \sqrt{(0.44)^2 + (8.96)^2} = \sqrt{80.5} = 8.97$$

Once the above calculations have been made, the overall measurement system can be evaluated.

1. The %EV = $100 \times (\sigma_e/\sigma_T)\% = 100 \times .44/8.97 = 4.91\%$
2. The %AV = $100 \times (\sigma_o/\sigma_T)\% = 100 \times 0/8.97 = 0\%$
3. The percent R&R is $100 \times (\sigma_m/\sigma_t)\%$

$$100\frac{\sigma_m}{\sigma_t}\% = 100\frac{0.44}{8.97} = 4.91\%$$

4. The number of distinct data categories that can be created with this measurement system is $1.41 \times (\text{PV}/\text{R\&R})$

$$1.41 \times \frac{46.15}{2.27} = 28.67 = 28$$

Since the minimum number of categories is five, the analysis indicates that this measurement system is more than adequate for process analysis or process control.

Gage R&R Analysis Using Minitab

Minitab has a built-in capability to perform gage repeatability and reproducibility studies. To illustrate these capabilities, the previous analysis will be repeated using Minitab. To begin, the data must be rearranged into the format expected by Minitab (Fig. 9.10). For reference purposes, columns C1–C4 contain the data in our original format and columns C5–C8 contain the same data in Minitab's preferred format.

Minitab offers two different methods for performing gage R&R studies: crossed and nested. Use gage R&R nested when each part can be measured by only one operator, as with destructive testing. Otherwise, choose gage R&R crossed. To do this, select Stat > Quality Tools > Gage R&R Study (Crossed) to reach the Minitab dialog box for our analysis (Fig. 9.11). In addition to choosing whether the study is crossed or nested, Minitab also offers both the ANOVA and the *Xbar* and *R* methods. You must choose the ANOVA option to obtain a breakdown of reproducibility by operator and operator by part. If the ANOVA method is selected, Minitab still displays the *Xbar* and *R* charts so you won't lose the information contained in the graphics. We will use ANOVA in this example. Note that the results of the calculations will differ slightly from those we obtained using the *Xbar* and *R* methods.

There is an option in gage R&R to include the process tolerance. This will provide comparisons of gage variation with respect to the specifications in addition to the variability with respect to process variation. This is useful information if the gage is to be used to make product acceptance decisions. If the process is "capable" in the sense that the total variability is less than the tolerance, then any gage that meets the criteria for

↓	C1	C2	C3	C4	C5	C6-T	C7	C8
	Part	Reading1	Reading2	Inspector	StackedData	Subscripts	Operator	PartNum
1	1	111.9	112.3	1	111.9	Reading1	1	1
2	2	108.1	108.1	1	108.1	Reading1	1	2
3	3	124.9	124.6	1	124.9	Reading1	1	3
4	4	118.6	118.7	1	118.6	Reading1	1	4
5	5	130.0	130.7	1	130.0	Reading1	1	5
6	1	111.4	112.9	2	111.4	Reading1	2	1
7	2	107.7	108.4	2	107.7	Reading1	2	2
8	3	124.6	124.2	2	124.6	Reading1	2	3
9	4	120.0	119.3	2	120.0	Reading1	2	4
10	5	130.4	130.1	2	130.4	Reading1	2	5
11					112.3	Reading2	1	1
12					108.1	Reading2	1	2
13					124.6	Reading2	1	3
14					118.7	Reading2	1	4
15					130.7	Reading2	1	5
16					112.9	Reading2	2	1
17					108.4	Reading2	2	2
18					124.2	Reading2	2	3
19					119.3	Reading2	2	4
20					130.1	Reading2	2	5

FIGURE 9.10 Data formatted for Minitab input.

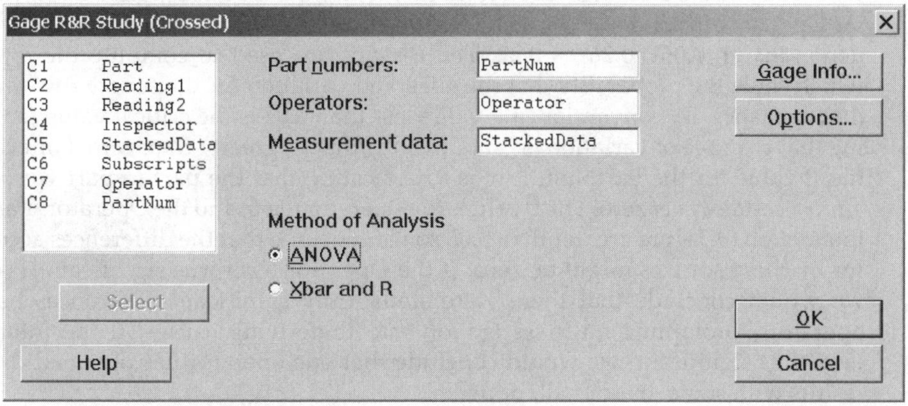

Gage R&R Study (Crossed)

C1 Part
C2 Reading1
C3 Reading2
C4 Inspector
C5 StackedData
C6 Subscripts
C7 Operator
C8 PartNum

Part numbers: PartNum
Operators: Operator
Measurement data: StackedData

Gage Info...
Options...

Method of Analysis
ANOVA
Xbar and R

Select
Help
OK
Cancel

FIGURE 9.11 Minitab gage R&R (crossed) dialog box.

checking the process can also be used for product acceptance. However, if the process is not capable, then its output will need to be sorted and the gage used for sorting may need more discriminatory power than the gage used for process control. For example, a gage capable of five distinct data categories for the process may have four or fewer for the product. For the purposes of illustration, we entered a value of 40 in the process tolerance box in the Minitab options dialog box (Fig. 9.12).

Gage R&R Study (Crossed) - Options

Study variation: 5.15 (number of standard deviations)

Process tolerance: 40

Historical sigma:

Do not display percent contribution

Do not display percent study variation

Draw plots on separate pages, one plot per page

Title: Gage R&R Example

Help OK Cancel

FIGURE 9.12 Minitab gage R&R (crossed) options dialog box.

Output

Minitab produces copious output, including six separate graphs, multiple tables, etc. Much of the output is identical to what has been discussed earlier in this chapter and won't be shown here.

Table 9.3 shows the analysis of variance for the R&R study. In the ANOVA the MS for repeatability (0.212) is used as the denominator or error term for calculating the F-ratio of the Operator*PartNum interaction; 0.269/0.212 = 1.27. The F-ratio for the Operator effect is found by using the Operator*PartNum interaction MS term as the denominator, 0.061/0.269 = 0.22. The F-ratios are used to compute the P values, which show the probability that the observed variation for the source row might be due to chance. By convention, a P value less than 0.05 is the critical value for deciding that a source of variation is "significant," that is, greater than zero. For example, the P value for the PartNum row is 0, indicating that the part-to-part variation is almost certainly *not* zero. The P values for Operator (0.66) and the Operator*PartNum interaction (0.34) are greater than 0.05 so we conclude that the differences accounted for by these sources might be zero. If the Operator term was significant ($P < 0.05$) we would conclude that there were statistically significant differences between operators, prompting an investigation into underlying causes. If the interaction term was significant, we would conclude that one operator has obtained different results with some, but not all, parts.

Source	DF	SS	MS	F	P
PartNum	4	1301.18	325.294	1208.15	0
Operator	1	0.06	0.061	0.22	0.6602
Operator*PartNum	4	1.08	0.269	1.27	0.34317
Repeatability	10	2.12	0.212		
Total	19	1304.43			

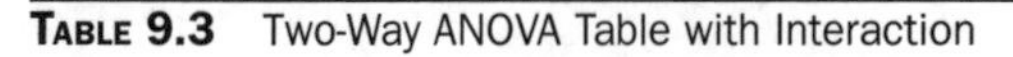

TABLE 9.3 Two-Way ANOVA Table with Interaction

Source	DF	SS	MS	F	P
PartNum	4	1301.18	325.294	1426.73	0
Operator	1	0.06	0.061	0.27	0.6145
Repeatability	14	3.19	0.228		
Total	19	1304.43			

TABLE 9.4 Two-Way ANOVA Table without Interaction

Minitab's next output is shown in Table 9.4. This analysis has removed the interaction term from the model, thereby gaining four degrees of freedom for the error term and making the test more sensitive. In some cases this might identify a significant effect that was missed by the larger model, but for this example the conclusions are unchanged.

Minitab also decomposes the total variance into components, as shown in Table 9.5. The VarComp column shows the variance attributed to each source, while the percentage of VarComp shows the percentage of the total variance accounted for by each source. The analysis indicates that nearly all of the variation is between parts.

The variance analysis shown in Table 9.5, while accurate, is not in original units. (Variances are the squares of measurements.) Technically, this is the correct way to analyze information on dispersion because variances are additive, while dispersion measurements expressed in original units are not. However, there is a natural interest in seeing an analysis of dispersion in the original units so Minitab provides this. Table 9.6

Source	VarComp	% of VarComp
Total gage R&R	0.228	0.28
Repeatability	0.228	0.28
Reproducibility	0	0
Operator	0	0
Part-to-Part	81.267	99.72
Total Variation	81.495	100

TABLE 9.5 Components of Variance Analysis

Source	StdDev	Study Var (5.15*SD)	% Study Var (% SV)	% Tolerance (SV/Toler)
Total gage R&R	0.47749	2.4591	5.29	6.15
Repeatability	0.47749	2.4591	5.29	6.15
Reproducibility	0	0	0	0
Operator	0	0	0	0
Part-to-Part	9.0148	46.4262	99.86	116.07
Total variation	9.02743	46.4913	100	116.23

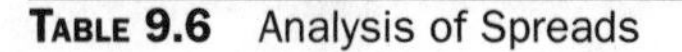

TABLE 9.6 Analysis of Spreads

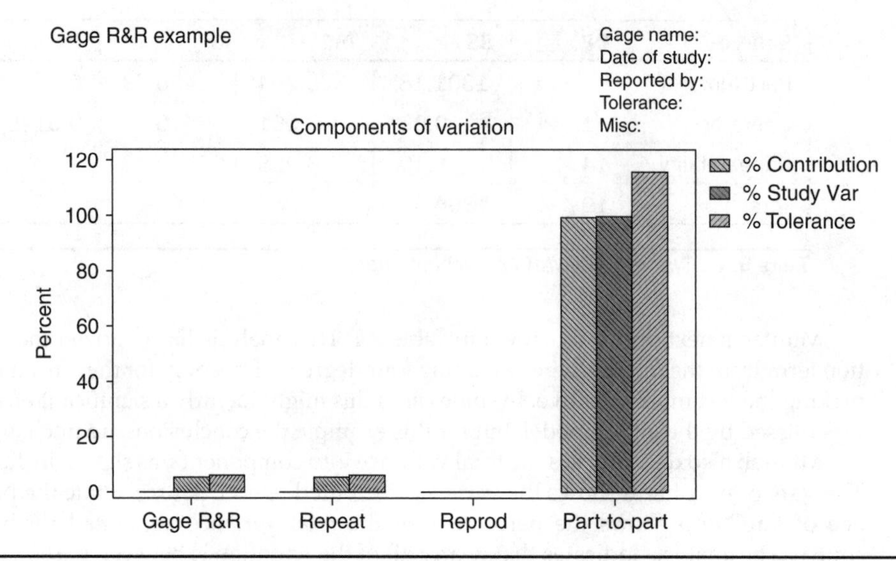

FIGURE 9.13 Graphical analysis of components of variation.

shows the spread attributable to the different sources. The StdDev column is the standard deviation, or the square root of the VarComp column in Table 9.5. The Study Var column shows the 99% confidence interval using the StdDev. The % Study Var column is the Study Var column divided by the total variation due to all sources. And the percent Tolerance is the Study Var column divided by the tolerance. It is interesting that the percent Tolerance column total is greater than 100%. This indicates that the measured process spread exceeds the tolerance. Although this isn't a process capability analysis, the data do indicate a possible problem meeting tolerances. The information in Table 9.6 is presented graphically in Fig. 9.13.

Linearity

Linearity can be determined by choosing parts or standards that cover all or most of the operating range of the measurement instrument. Bias is determined at each point in the range and a linear regression analysis is performed.

Linearity is defined as the slope times the process variance or the slope times the tolerance, whichever is greater. A scatter diagram should also be plotted from the data.

Linearity Example The following example is taken from *Measurement Systems Analysis,* published by the Automotive Industry Action Group.

A plant foreman was interested in determining the linearity of a measurement system. Five parts were chosen throughout the operating range of the measurement system based upon the process variation. Each part was measured by a layout inspection to determine its reference value. Each part was then measured 12 times by a single appraiser. The parts were selected at random. The part average and bias

Part →	1	2	3	4	5
Average	2.49	4.13	6.03	7.71	9.38
Ref. Value	2.00	4.00	6.00	8.00	10.00
Bias	+0.49	+0.13	+0.03	-0.29	-0.62

FIGURE 9.14 Gage data summary.

average were calculated for each part as shown in Fig. 9.14. The part bias was calculated by subtracting the part reference value from the part average.

A linear regression analysis was performed. In the regression, x is the reference value and y is the bias. The results are shown in Fig. 9.15.

The *P*-values indicate that the result is statistically significant, that is, there is actually a bias in the gage. The slope of the line is –0.132, and the intercept is 0.74. $R^2 = 0.98$, indicating that the straight line explains about 98% of the variation in the bias readings. The results can be summarized as follows:

Bias	$b + ax$ = 0:74 – 0.132 (Reference value)
Linearity	|slope| × process variation = 0.132 × 6 = 0.79, where 6 is the tolerance
% Linearity	100% × |slope| = 13.2%

Summary output					
Regression statistics					
Multiple R	0.98877098				
R Square	0.97766805				
Adjusted R Square	0.97022407				
Standard Error	0.07284687				
Observations	5				
ANOVA					
	df	*SS*	*ms*	*F*	*Significance F*
Regression	1	0.69696	0.69696	131.336683	0.00142598
Residual	3	0.01592	0.00530667		
Total	4	0.71288			
	Coefficients	*Standard error*	*t Stat*	*P-value*	
Intercept	0.74	0.07640244	9.68555413	0.0023371	
Ref. Value	-0.132	0.0115181	-11.460222	0.00142598	

FIGURE 9.15 Regression analysis of linearity summary data.

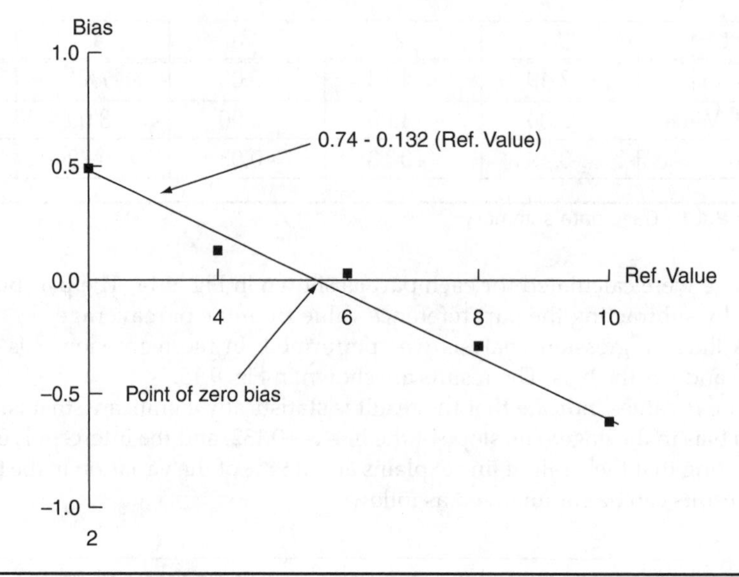

FIGURE 9.16 Graphical analysis of linearity.

Note that the zero bias point is found at

$$x = -\left(\frac{\text{intercept}}{\text{slope}}\right) = -\left(\frac{0.74}{-0.132}\right) = 5.61$$

In this case, this is the point of least bias. Greater bias exists as you move further from this value.

This information is summarized graphically in Fig. 9.16.

Linearity Analysis Using Minitab

Minitab has a built-in capability to perform gage linearity analysis. Figure 9.17 shows the data layout and dialog box. Figure 9.18 shows the Minitab output.

Note that Minitab doesn't show the *P*-values for the analysis so it is necessary to perform a supplementary regression analysis anyway to determine the statistical significance of the results. For this example, it is obvious from the scatter plot that the slope of the line isn't zero, so a *P*-value isn't required to conclude that non-linearity exists. The results aren't so clear for bias, which is only 0.867%. In fact, if we perform a one-sample t-test of the hypothesis that the mean bias is 0, we get the results shown in Fig. 9.19, which indicate the bias could be 0 ($P = 0.797$).*

*A problem with this analysis is that the datum for each part is an average of 12 measurements, not individual measurements. If we could obtain the 60 actual measurements the *P*-value would probably be different because the standard error would be based on 60 measurements rather than 5. On the other hand, the individual measurements would also be more variable, so the exact magnitude of the difference is impossible to determine without the raw data.

C1	C2	C3	C4	C5	C6	C7	C8	C9	C10
PART	Average	Reference	bias						
1	2.49	2	0.49						
2	4.13	4	0.13						
3	6.03	6	0.03						

Gage Linearity and Bias Study

C1 PART
C2 Average
C3 Reference
C4 bias

Part numbers: PART
Master measurements: Reference
Measurement data: Average
Process Variation: 6
(study variation from Gage RR)
or
(6*historical sigma)

Select | Help | Gage Info... | Options... | OK | Cancel

Figure 9.17 Minitab gage linearity dialog box.

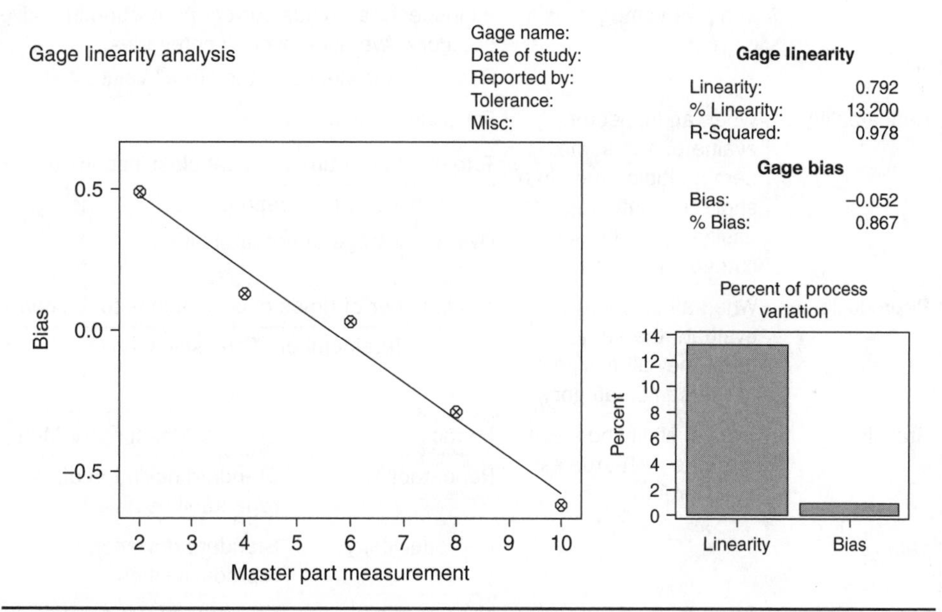

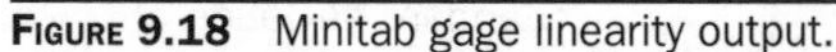

Figure 9.18 Minitab gage linearity output.

One-Sample T: bias

```
Test of mu = 0 vs mu not = 0

Variable          N      Mean     StDev   SE Mean
bias              5    -0.052     0.422     0.189

Variable            95.0% CI              T       P
bias          ( -0.576,   0.472)      -0.28   0.797
```

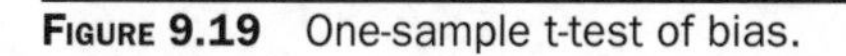

Figure 9.19 One-sample t-test of bias.

Attribute Measurement Error Analysis

Attribute data consist of classifications rather than measurements. Attribute inspection involves determining the classification of an item, for example, is it "good" or "bad"? The principles of good measurement for attribute inspection are the same as for measurement inspection (Table 9.7). Thus, it is possible to evaluate attribute measurement systems in

<table>
<tr><th>Measurement Concept</th><th>Interpretation for Attribute Data</th><th colspan="2">Suggested Metrics and Comments</th></tr>
<tr><td>Accuracy</td><td>Items are correctly categorized.</td><td colspan="2">$\frac{\text{Number of times correctly classified by all}}{\text{Totalnumber of evaluations by all}}$
Requires knowledge of the "true" value</td></tr>
<tr><td>Bias</td><td>The proportion of items in a given category is correct.</td><td colspan="2">Overall average proportion in a given category (for all inspectors) minus correct proportion in a given category. Averaged over all categories.
Requires knowledge of the "true" value.</td></tr>
<tr><td>Repeatability</td><td>When an inspector evaluates the same item multiple times in a short time interval, she assigns it to the same category every time.</td><td colspan="2">For a given inspector:
$\frac{\text{Totalnumber of times repeat classications agree}}{\text{Totalnumber of repeat classifications}}$
Overall: Average of repeatabilities</td></tr>
<tr><td>Reproducibility</td><td>When all inspectors evaluate the same item, they all assign it to the same category.</td><td colspan="2">$\frac{\text{Totalnumber of times classifications for all concur}}{\text{Totalnumber of classification}}$</td></tr>
<tr><td rowspan="5">Stability</td><td rowspan="5">The variability between attribute R&R studies at different times.</td><td>Metric</td><td>Stability Measure for Metric</td></tr>
<tr><td>Repeatability</td><td>Standard deviation of repeatabilities</td></tr>
<tr><td>Reproducibility</td><td>Standard deviation of reproducibilities</td></tr>
<tr><td>Accuracy</td><td>Standard deviation of accuracies</td></tr>
<tr><td>Bias</td><td>Average bias</td></tr>
<tr><td>Linearity</td><td>When an inspector evaluates items covering the full set of categories, her classifications are consistent across the categories.</td><td colspan="2">Range of inaccuracy and bias across all categories.
Requires knowledge of the "true" value.
Note: Because there is no natural ordering for nominal data, the concept of linearity doesn't really have a precise analog for attribute data on this scale. However, the suggested metrics will highlight interactions between inspectors and specific categories.</td></tr>
</table>

Table 9.7 Attribute Measurement Concepts

much the same way as we evaluate variable measurement systems. Much less work has been done on evaluating attribute measurement systems. The proposals provided in this book are those I've found to be useful for my employers and clients. The ideas are not part of any standard and you are encouraged to think about them critically before adopting them. I also include an example of Minitab's attribute gage R&R analysis.

Operational Definitions

An operational definition is defined as a requirement that includes a means of measurement. "High quality solder" is a requirement that must be operationalized by a clear definition of what "high quality solder" means. This might include verbal descriptions, magnification power, photographs, physical comparison specimens, and many more criteria.

Examples of Operational Definitions

1. Operational definition of the Ozone Transport Assessment Group's (OTAG) goal

   Goal: To identify reductions and recommend transported ozone and its precursors which, in combination with other measures, will enable attainment and maintenance of the ozone standard in the OTAG region.

   Suggested operational definition of the goal:

   1. A general modeled reduction in ozone and ozone precursors aloft throughout the OTAG region; and
   2. A reduction of ozone and ozone precursors both aloft and at ground level at the boundaries of non-attainment area modeling domains in the OTAG region; and
   3. A minimization of increases in peak ground level ozone concentrations in the OTAG region (This component of the operational definition is in review.).

2. Wellesley College Child Care Policy Research Partnership operational definition of unmet need

   1. Standard of comparison to judge the adequacy of neighborhood services: the median availability of services in the larger region (Hampden County).
   2. Thus, our definition of unmet need: The difference between the care available in the neighborhood and the median level of care in the surrounding region (stated in terms of child care slots indexed to the age-appropriate child population—"slots-per-tots").

3. Operational definitions of acids and bases

   1. An *acid* is any substance that increases the concentration of the H^+ ion when it dissolves in water.
   2. A *base* is any substance that increases the concentration of the OH^- ion when it dissolves in water.

4. Operational definition of "intelligence"

   1. Administer the Stanford-Binet IQ test to a person and score the result. The person's intelligence is the score on the test.

5. Operational definition of "dark blue carpet"

A carpet will be deemed to be dark blue if

1. Judged by an inspector medically certified as having passed the U.S. Air Force test for color-blindness
   1.1. It matches the PANTONE color card 7462 C when both carpet and card are illuminated by GE "cool white" fluorescent tubes;
   1.2. Card and carpet are viewed at a distance between 16 and 24 inches.

How to Conduct Attribute Inspection Studies

Some commonly used approaches to attribute inspection analysis are shown in Table 9.8.

Example of Attribute Inspection Error Analysis

Two sheets with identical lithographed patterns are to be inspected under carefully controlled conditions by each of the three inspectors. Each sheet has been carefully examined multiple times by journeymen lithographers and they have determined that one of the sheets should be classified as acceptable, the other as unacceptable. The inspectors sit on a stool at a large table where the sheet will be mounted for inspection. The inspector can adjust the height of the stool and the angle of the table. A lighted magnifying glass is mounted to the table with an adjustable arm that lets the inspector move it to any part of the sheet (see Fig. 9.20).

Each inspector checks each sheet once in the morning and again in the afternoon. After each inspection, the inspector classifies the sheet as either acceptable or unacceptable. The entire study is repeated the following week. The results are shown in Table 9.9.

In Table 9.9 the part column identifies which sheet is being inspected, and the standard column is the classification for the sheet based on the journey-men's evaluations. A 1 indicates that the sheet is acceptable, a 0 that it is unacceptable. The columns labeled InspA, InspB, and InspC show the classifications assigned by the three inspectors respectively. The reproducible column is a 1 if all three inspectors agree on the classification, whether their classification agrees with the standard or not. The accurate column is a 1 if *all three* inspectors classify the sheet correctly as shown in the standard column.

Individual Inspector Accuracy

Individual inspector accuracy is determined by comparing each inspector's classification with the standard. For example, in cell C2 of Table 9.9 inspector A classified the unit as acceptable, and the standard column in the same row indicates that the classification is correct. However, in cell C3 the unit is classified as unacceptable when it actually is acceptable. Continuing this evaluation shows that inspector A made the correct assessment 7 out of 8 times, for an accuracy of 0.875 or 87.5%. The results for all inspectors are given in Table 9.10.

Repeatability and Pairwise Reproducibility

Repeatability is defined in Table 9.7 as the same inspector getting the same result when evaluating the same item more than once within a short time interval. Looking at InspA

True Value	Method of Evaluation	Comments
	Expert Judgment: An expert looks at the classifications after the operator makes normal classifications and decides which are correct and which are incorrect.	☐ *Metrics*: 1. Percent correct ☐ Quantifies the accuracy of the classifications ☐ Simple to evaluate ☐ Who says the expert is correct? ☐ Care must be taken to include all types of attributes ☐ Difficult to compare operators since different units are classified by different people ☐ Acceptable level of performance must be decided upon. Consider cost, impact on customers, etc
Known	*Round Robin Study*: A set of carefully identified objects is chosen to represent the full range of attributes. 1. Each item is evaluated by an expert and its condition recorded. 2. Each item is evaluated by every inspector at least twice.	☐ *Metrics*: 1. Percent correct by inspector 2. Inspector repeatability 3. Inspector reproducibility 4. Stability 5. Inspector "linearity" ☐ Full range of attributes included ☐ All aspects of measurement error quantified ☐ People know they're being watched, may affect performance ☐ Not routine conditions ☐ Special care must be taken to insure rigor ☐ Acceptable level of performance must be decided upon for each type of error. Consider cost, impact on customers, etc ☐ *Metrics:* 1. Inspector repeatability 2. Inspector reproducibility 3. Stability 4. Inspector "linearity"
Unknown	*Inspector Concurrence Study*: A set of carefully identified objects is chosen to represent the full range of attributes, to the extent possible. 1. Each item is evaluated by every inspector at least twice.	☐ Like a round robin, except true value isn't known ☐ No measures of accuracy or bias are possible. Can only measure agreement between equally qualified people ☐ Full range of attributes included ☐ People know they're being watched, may affect performance ☐ Not routine conditions ☐ Special care must be taken to insure rigor ☐ Acceptable level of performance must be decided upon for each type of error. Consider cost, impact on customers, etc

Table 9.8 Methods of Evaluating Attribute Inspection

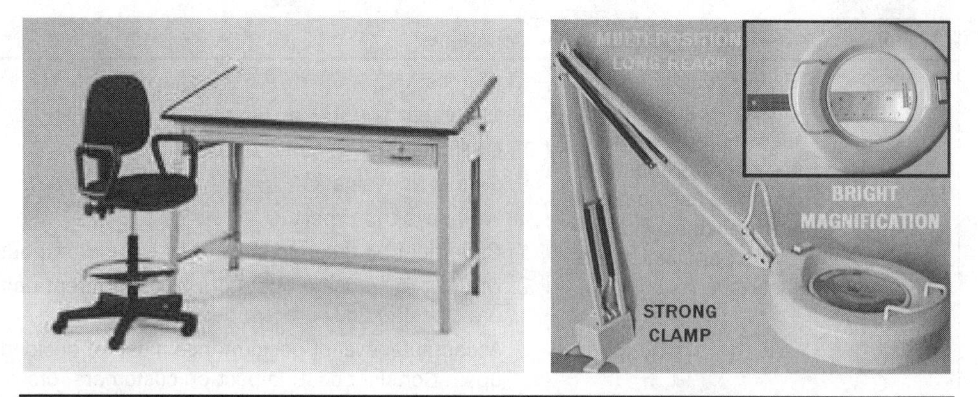

FIGURE 9.20 Lithography inspection station table, stool and magnifying glass.

	A	B	C	D	E	F	G	H	I
1	**Part**	**Standard**	**InspA**	**InspB**	**InspC**	**Date**	**Time**	**Reproducible**	**Accurate**
2	1	1	1	1	1	Today	Morning	1	1
3	1	1	0	1	1	Today	Afternoon	0	0
4	2	0	0	0	0	Today	Morning	1	1
5	2	0	0	0	1	Today	Afternoon	0	0
6	1	1	1	1	1	Last Week	Morning	1	1
7	1	1	1	1	0	Last Week	Afternoon	0	0
8	2	0	0	0	1	Last Week	Morning	0	0
9	2	0	0	0	0	Last Week	Afternoon	1	1

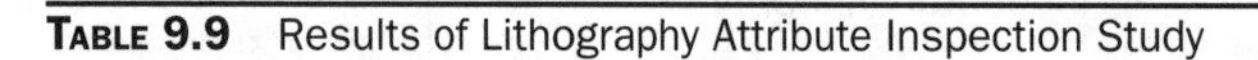

TABLE 9.9 Results of Lithography Attribute Inspection Study

Inspector	A	B	C
Accuracy	87.5%	100.0%	62.5%

TABLE 9.10 Inspector Accuracies

we see that when she evaluated Part 1 in the morning of "Today" she classified it as acceptable (1), but in the afternoon she said it was unacceptable (0). The other three morning/afternoon classifications matched each other. Thus, her repeatability is 3/4 or 75%.

	Overall			Today				Last Week		
	A	B	C		A	B	C		A	B	C
A	0.75	0.88	0.50	A	0.50	0.75	0.50	A	1.00	1.00	0.50
B		1.00	0.50	B		1.00	0.75	B		1.00	0.50
C			0.25	C			0.50	C			0.00

TABLE 9.11 Repeatability and Pairwise Reproducibility for Both Days Combined

Pairwise reproducibility is the comparison of each inspector with every other inspector when checking the same part at the same time on the same day. For example, on Part 1/Morning/Today, InspA's classification matched that of InspB. However, for Part 1/*Afternoon*/Today InspA's classification was different than that of InspB. There are eight such comparisons for each pair of inspectors. Looking at InspA versus InspB we see that they agreed 7 of the 8 times, for a pairwise repeatability of $7/8 = 0.875$.

In Table 9.11 the diagonal values are the repeatability scores and the off-diagonal elements are the pairwise reproducibility scores. The results are shown for "Today," "Last Week" and both combined.

Overall Repeatability, Reproducibility, Accuracy and Bias

Information is always lost when summary statistics are used, but the data reduction often makes the tradeoff worthwhile. The calculations for the overall statistics are operationally defined as follows:

- Repeatability is the average of the repeatability scores for the 2 days combined; that is, $(0.75 + 1.00 + 0.25)/3 = 0.67$.
- *Reproducibility* is the average of the reproducibility scores for the 2 days combined (see Table 9.9); that is,

$$\left(\frac{1+0+1+0}{4}+\frac{1+0+0+1}{4}\right)\Bigg/2 = 0.50$$

- *Accuracy* is the average of the accuracy scores for the 2 days combined (see Table 9.9); that is,

$$\left(\frac{1+0+0+0}{4}+\frac{1+0+0+0}{4}\right)\Bigg/2 = 0.25$$

- *Bias* is the estimated proportion in a category minus the true proportion in the category. In this example the true percent defective is 50% (1 part in 2). Of the 24 evaluations, 12 evaluations classified the item as defective. Thus, the bias is $0.5 - 0.5 = 0$

Overall Stability

Stability is calculated for each of the above metrics separately, as shown in Table 9.12.

Stability of...	Operational Definition of Stability	Stability Result
Repeatability	Standard deviation of the six repeatabilities (0.5, 1, 0.5, 1, 1, 1)	0.41
Reproducibility	Standard deviation of the average repeatabilities. For data in Table 9.9, STDEV [(VERAGE (H2:H5), AVERAGE (H6:H9)]	0.00
Accuracy	Standard deviation of the average accuracies. For data in Table 9.9, = STDEV [AVERAGE (2:5), AVERAGE (6:9)]	0.00
Bias	Average of bias over the 2 weeks	0.0

Table 9.12 Stability Analysis

Interpretation of Results

1. The system *overall* appears to be unbiased and accurate. However, the evaluation of individual inspectors indicates that there is room for improvement.
2. The results of the individual accuracy analysis indicate that inspector C has a problem with accuracy, see Table 9.10.
3. The results of the R&R (pairwise) indicate that inspector C has a problem with both repeatability and reproducibility, see Table 9.11.
4. The repeatability numbers are not very stable (Table 9.12). Comparing the diagonal elements for Today with those of Last Week in Table 9.11, we see that inspectors A and C tended to get different results for the different weeks. Otherwise the system appears to be relatively stable.
5. Reproducibility of inspectors A and B is not perfect. Some benefit might be obtained from looking at reasons for the difference.
6. Since inspector B's results are more accurate and repeatable, studying her might lead to the discovery of best practices.

Minitab Attribute Gage R&R Example

Minitab includes a built-in capability to analyze attribute measurement systems, known as "attribute gage R&R." We will repeat the above analysis using Minitab.

Minitab can't work with the data as shown in Table 9.9, it must be rearranged. Once the data are in a format acceptable to Minitab, we enter the Attribute Gage R&R Study dialog box by choosing Stat > Quality Tools > Attribute Gage R&R Study (see Fig. 9.21). Note the checkbox "Categories of the attribute data are ordered." Check this box if the data are ordinal and have more than two levels. Ordinal data means, for example, a 1 is in some sense "bigger" or "better" than a 0. For example, if we ask raters in a taste test a question like the following: "Rate the flavor as 0 (awful), 1 (OK), or 2 (delicious)." Our data are ordinal (acceptable is better than unacceptable), but there are only two levels, so we will not check this box.

Minitab evaluates the repeatability of appraisers by examining how often the appraiser "agrees with him/herself across trials." It does this by looking at all of the classifications for each part and counting the number of *parts* where *all* classifications agreed. For our example each appraiser looked at two parts four times each. Minitab's output, shown in Fig. 9.22, indicates that InspA rated 50% of the parts consistently,

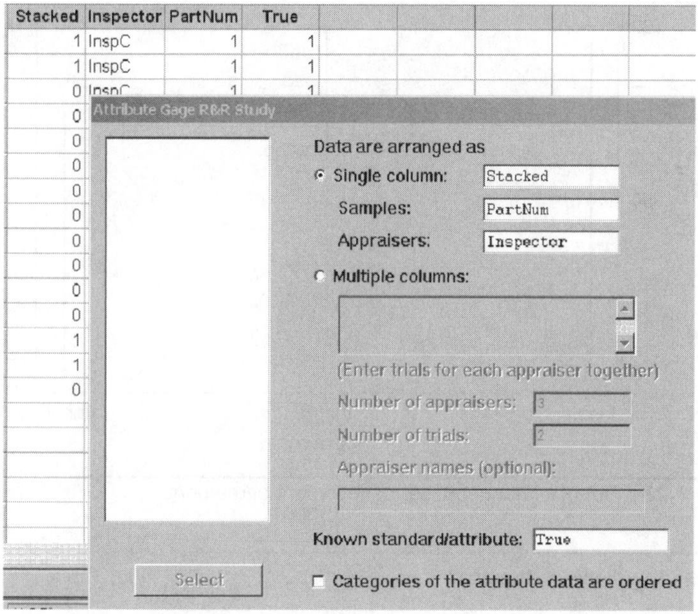

Stacked	Inspector	PartNum	True
1	InspC	1	1
1	InspC	1	1
0	InspC	1	1
0			
0			
0			
0			
0			
0			
0			
0			
0			
1			
1			
0			

Attribute Gage R&R Study

Data are arranged as

Single column: Stacked

Samples: PartNum

Appraisers: Inspector

Multiple columns:

(Enter trials for each appraiser together)

Number of appraisers: 3

Number of trials: 2

Appraiser names (optional):

Known standard/attribute: True

Select

Categories of the attribute data are ordered

Figure 9.21 Attribute gage R&R dialog box and data layout within appraiser analysis.

Within Appraiser

```
Assessment Agreement

Appraiser # Inspected # Matched Percent (%)        95.0% CI
InspA             2         1         50.0 (  1.3,   98.7)
InspB             2         2        100.0 ( 22.4,  100.0)
InspC             2         0          0.0 (  0.0,   77.6)

# Matched: Appraiser agrees with him/herself across trials.
```

Figure 9.22 Minitab within appraiser assessment agreement.

InspB 100%, and InspC 0%. The 95% confidence interval on the percentage agreement is also shown. The results are displayed graphically in Fig. 9.23.

Accuracy Analysis

Minitab evaluates accuracy by looking at how often all of an appraiser's classifications for a given part agree with the standard. Figure 9.24 shows the results for our example. As before, Minitab combines the results for both days. The plot of these results is shown in Fig. 9.25.

Minitab also looks at whether or not there is a distinct pattern in the disagreements with the standard. It does this by counting the number of times the appraiser classified an item as a 1 when the standard said it was a 0 (the 1/0 Percent column), how often the appraiser classified an item as a 0 when it was a 1 (the 0/1 Percent column), and how often the

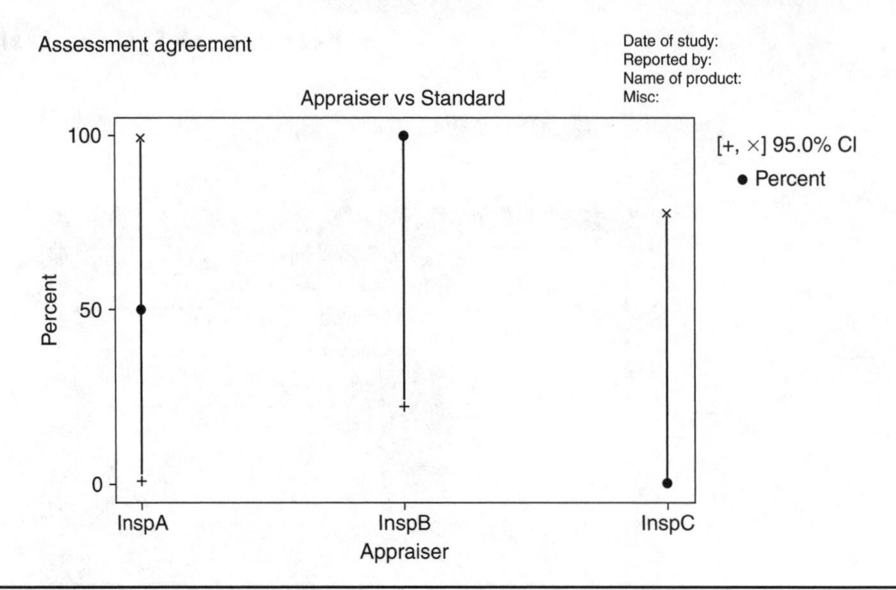

FIGURE 9.23 Plot of within appraiser assessment agreement.

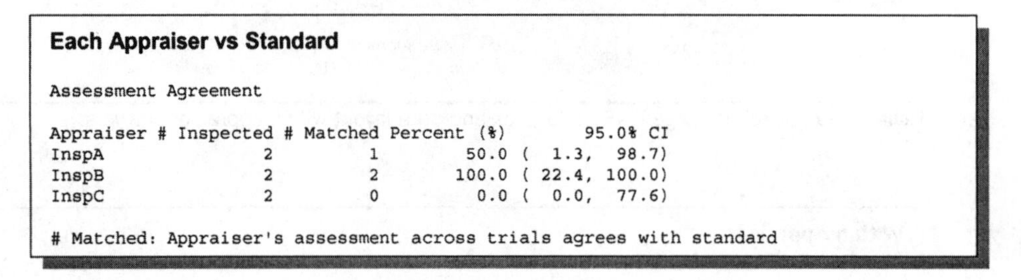

Each Appraiser vs Standard

Assessment Agreement

Appraiser	# Inspected	# Matched	Percent (%)	95.0% CI
InspA	2	1	50.0	(1.3, 98.7)
InspB	2	2	100.0	(22.4, 100.0)
InspC	2	0	0.0	(0.0, 77.6)

Matched: Appraiser's assessment across trials agrees with standard

FIGURE 9.24 Minitab appraiser versus standard agreement.

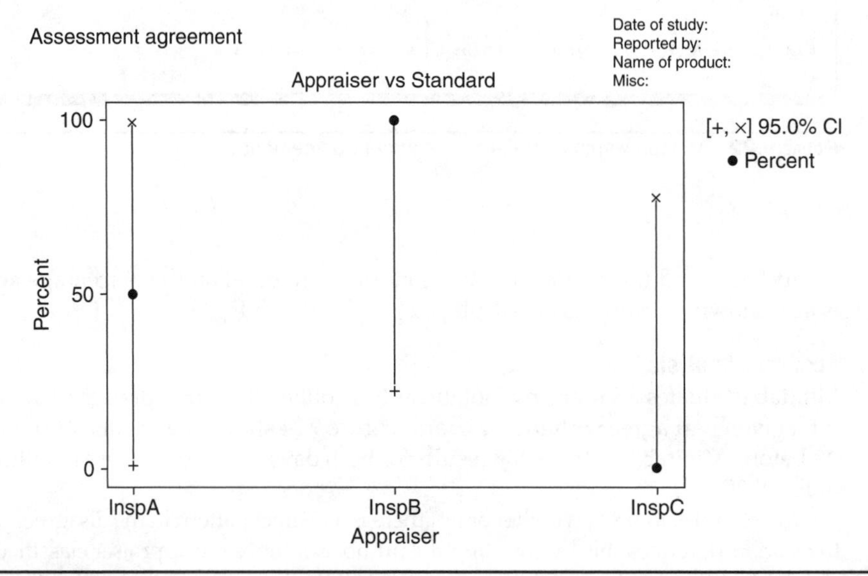

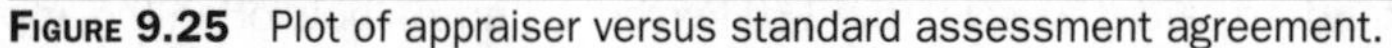
FIGURE 9.25 Plot of appraiser versus standard assessment agreement.

```
Assessment Disagreement

Appraiser    # 1/0 Percent (%)    # 0/1 Percent (%)   # Mixed Percent (%)
InspA            0         0.0        0         0.0         1        50.0
InspB            0         0.0        0         0.0         0         0.0
InspC            0         0.0        0         0.0         2       100.0

# 1/0: Assessments across trials = 1 / standard = 0.
# 0/1: Assessments across trials = 0 / standard = 1.
# Mixed: Assessments across trials are not identical.
```

Figure 9.26 Minitab appraiser assessment disagreement analysis.

```
Between Appraisers
Assessment Agreement

# Inspected # Matched Percent (%)       95.0% CI
          2         0           0.0 (  0.0,  77.6)

# Matched: All appraisers' assessments agree with each other.
```

Figure 9.27 Minitab between appraisers assessment agreement.

```
All Appraisers vs Standard
Assessment Agreement

# Inspected # Matched Percent (%)       95.0% CI
          2         0           0.0 (  0.0,  77.6)

# Matched: All appraisers' assessments agree with standard.
```

Figure 9.28 Minitab assessment versus standard agreement across all appraisers.

appraiser's classifications were mixed, i.e., is not repeatable (the # Mixed Percent column). The results are shown in Fig. 9.26. The results indicate that there is no consistent bias, defined as consistently putting a unit into the same wrong category. The problem, as was shown in the previous analysis, is that appraisers A and C are not repeatable.

Between Appraiser Assessments

Next, Minitab looks at all of the appraiser assessments for each part and counts how often every appraiser agrees on the classification of the part. The results, shown in Fig. 9.27, indicate that this never happened during our experiment. The 95% confidence interval is also shown.

All Appraisers versus Standard

Finally, Minitab looks at all of the appraiser assessments for each part and counts how often every appraiser agrees on the classification of the part and their classification agrees with the standard. This can't be any better than the between appraiser assessment agreement shown in Fig. 9.27. Unsurprisingly, the results, shown in Fig. 9.28, indicate that this never happened during our experiment. The 95% confidence interval is also shown.

CHAPTER 10

Analyze Phase

The key objectives of the Analyze phase include:

- For existing processes, analyze the value stream to identify ways to eliminate the gap between the current performance and the desired performance.
- Analyze the sources of variation that contribute to the gap (for DMAIC) or that will contribute to the design performance (for DMADV).
- Determine the drivers, the little *x*'s that correlate to the customer requirements (CTQ, CTS, CTC) and significantly influence the process or design.
- Use benchmarking techniques described in Chap. 3 to evaluate best in class for similar products or services.

Value Stream Analysis

A value stream consists of all activities, both value added and non-value added, required to bring a product from raw material into the hands of the customer, a customer requirement from order to delivery, and a design from concept to launch. Value stream improvement usually begins at the door-to-door level within a facility, and then expands outward to eventually encompass the full value stream (Womack and Jones, 1996, p. 311). A value stream consists of product and service flows, as well as information flows.

Lean principles are used to analyze the value stream. Lean, also known as the Lean Production System, has its origins in the post-World War II era in Japan. It was developed by Taiichi Ohno, a Toyota production executive, in response to a number of problems that plagued Japanese industry. The main problem was that of high-variety production required to serve the domestic Japanese market. Mass production techniques, which were developed by Henry Ford to economically produce long runs of identical product, were ill-suited to the situation faced by Toyota. Today the conditions faced by Toyota in the late 1940s are common throughout industry and Lean is being adopted by businesses all over the world as a way to improve efficiency and to better serve customers.

The Lean approach (the term Lean was coined in the early 1990s by MIT researchers) systematically minimizes waste—called *muda*—in the value stream. *Muda* includes all types of defective work, not just defective products. Wasted time, motion, and materials are all *muda*. Ohno (1988) identified the following types of *muda* in business:

1. Defects
2. Overproduction
3. Inventories (in process or finished goods)

4. Unnecessary processing
5. Unnecessary movement of people
6. Unnecessary transport of goods
7. Waiting

Womack and Jones (1996) added another type of *muda*:

8. Designing goods and services that don't meet customers' needs

Value is what customers want or need, and are willing and able to pay for. Waste is any activity that consumes resources but creates no value for the customer, thus waste activities are called "non-value added." Differentiating between the two may not be easy, especially for new products or services, but it must be done. For existing products use focus groups, surveys, and other methods described in this text. For new products, consider the DFSS methods. Most importantly, DO NOT RELY ON INTERNAL SOURCES! Most companies start with what they already know and go from there, tweaking their existing offering in some way. Customer input involves asking customers what they like or don't like about the existing offering, or what they'd like to see added or changed. The result is incremental change that may or may not address what the customers are really after. The definition of value must begin with the producer and customer jointly analyzing value and challenging old beliefs.

Consider a team with the task of reducing defects on supermarket shelving. The number one problem was "weld dents," a condition caused when brackets were welded to the shelves. A great deal of effort went into inspecting shelves for this condition, running laboratory tests to determine the impact of weld dents on the durability of the shelves, reworking shelves that had weld dents, etc. Scrap costs were very high. When the team met with customers to try to operationally define unacceptable weld dents they made an amazing discovery: customers didn't know what weld dents were! Even more strange, when shown shelves with no weld dents and those with "extreme" weld dents, customers couldn't care less. However, customers *did* care about the shape of the front of the shelves. They wanted nice, straight looking shelf fronts that looked streamlined when lined up in long supermarket aisles. They were not happy at all with what was being delivered. No one inside the company knew that this was important to customers, and no efforts were underway to improve this aspect of the product.

If the supermarket manager was asked to define value, chances are he wouldn't say "Shelves that have straight fronts that line up." Instead he might say "Shelves that look good to my customers when they look down the aisle." The importance of obtaining the voice of the customer, and using this voice to drive business processes, was discussed in Chaps. 2 and 3. Those vital Six Sigma lessons need to be integrated into Lean as well.

With your definition of value in hand, you can now begin to evaluate which activities add value and which activities are *muda*. The results are often surprising. In some cases most activities are not value added. For example, one Six Sigma team working on improving purchase order (PO) cycle time (defined as the time from receiving a request for a PO to the time the requestor received the PO) conducted a little test. They made a list of all the people whose signature was needed for PO approval. Then the team members (with the approval of the director of purchasing) hand-carried 10 POs through the process. Each purchasing agent was to treat the team member's request as their number 1 priority, dropping every other activity until it was completed. The team discovered that it took an average of about 6 hours to process a PO. The average processing time in the real world was 6 weeks. Assuming a 40-hour work-week, the value-added time accounted for only 2.5% of the total time a PO was in the system. The remaining 97.5% was *muda*.

Even that's not the full extent of the *muda*. During the walk-throughs the team also began to question why some of the approvals were needed. In some cases, such as POs for standard hardware or basic supplies, the requestor could be empowered to place the order. Many POs could be eliminated completely with automatic *pull* ordering systems (see Chap. 11 for more on pull systems). The value-added portion of the purchase order approval process was tiny indeed.

The immediate impact of such discoveries is fear. Imagine yourself as the director of purchasing or a purchasing agent. Along comes a team with data that indicate that most of your department is non-value added. Is it any wonder that change agents so often talk about "resistance to change"? Who wouldn't resist change when that change is you losing your job? Yet this is often the case and the leadership needs to face up to this reality and to plan for it. They have a responsibility to the shareholders that dictates that they reduce *muda*. They have a responsibility to customers to produce value. But they also have a responsibility to the employees to treat them fairly. Unless all the leadership makes it clear that fair treatment is guaranteed, you can expect strong resistance from people threatened by the change.

The purchasing department needs to rethink the value they add, that is, their mission. If their job isn't bureaucratic paper-shuffling, then what is it? Perhaps it is better defined as improving the integration of the supply chain with the rest of the value stream (see the following section for a discussion of the value stream). This might involve looking at how suppliers can help design easier-to-produce parts, how they can deliver to precisely the right place and at precisely the right time, what they can do to help your customers succeed, etc. This is easier to do in the process enterprise, where core business processes control the definition of work (see Chap. 2). In the end the transformed "purchasing department" will probably look much different than it did at the beginning. But if people feel that management treated everyone fairly chances are morale will improve even while *muda* is eliminated. After all, who wants to be *muda*?

The good news is that when Lean organizations redefine value, they often find that they have discovered the key to finding more customers (and more sales) very quickly. The increased demand often outpaces the rate at which resources are converted from *muda* to value creation. Although this isn't guaranteed, it happens often enough to provide a measure of comfort to employees, especially if they see it happening in their own organization. They may still need to acquire new skills to do a different kind of work, but they are usually able to adapt to this.

When trying to identify *muda* it may be helpful to think of certain categories of waste. One handy mnemonic is CLOSEDMITTS (Spencer, 1999) (Table 10.1).

Value Stream Mapping

Value stream mapping, also known as material and information flow mapping, is a variation of process mapping that looks at how value flows into and through a process and to the customer, and how information flow facilitates the work flow. One way to view a process is the logical flow of work. Another view is the physical flow of work. Figure 10.1 shows the logical flow of work for a technical support process. The team determined that the value-added steps were the ones shown with a drop-shadow box. The process map makes it obvious that the bulk of the work in the process is not part of

Type of Waste	Example
Complexity	Unnecessary steps, excessive documentation, too many permissions needed
Labor	Inefficient operations, excess headcount
Overproduction	Producing more than the customer demands. Producing before the customer needs it
Space	Storage for inventory, parts awaiting disposition, parts awaiting rework and scrap storage. Excessively wide aisles. Other wasted space
Energy	Wasted power or human energy
Defects	Repair, rework, repeated service, multiple calls to resolve problems
Materials	Scrap, ordering more than is needed
Idle materials	Material that just sits, inventory
Time	Waste of time
Transportation	Movement that adds no value
Safety hazards	Unsafe or accident-prone environments

TABLE 10.1 CLOSEDMITTS

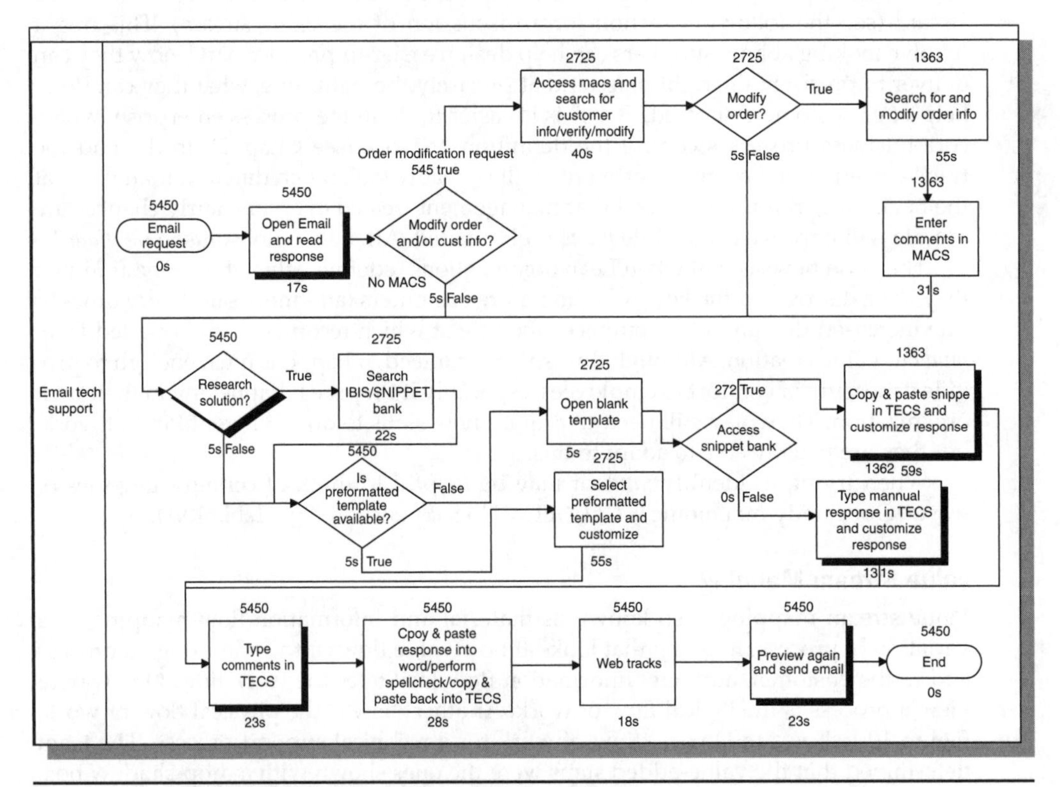

FIGURE 10.1 Flow of work through email technical support.

the value stream, it's *muda*. Based on the times shown for each activity and the path for a given support issue, somewhere between 38 and 49% of the total time is used for value-added activities.

Unfortunately, in many cases all of the non-value added steps cannot be immediately eliminated. The non-value added and unnecessary steps that can be eliminated without consequence to the business or the customer are sometimes referred to as Type II *muda*. A few of the steps involve recording information that can be used in the future to make it faster and easier to find the right answers. This is an example of a non-value added task that is necessary, based on the internal policies. Sometimes referred to as business value added (BVA) tasks, since they could be justified as necessary for business operations, these tasks cannot be eliminated immediately without consequence. Other examples of typical BVA activities include most quality functions, such as inspections, audits, and SPC, as well as related management approvals. These functions only exist because of poor quality levels: customers would not be willing to pay for these if they had alternate suppliers that could guarantee perfect quality without the cost and risk of these activities. In some cases, BVA activities are related to regulations, such as in the production of pharmaceuticals. In these cases, at least some of the BVA activities can be reduced when justified as unnecessary by sufficient analysis. Many times, BVA activities are excellent targets for significant cost reduction through process improvement and redesign. Of course, in the absence of process improvement and process redesign, the business need still exists.

How Do We Make Value Flow?

The key to value flow is the customer's requirements. What the customer needs and when he needs it drives all activity. This concept is often called *Takt time*. The formula for *Takt* time is shown in Eq. (10.1).

$$\text{Takt time} = \frac{\text{available work time}}{\text{customer required volume}} \tag{10.1}$$

Work time does not include lunches, breaks, or other process downtime. Generally, *Takt* time is used to create short-term (daily, weekly) work schedules.

Example of Takt Time Calculation

A satellite manufacturer receives orders for 26 satellites per year.

$$\text{Takt time} = \frac{\text{26 satellites}}{\text{260 work days/year}} = \text{1 satellite every 10 days} \tag{10.2}$$

This means that every work cell and operation has to move one system's worth of work through every 10 work days, no less and no more. For example, if an average satellite requires 10 batteries, then the battery operation needs to produce one battery per work day, if a satellite needs 1,000 circuit boards, then 100 boards need to be completed every work day.

If the historical process average is 20 satellites per year, then the time to produce a satellite is 13 work days, substantially short of the 10 day *Takt* time. In this case efforts need to focus on improving cycle time. On the other hand, if the historical average is 30 satellites per year, then production time is only 8.67 days per satellite and focus should be on increasing sales and reducing resources to the level dictated by customer demand.

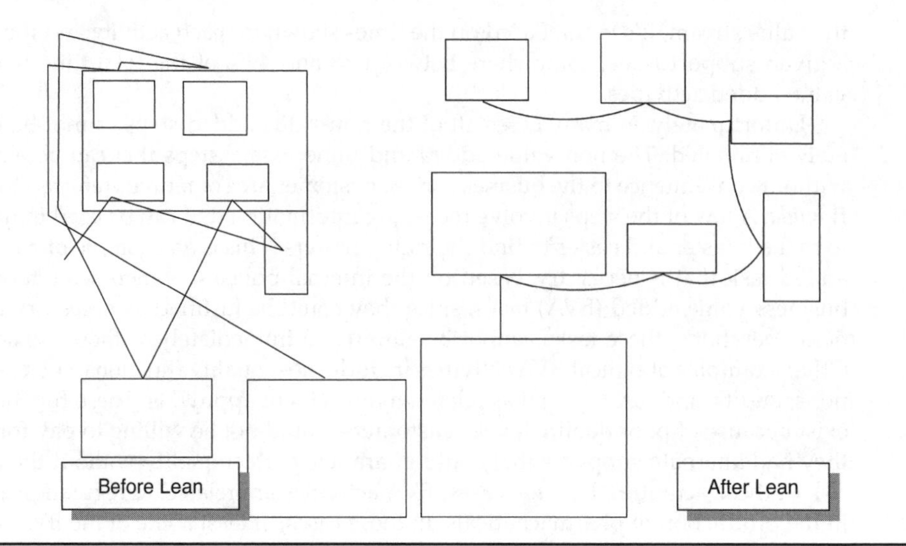

Figure 10.2 Spaghetti chart versus Lean flow.

Spaghetti Charts

Current state physical work flow is often depicted on *spaghetti charts*. A spaghetti chart is a map of the path taken by a specific product as it travels down the value stream in a mass-production organization, so-called because the product's route typically looks like a plate of spaghetti. To create a spaghetti chart, like the one shown on the left in Fig. 10.2, tell a person to "be the part" and to physically walk through the process as the part would go through it. Sometimes a part travels *miles* in the original process configuration, and only a few feet in the Lean layout. The Lean layout is shown on the right in Fig. 10.2. The difference between the current state layout and the Lean layout is *muda*.

When setting goals for a future state process, it is often helpful to stretch the mind. One way to become inspired is to identify the absolute best in class performance for a particular activity. For example, the quick lube joints' claim to exceptional value is that they can get you in and out in 15 minutes or less, much quicker than the corner "service station" which often took a couple of hours or more. But consider the pit crew of a Nascar racing team, which can perform maintenance on a car so fast (14 seconds or less) they make your local Quickie Lube look like they're working at a crawl. And during those 14 seconds they do a great deal more than change the car's fluids. They gas it up, wash the windows, change all of the tires, etc. (Fig. 10.3). There are many published examples of Lean achievements that can serve to educate and inspire. At the CAMI factory operated by GM and Suzuki, machine changeover time was reduced from 36 hours to 6 minutes.

Analyzing the Sources of Variation

Whether the focus of the Six Sigma project relates directly to the mechanics of the value stream, such as cycle time and cost/capacity-related resource allocation, or indirectly, as in the relative quality of its output, a critical piece of the analysis is the understanding of the relative contributions of the sources of variation impacting the value stream. A review of the statistical control chart constructed in the Measure phase will provide input as to the type of variation: common cause intrinsic to the process or special causes

Figure 10.3 UPS racing team pit crew.

that occur sporadically under specific conditions. As discussed in the Measure stage, the response to each of these types of variation differs significantly.

The potential sources of process variation may be brainstormed by the Six Sigma team using a cause and effect diagram. These potential causes must then be analyzed for their significance using more advanced statistical tools, including designed experiments and their associated enumerative methods.

The basic statistical methods, including confidence intervals and hypothesis tests, are discussed in the sections that follow. While these rather simple statistical methods may be used to directly compare a sample to its desired properties, or one sample to another, designed experiments will build on these concepts in applying ANOVA (analysis of variance) techniques to multiple sources of variation, allowing quantification of the relative contribution of each source to the total error. General regression and correlation analysis is presented as a precursor to the designed experiments, to aid in the understanding of the experimental analysis techniques.

Cause and Effect Diagrams

With most practical applications, the number of possible causes for any given problem can be huge. Dr. Kaoru Ishikawa developed a simple method of graphically displaying the causes of any given quality problem. His method is called by several names, the Ishikawa diagram, the fishbone diagram, and the cause and effect diagram.

Cause and effect diagrams are tools that are used to organize and graphically display all of the knowledge a group has relating to a particular problem.

Constructing the cause and effect diagram is very simple. The steps are:

1. Draw a box on the far right-hand side of a large sheet of paper and draw a horizontal arrow that points to the box. Inside of the box, write the description of the problem you are trying to solve.

2. Write the names of the categories above and below the horizontal line. Think of these as branches from the main trunk of the tree.
3. Draw in the detailed cause data for each category. Think of these as limbs and twigs on the branches.

A good cause and effect diagram will have many "twigs," as shown in Fig. 10.4. If your cause and effect diagram doesn't have a lot of smaller branches and twigs, it shows that the understanding of the problem is superficial. Chances are that you need the help of someone outside of your group to aid in the understanding, perhaps someone more closely associated with the problem.

Cause and effect diagrams come in several basic types. The dispersion analysis type is created by repeatedly asking "why does this dispersion occur?" For example, we might want to know why all of our fresh peaches don't have the same color.

The production process class cause and effect diagram uses production processes as the main categories, or branches of the diagram. The processes are shown joined by the horizontal line. Figure 10.5 is an example of this type of diagram.

The cause enumeration cause and effect diagram simply displays all possible causes of a given problem grouped according to rational categories. This type of cause and effect diagram lends itself readily to the brainstorming approach we are using.

A variation of the basic cause and effect diagram, developed by Dr. Ryuji Fukuda of Japan, is cause and effect diagrams with the addition of cards, or CEDAC. The main difference is that the group gathers ideas outside of the meeting room on small cards, as well as in group meetings. The cards also serve as a vehicle for gathering input from people who are not in the group; they can be distributed to anyone involved with the process. Often the cards provide more information than the brief entries on a standard cause and effect diagram. The cause and effect diagram is built by actually placing the cards on the branches.

Boxplots

A *boxplot* displays summary statistics for a set of distributions. It is a plot of the 25th, 50th, and 75th percentiles, as well as values far removed from the rest.

Figure 10.6 shows an annotated sketch of a boxplot. The lower boundary of the box is the 25th percentile. Tukey refers to the 25th and 75th percentile "hinges." Note that the 50th percentile is the median of the overall data set, the 25th percentile is the median of those values below the median, and the 75th percentile is the median of those values above the median. The horizontal line inside the box represents the median. Fifty percent of the cases are included within the box. The box length corresponds to the interquartile range, which is the difference between the 25th and 75th percentiles.

The boxplot includes two categories of cases with outlying values. Cases with values that are more than 3 box-lengths from the upper or lower edge of the box are called extreme values. On the boxplot, these are designated with an asterisk (*). Cases with values that are between 1.5 and 3 box-lengths from the upper or lower edge of the box are called outliers and are designated with a circle. The largest and smallest observed values that aren't outliers are also shown. Lines are drawn from the ends of the box to these values. (These lines are sometimes called whiskers and the plot is then called a box-and-whiskers plot.)

Despite its simplicity, the boxplot contains an impressive amount of information. From the median you can determine the central tendency, or location. From the length

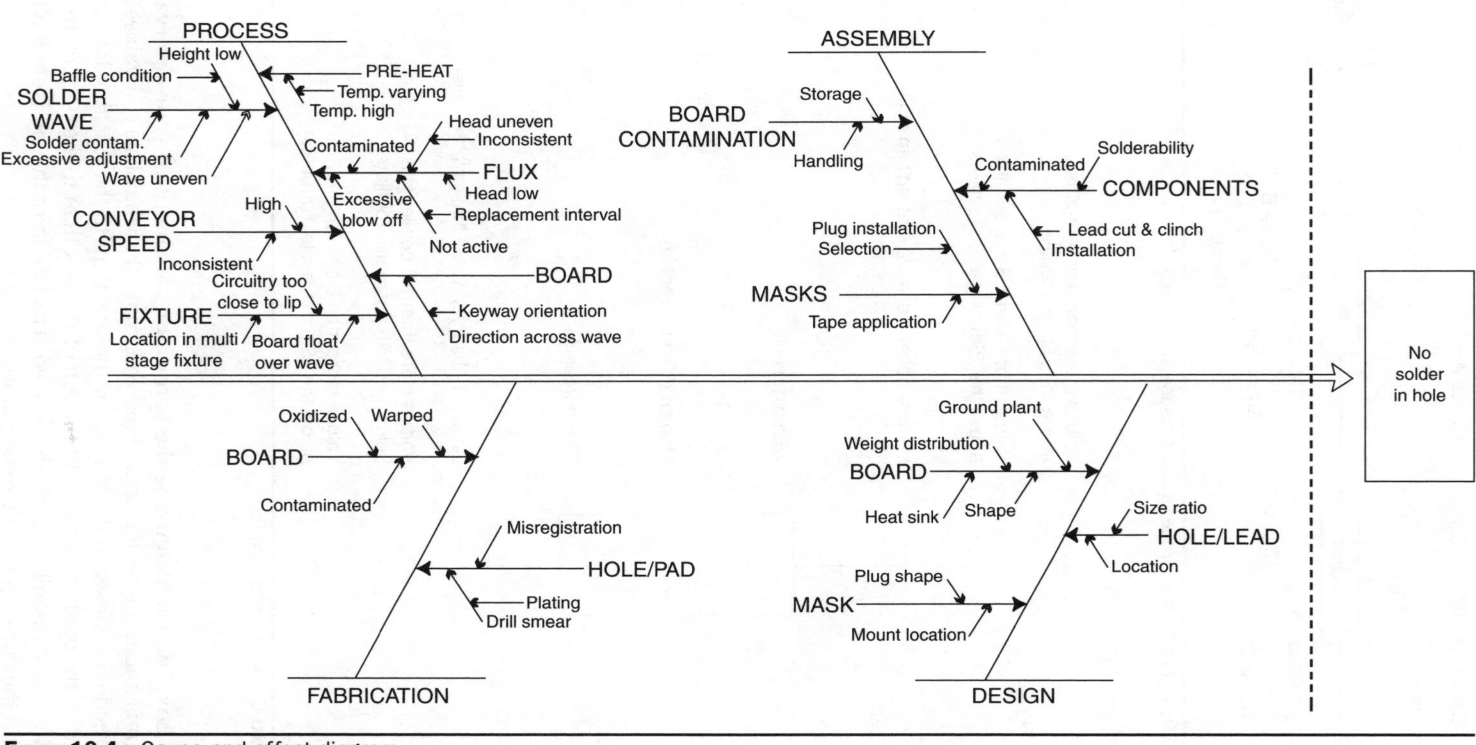

FIGURE 10.4 Cause and effect diagram.

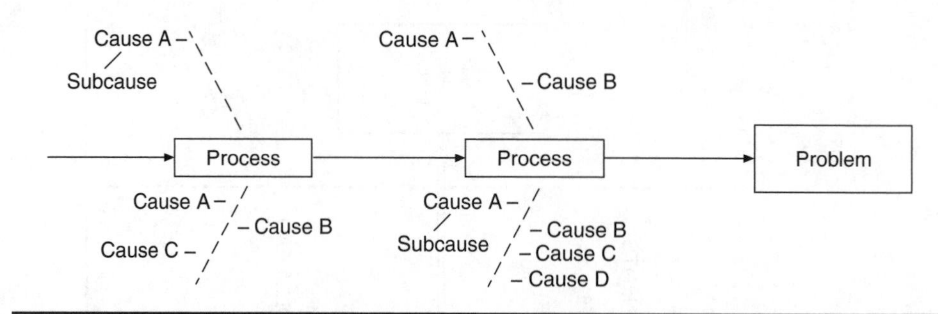

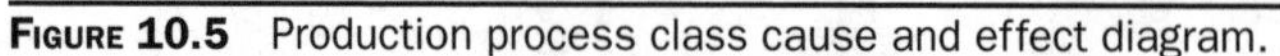

FIGURE 10.5 Production process class cause and effect diagram.

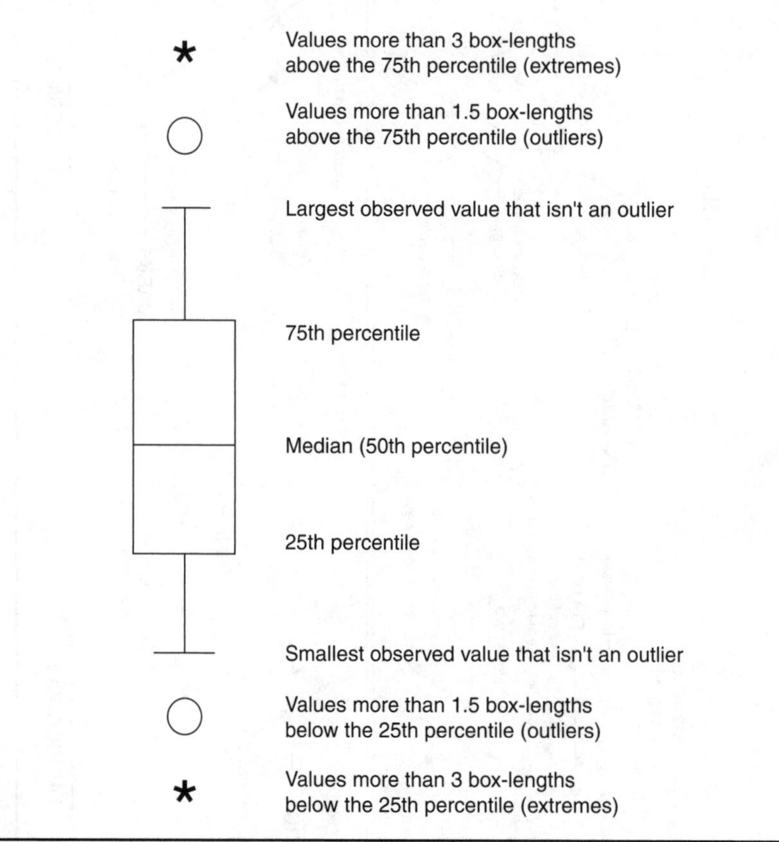

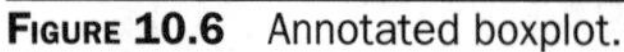

FIGURE 10.6 Annotated boxplot.

of the box, you can determine the spread, or variability, of your observations. If the median is not in the center of the box, you know that the observed values are skewed. If the median is closer to the bottom of the box than to the top, the data are positively skewed. If the median is closer to the top of the box than to the bottom, the opposite is true: the distribution is negatively skewed. The length of the tail is shown by the whiskers and the outlying and extreme points.

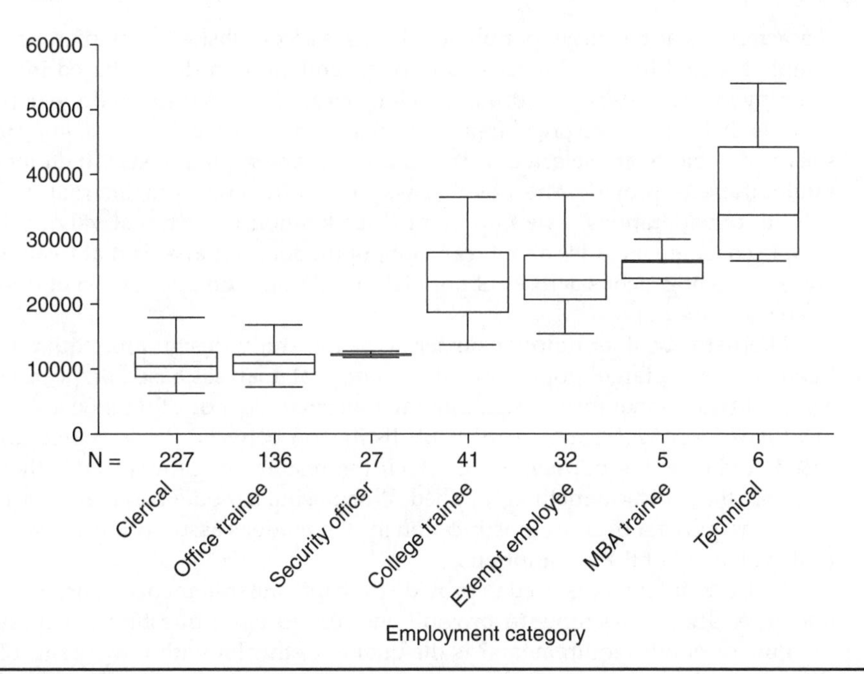

FIGURE 10.7 Boxplots of salary by job category.

Boxplots are particularly useful for comparing the distribution of values in several groups. Figure 10.7 shows boxplots for the salaries for several different job titles.

The boxplot makes it easy to see the different properties of the distributions. The location, variability, and shapes of the distributions are obvious at a glance. This ease of interpretation is something that statistics alone cannot provide.

Statistical Inference

This section discusses the basic concept of statistical inference. The reader should also consult the glossary in the Appendix for additional information. Inferential statistics belong to the enumerative class of statistical methods. All statements made in this section are valid only for stable processes, that is, processes in statistical control. Although most applications of Six Sigma are analytic, there are times when enumerative statistics prove useful. The term *inference* is defined as (1) the act or process of deriving logical conclusions from premises known or assumed to be true, or (2) the act of reasoning from factual knowledge or evidence. Inferential statistics provide information that is used in the process of inference. As can be seen from the definitions, inference involves two domains: the premises and the evidence or factual knowledge. Additionally, there are two conceptual frameworks for addressing premises questions in inference: the design-based approach and the model-based approach.

As discussed by Koch and Gillings (1983), a statistical analysis whose only assumptions are random selection of units or random allocation of units to experimental conditions results in *design-based inferences*; or, equivalently, randomization-based inferences. The objective is to structure sampling such that the sampled population has the same characteristics as the target population. If this is accomplished then inferences from the sample are said to have internal validity. A limitation on design-based inferences for experimental studies is that formal conclusions are restricted to the finite population of subjects that actually received treatment, that is, they lack *external validity*. However, if sites and subjects are selected at random from larger eligible sets, then models with random effects provide one possible way of addressing both internal and external validity considerations. One important consideration for external validity is that the sample coverage includes all relevant subpopulations; another is that treatment differences be homogeneous across subpopulations. A common application of design-based inference is the survey.

Alternatively, if assumptions external to the study design are required to extend inferences to the target population, then statistical analyses based on postulated probability distributional forms (e.g., binomial, normal, etc.) or other stochastic processes yield *model-based inferences*. A focus of distinction between design-based and model-based studies is the population to which the results are generalized rather than the nature of the statistical methods applied. When using a model-based approach, external validity requires substantive justification for the model's assumptions, as well as statistical evaluation of the assumptions.

Statistical inference is used to provide probabilistic statements regarding a scientific inference. Science attempts to provide answers to basic questions, such as can this machine meet our requirements? Is the quality of this lot within the terms of our contract? Does the new method of processing produce better results than the old? These questions are answered by conducting an experiment, which produces data. If the data vary, then statistical inference is necessary to interpret the answers to the questions posed. A statistical model is developed to describe the probabilistic structure relating the observed data to the quantity of interest (the *parameters*), that is, a scientific hypothesis is formulated. Rules are applied to the data and the scientific hypothesis is either rejected or not. In formal tests of a hypothesis, there are usually two mutually exclusive and exhaustive hypotheses formulated: a *null hypothesis* and an *alternate hypothesis*.

Chi-Square, Student's *T*, and *F* Distributions

In addition to the distributions present earlier in the Measure phase, these three distributions are used in Six Sigma to test hypotheses, construct confidence intervals, and compute control limits.

Chi-Square

Many characteristics encountered in Six Sigma have normal or approximately normal distributions. It can be shown that in these instances the distribution of sample variances has the form (except for a constant) of a chi-square distribution, symbolized χ^2. Tables have been constructed giving abscissa values for selected ordinates of the cumulative χ^2 distribution. One such table is given in Appendix 4.

The χ^2 distribution varies with the quantity υ, which for our purposes is equal to the sample size minus 1. For each value of υ there is a different χ^2 distribution. Equation (10.3) gives the pdf for the χ^2.

$$f(\chi^2) = \frac{e^{-\chi^2/2}(\chi^2)^{(\upsilon-2)/2}}{2^{\upsilon/2}\left(\frac{\upsilon-2}{2}\right)!} \tag{10.3}$$

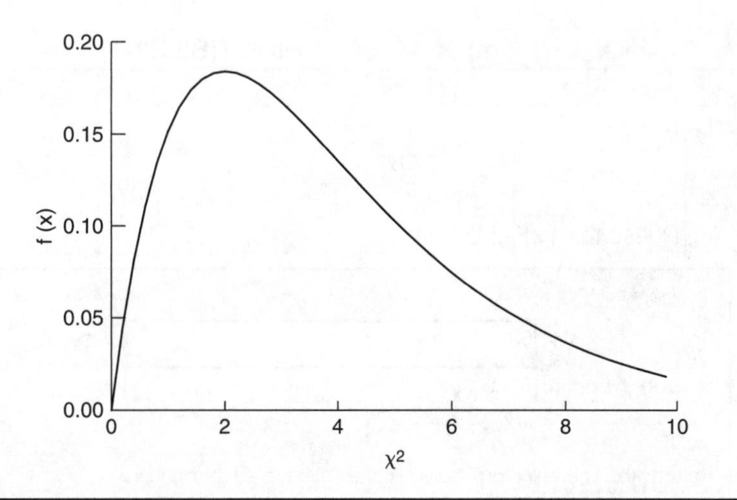

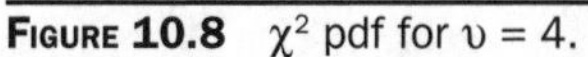
FIGURE 10.8 χ^2 pdf for $\upsilon = 4$.

Figure 10.8 shows the pdf for $\upsilon = 4$.

Example

The use of χ^2 is illustrated in this example to find the probability that the variance of a sample of n items from a specified normal universe will equal or exceed a given value s^2; we compute $\chi^2 = (n-1)\, s^2/\sigma^2$. Now, let's suppose that we sample $n = 10$ items from a process with $\sigma^2 = 25$ and wish to determine the probability that the sample variance will exceed 50. Then

$$\frac{(n-1)s^2}{\sigma^2} = \frac{9(50)}{25} = 18$$

We enter Appendix 4 (χ^2) at the line for $\upsilon = 10 - 1 = 9$ and note that 18 falls between the columns for the percentage points of 0.025 and 0.05. Thus, the probability of getting a sample variance in excess of 50 is about 3%.

It is also possible to determine the sample variance that would be exceeded only a stated percentage of the time. For example, we might want to be alerted when the sample variance exceeded a value that should occur only once in 100 times. Then we set up the χ^2 equation, find the critical value from Appendix 4, and solve for the sample variance. Using the same values as above, the value of s^2 that would be exceeded only once in 100 times is found as follows:

$$\frac{9s^2}{\sigma^2} = \frac{9s^2}{25} = 21.7 \Rightarrow s^2 = \frac{21.7 \times 25}{9} = 60.278$$

In other words, the variance of samples of size 10, taken from this process, should be less than 60.278, 99% of the time.

Example of Chi-Squared Probability Calculations Using Microsoft Excel

Microsoft Excel has a built-in capability to calculate chi-squared probabilities. To solve the above problem using Excel, enter the n and x values as shown in Fig. 10.9. Note that

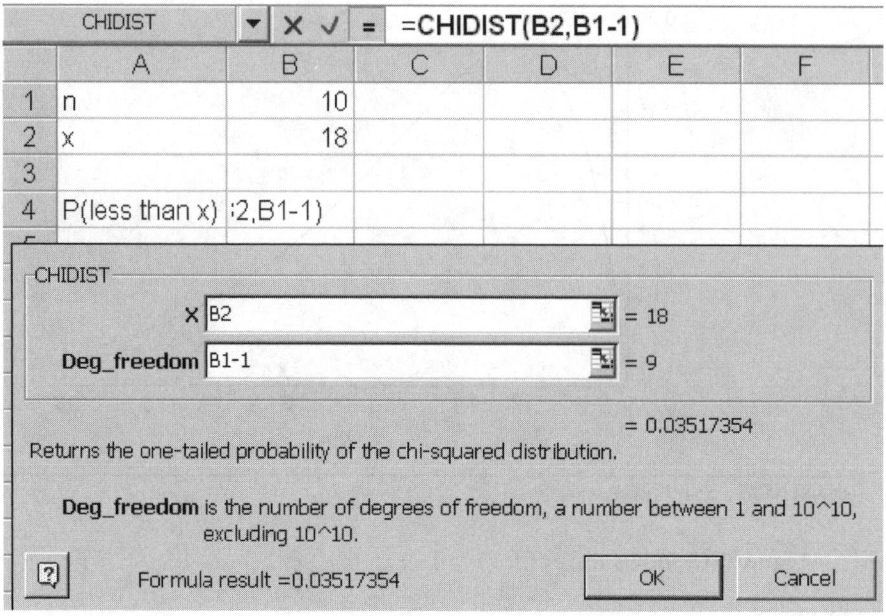

CHIDIST | =CHIDIST(B2,B1-1)

	A	B	C	D	E	F
1	n	10				
2	x	18				
3						
4	P(less than x)	:2,B1-1)				

CHIDIST

X B2 = 18

Deg_freedom B1-1 = 9

= 0.03517354

Returns the one-tailed probability of the chi-squared distribution.

Deg_freedom is the number of degrees of freedom, a number between 1 and 10^10, excluding 10^10.

Formula result =0.03517354 OK Cancel

FIGURE 10.9 Example of finding chi-squared probability using Microsoft Excel.

Excel uses degrees of freedom rather than the sample size in its calculations; degrees of freedom is the sample size minus one, as shown in the Deg_freedom box in Fig. 10.9. The formula result near the bottom of the screen gives the desired probability.

Example of Inverse Chi-Squared Probability Calculations Using Microsoft Excel

Microsoft Excel has a built-in capability to calculate chi-squared probabilities, making it unnecessary to look up the probabilities in tables. To find the critical chi-squared value for the above problem using Excel, use the CHIINV function and enter the desired probability and degrees of freedom as shown in Fig. 10.10. The formula result near the bottom of the screen gives the desired critical value.

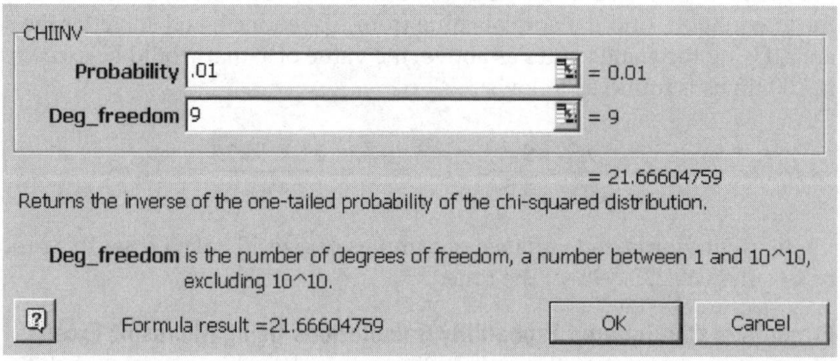

CHIINV

Probability .01 = 0.01

Deg_freedom 9 = 9

= 21.66604759

Returns the inverse of the one-tailed probability of the chi-squared distribution.

Deg_freedom is the number of degrees of freedom, a number between 1 and 10^10, excluding 10^10.

Formula result =21.66604759 OK Cancel

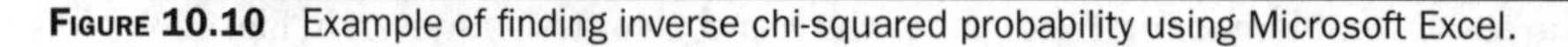

FIGURE 10.10 Example of finding inverse chi-squared probability using Microsoft Excel.

Student's *T* Distribution

The *t* statistic is commonly used to test hypotheses regarding means, regression coefficients and a wide variety of other statistics used in quality engineering. "Student" was the pseudonym of W.S. Gosset, whose need to quantify the results of small scale experiments motivated him to develop and tabulate the probability integral of the ratio which is now known as the *t* statistic and is shown in Eq. (10.4).

$$t = \frac{\overline{X} - \mu}{s / \sqrt{n}} \tag{10.4}$$

In Eq. (10.4), the denominator is the standard deviation of the sample mean. Percentage points of the corresponding distribution function of *t* may be found in Appendix 3. There is a *t* distribution for each sample size of $n > 1$. As the sample size increases, the *t* distribution approaches the shape of the normal distribution, as shown in Fig. 10.11.

One of the simplest (and most common) applications of the student's *t* test involves using a sample from a normal population with mean μ and variance σ^2. This is demonstrated in the hypothesis testing section later in this chapter.

F Distribution

Suppose we have two random samples drawn from a normal population. Let s_1^2 be the variance of the first sample and s_1^2 be the variance of the second sample. The two samples need not have the same sample size. The statistic *F* given by

$$F = \frac{s_1^2}{s_2^2} \tag{10.5}$$

has a sampling distribution called the *F distribution*. There are two sample variances involved and two sets of degrees of freedom, $n_1 - 1$ in the numerator and $n_2 - 1$ in the

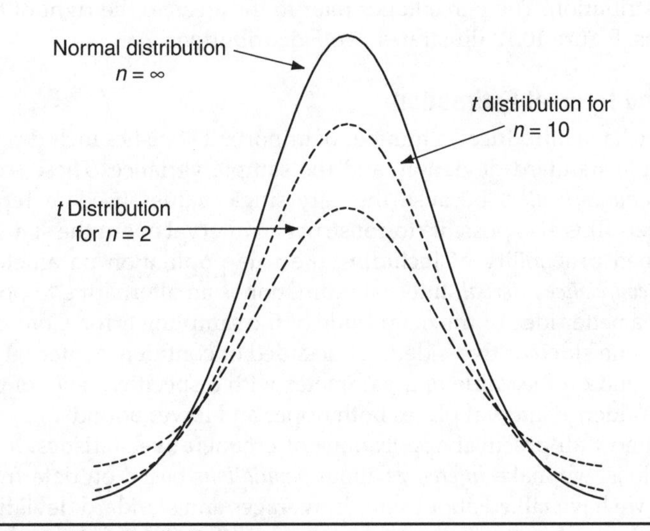

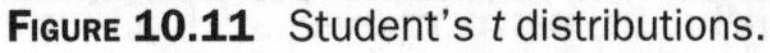
FIGURE 10.11 Student's *t* distributions.

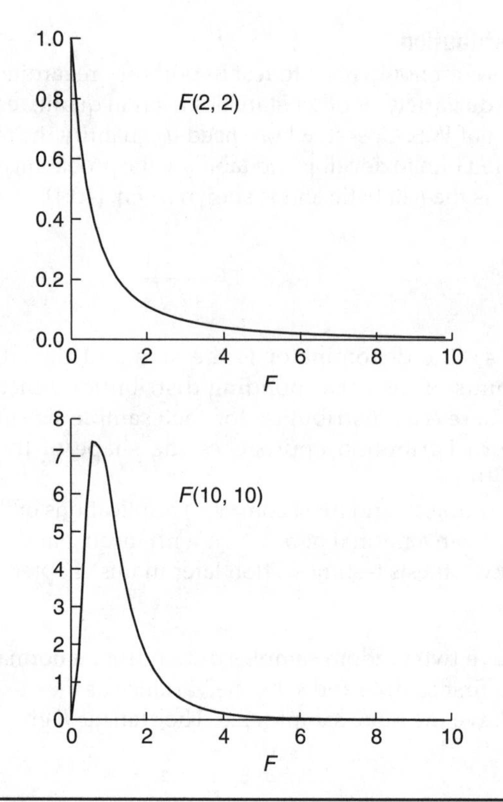

Figure 10.12 *F* distributions.

denominator. Appendix 5 and 6 provide values for the 1 and 5% percentage points for the *F* distribution. The percentages refer to the areas to the right of the values given in the tables. Figure 10.12 illustrates two *F* distributions.

Point and Interval Estimation

So far, we have introduced a number of important statistics including the sample mean, the sample standard deviation, and the sample variance. These sample statistics are called *point estimators* because they are single values used to represent population parameters. It is also possible to construct an interval about the statistics that has a predetermined probability of including the true population parameter. This interval is called a *confidence interval*. Interval estimation is an alternative to point estimation that gives us a better idea of the magnitude of the sampling error. Confidence intervals can be either one-sided or two-sided. A one-sided or confidence interval places an upper or lower bound on the value of a parameter with a specified level of confidence. A two-sided confidence interval places both upper and lower bounds.

In almost all practical applications of enumerative statistics, including Six Sigma applications, we make *inferences* about *populations* based on data from *samples*. In this chapter, we have talked about sample averages and standard deviations; we have even used these numbers to make statements about future performance, such as long term yields or potential failures. A problem arises that is of considerable practical importance: any estimate that is based on a sample has some amount of sampling error. This is true even though the sample estimates are the "best estimates" in the sense that they are (usually) unbiased estimators of the population parameters.

Estimates of the Mean

For random samples with replacement, the sampling distribution of $\bar{X}$ has a mean μ and a standard deviation equal to $\sigma/\sqrt{n}$. For large samples the sampling distribution of $\bar{X}$ is approximately normal and normal tables can be used to find the probability that a sample mean will be within a given distance of μ.

For example, in 95% of the samples we will observe a mean within $\pm 1.96\sigma/\sqrt{n}$ of μ. In other words, in 95% of the samples the interval from $\bar{X} - 1.96\sigma/\sqrt{n}$ to $\bar{X} + 1.96\sigma/\sqrt{n}$ will include μ. This interval is called a "95% confidence interval for estimating μ." It is usually shown using inequality symbols:

$$\bar{X} - 1.96\sigma/\sqrt{n} < \mu\bar{X} + 1.96\sigma/\sqrt{n}$$

The factor 1.96 is the Z value obtained from the normal in the Appendix 2. It corresponds to the Z value beyond which 2.5% of the population lie. Since the normal distribution is symmetric, 2.5% of the distribution lies above Z and 2.5% below –Z. The notation commonly used to denote Z values for confidence interval construction or hypothesis testing is $Z_{\alpha/z}$ where $100(1 - \alpha)$ is the desired confidence level in percent. For example, if we want 95% confidence, $\alpha = 0{:}05$, $100(1 - \alpha) = 95\%$, and $Z_{0.025} = 1.96$. In hypothesis testing the value of α is known as the *significance level*.

Example: Estimating μ When σ Is Known

Suppose that σ is known to be 2.8. Assume that we collect a sample of $n = 16$ and compute $\bar{X} = 15.7$. Using the e equation mentioned in previous section we find the 95% confidence interval for μ as follows:

$$\bar{X} - 1.96\sigma/\sqrt{n} < \mu < \bar{X} + 1.96\sigma/\sqrt{n}$$

$$15.7 - 1.96(2.8/\sqrt{16}) < \mu < 15.7 + 1.96(2.8/\sqrt{16})$$

$$14.33 < \mu < 17.07$$

There is a 95% *level of confidence* associated with this interval. The numbers 14.33 and 17.07 are sometimes referred to as the *confidence limits*.

Note that this is a two-sided confidence interval. There is a 2.5% probability that 17.07 is lower than μ and a 2.5% probability that 14.33 is greater than μ. If we were only interested in, say, the probability that μ were greater than 14.33, then the one-sided confidence interval would be $\mu > 14.33$ and the one-sided confidence level would be 97.5%.

Example of Using Microsoft Excel to Calculate the Confidence Interval for the Mean When Sigma Is Known

Microsoft Excel has a built-in capability to calculate confidence intervals for the mean. The dialog box in Fig. 10.13 shows the input. The formula result near the bottom of

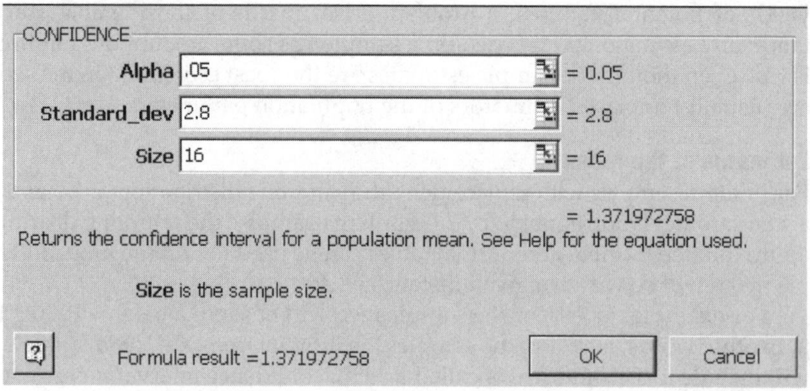

FIGURE 10.13 Example of finding the confidence interval when sigma is known using Microsoft Excel.

the screen gives the interval width as 1.371972758. To find the lower confidence limit subtract the width from the mean. To find the upper confidence limit add the width to the mean.

Example: Estimating μ When σ Is Unknown

When σ is not known and we wish to replace σ with s in calculating confidence intervals for μ, we must replace $Z_{\alpha/2}$ with $t_{\alpha/2}$ and obtain the percentiles from tables for student's t distribution instead of the normal tables. Let's revisit the example above and assume that instead of knowing σ, it was estimated from the sample, that is, based on the sample of $n = 16$, we computed $s = 2.8$ and $\bar{X} = 15.7$. Then the 95% confidence interval becomes:

$$\bar{X} - 2.131s/\sqrt{n} < \mu < \bar{X} + 2.131s/\sqrt{n}$$

$$15.7 - 2.131(2.8/\sqrt{16}) < \mu < 15.7 + 2.131(2.8/\sqrt{16})$$

$$14.21 < \mu < 17.19$$

It can be seen that this interval is wider than the one obtained for known σ. The $t_{\alpha/2}$ value found for 15 df is 2.131 (see Table 3 in the Appendix), which is greater than $Z_{\alpha/2} =$ 1.96 above.

Example of Using Microsoft Excel to Calculate the Confidence Interval for the Mean When Sigma Is Unknown

Microsoft Excel has no built-in capability to calculate confidence intervals for the mean when sigma is not known. However, it does have the ability to calculate t-values when given probabilities and degrees of freedom. This information can be entered into an equation and used to find the desired confidence limits. Figure 10.14 illustrates the approach. The formula bar shows the formula for the 95% upper confidence limit for the mean in cell B7.

B7		=$B$1+TINV($B$4,$B$3-1)* $B$2/SQRT($B$3)
	A	B
1	Mean	15.7
2	sigma	2.8
3	n	16
4	Alpha	0.05
5		
6	Lower Confidence Limit	14.21
7	Upper Confidence Limit	17.19

FIGURE 10.14 Example of finding the confidence interval when sigma is unknown using Microsoft Excel.

Hypothesis Testing

Statistical inference generally involves four steps:

1. Formulating a hypothesis about the population or "state of nature"
2. Collecting a sample of observations from the population
3. Calculating statistics based on the sample
4. Either accepting or rejecting the hypothesis based on a predetermined acceptance criterion

There are two types of error associated with statistical inference:

Type I error (α error)—The probability that a hypothesis that is actually true will be rejected. The value of α is known as the significance level of the test.

Type II error (β error)—The probability that a hypothesis that is actually false will be accepted.

Type II errors are often plotted in what is known as an operating characteristics curve.

Confidence intervals are usually constructed as part of a *statistical test of hypotheses*. The hypothesis test is designed to help us make an inference about the true population value at a desired level of confidence. We will look at a few examples of how hypothesis testing can be used in Six Sigma applications.

Example: Hypothesis Test of Sample Mean

Experiment: The nominal specification for filling a bottle with a test chemical is 30 cc. The plan is to draw a sample of $n = 25$ units from a stable process and, using the sample mean and standard deviation, construct a two-sided confidence interval (an interval that extends on either side of the sample average) that has a 95% probability of including the true population mean. If the interval includes 30, conclude that the lot mean is 30, otherwise conclude that the lot mean is not 30.

Result: A sample of 25 bottles was measured and the following statistics computed

$$\bar{X} = 28 \text{ cc}$$
$$s = 6 \text{ cc}$$

The appropriate test statistic is t, given by the formula

$$t = \frac{\bar{X} - \mu}{s/\sqrt{n}} = \frac{28 - 30}{6/\sqrt{25}} = -1.67$$

Table 3 in the Appendix gives values for the t statistic at various degrees of freedom. There are $n - 1$ degrees of freedom (df). For our example we need the $t_{.975}$ column and the row for 24 df. This gives a t value of 2.064. Since the absolute value of this t value is greater than our test statistic, we fail to reject the hypothesis that the lot mean is 30 cc. Using statistical notation this is shown as:

H_0:μ = 30 cc (the *null hypothesis*)
H_1:μ is not equal to 30 cc (the *alternate hypothesis*)
$\alpha = .05$ (*Type I error or level of significance*)
Critical region: $-2.064 \leq t_0 \leq +2.064$
Test statistic: $t = -1.67$.

Since t lies inside the critical region, fail to reject H_0, and accept the hypothesis that the lot mean is 30 cc for the data at hand.

Example: Hypothesis Test of Two Sample Variances

The variance of machine X's output, based on a sample of $n = 25$ taken from a stable process, is 100. Machine Y's variance, based on a sample of 10, is 50. The manufacturing representative from the supplier of machine X contends that the result is a mere "statistical fluke." Assuming that a "statistical fluke" is something that has less than 1 chance in 100, test the hypothesis that both variances are actually equal.

The test statistic used to test for equality of two sample variances is the F statistic, which, for this example, is given by the equation

$$F = \frac{s_1^2}{s_2^2} = \frac{100}{50} = 2, \text{numerator df} = 24, \text{denominator df} = 9$$

Using Table 5 in the Appendix for $F_{.99}$ we find that for 24 df in the numerator and 9 df in the denominator $F = 4.73$. Based on this we conclude that the manufacturer of machine X could be right, the result could be a statistical fluke. This example demonstrates the volatile nature of the sampling error of sample variances and standard deviations.

Example: Hypothesis Test of a Standard Deviation Compared to a Standard Value

A machine is supposed to produce parts in the range of 0.500 inch plus or minus 0.006 inch. Based on this, your statistician computes that the absolute worst standard deviation tolerable is 0.002 inch. In looking over your capability charts you find that the best machine in the shop has a standard deviation of 0.0022, based on a sample of 25 units.

In discussing the situation with the statistician and management, it is agreed that the machine will be used if a one-sided 95% confidence interval on sigma includes 0.002.

The correct statistic for comparing a sample standard deviation with a standard value is the chi-square statistic. For our data we have $s = 0.0022$, $n = 25$, and $\sigma_0 = 0{:}002$. The χ^2 statistic has $n - 1 = 24$ degrees of freedom. Thus,

$$\chi^2 = \frac{(n-1)s^2}{\sigma^2} = \frac{24 \times (0.0022)^2}{(0.002)^2} = 29.04$$

Appendix 4 gives, in the 0.05 column (since we are constructing a one-sided confidence interval) and the df = 24 row, the critical value $\chi^2 = 36.42$. Since our computed value of χ^2 is less than 36.42, we use the machine. The reader should recognize that all of these exercises involved a number of assumptions, for example, that we "know" that the best machine has a standard deviation of 0.0022. In reality, this knowledge must be confirmed by a stable control chart.

Resampling (Bootstrapping)

A number of criticisms have been raised regarding the methods used for estimation and hypothesis testing:

- They are not intuitive.
- They are based on strong assumptions (e.g., normality) that are often not met in practice.
- They are difficult to learn and to apply.
- They are error-prone.

In recent years a new method of performing these analyses has been developed. It is known as resampling or bootstrapping. The new methods are conceptually quite simple: using the data from a sample, calculate the statistic of interest repeatedly and examine the distribution of the statistic. For example, say you obtained a sample of $n = 25$ measurements from a lot and you wished to determine a confidence interval on the statistic C_{PK}.* Using resampling, you would tell the computer to select a sample of $n = 25$ *from the sample results*, compute C_{PK}, and repeat the process many times, say 10,000 times. You would then determine whatever percentage point value you wished by simply looking at the results. The samples would be taken "with replacement," that is, a particular value from the original sample might appear several times (or not at all) in a resample.

Resampling has many advantages, especially in the era of easily available, low-cost computer power. Spreadsheets can be programmed to resample and calculate the statistics of interest. Compared with traditional statistical methods, resampling is easier for most people to understand. It works without strong assumptions, and it is simple. Resampling doesn't impose as much baggage between the engineering problem and the statistical result as conventional methods. It can also be used for more advanced problems, such as modeling, design of experiments, etc.

For a discussion of the theory behind resampling, see Efron (1982). For a presentation of numerous examples using a resampling computer program see Simon (1992).

*See Chap. 6.

Regression and Correlation Analysis

The simplest tool used in regression analysis is often called a *Scatter plot* or *Scatter diagram.* A scatter diagram is a plot of one variable versus another. One variable is called the *independent variable* and it is usually shown on the horizontal (bottom) axis. The other variable is called the *dependent variable* and it is shown on the vertical (side) axis.

Scatter diagrams are used to evaluate cause and effect relationships. The assumption is that the independent variable is causing a change in the dependent variable. Scatter plots are used to answer such questions as "Does the length of training have anything to do with the amount of scrap an operator makes?"

For example, an orchard manager has been keeping track of the weight of peaches on a day by day basis. The data are provided in Table 10.2.

The scatter diagram is shown in Fig. 10.15.

Pointers for Using Scatter Diagrams

- Scatter diagrams display different patterns that must be interpreted; Fig. 10.16 provides a scatter diagram interpretation guide.

Number	Days on Tree	Weight (Ounces)
1	75	4.5
2	76	4.5
3	77	4.4
4	78	4.6
5	79	5.0
6	80	4.8
7	80	4.9
8	81	5.1
9	82	5.2
10	82	5.2
11	83	5.5
12	84	5.4
13	85	5.5
14	85	5.5
15	86	5.6
16	87	5.7
17	88	5.8
18	89	5.8
19	90	6.0
20	90	6.1

From Pyzdek, 1990, P. 67.

TABLE 10.2 Raw Data for Scatter Diagram

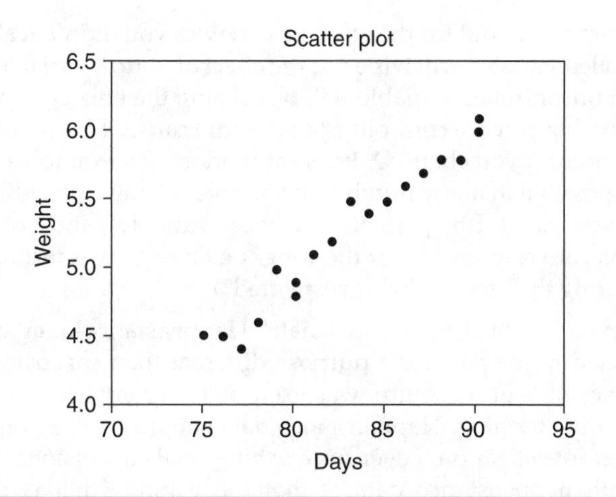

FIGURE 10.15 Completed scatter diagram. (*From Pyzdek (1990), P. 68. Copyright © 1990 by Quality Publishing.*)

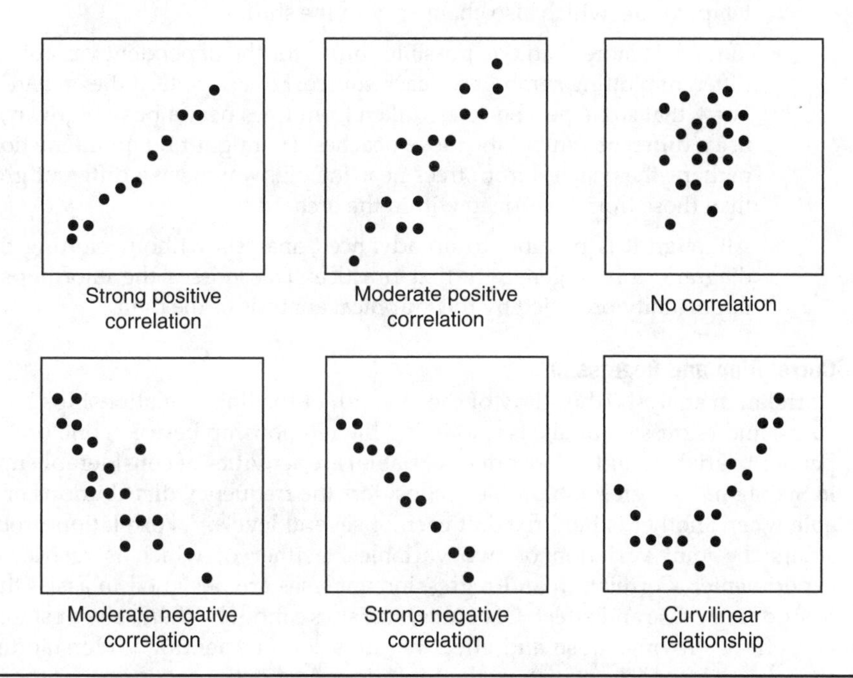

FIGURE 10.16 Scatter diagram interpretation guide. (*From Pyzdek, 1990 . P.69.*)

- Be sure that the independent variable, *X*, is varied over a sufficiently large range. When *X* is changed only a small amount, you may not see a correlation with *Y*, even though the correlation really does exist.
- If you make a prediction for *Y*, for an *X* value that lies outside of the range you tested, be advised that the prediction is highly questionable and should be tested thoroughly. Predicting a *Y* value beyond the *X* range actually tested is called extrapolation.

- Keep an eye out for the effect of variables you didn't evaluate. Often, an uncontrolled variable will wipe out the effect of your *X* variable. It is also possible that an uncontrolled variable will be causing the effect and you will mistake the *X* variable you are controlling as the true cause. This problem is much less likely to occur if you choose *X* levels at random. An example of this is our peaches. It is possible that any number of variables changed steadily over the time period investigated. It is possible that these variables, and not the independent variable, are responsible for the weight gain (e.g., was fertilizer added periodically during the time period investigated?).
- Beware of "happenstance" data! Happenstance data are data that were collected in the past for a purpose different than constructing a scatter diagram. Since little or no control was exercised over important variables, you may find nearly anything. Happenstance data should be used only to get ideas for further investigation, never for reaching final conclusions. One common problem with happenstance data is that the variable that is truly important is not recorded. For example, records might show a correlation between the defect rate and the shift. However, perhaps the real cause of defects is the ambient temperature, which also changes with the shift.
- If there is more than one possible source for the dependent variable, try using different plotting symbols for each source. For example, if the orchard manager knew that some peaches were taken from trees near a busy highway, he could use a different symbol for those peaches. He might find an interaction, that is, perhaps the peaches from trees near the highway have a different growth rate than those from trees deep within the orchard.
- Although it is possible to do advanced analysis without plotting the scatter diagram, this is generally bad practice. This misses the enormous learning opportunity provided by the graphical analysis of the data.

Correlation and Regression

Correlation analysis (the study of the strength of the linear relationships among variables) and regression analysis (modeling the relationship between one or more independent variables and a dependent variable) are activities of considerable importance in Six Sigma. A regression problem considers the frequency distributions of one variable when another is held fixed at each of several levels. A correlation problem considers the joint variation of two variables, neither of which is restricted by the experimenter. Correlation and regression analyses are designed to assist the analyst in studying cause and effect. Of course, statistics cannot by themselves establish cause and effect. Proving cause and effect requires sound scientific understanding of the situation at hand. The statistical methods described in this section assist the analyst in performing this task.

Linear Models

A linear model is simply an expression of a type of association between two variables, x and y. A *linear relationship* simply means that a change of a given size in x produces a proportionate change in y. Linear models have the form:

$$y = a + \mathrm{b}x \tag{10.6}$$

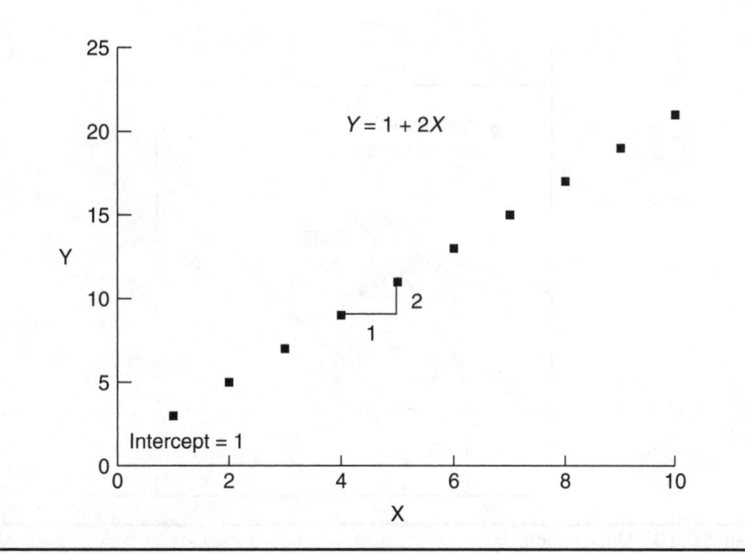

FIGURE 10.17 Scatter diagram of a linear relationship.

where a and b are constants. The equation simply says that when x changes by one unit, y will change by b units. This relationship can be shown graphically.

In Fig. 10.17, $a = 1$ and $b = 2$. The term a is called the intercept and b is called the slope. When $x = 0$, y is equal to the intercept, Fig. 10.17 depicts a perfect linear fit, for example, if x is known we can determine y exactly. Of course, perfect fits are virtually unknown when real data are used. In practice we must deal with error in x and y. These issues are discussed below.

Many types of associations are nonlinear, but can be converted to linear for ease of analysis, as described later in this chapter.

When conducting regression and correlation analysis we can distinguish two main types of variables. One type we call *predictor variables* or *independent variables*; the other, *response variables* or *dependent variables*. By predictor independent variables we usually mean variables that can either be set to a desired variable (e.g., oven temperature) or else take values that can be observed but not controlled (e.g., outdoors ambient humidity). As a result of changes that are deliberately made, or simply take place in the predictor variables, an effect is transmitted to the response variables (e.g., the grain size of a composite material). We are usually interested in discovering how changes in the predictor variables affect the values of the response variables. Ideally, we hope that a small number of predictor variables will "explain" nearly all of the variation in the response variables.

In practice, it is sometimes difficult to draw a clear distinction between independent and dependent variables. In many cases it depends on the objective of the investigator. For example, an analyst may treat ambient temperature as a predictor variable in the study of paint quality, and as the response variable in a study of clean room particulates. However, the above definitions are useful in planning Six Sigma studies.

Another idea important to studying cause and effect is that of the *data space* of the study. The data space of a study refers to the region bounded by the range of the independent variables under study. In general, predictions based on values outside the data

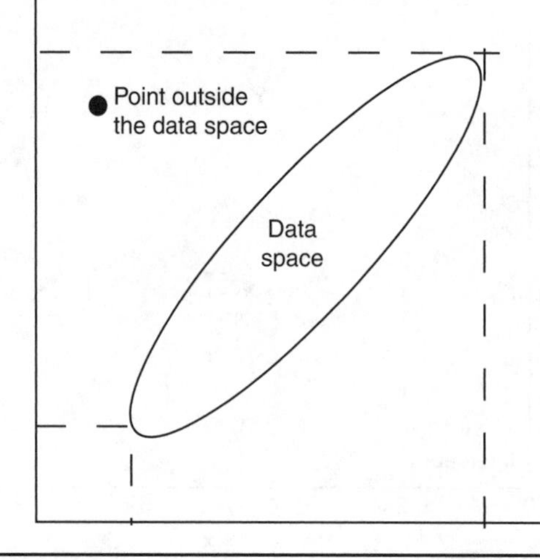

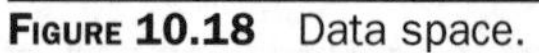
FIGURE 10.18 Data space.

space studied, called *extrapolations*, are little more than speculation and not advised. Figure 10.18 illustrates the concept of data space for two independent variables. Defining the data space can be quite tricky when large numbers of independent variables are involved.

While the numerical analysis of data provides valuable information, it should always be supplemented with graphical analysis as well. Scatter diagrams are one very useful supplement to regression and correlation analysis. Figure 10.19 illustrates the value of supplementing numerical analysis with scatter diagrams.

In other words, although the scatter diagrams clearly show four distinct processes, the statistical analysis does not. In Six Sigma, numerical analysis alone is not enough.

Least-Squares Fit

If all data fell on a perfectly straight line it would be easy to compute the slope and intercept given any two points. However, the situation becomes more complicated when there is "scatter" around the line. That is, for a given value of x, more than one value of y appears. When this occurs, we have error in the model. Figure 10.20 illustrates the concept of error.

The model for a simple linear regression with error is:

$$y = a + bx + \varepsilon \tag{10.7}$$

where ε represents error. Generally, assuming the model adequately fits the data, errors are assumed to follow a normal distribution with a mean of 0 and a constant standard deviation. The standard deviation of the errors is known as the *standard error*. We discuss ways of verifying our assumptions below.

When error occurs, as it does in nearly all "real-world" situations, there are many possible lines which might be used to model the data. Some method must be found which provides, in some sense, a "best-fit" equation in these everyday situations.

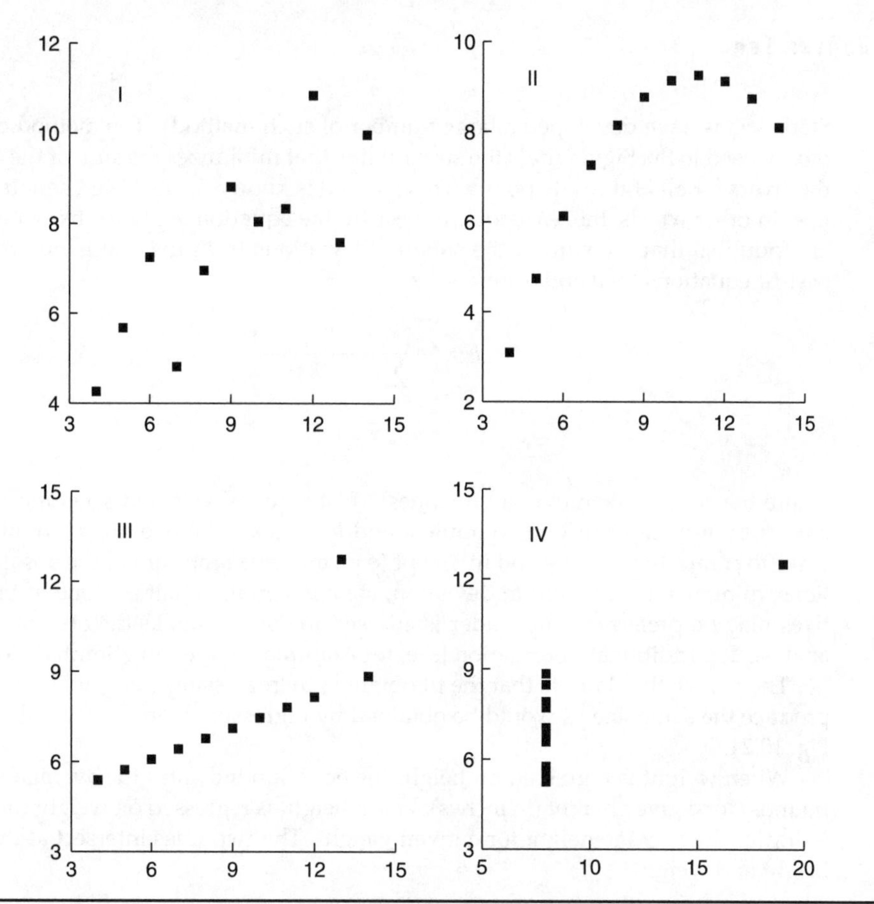

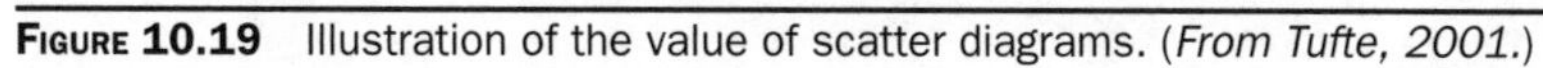

FIGURE 10.19 Illustration of the value of scatter diagrams. (*From Tufte, 2001.*)

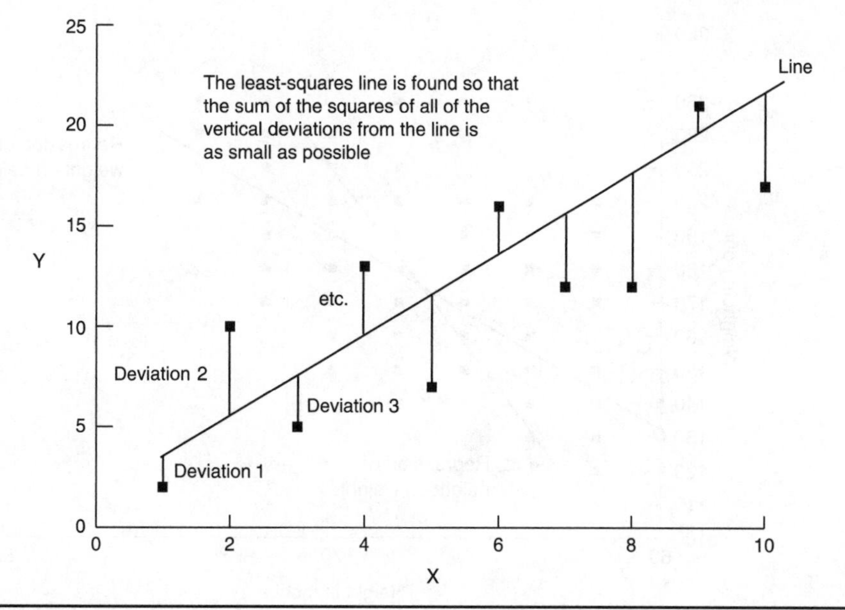

FIGURE 10.20 Error in the linear model.

Statisticians have developed a large number of such methods. The method most commonly used in Six Sigma finds the straight line that minimizes the sum of the squares of the errors for all of the data points. This method is known as the "least-squares" best-fit line. In other words, the least-squares best-fit line equation is $y_i' = a + bx_i$ where a and b are found so that the sum of the squared deviations from the line is minimized. The best-fit equations for a and b are:

$$b = \frac{\sum(X_i - \bar{X})(Y_i - \bar{Y})}{\sum(X_i - \bar{X})^2} \tag{10.8}$$

$$a = \bar{Y} - b\bar{X} \tag{10.9}$$

where the sum is taken over all n values. Most spreadsheets and scientific calculators have a built-in capability to compute a and b. As stated above, there are many other ways to compute the slope and intercept (e.g., minimize the sum of the absolute deviations, minimize the maximum deviation, etc.); in certain situations one of the alternatives may be preferred. The reader is advised to consult books devoted to regression analysis for additional information [see, for example, Draper and Smith (1981)].

The reader should note that the fit obtained by regressing x on y will not in general produce the same line as would be obtained by regressing y on x. This is illustrated in Fig. 10.21.

When weight is regressed on height the equation indicates the average weight (in pounds) for a given height (in inches). When height is regressed on weight the equation indicates the average height for a given weight. The two lines intersect at the average height and weight.

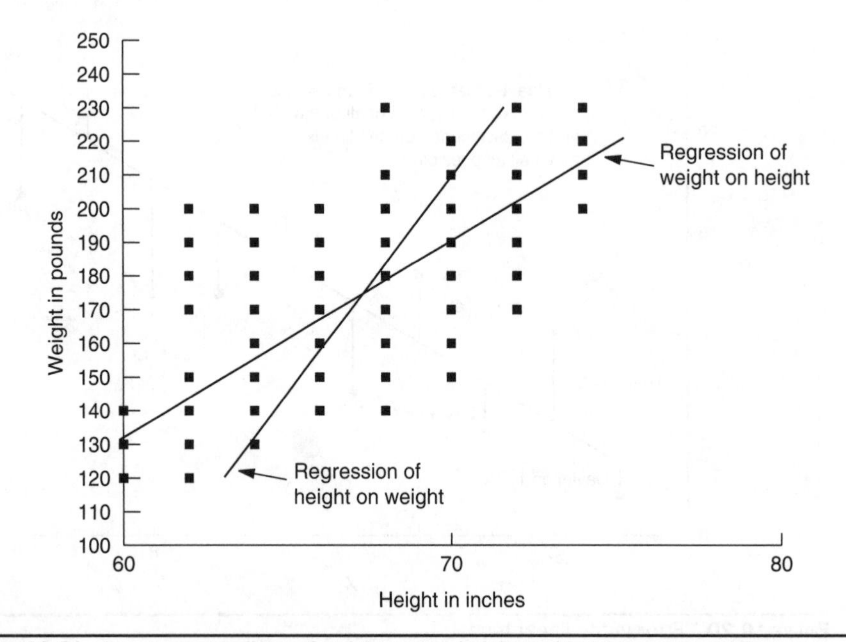

FIGURE 10.21 Least-squares lines of weight versus height and height versus weight.

These examples show how a single independent variable is used to model the response of a dependent variable. This is known as *simple linear regression*. It is also possible to model the dependent variable in terms of two or more independent variables; this is known as *multiple linear regression*. The mathematical model for multiple linear regression has additional terms for the additional independent variables. Equation (10.10) shows a linear model when there are two independent variables.

$$\hat{y} = a + b_1 x_1 + b_2 x_2 + \varepsilon \tag{10.10}$$

where x_1, x_2 are independent variables, b_1 is the coefficient for x_1 and b_2 is the coefficient for x_2.

Example of Regression Analysis

A restaurant conducted surveys of 42 customers, obtaining customer ratings on staff service, food quality, and overall satisfaction with their visit to the restaurant. Figure 10.22 shows the regression analysis output from a spreadsheet regression function (Microsoft Excel).

The data consist of two independent variables, staff and food quality, and a single dependent variable, overall satisfaction. The basic idea is that the quality of staff service and the food are *causes* and the overall satisfaction score is an *effect*. The regression output is interpreted as follows:

- Multiple R—the multiple correlation coefficient. It is the correlation between y and $\hat{y}$. For the example: multiple $R = 0.847$, which indicates that y and $\hat{y}$ are

Summary output						
Regression statistics						
Multiple R	0.847					
R square	0.717					
Adjusted R square	0.703					
Standard error	0.541					
Observations	42					
ANOVA						
	df	*SS*	*ms*	*F*	*Significance F*	
Regression	2	28.97	14.49	49.43	0.00	
Residual	39	11.43	0.29			
Total	41	40.40				
	Coefficients	*Standard error*	*t Stat*	*P-value*	*Lower 95%*	*Upper 95%*
Intercept	-1.188	0.565	-2.102	0.042	-2.331	-0.045
Staff	0.902	0.144	6.283	0.000	0.611	1.192
Food	0.379	0.163	2.325	0.025	0.049	0.710

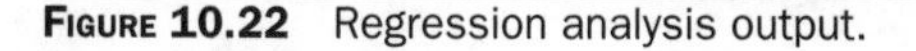

Figure 10.22 Regression analysis output.

highly correlated, which implies that there is an association between overall satisfaction and the quality of the food and service.

- R square—the square of multiple R, it measures the proportion of total variation about the mean $\bar{Y}$ explained by the regression. For the example: $R^2 = 0.717$, which indicates that the fitted equation explains 71.7% of the total variation about the average satisfaction level.
- Adjusted R square—a measure of R^2 "adjusted for degrees of freedom." The equation is

$$\text{Adjusted } R^2 = 1-(1-R^2)\left(\frac{n-1}{n-p}\right) \tag{10.11}$$

where p is the number of parameters (coefficients for the xs) estimated in the model. For the example: $p = 2$, since there are two x terms. Some experimenters prefer the adjusted R^2 to the unadjusted R^2, while others see little advantage to it (e.g., Draper and Smith, 1981, p. 92).

- Standard error—the standard deviation of the residuals. The *residual* is the difference between the observed values of y and the predicted values based on the regression equation.
- Observations—refer to the number of cases in the regression analysis, or n.
- ANOVA, or analysis of variance a table examining the hypothesis that the variation explained by the regression is zero. If this is so, then the observed association could be explained by chance alone. The rows and columns are those of a standard one-factor ANOVA table (discussed in more detail later in this chapter). For this example, the important item is the column labeled "Significance *F*." The value shown, 0.00, indicates that the probability of getting these results due to chance alone is less than 0.01; that is, the association is probably not due to chance alone. Note that the ANOVA applies to the entire *model*, not to the individual variables.

    The next table in the output examines each of the terms in the linear model separately. The *intercept* is as described above, and corresponds to our term *a* in the linear equation. Our model uses two independent variables. In our terminology staff = b_1, food = b_2. *Thus, reading from the coefficients* column, the linear model is: $\bar{y} = -1.188 + 0.902 \times$ staff score $+ 0.379 \times$ food score. The remaining columns test the hypotheses that each coefficient in the model is actually zero.

- Standard error column—gives the standard deviations of each term, i.e., the standard deviation of the intercept = 0.565, etc.
- t Stat column—the coefficient divided by the standard error, i.e., it shows how many standard deviations the observed coefficient is from zero.
- *P*-value—shows the area in the tail of a t distribution beyond the computed t value. For most experimental work, a *P*-value less than 0.05 is accepted as an indication that the coefficient is significantly different than zero. All of the terms in our model have significant *P*-values.
- Lower 95 and Upper 95% columns—a 95% confidence interval on the coefficient. If the confidence interval does not include zero, we will fail to reject the hypothesis that the coefficient is zero. None of the intervals in our example include zero.

Correlation Analysis

As mentioned earlier, a correlation problem considers the joint variation of two variables, neither of which is restricted by the experimenter. Unlike regression analysis, which considers the effect of the independent variable(s) on a dependent variable, correlation analysis is concerned with the joint variation of one independent variable with another. In a correlation problem, the analyst has two measurements for each individual item in the sample. Unlike a regression study where the analyst controls the values of the x variables, correlation studies usually involve spontaneous variation in the variables being studied. Correlation methods for determining the strength of the linear relationship between two or more variables are among the most widely applied statistical techniques. More advanced methods exist for studying situations with more than two variables (e.g., canonical analysis, factor analysis, principal components analysis, etc.), however, with the exception of multiple regression, our discussion will focus on the linear association of two variables at a time.

In most cases, the measure of correlation used by analysts is the statistic r, sometimes referred to as *Pearson's product-moment correlation*. Usually x and y are assumed to have a bivariate normal distribution. Under this assumption r is a sample statistic which estimates the population correlation parameter ρ. One interpretation of r is based on the linear regression model described earlier, namely that r^2 is the proportion of the total variability in the y data which can be explained by the linear regression model. The equation for r is:

$$r = \frac{s_{xy}}{s_x s_y} = \frac{n\sum xy - \sum x \sum y}{\sqrt{\left[n\sum x^2 - \left(\sum x\right)^2\right]\left[n\sum y^2 - \left(\sum y\right)^2\right]}} \tag{10.12}$$

and, of course, r^2 is simply the square of r. r is bounded at -1 and $+1$. When the assumptions hold, the significance of r is tested by the regression ANOVA.

Interpreting r can become quite tricky, so scatter plots should always be used (see above). When the relationship between x and y is nonlinear, the "explanatory power" of r is difficult to interpret in precise terms and should be discussed with great care. While it is easy to see the value of very high correlations such as $r = 0{:}99$, it is not so easy to draw conclusions from lower values of r, even when they are statistically significant (i.e., they are significantly different than 0.0). For example, $r = 0.5$ does *not* mean the data show half as much clustering as a perfect straight-line fit. In fact, $r = 0$ does *not* mean that there is no relationship between the x and y data, as Fig. 10.23 shows. When $r > 0$, y tends to increase when x increases. When $r < 0$, y tends to decrease when x increases.

Although $r = 0$, the relationship between x and y is perfect, albeit nonlinear. At the other extreme, $r = 1$, a "perfect correlation," does not mean that there is a cause and effect relationship between x and y. For example, both x and y might be determined by a third variable, z. In such situations, z is described as a *lurking variable* which "hides" in the background, unknown to the experimenter. Lurking variables are behind some of the infamous silly associations, such as the association between teacher's pay and liquor sales (the lurking variable is general prosperity).*

*It is possible to evaluate the association of x and y by removing the effect of the lurking variable. This can be done using regression analysis and computing partial correlation coefficients. This advanced procedure is described in most texts on regression analysis.

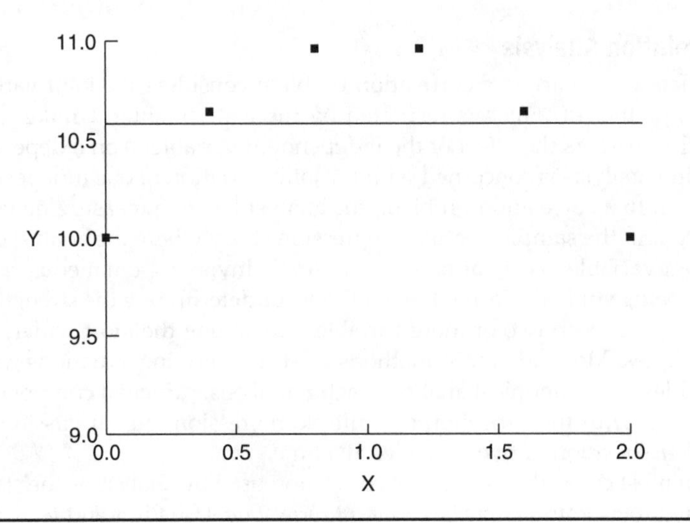

Figure 10.23 Interpreting $r = 0$ for curvilinear data.

Establishing causation requires solid scientific understanding. Causation cannot be "proven" by statistics alone. Some statistical techniques, such as path analysis, can help determine if the correlations between a number of variables are consistent with causal assumptions. However, these methods are beyond the scope of this book.

Designed Experiments

Designed experiments play an important role in quality improvement. While the confidence intervals and hypothesis tests previously discussed are limited to rather simple comparisons between one sample and requirements or between two samples, the designed experiments will use ANOVA (analysis of variance) techniques to partition the variation in a response amongst the potential sources of variation. This section will introduce the basic concepts involved and it will contrast the statistically designed experiment with the "one factor at a time" (OFAT) approach used traditionally. Also briefly discussed are the concepts involved in Taguchi methods, statistical methods named after their creator, Dr. Genichi Taguchi.

The Traditional Approach versus Statistically Designed Experiments The traditional approach, which most of us learned in high school science class, is to hold all factors constant except one. When this approach is used we can be sure that the variation is due to a cause and effect relationship or so we are told. However, this approach suffers from a number of problems:

- It usually isn't possible to hold all other variables constant.
- There is no way to account for the effect of joint variation of independent variables, such as interaction.
- There is no way to account for experimental error, including measurement variation.

The statistically designed experiment usually involves varying two or more variables simultaneously and obtaining multiple measurements under the same experimental conditions. The advantage of the statistical approach is threefold:

1. Interactions can be detected and measured. Failure to detect interactions is a major flaw in the OFAT approach.
2. Each value does the work of several values. A properly designed experiment allows you to use the same observation to estimate several different effects. This translates directly to cost savings when using the statistical approach.
3. Experimental error is quantified and used to determine the confidence the experimenter has in his conclusions.

Terminology

Much of the early work on the design of experiments involved agricultural studies. The language of experimental design still reflects these origins. The experimental area was literally a piece of ground. A block was a smaller piece of ground with fairly uniform properties. A plot was smaller still and it served as the basic unit of the design. As the plot was planted, fertilized and harvested, it could be split simply by drawing a line. A treatment was actually a treatment, such as the application of fertilizer. Unfortunately for the Six Sigma analyst, these terms are still part of the language of experiments. The analyst must do his or her best to understand quality improvement experimenting using these terms. Natrella (1963) recommends the following:

Experimental area can be thought of as the scope of the planned experiment. For us, a block can be a group of results from a particular operator, or from a particular machine, or on a particular day—any planned natural grouping which should serve to make results from one block more alike than results from different blocks. For us, a treatment is the factor being investigated (material, environmental condition, etc.) in a single factor experiment. In factorial experiments (where several variables are being investigated at the same time) we speak of a treatment combination and we mean the prescribed levels of the factors to be applied to an experimental unit. For us, a yield is a measured result and, happily enough, in chemistry it will sometimes be a yield.

Definitions

A designed experiment is an experiment where one or more factors, called independent variables, believed to have an effect on the experimental outcome are identified and manipulated according to a predetermined plan. Data collected from a designed experiment can be analyzed statistically to determine the effect of the independent variables, or combinations of more than one independent variable. An experimental plan must also include provisions for dealing with extraneous variables, that is, variables not explicitly identified as independent variables.

Response variable—The variable being investigated, also called the *dependent variable*, sometimes called simply *response*.

Primary variables—The controllable variables believed most likely to have an effect. These may be quantitative, such as temperature, pressure, or speed, or they may be qualitative such as vendor, production method, and operator.

Background variables—Variables, identified by the designers of the experiment, which may have an effect but either cannot or should not be deliberately manipulated

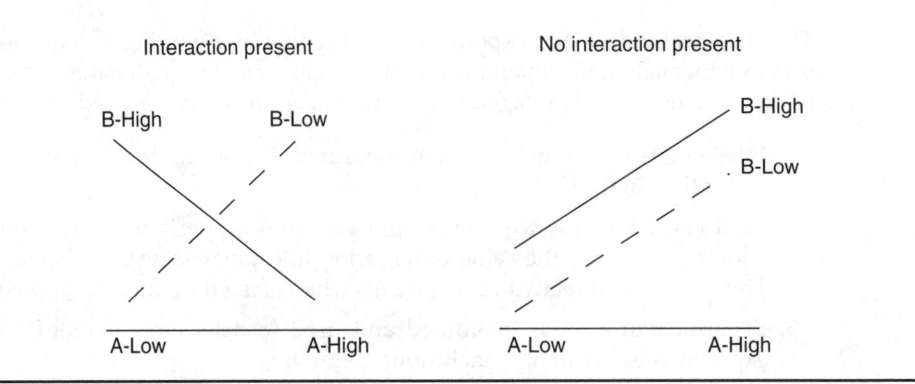

Figure 10.24 Illustration of interaction.

or held constant. The effect of background variables can contaminate primary variable effects unless they are properly handled. The most common method of handling background variables is blocking (blocking is described later in this chapter).

Experimental error—In any given experimental situation, a great many variables may be potential sources of variation. So many, in fact, that no experiment could be designed that deals with every possible source of variation explicitly. Those variables that are not considered explicitly are analogous to common causes of variation. They represent the "noise level" of the process and their effects are kept from contaminating the primary variable effects by *randomization*. Randomization is a term meant to describe a procedure that assigns test units to test conditions in such a way that any given unit has an equal probability of being processed under a given set of test conditions.

Interaction—A condition where the effect of one factor depends on the level of another factor. Interaction is illustrated in Fig. 10.24.

Design Characteristics

Good experiments don't just happen, they are a result of careful planning. A good experimental plan depends on (Natrella 1963):

- The purpose of the experiment
- Physical restrictions on the process of taking measurements
- Restrictions imposed by limitations of time, money, material, and personnel

The analyst must explain clearly why the experiment is being done, why the experimental treatments were selected, and how the completed experiment will accomplish the stated objectives. The experimental plan should be in writing and it should be endorsed by all key participants. The plan will include a statement of the objectives of the experiment, the experimental treatments to be applied, the size of the experiment, the time frame, and a brief discussion of the methods to be used to analyze the results. Two concepts are of particular interest to the Six Sigma analyst: replication and randomization.

Replication—The collection of more than one observation for the same set of experimental conditions. Replication allows the experimenter to estimate experimental error. If variation exists when all experimental conditions are held constant, the cause must be something other than the variables being controlled by the experimenter. Experimental error can be estimated without replicating the entire experiment. If a process has been in statistical control for a period of time, experimental error can be estimated from the control chart. Replication also serves to decrease bias due to uncontrolled factors.

Randomization—In order to eliminate bias from the experiment, variables not specifically controlled as factors should be randomized. This means that allocations of specimens to treatments should be made using some mechanical method of randomization, such as a random numbers table. Randomization also ensures valid estimates of experimental error.

Types of Design

Experiments can be designed to meet a wide variety of experimental objectives. A few of the more common types of experimental designs are defined here.

Fixed-effects model: An experimental model where all possible factor levels are studied. For example, if there are three different materials, all three are included in the experiment.

Random-effects model: An experimental model where the levels of factors evaluated by the experiment represent a sample of all possible levels. For example, if we have three different materials but only use two materials in the experiment.

Mixed model: An experimental model with both fixed and random effects.

Completely randomized design: An experimental plan where the order in which the experiment is performed is completely random, for example,

Level	Test Sequence Number
A	7, 1, 5
B	2, 3, 6
C	8, 4

Randomized-block design: An experimental design is one where the experimental observations are divided into "blocks" according to some criterion. The blocks are filled sequentially, but the order within the block is filled randomly. For example, assume we are conducting a painting test with different materials, material A and material B. We have four test pieces of each material. Ideally we would like to clean all of the pieces at the same time to ensure that the cleaning process doesn't have an effect on our results; but what if our test requires that we use a cleaning tank that cleans two test pieces at a time? The tank load then becomes a "blocking factor." We will have four blocks, which might look like this:

Material	Tank Load	Test Piece Number
A	1	7
B		1
B	2	5
A		2
B	3	3
A		6
B	4	4
A		8

Since each material appears exactly once per cleaning tank load we say the design is *balanced*. The material totals or averages can be compared directly. The reader should be aware that statistical designs exist to handle more complicated "unbalanced designs."

Latin-square designs: Designs where each treatment appears once and only once in each row and column. A Latin-square plan is useful when it is necessary or desirable to allow for two specific sources of nonhomogeneity in the conditions affecting test results. Such designs were originally applied in agricultural experimentation when the two sources of nonhomogeneity were the two directions on the field and the "square" was literally a square piece of ground. Its usage has been extended to many other applications where there are two sources of nonhomogeneity that may affect experimental results—for example, machines, positions, operators, runs, and days. A third variable is then associated with the other two in a prescribed fashion. The use of Latin squares is restricted by two conditions:

1. The number of rows, columns and treatments must all be the same.
2. There must be no interactions between row and column factors.

Natrella (1963, pp. 13–30) provides the following example of a Latin square. Suppose we wish to compare four materials with regard to their wearing qualities. Suppose further that we have a wear-testing machine which can handle four samples simultaneously. Two sources of inhomogeneity might be the variations from run to run, and the variation among the four positions on the wear machine. In this situation, a 4 × 4 Latin square will enable us to allow for both sources of inhomogeneity if we can make four runs. The Latin square plan is as in Fig. 10.25 (the four materials are labeled A, B, C, D).

The procedure to be followed in using a given Latin square is as follows:

1. Permute the columns at random
2. Permute the rows at random
3. Assign letters randomly to the treatments

One-Factor ANOVA

The following example will be used to illustrate the interpretation of a single factor analysis of variance. With the widespread availability of computers, few people actually perform such complex calculations by hand. The analysis below was performed using Microsoft Excel. Commonly used statistical methods such as regression and ANOVA are included in most high-end spreadsheets.

The coded results in Table 10.3 were obtained from a single factor, completely randomized experiment, in which the production outputs of three machines (A, B, and C) were to be compared.

	Position number			
Run	(1)	(2)	(3)	(4)
1	A	B	C	D
2	B	C	D	A
3	C	D	A	B
4	D	A	B	C

Figure 10.25 A 4 × 4 Latin square.

A	B	C
4	2	−3
8	0	1
5	1	−2
7	2	−1
6	4	0

TABLE 10.3 Experimental Raw Data (Coded)

ANOVA: SINGLE FACTOR						
SUMMARY						
Groups	*Count*	*Sum*	*Average*	*Variance*		
A	5	30.000	6.000	2.500		
B	5	9.000	1.800	2.200		
C	5	−5.000	−1.000	2.500		
ANOVA						
Source of Variation	*SS*	*df*	*MS*	*F*	*P*-value	*F* crit
Between groups	124.133	2	62.067	25.861	0.000	3.885
Within groups	28.800	12	2.400			
Total	152.933	14				

TABLE 10.4 Results of the Analysis

An ANOVA of these results produced the results shown in Table 10.4.

The first part of Table 10.4 shows descriptive statistics for the data; the analyst should always look carefully at these easily understood results to check for obvious errors. The results show that the means vary from a low of −1 for machine C to a high of 6 for machine A.

ANOVA Procedure

ANOVA proceeds as follows:

1. State the null and alternative hypotheses: the ANOVA table tests the hypotheses: H_0 (all means are equal) versus H_a (at least two of the means are different).
2. Choose the level of significance. For this analysis a significance level $\alpha = 0.05$ was selected.
3. Compute the *F* statistic, the ratio of the mean square between groups to the mean square within groups.

4. Assuming that the observations are random samples from normally distributed populations with equal variances, and that the hypothesis is true, the critical value of F is found in the Appendix 5 or 6. The numerator will have the degrees of freedom shown in the degrees of freedom column for the between groups row. The denominator will have the degrees of freedom shown in the degrees of freedom column for the within groups row.
5. If the computed $F > F_{1-\alpha}$ then reject the null hypothesis and conclude the alternate hypothesis. Otherwise fail to reject the null hypothesis.

The ANOVA table shows that for these data F computed is 62.067/2.4 = 25.861 and F critical at $\alpha = 0.05$ with numerator df = 2 and denominator df = 12 is 3.885.* Since 25.861 > 3.885 we reject the null hypothesis and conclude that the machines produce different results. Note that all we know is that at least the two extreme machines (A and C) are different. The ANOVA does *not* tell us if A and B or B and C are significantly different. There are methods which can make this determination, such as *contrasts*. The reader is referred to a text on design of experiments, for example, Montgomery (1984) for additional information.

Performing ANOVA Manually

On rare occasions (such as taking a Black Belt exam), the analyst may find that computers are not available and the analysis must be performed "by hand." The analysis is illustrated below.

		Total	*N*	Sum of Squares
Treatment A	4, 8, 5, 7, 6	30	5	190
Treatment B	2, 0, 1, 2, 4	9	5	25
Treatment C	−3, 1, −2, −1, 0	−5	5	15
Totals		34	15	230

$$\text{Total sum of square} = 230 - \frac{(34)^2}{15} = 152.933$$

$$\text{Treatment sum of square} = \frac{(30)^2}{5} + \frac{(9)^2}{5} + \frac{(-5)^2}{5} - \frac{(34)^2}{15} = 124.133$$

$$\begin{aligned}\text{Error sum of squares} &= \text{total sum of squares} - \text{treatment sum of squares}\\ &= 152.933 - 124.133 = 28.8\end{aligned}$$

These values are placed in the sum of squares (SS) column in the ANOVA table (Table 10.4). The remainder of the ANOVA table is obtained through simple division.

Examples of Applying Common DOE Methods Using Software

This section includes examples of the most commonly used design of experiment methods using software. Whenever possible the examples employ popular software, such as

*Referring to the critical value is actually unnecessary; the *P*-value of 0.000 indicates that the probability of getting an *F* value as large as that computed is less than 1 in 1,000.

Microsoft Excel. For detailed mathematical background on these methods, the reader is referred to any of the many fine books on the subject (e.g. Box et al., 1978; Hicks, 1993; Montgomery, 1996). DOE PC, a full-featured commercial software for design and analysis of experiments is available from http://www.qualityamerica.com. A statistical analysis shareware package for Windows operating systems can be downloaded from http://www.dagonet.com/scalc.htm. MINITAB includes DOE capabilities.

Two-Way ANOVA with No Replicates

When experiments are conducted which involve two factors, and it is not possible to obtain repeat readings for a given set of experimental conditions, a two-way analysis of variance may be used. The following example assumes that experimental treatments are assigned at random. Note that if the factors involved are each tested at only two levels, the full factorial analysis method described below could also be used.

Example of Two-Way ANOVA with No Replicates

An experiment was conducted to evaluate the effect of different detergents and water temperatures on the cleanliness of ceramic substrates. The experimenter selected three different detergents based on their pH levels, and conducted a series of experiments at four different water temperatures. Cleanliness was quantified by measuring the contamination of a distilled water beaker after rinsing the parts cleaned using each treatment combination. The coded data are shown in Table 10.5.

Part one of the Excel output (Table 10.6) provides descriptive statistics on the different treatment levels. The ANOVA table is shown in part two. Note that in the previously presented raw data table the rows represent the different temperatures and the columns the different detergents. Because there are no replicates, Excel is not able to provide an estimate of the interaction of detergent and water temperature. If you suspect that an interaction may be present, then you should try to replicate the experiment to estimate this effect. For this experiment, any *P*-value less than 0.05 would indicate a significant effect. The ANOVA table indicates that there are significant differences between the different detergents and the different water temperatures. To identify *which* differences are significant the experimenter can examine the means of the different detergents and water temperatures using *t*-tests. (Excel's data analysis tools add-in includes these tests.) Be aware that the Type I error is affected by conducting multiple *t*-tests. If the Type I error on a single *t*-test is α, then the overall Type I error for k such tests is $1 - (1 - \alpha)^k$. For example, if $\alpha = 0.01$ and three pairs of means are examined, then the combined Type I error for all three *t*-tests is $1 - (1 - 0.01)^3 = 1 - (0.99)^3 = 0.03$. Statistical methods exist that guarantee an overall level of Type I error for simultaneous comparisons (Hicks, 1973, pp. 31–38).

	Detergent A	Detergent B	Detergent C
Cold	15	18	10
Cool	12	14	9
Warm	10	18	7
Hot	6	12	5

Table 10.5 Cleaning Experiment Raw Data

SUMMARY OUTPUT						
	Count	Sum	Average	Variance		
Cold water	3	43	4.333333	16.33333		
Cool water	3	35	11.666667	6.333333		
Warm water	3	35	11.666667	32.33333		
Hot water	3	23	7.6666667	14.33333		
Detergent A	4	43	10.75	14.25		
Detergent B	4	62	15.5	9		
Detergent C	4	31	7.75	4.916667		
ANOVA						
Source of variation	SS	df	MS	F	P-value	F crit
Rows	68	3	22.666667	8.242424	0.015043179	4.757055
Columns	122.1666667	2	61.083333	22.21212	0.001684751	5.143249
Error	16.5	6	2.75			
Total	206.6666667	11				

TABLE 10.6 Cleaning Experiment Two-Way ANOVA Output from Microsoft Excel (Two-Factor without Replication)

Two-Way ANOVA with Replicates

If you are investigating two factors which might interact with one another, and you can obtain more than one result for each combination of experimental treatments, then two-way analysis of variance with replicates may be used for the analysis. Spreadsheets such as Microsoft Excel include functions that perform this analysis.

Example of Two-Way ANOVA with Replicates

An investigator is interested in improving a process for bonding photoresist to copper clad printed circuit boards. Two factors are to be evaluated: the pressure used to apply the photoresist material and the preheat temperature of the photoresist. Three different pressures and three different temperatures are to be evaluated; the number of levels need not be the same for each factor and there is no restriction on the total number of levels. Each experimental combination of variables is repeated 5 times. Note that while Excel requires equal numbers of replicates for each combination of treatments, most statistical analysis packages allow different sample sizes to be used. The experimenter recorded the number of photoresist defects per batch of printed wiring boards. The coded data are shown in Table 10.7.

These data were analyzed using Excel's two-way ANOVA with replicates function. The results are shown in Table 10.8.

	High Pressure	Med Pressure	Low Pressure
High temp	39	32	18
	30	31	20
	35	28	21
	43	28	25
	25	29	26
Med temp	38	10	22
	31	15	28
	31	25	29
	30	31	26
	35	36	20
Low temp	30	21	25
	35	22	24
	36	25	20
	37	24	21
	39	27	21

TABLE 10.7 Photoresist Experiment Raw Data ANOVA Results

As before, part one of the Excel output provides descriptive statistics on the different treatment levels. The ANOVA table is shown in part two. Because there are now replicates, Excel is able to provide an estimate of the interaction of pressure and temperature. For this experiment, the experimenter decided that any *P*-value less than 0.05 would indicate a significant effect. The ANOVA table *P*-value of less than 0.001 indicates that there are significant differences between the different columns (pressure), but the *P*-value of 0.6363 indicates that there is not a significant difference between the rows (temperature). The interaction of pressure and temperature is also not significant, as indicated by the *P*-value of 0.267501.

Since the *P*-value indicates that at least one difference is significant, we know that the largest difference of $34.26666667 - 23.06666667 = 11.2$ is significant. To identify *which other* differences are significant the experimenter can examine the means of the different pressures using *t*-tests. (Excel's data analysis tools add-in includes these tests.) Be aware that the Type I error is affected by conducting multiple *t*-tests. If the Type I error on a single *t*-test is α, then the overall Type I error for k such tests is $1 - (1 - \alpha)^k$. For example, if $\alpha = 0.01$ and three pairs of means are examined, then the combined Type I error for all three *t*-tests is $1 - (1 - 0.01)^3 = 1 - (0.99)^3 = 0.03$.

Full and Fractional Factorial

Full factorial experiments are those where at least one observation is obtained for every possible combination of experimental variables. For example, if A has 2 levels, B has 3 levels and C has 5 levels, a full factorial experiment would have at least $2 \times 3 \times 5 = 30$ observations.

SUMMARY OUTPUT				
	High pressure	Med pressure	Low pressure	Total
High temp				
Count	5	5	5	15
Sum	172	148	110	430
Average	34.4	29.6	22	28.66667
Variance	50.8	3.3	11.5	46.66667
Med temp				
Count	5	5	5	15
Sum	165	117	125	407
Average	33	23.4	25	27.13333
Variance	11.5	117.3	15	59.98095
Low temp				
Count	5	5	5	15
Sum	177	119	111	407
Average	35.4	23.8	22.2	27.13333
Variance	11.3	5.7	4.7	43.26667
Total				
Count	15	15	15	
Sum	514	384	346	
Average	34.26666667	25.6	23.06666667	
Variance	22.06666667	44.68571429	10.92380952	

ANOVA						
Source of variation	*SS*	*df*	*MS*	*F*	*P-value*	*F crit*
Sample	23.5111111	2	11.7555556	0.45781	0.6363	3.259444
Columns	1034.84444	2	517.422222	20.1506	1.34E-06	3.259444
Interaction	139.555556	4	34.8888889	1.35872	0.267501	2.633534
Within	924.4	36	25.6777778			
Total	2122.31111	44				

Table 10.8 Photoresist Experiment Two-Way ANOVA Output from Microsoft Excel (Two-factor with Replication)

Fractional factorial or fractional replicate are experiments where there are some combinations of experimental variables where observations were not obtained. Such experiments may not allow the estimation of every interaction. However, when carefully planned, the experimenter can often obtain all of the information needed at a significant saving.

Analyzing Factorial Experiments

A simple method exists for analyzing the common 2^n experiment. The method, known as the Yates method, can be easily performed with a pocket calculator or programmed into a spreadsheet. It can be used with any properly designed 2^n experiment, regardless of the number of factors being studied.

To use the Yates algorithm, the data are first arranged in standard order (of course, the actual running order is random). The concept of standard order is easier to understand if demonstrated. Assume that we have conducted an experiment with three factors, A, B, and C. Each of the three factors is evaluated at two levels, which we will call low and high. A factor held at a low level will be identified with a "–" sign, one held at a high level will be identified with a "+" sign. The eight possible combinations of the three factors are identified using the scheme shown in the table below.

ID	A	B	C
(1)	–	–	–
a	+	–	–
b	–	+	–
ab	+	+	–
c	–	–	+
ac	+	–	+
bc	–	+	+
abc	+	+	+

Note that the table begins with all factors at their low level. Next, the first factor is high and all others are low. When a factor is high, it is shown in the ID column, otherwise it is not. For example, whenever *a* appears it indicates that factor A is at its high level. To complete the table you simply note that as each factor is added to the table it is "multiplied" by each preceding row. Thus, when *b* is added it is multiplied by *a*, giving the row *ab*. When *c* is added it is multiplied by, in order, *a*, *b*, and *ab*, giving the remaining rows in the table. (As an exercise, the reader should add a fourth factor D to the above table. Hint: the result will be a table with eight more rows.) Once the data are in standard order, add a column for the data and one additional column for each variable, for example, for our three variables we will add four columns.

ID	A	B	C	Data	1	2	3
(1)	–	–	–				
a	+	–	–				
b	–	+	–				
ab	+	+	–				
c	–	–	+				
ac	+	–	+				
bc	–	+	+				
abc	+	+	+				

Record the observations in the data column (if the experiment has been replicated, record the totals). Now record the sum of the data values in the first two rows that is, (1) + *a* in the first cell of the column labeled column 1. Record the sum of the next two rows in the second cell (i.e., *b* + *ab*). Continue until the top half of column 1 is completed. The lower half of column 1 is completed by subtracting one row from the next, for example, the fifth value in column 1 is found by subtracting −5 − 2 = −3. After completing column 1 the same process is completed for column 2, using the values in column 1. Column 3 is created using the values in column 2. The result is shown below.

ID	A	B	C	Data	1	2	3
(1)	–	–	–	−2	−7	21	−17
a	+	–	–	−5	28	−38	−15
b	–	+	–	15	−29	−5	55
ab	+	+	–	13	−9	−10	1
c	–	–	+	−12	−3	35	−59
ac	+	–	+	−17	−2	20	−5
bc	–	+	+	−2	−5	1	−15
abc	+	+	+	−7	−5	0	−1

Example of Yates Method

The table below shows sample data from an actual experiment. The experiment involved a target shooter trying to improve the number of targets hit per box of 25 shots. Three variables were involved: *a* = the gauge of the shotgun (12-gauge and 20-gauge), *b* = the shot *size* (6 shot and 8 shot), and *c* = the length of the handle on the target thrower (short or long). The shooter ran the experiment twice. The column labeled "1st" is the number of hits the first time the combination was tried. The column labeled "2nd" is the number of hits the second time the combination was tried. The Yates analysis begins with the sums shown in the column labeled Sum.

ID	1st	2nd	Sum	1	2	3	Effect	df	SS	MS	*F* Ratio
1	22	19	41	86	167	288	18	Avg.			
a	21	24	45	81	121	20	2.5	1	25.00	25.00	3.64
b	20	18	38	58	9	0	0	1	0.00	0.00	0.00
ab	21	22	43	63	11	4	0.5	1	1.00	1.00	0.15
c	12	15	27	4	−5	−46	−5.75	1	132.25	132.25	19.24
ac	12	19	31	5	5	2	0.25	1	0.25	0.25	0.04
bc	13	15	28	4	1	10	1.25	1	6.25	6.25	0.91
abc	20	15	35	7	3	2	0.25	1	0.25	0.25	0.04
Error								8	55.00	6.88	
Total	141	147						15	220.00		

The first row in the Effect column is simply the first row of column 3 (288) divided by the count ($r \times 2^n$); this is simply the average. Subsequent rows in the Effect column are found by dividing the numbers in column 3 by $r \times 2^{n-1}$. The Effect column provides the impact of the given factor on the response; thus, the shooter hit, on average, 2.5 more targets per box when shooting a 12-gauge than he did when shooting a 20-gauge.

The next question is whether or not these differences are statistically significant, that is, could they be due to chance alone? To answer this question we will use the F ratio of the effect MS for each factor to the error MS. The degrees of freedom (df) for each effect is simply 1 (the number of factor levels minus 1), the total df is $N - 1$, and the error df is the total df minus the sum of the factor dfs. The sum of squares (SS) for each factor is the column 3 value squared divided by $r \times 2^n$; for example., $SS_A = 20^2/16 = 25$. The total SS is the sum of the individual values squared minus the first row in column 3 squared divided by $r \times 2^n$; for example,

$$(22^2 + 21^2 + \cdots + 15^2) - \frac{288^2}{16} = 220$$

The error SS is the total SS minus the factor SS. The MS and F columns are computed using the same approach as shown above for one-way ANOVA. For the example the F ratio for factor c (thrower) is significant at $\alpha < 0.01$ and the F ratio for factor a (gauge) is significant at $\alpha < 0.10$; no other F ratios are significant.

Screening Experiments. In many cases, there are a sufficient number of variables to make even a fractional factorial design, with interactions, to be quite large. In those cases, a screening design is useful as an initial pass to filter out factors that are highly insignificant.

Consider this example of solder defects in electronics manufacturing. The solder team decided to list as many items as possible that might be causing solder problems. Since many variables had already been studied in earlier parts of the project, the list was not unreasonably long. The team looked at ways to control the variables listed and was able to develop methods for eliminating the effects of many variables on their list. The remaining list included the following factors:

Variable	Low Level (–)	High Level (+)
A: Prebaking of boards in an oven	No	Yes
B: Preheat time	10 s	20 s
C: Preheat temperature	150°F	200°F
D: Distance from preheat element to board surface	25 cm	50 cm
E: Line speed	3 fpm	5 fpm
F: Solder temperature	495°F	505°F
G: Circuit density	Low	High
H: Was the board in a fixture?	No	Yes

This information was used to create an experimental design using a statistical software package. There are many packages on the market that perform similar analyses to the one shown here.

Run	A	B	C	D	E	F	G	H	Response
1	+	−	−	−	−	+	+	+	65
2	+	−	+	+	−	+	−	−	85
3	+	+	−	−	+	+	−	−	58
4	−	+	−	−	+	−	+	+	57
5	−	−	−	−	−	−	−	−	63
6	+	+	+	+	+	+	+	+	75
7	−	+	−	+	−	+	+	−	77
8	−	+	+	−	−	+	−	+	60
9	+	−	+	−	+	−	−	+	67
10	+	+	+	−	−	−	+	−	56
11	−	−	+	−	+	+	+	−	63
12	−	−	−	+	+	+	−	+	81
13	+	+	−	+	−	−	−	+	73
14	+	−	−	+	+	−	+	−	87
15	−'	+	+	+	+	−	−	−	75
16	−	−	+	+	−	−	+	+	84

Table 10.9 Screening Experiment Layout. Data Matrix (Randomized)

Since this is only to be a screening experiment, the team was not interested in obtaining estimates of factor interactions. The focus was to identify important main effects. The software allows selection from among several designs. The Black Belt decided upon the design which would estimate the main effects with the smallest number of test units. This design involved testing 16 units. The data matrix produced by the computer is shown in Table 10.9. The run order has been randomized by the computer. If the experiment cannot be conducted in that particular order, the computer software would allow the data to be run in blocks and it would adjust the analysis accordingly. The program also tells us that the design is of resolution IV, which means that main effects are not confounded with each other or any two-factor interactions.

In Table 10.9 the "−" indicates that the variable is run at its low level, while a "+" sign indicates that it is to be run at its high level. For example, the unit for run 16 was processed as follows:

- Prebaking = No
- Preheat time = 10 seconds
- Preheat temperature = 200°F
- Distance from preheat element to board surface = 50 cm
- Line speed = 3 fpm
- Solder temperature = 495°F

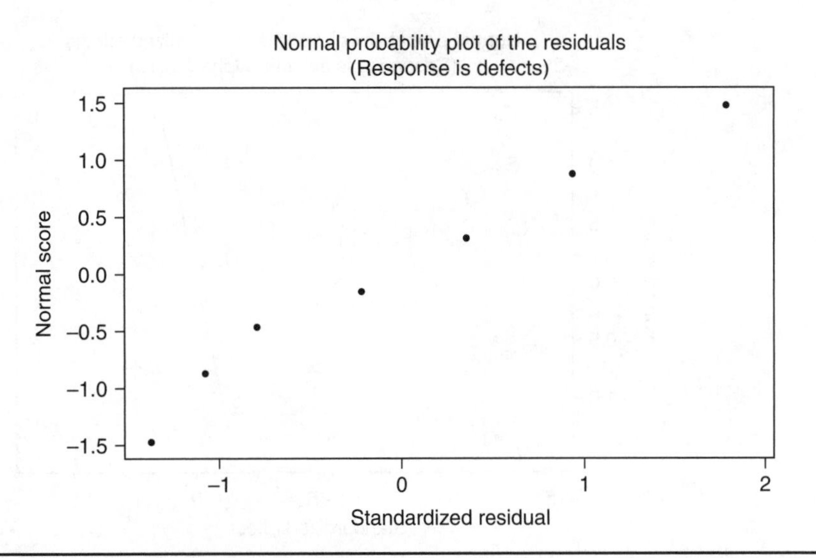

FIGURE 10.26 Residuals from experimental model.

- Circuit density = High
- Fixture used = Yes
- Defects per standard unit = 84

Experimental data were collected using the randomized run order recommended by the software. The "response" column are data that were recorded in terms of defective solder joints per "standard unit," where a standard unit represented a circuit board with a median number of solder joints.* The results are shown in Table 10.10.

A model that fits the data well would produce residuals that fall along a straight line. The Black Belt concluded that the fit of the model was adequate.

The analysis indicates that factors B (preheat time) and D (distance from preheat element to board surface) produce significant effects. Figure 10.26 shows a normal probability plot of the experimental effects. This figure plots the coefficients column from Table 10.10 on a normal probability scale. If the factor's effect was due to chance variation it would plot close to the line representing normal variation. In Fig. 10.27 the effects of B and D are shown to be further from the line than can be accounted for by random variation.

The effects of the significant factors are graphed in response units in Fig. 10.27.

Since the response is a defect count, the graph indicates that the low level of factor D gives better results, while the high level of factor B gives the better results. This can also be seen by examination of the coefficients for the variables. When D is low the average defect rate is 18.5 defects per unit better than when D is high; when B is high the average defect rate is 8 defects per unit better than when B is low.

*Technically, a Poisson model would be the correct choice here. However, use of a normal model, which the analysis assumes, is reasonably accurate for defect counts of this magnitude. The team also evaluated the variance, more specifically, the log of the variance. The variances at each factor combination did not differ significantly and are not shown here.

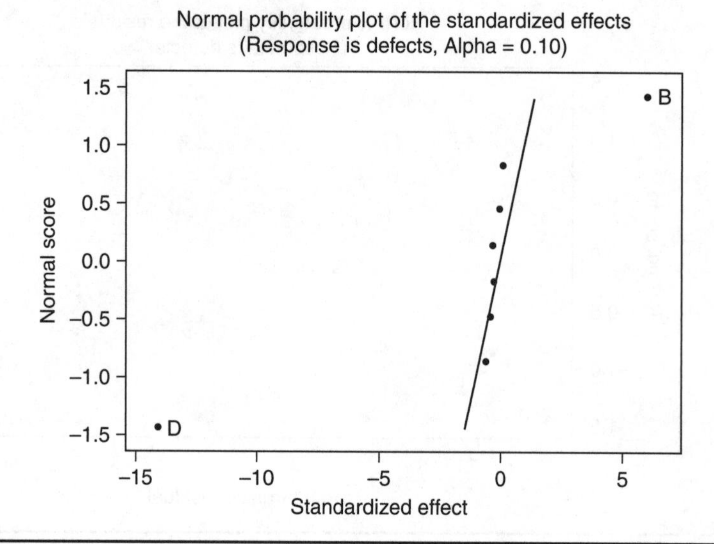

FIGURE 10.27 Significant factor effects.

Estimated effects and coefficients for response (coded units)						
Term	Effect	Coef	StDev coef	T	P	
Constant		70.375	0.6597	106.67	0.000	
A	−0.750	−0.375	0.6597	−0.57	0.588	
B	8.000	4.000	0.6597	6.06	0.001	
C	−0.500	−0.250	0.6597	−0.38	0.716	
D	−18.500	−9.250	0.6597	−14.02	0.000	
E	0.000	0.000	0.6597	0.00	1.000	
F	−0.250	−0.125	0.6597	−0.19	0.855	
G	−0.250	−0.125	0.6597	−0.19	0.855	
H	0.250	0.125	0.6597	0.19	0.855	
ANOVA for defects (coded units)						
Source of variation	*df*	*Seq. SS*	*Adj. SS*	*Adj. MS*	*F*	*P-value*
Main effects	8	1629.00	1629.00	203.625	29.24	0.000
Residual error	7	48.75	48.75	6.964		
Total	15	1677.75				

TABLE 10.10 Results of Experimental Data Analysis. Fractional factorial fit

The team met to discuss these results. They decided to set all factors that were not found to be statistically significant to the levels that cost the least to operate, and factors B and D at their midpoints. The process would be monitored at these settings for a while to determine that the results were similar to what the team expected based on the experimental analysis. While this was done, another series of experiments would be planned to further explore the significant effects uncovered by the screening experiment.

Based on the screening experiment, the linear model for estimating the defect rate was found from the coefficients in Table 10.10 to be

$$\text{Defect rate} = 70.375 + 4\text{B} - 9.25\text{D}$$

Power and Sample Size

The term *power* of a statistical test refers to the probability that the will lead to correctly rejecting a false Null Hypothesis, that is, $1-\beta$, where beta is the probability of failing to reject the false Null Hypothesis. Generally, the power of a statistical test is improved when:

- There is a large difference between the null and alternative conditions,
- The population sigma is small,
- The sample size is large; or,
- The significance (α) is large.

Many statistical software packages provide Power and Sample Size calculations. Minitab's *Power and Sample Size* option in the *Stat* menu can estimate these for a variety of test formats.

Example

Consider a one-way ANOVA test of the hypothesis that four populations have equal means. A sample of $n = 5$ is taken from each population whose historical standard deviation is 2.0. If we are interested in detecting a difference of 3 units in the means, the software can estimate the power of the test after completing the *Power and Sample Size for one-way ANOVA* dialog box as:

- Number of levels: 4
- Sample sizes: 5
- Values of the maximum difference between means: 3
- Standard deviation: 2
- Significance level (in the *Options dialog*): 0.05

The probability the assumption of equal means is rejected is found to be about 39% in this case. Note that if the sample size is increased to 10 the power is improved to 77%.

Testing Common Assumptions

Many statistical tests are only valid if certain underlying assumptions are met. In most cases, these assumptions are stated in the statistical textbooks along with the descriptions of the particular statistical technique. This chapter describes some of the more common assumptions encountered in Six Sigma project work and how to test for them. However, the subject of testing underlying assumptions is a big one and you might wish to explore it further with a Master Black Belt.

Continuous versus Discrete Data

Data come in two basic flavors: continuous and discrete, as discussed in Chap. 7. To review the basic idea, continuous data are numbers that can be expressed to any desired level of precision, at least in theory. For example, using a mercury thermometer I can say that the temperature is 75 degrees Fahrenheit. With a home digital thermometer I could say it's 75.4 degrees. A weather bureau instrument could add additional decimal places. Discrete data can only assume certain values. For example, the counting numbers can only be integers. Some survey responses force the respondent to choose a particular number from a list (pick a rating on a scale from 1 to 10).

Some statistical tests assume that you are working with either continuous or discrete data. For example, ANOVA assumes that continuous data are being analyzed, while chi-square and correspondence analysis assume that your data are counts. In many cases the tests are insensitive to departures from the data-type assumption. For example, expenditures can only be expressed to two decimal places (dollars and cents), but they can be treated as if they are continuous data. Counts can usually be treated as continuous data if there are many different counts in the data set. For example, if the data are defect counts ranging from 10 to 30 defects with all 21 counts showing up in the data (10, 11, 12, 28, 29, 30).

You Have Discrete Data But Need Continuous Data In some cases, however, the data type matters. For example, if discrete data are plotted on control charts intended for continuous data the control limit calculations will be incorrect. Run tests and other nonparametric tests will also be affected by this. The problem of "discretized" data is often caused by rounding the data to too few decimal places when they are recorded. This rounding can be human caused, or it might be a computer program not recording or displaying enough digits. The simple solution is to record more digits. The problem may be caused by an inadequate measurement system. This situation can be identified by a measurement system analysis (see Chap. 9). The problem can be readily detected by creating a dot plot of the data.

You Have Continuous Data But Need Discrete Data Let's say you want to determine if operator experience has an impact on the defects. One way to analyze this is to use a technique such as regression analysis to regress X = years of experience on Y = defects. Another would be to perform a chi-square analysis on the defects by experience level. To do this you need to put the operators into discrete categories, then analyze the defects in each category. This can be accomplished by "discretizing" the experience variable. For example, you might create the following discrete categories:

Experience (years)	Experience Category
Less than 1	New
1 to 2	Moderately experienced
3 to 5	Experienced
More than 5	Very experienced

The newly classified data are now suitable for chi-square analysis or other techniques that require discrete data.

Independence Assumption

Statistical independence means that two values are not related to one another. In other words, knowing what one value provides no information as to what the other value is. If you throw two dice and I tell you that one of them is a 4, that information doesn't help you predict the value on the other die. Many statistical techniques assume that the data are independent. For example, if a regression model fits the data adequately, then the residuals will be independent. Control charts assume that the individual data values are independent; that is, knowing the diameter of piston 100 doesn't help me predict the diameter of piston 101, nor does it tell me what the diameter of piston 99 was. If I don't have independence, the results of my analysis will be wrong. I will believe that the model fits the data when it does not. I will tamper with controlled processes.

Independence can be tested in a variety of ways. If the data are normal (testing the normality assumption is discussed below) then the run tests described for control charts can be used.

A scatter plot can also be used. Let $y = X_{t-1}$ and plot X versus Y. You will see random patterns if the data are independent. Software such as Minitab offer several ways of examining independence in time series data. Note: lack of independence in time series data is called *autocorrelation*.

If you don't have independence you have several options. In many cases the best course of action is to identify the reason why the data are not independent and fix the underlying cause. If the residuals are not independent, add terms to the model. If the process is drifting, add compensating adjustments.

If fixing the root cause is not a viable option, an alternative is to use a statistical technique that accounts for the lack of independence. For example, the EWMA control chart or a time series analysis that can model autocorrelated data. Another is to modify the technique to work with your autocorrelated data, such as using sloped control limits on the control chart. If data are cyclical you can create uncorrelated data by using a sampling interval equal to the cycle length. For example, you can create a control chart comparing performance on Monday mornings.

Normality Assumption

Statistical techniques such as t-tests, Z-tests, ANOVA, and many others assume that the data are at least approximately normal. This assumption is easily tested using software. There are two approaches to testing normality: graphical and statistical.

Graphical Evaluation of Normality One graphical approach involves plotting a histogram of the data, then superimposing a normal curve over the histogram. This approach works best if you have at least 200 data points, and the more the merrier. For small data sets the interpretation of the histogram is difficult; the usual problem is seeing a lack of fit when none exists. In any case, the interpretation is subjective and two people often reach different conclusions when viewing the same data. Figure 10.28 shows four histograms for normally distributed data with mean = 10, sigma = 1 and sample sizes ranging from 30 to 500.

An alternative to the histogram/normal curve approach is to calculate a "goodness-of-fit" statistic and a P-value. This gives an unambiguous acceptance criterion; usually the researcher rejects the assumption of normality if $P < 0.05$.

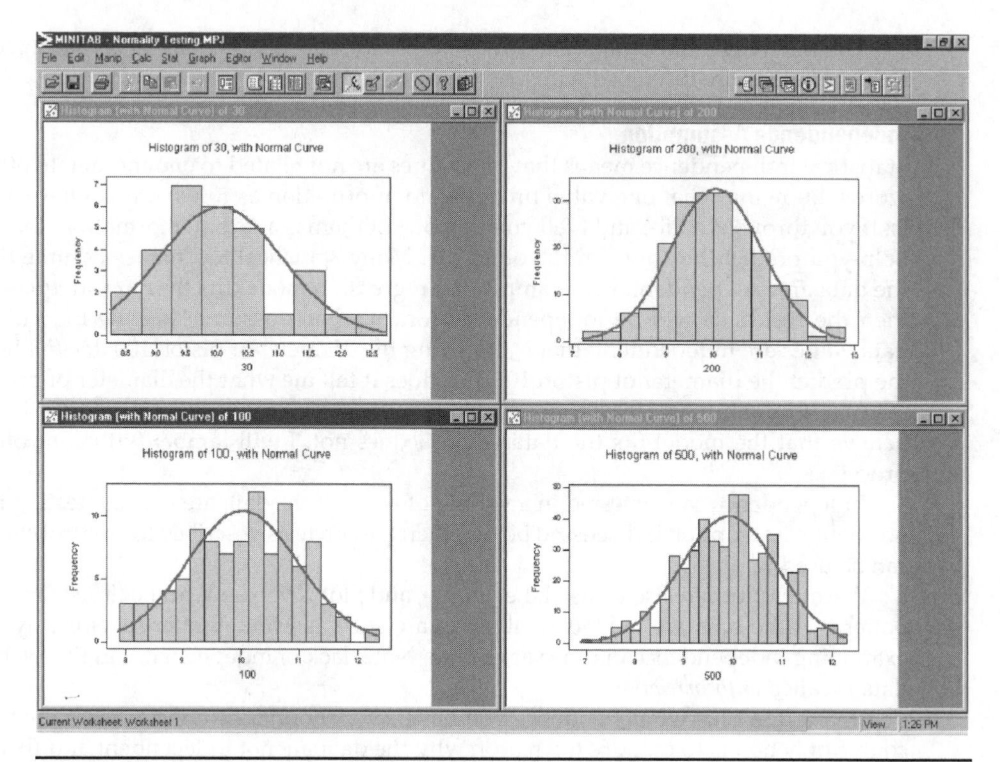

FIGURE 10.28 Histograms with normal curves for different sample sizes.

However, it has the disadvantage of being nongraphical. This violates the three rules of data analysis:

1. Plot the data
2. Plot the data
3. Plot the data

To avoid violating these important rules, the usual approach is to supplement the statistical analysis with a probability plot. The probability plot is scaled so that normally distributed data will plot as a straight line. Figure 10.29 shows the probability plots that correspond to the histograms and normal curves in Fig. 10.28. The table below Fig. 10.29 shows that the *P*-values are all comfortably above 0.05, leading us to conclude that the data are reasonably close to the normal distribution.

N	30	100	200	500
P-Value	0.139	0.452	0.816	0.345

What to Do If the Data Aren't Normal When data are not normal, the following steps are usually pursued:

- *Do nothing*—Often the histogram or probability plot shows that the normal model fits the data well "where it counts." If the primary interest is in the tails,

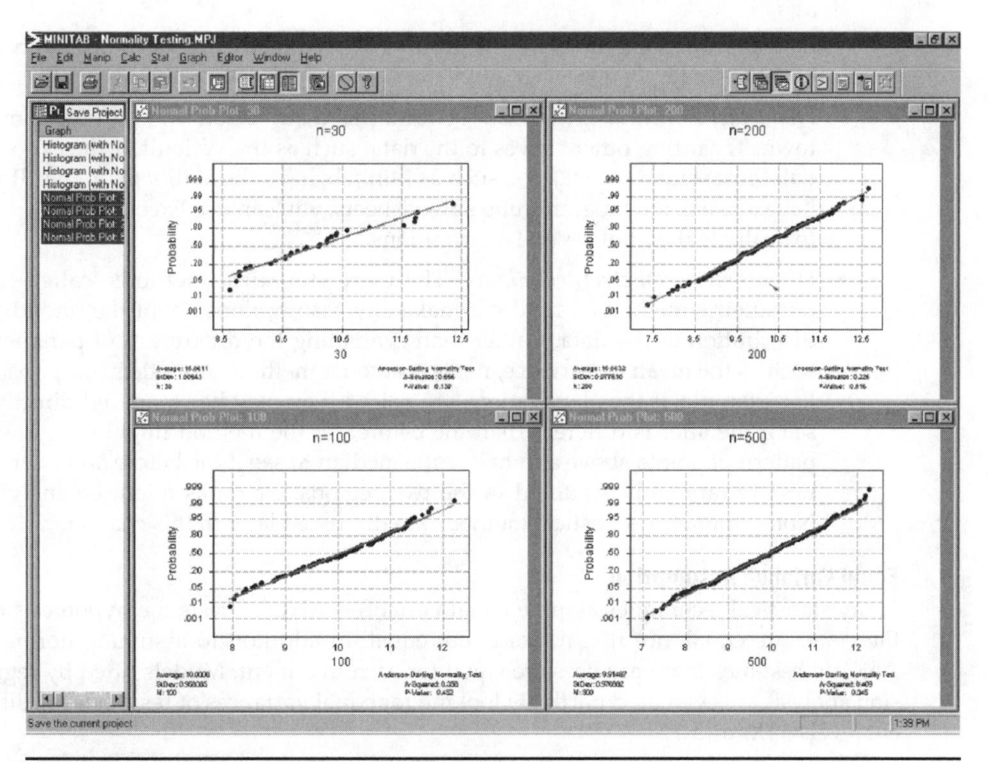

FIGURE 10.29 Normal probability plots and goodness of fit tests.

for example, and the curve fits the data well there, then proceed to use the normal model despite the fact that the *P*-value is less than 0.05. Or if the model fits the middle of the distribution well and that's your focus, go with it. Likewise, if you have a very large sample you may get *P*-values greater than 0.05 even though the model appears to fit well *everywhere*. I work with clients who routinely analyze data sets of 100,000+ records. Samples this large will flag functionally and economically unimportant departures from normality as "statistically significant," but it isn't worth the time or the expense to do anything about it.

- *Transform the data*—It is often possible to make the data normal by performing a mathematical operation on the data. For example, if the data distribution has very long tails to the high side, taking the logarithm often creates data that are normally distributed. Minitab's control chart feature offers the Box-Cox normalizing power transformation that works with many data distributions encountered in Six Sigma work. The downside to transforming is that data have to be returned to the original measurement scale before being presented to non-technical personnel. Some statistics can't be directly returned to their original units; for example, if you use the log transform then you can't find the mean of the original data by taking the inverse log of the mean of the transformed data.
- *Use averages*—Averages are a special type of transformation because averages of subgroups always tend to be normally distributed, even if the underlying

data are not. Sometimes the subgroup sizes required to achieve normality can be quite small.

- *Fit another statistical distribution*—The normal distribution isn't the only game in town. Try fitting other curves to the data, such as the Weibull or the exponential. Most statistics packages, such as Minitab, have the ability to do this. If you have a knack for programming spreadsheets, you can use Excel's solver add-in to evaluate the fit of several distributions.
- *Use a non-parametric technique*—There are statistical methods, called non-parametric methods, that don't make any assumptions about the underlying distribution of the data. Rather than evaluating the differences of parameters such as the mean or variance, non-parametric methods use other comparisons. For example, if the observations are paired they may be compared directly to see if the after is different than the before. Or the method might examine the pattern of points above and below the median to see if the before and after values are randomly scattered in the two regions. Or ranks might be analyzed. Non-parametric statistical methods are discussed later in this chapter.

Equal Variance Assumption

Many statistical techniques assume equal variances. ANOVA tests the hypothesis that the *means* are equal, not that *variances* are equal. In addition to assuming normality, ANOVA assumes that variances are equal for each treatment. Models fitted by regression analysis are evaluated partly by looking for equal variances of residuals for different levels of *X*s and *Y*.

Minitab's test for equal variances is found in Stat > ANOVA > Test for Equal Variances. You need a column containing the data and one or more columns specifying the factor level for each data point. If the data have already passed the normality test, use the *P*-value from Bartlett's test to test the equal variances assumption. Otherwise, use the *P*-value from Levene's test. The test shown in Fig. 10.30 involved five factor levels and Minitab shows a confidence interval bar for sigma of each of the five samples; the tick mark in the center of the bar represents the sample sigma. These are the data from the sample of 100 analyzed earlier and found to be normally distributed, so Bartlett's test can be used. The *P*-value from Bartlett's test is 0.182, indicating that we can expect this much variability from populations with equal variances 18.2% of the time. Since this is greater than 5%, we fail to reject the null hypothesis of equal variances. Had the data not been normally distributed we would've used Levene's test, which has a *P*-value of 0.243 and leads to the same conclusion.

Linear Model Assumption

Many types of associations are nonlinear. For example, over a given range of x values, y might increase, and for other x values, y might decrease. This *curvilinear relationship* is shown in Fig. 10.31.

Here we see that y increases when x increases and is less than 1, and decreases as x increases when x is greater than 1. Curvilinear relationships are valuable in the design of robust systems. A wide variety of processes produces such relationships.

It is often helpful to convert these nonlinear forms to linear form for analysis using standard computer programs or scientific calculators. Several such transformations are shown in Table 10.11.

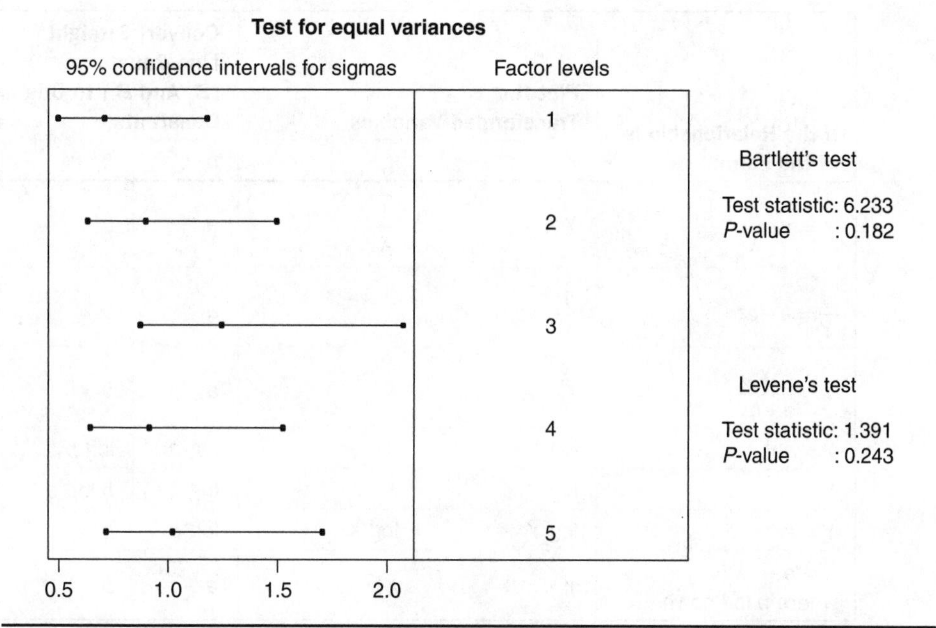

FIGURE 10.30 Output from Minitab's test for equal variances.

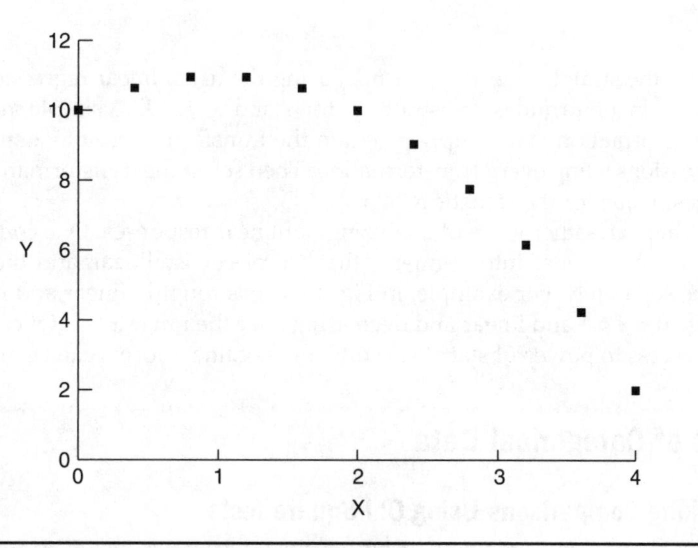

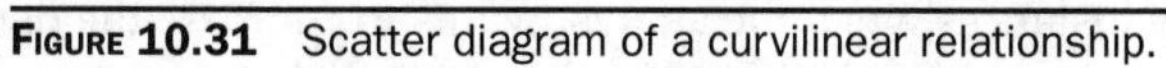
FIGURE 10.31 Scatter diagram of a curvilinear relationship.

If the Relationship Is of the Form:	Plot the Transformed Variables		Convert Straight Line Constants (B_0 And B_1) to Original Constants	
	Y_T	X_T	b_0	b_1
$Y = a + \frac{b}{X}$	Y	$\frac{1}{X}$	a	b
$\frac{1}{Y} = a + bX$	$\frac{1}{Y}$	X	a	b
$Y = \frac{X}{a + bX}$	$\frac{X}{Y}$	X	a	b
$Y = ab^X$	$\log Y$	X	$\log a$	$\log b$
$Y = ae^{bx}$	$\log Y$	X	$\log a$	$b \log e$
$Y = aX^b$	$\log Y$	$\log X$	$\log a$	b
$Y = a + bX^n$ where n is known	Y	X^n	a	b

(From Natrella (1963), pp. 5–31)

TABLE 10.11 Some Linearizing Transformations

Fit the straight line $Y_T = b_0 + b_1X_T$ using the usual linear regression procedures (see below). In all formulas, substitute Y_T for Y and X_T for X. A simple method for selecting a transformation is to simply program the transformation into a spreadsheet and run regressions using every transformation. Then select the transformation which gives the largest value for the statistic R^2.

There are other ways of analyzing nonlinear responses. One common method is to break the response into segments that are piecewise linear, and then to analyze each piece separately. For example, in Fig. 10.31 y is roughly linear and increasing over the range $0 < x < 1$ and linear and decreasing over the range $x > 1$. Of course, if the analyst has access to powerful statistical software, nonlinear forms can be analyzed directly.

Analysis of Categorical Data

Making Comparisons Using Chi-Square Tests

In Six Sigma, there are many instances when the analyst wants to compare the percentage of items distributed among several categories. The things might be operators, methods, materials, or any other grouping of interest. From each of the groups a sample is taken, evaluated, and placed into one of several categories (e.g., high quality, marginal quality, reject quality). The results can be presented as a table with m rows representing the groups of interest and k columns representing the categories. Such tables can be analyzed to answer the question "Do the groups *differ* with regard to the proportion of items in the categories?" The chi-square statistic can be used for this purpose.

Example of Chi-Square Test

The following example is from Natrella (1963):

Rejects of metal castings were classified by cause of rejection for three different weeks, as given in the following tabulation. The question to be answered is: Does the distribution of rejects differ from week to week?

	Cause of Rejection							
	Sand	Misrun	Shift	Drop	Corebreak	Broken	Other	Total
Week 1	97	8	18	8	23	21	5	180
Week 2	120	15	12	13	21	17	15	213
Week 3	82	4	0	12	38	25	19	180
Total	299	27	30	33	82	63	39	573

Chi-square (X^2) is computed by first finding the expected frequencies in each cell. This is done using the equation:

$$\text{Frequency expected} = f_e = \frac{\text{row sum} \times \text{column sum}}{\text{overall sum}}$$

For example, for week 1, the frequency expected of sand rejects is $(180 \times 299)/573 = 93.93$. The table below shows the frequency expected for the remainder of the cells.

	Sand	Misrun	Shift	Drop	Corebreak	Broken	Other
Week 1	93.93	8.48	9.42	10.37	25.76	19.79	12.25
Week 2	111.15	10.04	11.15	12.27	30.48	23.42	14.50
Week 3	93.93	8.48	9.42	10.37	25.76	19.79	12.25

The next step is to compute X^2 as follows:

$$X^2 = \sum_{\text{over all cells}} \frac{(\text{Frequency expected} - \text{Frequency observed})^2}{\text{Frequency expected}}$$

$$= \frac{(93.93-97)^2}{93.93} + \cdots + \frac{(12.25-19)^2}{12.25} = 45.60$$

Next choose a value for α; we will use $\alpha = 0.10$ for this example. The degrees of freedom for the X^2 test are $(k-1)(m-1) = 12$. Referring to Appendix 4 we find the critical value of $X^2 = 18.55$ for our values. Since our computed value of X^2 exceeds the critical value, we conclude that the weeks differ with regard to proportions of various types of defectives.

ResponseVariable and LogisticRegression Type	Number of ResponseCategories	ResponseCharacteristics	Examples
Binary	2	Two levels	Go/not-go, pass/fail, buy/doesn't buy, yes/no, recovers/dies, male/female
Ordinal	3 or more	Natural ordering of the levels	Dissatisfied/neutral/satisfied, none/mild/severe, fine/medium/coarse
Nominal	3 or more	No natural ordering of the levels	Black/white/Hispanic, black hair/brown hair/blonde hair, sunny/rainy/cloudy

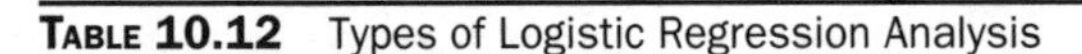
TABLE 10.12 Types of Logistic Regression Analysis

Logistic Regression

Logistic regression, like least squares regression, investigates the relationship between a response variable and one or more predictors. However, linear regression is used when response variables are continuous, while logistic regression techniques are used with categorical response variables. We will look at three different types of logistic regression, based on the type of response variable being analyzed (see Table 10.12).

The basic idea behind logistic regression is very simple, as shown in Fig. 10.32. *X* is a hypothetical "cause" of a response. *X* can be either continuous or categorical. *Y* is an event that we are interested in and it must be categorical. a model can have multiple *Xs*, but only one response variable. For example, *Y* might be whether a prospect purchased a magazine or not, and *Xs* might be the age and race of the prospect. The model would produce a prediction of the probability of a magazine being purchased based on the age

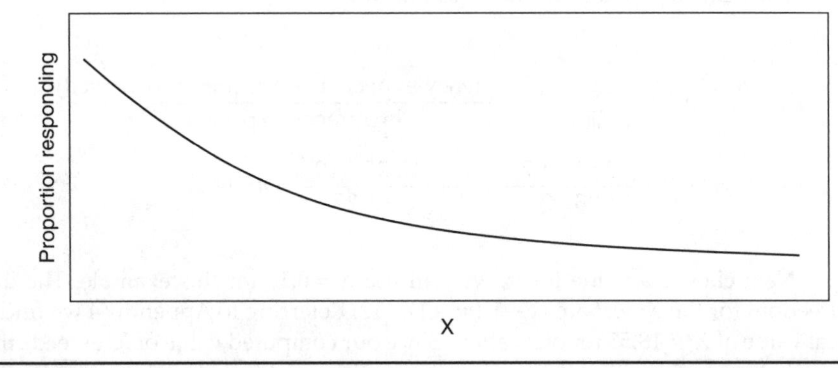

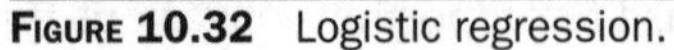
FIGURE 10.32 Logistic regression.

and race of the prospect, which might be used to prioritize a list for telemarketing purposes.

The Logit

Figure 10.32 illustrates a direct modeling of the proportion responding versus a predictor variable. The problem is that in the real world the response pattern can take on a wide variety of forms and a simple model of the proportion responding as a function of predictors isn't flexible enough to take on all of the various shapes. The solution to this is to use a mathematical function, called the logit, that makes it possible to develop versatile models. The formula for the logit is shown in Eq. (10.12). Although it looks intimidating, it is really very similar to the equation for a linear regression. Notice that e is raised to a power that is just a linear function of the Xs. In fact, the power term is just the multiple linear regression model. However, where linear regression can only model straight-line functions, the logit takes on a wide variety of curve shapes as the estimates of the parameters vary. Figure 10.33 shows logit curves for a few values of β, with α held constant at 0 (changing α would result in shifting the curves left or right).

$$P(x) = \frac{e^{\alpha+\beta_1 x_1+\beta_2 x_2+\cdots+\beta_n x_n}}{1+e^{\alpha+\beta_1 x_1+\beta_2 x_2+\cdots+\beta_n x_n}} \tag{10.13}$$

Odds Ratios

When the logit link is used (it's the default in most software packages, including Minitab), logistic regression evaluates the odds of some event of interest happening versus the odds of it not happening. This is done via *odds ratios*. "Odds" and probabilities are similar, but not identical. In a standard deck of cards there are 13 different card

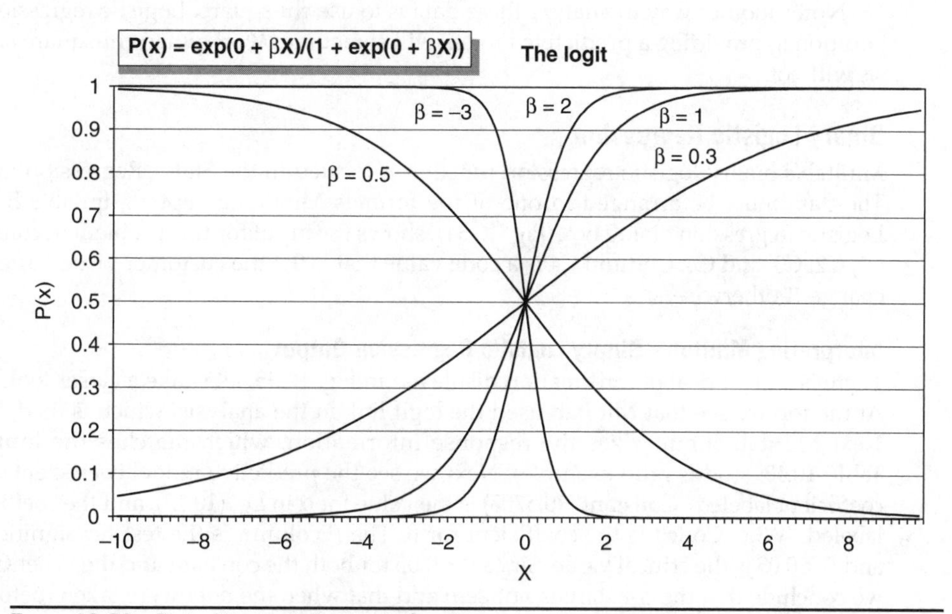

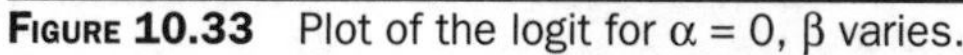
FIGURE 10.33 Plot of the logit for $\alpha = 0$, β varies.

Web Site Design	Found Answer	Didn't' Find Answer
Old	50	169
New	26	46

TABLE 10.13 Odds Ratio Example

values, ace, king, queen, and so on. The *odds* of a randomly selected card being an ace is 12-to-1, i.e., there are 12 nonaces to 1 ace. The *probability* of selecting an ace is 1-in-13, that is, there are 13 choices of which 1 is an ace. In most statistical analyses used in Six Sigma work we use probabilities, but logistic regression uses odds for its calculations.

Consider a Six Sigma project involving a Web site. The goal of the project is to make it easier for customers to find what they are looking for. A survey was administered to people who visited the Web site and the results in Table 10.13 were obtained. The Black Belt wants to know if the design change had an impact on the customer's ability to find an answer to their question.

The odds ratio for these data is calculated as follows:

$$\text{Odds of finding answer with old design} = 50/169 = 0.0296$$
$$\text{Odds of finding answer with new design} = 26/46 = 0.565$$
$$\text{Odds ratio} = 0.565/0.296 = 1.91$$

It can be seen that the odds of the customer finding the answer appears to be 91% better with the new design than with the old design. However, to interpret this result properly we must know if this improvement is statistically significant. We can determine this by using binary logistic regression.

Note: another way to analyze these data is to use chi-square. Logistic regression, in addition to providing a predictive model, will sometimes work when chi-square analysis will not.

Binary Logistic Regression

Minitab's binary logistic regression function is located in the Stat > Regression menu. The data must be arranged in one of the formats Minitab accepts. Minitab's Binary Logistic Regression dialog box (Fig. 10.34), shows the input for this problem in columns C1, C2, C3, and C4. Column C4 is a code value that is 0 if the customer visited after the change, 1 otherwise.

Interpreting Minitab's Binary Logistic Regression Output

There is a great deal of information displayed in Fig. 10.35; let's take a closer look at it. At the top we see that Minitab used the logit link in the analysis, which is its default. Next Minitab summarizes the response information, which matches the input in Table 10.13—(odds ratio example). Next we see the predictive model coefficients. The coefficient labeled "Constant" (0.5705) is the value for α in Eq. (10.13), and the coefficient labeled "WhenCode" is the coefficient for β. The P column is the test for significance and $P < 0.05$ is the critical value. Since $P < 0.05$ for both the constant and the WhenCode, we conclude that the constant is not zero and that when the data were taken (before or after the design change) made a difference.

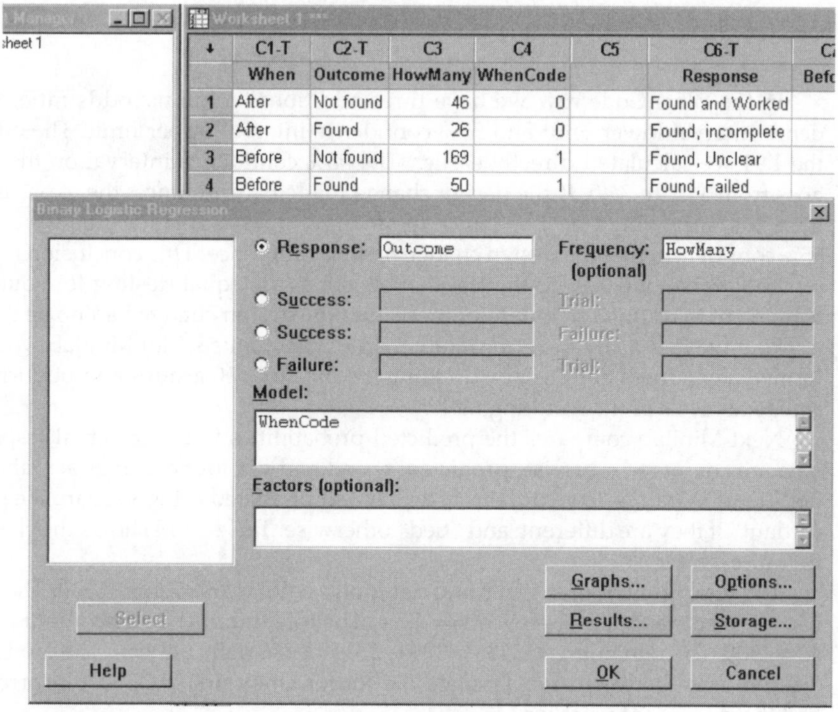

sheet 1

↓	C1-T	C2-T	C3	C4	C5	C6-T
	When	Outcome	HowMany	WhenCode		Response
1	After	Not found	46	0		Found and Worked
2	After	Found	26	0		Found, incomplete
3	Before	Not found	169	1		Found, Unclear
4	Before	Found	50	1		Found, Failed

Binary Logistic Regression

Response: Outcome
Frequency (optional): HowMany
Success:
Trial:
Success:
Failure:
Failure:
Trial:
Model:
WhenCode
Factors (optional):
Select
Graphs...
Options...
Results...
Storage...
Help
OK
Cancel

FIGURE 10.34 Minitab's Binary Logistic Regression dialog box.

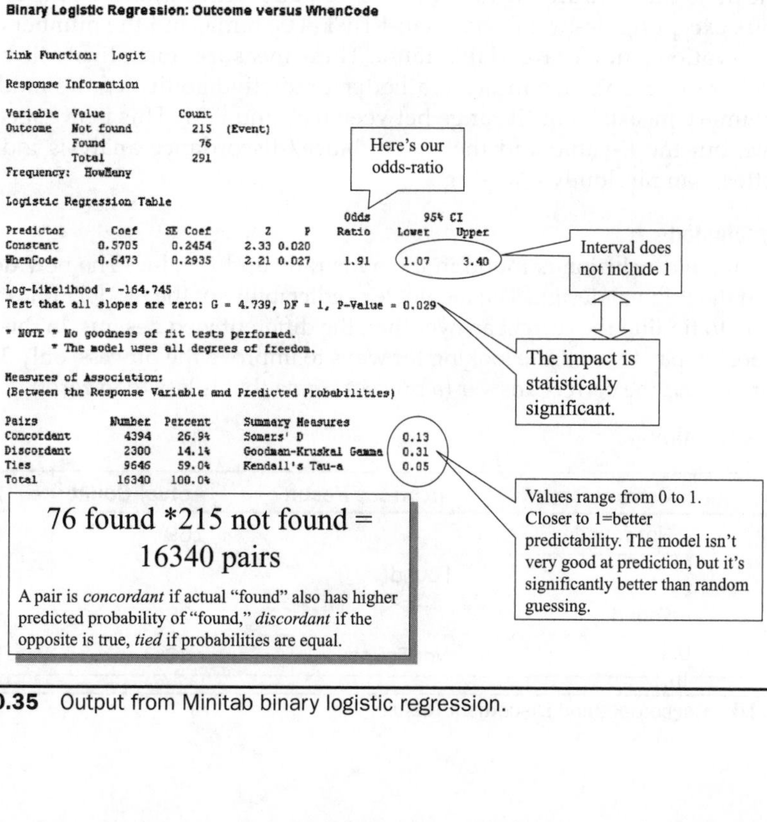

```
Binary Logistic Regression: Outcome versus WhenCode

Link Function:  Logit

Response Information

Variable  Value        Count
Outcome   Not found      215  (Event)
          Found           76
          Total          291
Frequency:  HowMany

Logistic Regression Table
                                                  Odds       95% CI
Predictor       Coef    SE Coef        Z      P   Ratio   Lower   Upper
Constant      0.5705     0.2454     2.33  0.020
WhenCode      0.6473     0.2935     2.21  0.027    1.91    1.07    3.40

Log-Likelihood = -164.745
Test that all slopes are zero: G = 4.739, DF = 1, P-Value = 0.029

* NOTE * No goodness of fit tests performed.
       * The model uses all degrees of freedom.

Measures of Association:
(Between the Response Variable and Predicted Probabilities)

Pairs         Number  Percent   Summary Measures
Concordant      4394    26.9%   Somers' D                 0.13
Discordant      2300    14.1%   Goodman-Kruskal Gamma     0.31
Ties            9646    59.0%   Kendall's Tau-a           0.05
Total          16340   100.0%
```

Here's our odds-ratio

Interval does not include 1

The impact is statistically significant.

Values range from 0 to 1. Closer to 1=better predictability. The model isn't very good at prediction, but it's significantly better than random guessing.

76 found *215 not found = 16340 pairs

A pair is *concordant* if actual "found" also has higher predicted probability of "found," *discordant* if the opposite is true, *tied* if probabilities are equal.

FIGURE 10.35 Output from Minitab binary logistic regression.

In the WhenCode row we have three additional columns: odds ratio, 95% confidence interval lower limit and 95% confidence interval upper limit. The odds ratio is the 1.91 we calculated directly earlier. The 95% confidence interval on the odds ratio goes from 1.07 to 3.40. If the design change made no difference, the expected value of the odds ratio would be 1.00. Since the interval doesn't include 1.00 we conclude (at 95% confidence) that the design change made a difference. This conclusion is confirmed by the *P*-value of 0.029 for the test that all slopes are equal (testing for equal slopes is equivalent to testing the null hypothesis that the design change had no effect).

Had we had a covariate term (an *X* on a continuous scale) Minitab would've performed a goodness of fit test by dividing the data into 10 groups and performing a chi-square analysis of the resulting table.

Next Minitab compares the predicted probabilities with the actual responses. The data are compared pairwise, predicted: found and not found versus actual: found and not found. A pair is "concordant" if actual and predicted categories are the same, "discordant" if they are different, and "tied" otherwise. Table 10.14 shows the classifications for our example.

The total number of found times not found pairs is $76 \times 215 = 16340$. The total number of concordant pairs is $169 \times 26 = 4394$. The total number of discordant pairs is $50 \times 46 = 2300$. The remaining $16340 - 4394 - 2300 = 9646$ pairs are ties. The model correctly discriminated between and classified the concordant pairs, or 27%. It incorrectly classified the discordant pairs, or 14%.

Somers' D, Goodman-Kruskal Gamma, and Kendall's Tau-a are summaries of the table of concordant and discordant pairs. The numbers have the same numerator: the number of concordant pairs minus the number of discordant pairs. The denominators are the total number of pairs with Somers' D, the total number of pairs excepting ties with Goodman-Kruskal Gamma, and the number of all possible observation pairs for Kendall's Tau-a. These measures most likely lie between 0 and 1 where larger values indicate a better predictive ability of the model. The three summary measures of fit range between 0.05 and 0.31. This isn't especially impressive, but the *P*-value and the concordance/discordance analysis indicate that it's better than randomly guessing.

Conclusion

The main conclusion is found in the odds ratio and *P*-value. The new design is better than the original design. The mediocre predictability of the model indicates that there's more to finding the correct answer than the different web designs. In this case it would probably pay to continue looking for ways to improve the process, only 36% of the customers find the correct answer (a process sigma that is less than zero!).

Design	Correct Result	Incorrect Result	Actual Count	Result
Old	Not found		169	Concordant
		Found	50	Discordant
New	Found		26	Concordant
		Not Found	46	Discordant

TABLE 10.14 Concordant and Discordant Results

Ordinal Logistic Regression

If the response variable has more than two categories, and if the categories have a natural order, then use ordinal logistic regression. Minitab's procedure for performing this analysis assumes parallel logistic regression lines. You may also want to perform a nominal logistic regression, which doesn't assume parallel regression lines, and compare the results. An advantage to using ordinal logistic regression is that the output includes estimated probabilities for the response variables as a function of the factors and covariates.

Ordinal Logistic Regression Example

A call center conducted a survey of its customers to determine the impact of various call center variables on overall customer satisfaction. Customers were asked to read a statement, then to respond by indicating the extent of their agreement with the statement. The two survey items we will analyze are:

Q3: The technical support representative was professional. (X)

Q17: I plan to use XXX in the future, should the need arise. (Y)

Customers were asked to choose one of the following responses to each question:

1. I strongly disagree with the statement.
2. I disagree with the statement.
3. I neither agree nor disagree with the statement.
4. I agree with the statement.
5. I strongly agree with the statement.

The results are shown in Table 10.15. Table 10.16 presents the first part of the Minitab worksheet for the data—note that this is the same information as in Table 10.15, just rearranged. There is one row for each combination of responses to Q3 and Q17.

Minitab's dialog box for this example is shown in Fig. 10.36. The storage dialog box allows you to tell Minitab to calculate the probabilities for the various responses. I also recommend telling Minitab to calculate the number of occurrences so that you can cross check your frequencies with Minitab's to ensure that you have the data in the correct format. When you tell Minitab to store results, the information is placed in new columns

Frequency Table					
	Q17 Response				
Q3 RESPONSE	1	2	3	4	5
1	7	6	7	12	9
2	5	2	8	18	3
3	4	2	20	42	10
4	7	5	24	231	119
5	0	2	14	136	303

Table 10.15 Survey Response Cross-Tabulation

Q3Response	Freq	Q17Response
1	7	1
2	5	1
3	4	1
4	7	1
5	0	1
1	6	2
2	2	2
Etc.	Etc.	Etc.

TABLE 10.16 Table 10.15 data reformatted for Minitab

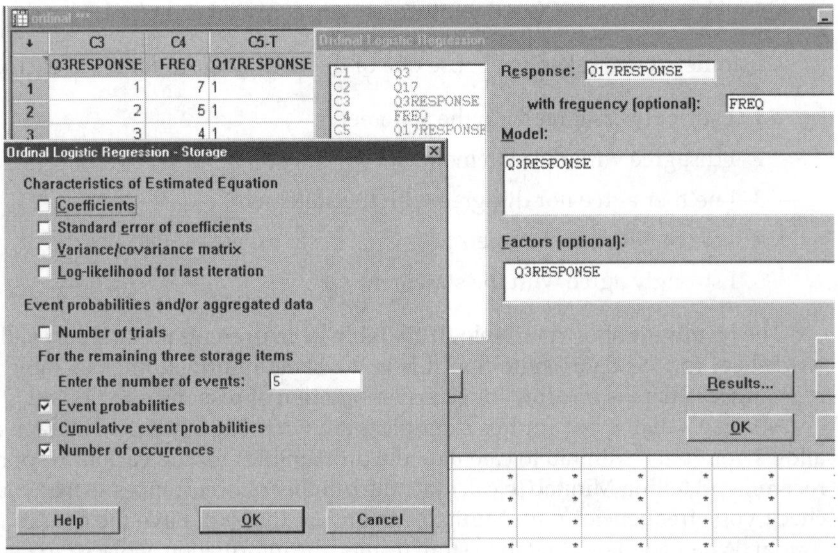

FIGURE 10.36 Ordinal Logistic Regression Minitab dialog boxes.

in your active worksheet, not in the session window. Note the data entries for the response, frequency, model, and factors.

Minitab's session window output is shown in Fig. 10.37. For simplicity only part of the output is shown. The goodness-of-fit statistics (concordance, discordance, etc.) have been omitted, but the interpretation is the same as for binary logistic regression. Minitab needs to designate one of the response values as the reference event. Unless you specifically choose a reference event, Minitab defines the reference event based on the data type:

- For numeric factors, the reference event is the greatest numeric value.
- For date/time factors, the reference event is the most recent date/time.
- For text factors, the reference event is the last in alphabetical order. A summary of the interpretation follows:

```
Ordinal Logistic Regression: Q17RESPONSE versus Q3RESPONSE

Link Function:  Logit

Response Information

Variable  Value        Count
Q17RESPO  1               23
          2               17
          3               73
          4              439
          5              444
          Total          996
Frequency:  FREQ

   24 cases were used
    1 cases contained missing values
      or was a case with zero frequency.

Logistic Regression Table
                                                      Odds       95% CI
Predictor        Coef     SE Coef        Z      P    Ratio    Lower    Upper
Const(1)      -2.0630      0.3358    -6.14  0.000
Const(2)      -1.4516      0.3094    -4.69  0.000
Const(3)      -0.1995      0.2926    -0.68  0.496
Const(4)       2.5735      0.3121     8.25  0.000
Q3RESPON
 2            -0.0905      0.4253    -0.21  0.831     0.91     0.40     2.10
 3            -0.5898      0.3638    -1.62  0.105     0.55     0.27     1.13
 4            -1.8408      0.3176    -5.80  0.000     0.16     0.09     0.30
 5            -3.2571      0.3254   -10.01  0.000     0.04     0.02     0.07

Log-likelihood = -941.733
Test that all slopes are zero: G = 246.547, DF = 4, P-Value = 0.000
```

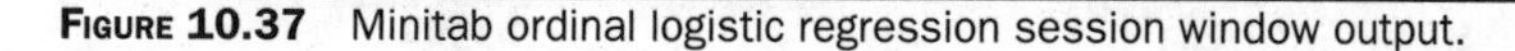

Figure 10.37 Minitab ordinal logistic regression session window output.

- The odds of a reference event is the ratio of P(event) to P(not event).
- The estimated coefficient can also be used to calculate the odds ratio, or the ratio between two odds. Exponentiating the parameter estimate of a factor yields the ratio of P(event)/P(not event) for a certain factor level compared to the reference level.

You can change the default reference event in the Options subdialog box. For our example, category 5 (strongly agree) is the reference event. The odds ratios are calculated as the probability of the response being a 5 versus the probability that it is not a 5. For factors, the *smallest* numerical value is the reference event. For the example, this is a Q3 response of 1.

The odds ratios and their confidence intervals are given near the bottom of the table. a negative coefficient and an odds ratio less than 1 indicate that higher responses to Q17 tend to be associated with higher responses to Q3. Odds ratios whose confidence intervals do not include 1.00 are statistically significant. For example, this applies to responses of 4 or 5 to Q3, that is, a customer who chooses a 4 or 5 in response to Q3 is more likely to choose a 5 in response to Q17.

The statistical probabilities stored by Minitab are plotted in Fig. 10.38. The lines for Q3 = 4 and Q3 = 5, the factor categories with significant odds ratios, are shown as bold lines. Note that the gap between these two lines and the other lines is greatest for Q17 = 5.

Nominal Logistic Regression

Nominal logistic regression, as indicated in Table 10.12, is used when the response is categorical, there are two or more response categories, and there is no natural ordering

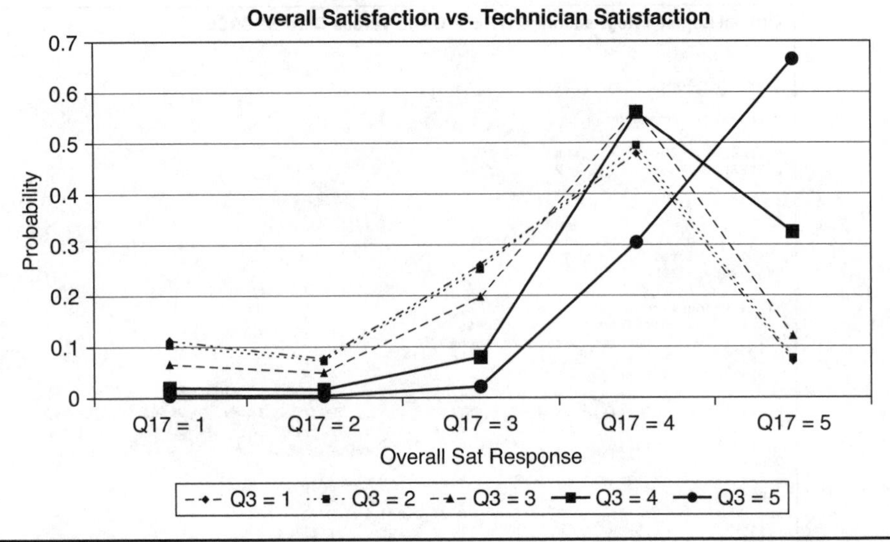

FIGURE 10.38 Minitab stored results.

of the response categories. It can also be used to evaluate whether the parallel line assumption of ordinal logistic regression is reasonable.

Example of Nominal Logistic Regression

Upon further investigation the Master Black Belt discovered that the Black Belt working on the Web site redesign project described in the binary logistic regression example section above had captured additional categories. Rather than just responding that the answer to their question was found or not found, there were several other response categories (Figs. 10.39 and 10.40). Since the various not found subcategories have no natural order, nominal logistic regression is the correct procedure for analyzing these data.

The result of Minitab's analysis, shown in Fig. 10.41, shows that only the odds ratio for found and worked versus not found is significant. The confidence interval for all other found subcategories compared with found and worked includes 1.00. The family *P*-value is a significance test for all comparisons simultaneously. Since we are making four comparisons, the significance level is higher than that of each separate test.

Comparison with Chi-Square

If a chi-square analysis is performed on the web redesign data Minitab produces the output shown in Fig. 10.42. Note that the chi-square procedure prints a warning that there are two cells with less than the recommended minimum expected frequency of 5.0. It also gives a *P*-value of 0.116, which is greater than the critical value of 0.05, leading to a somewhat different conclusion than the logistic regression analysis. The chi-square test only lets us look at the significance of the overall result, which is analogous to the "family *P*-value" test performed in the nominal logistic regression analysis. However, in this case we are primarily concerned with the improved odds of finding the *correct* answer with the new web design versus the old web design, which is provided by logit 4 of the logistic regression.

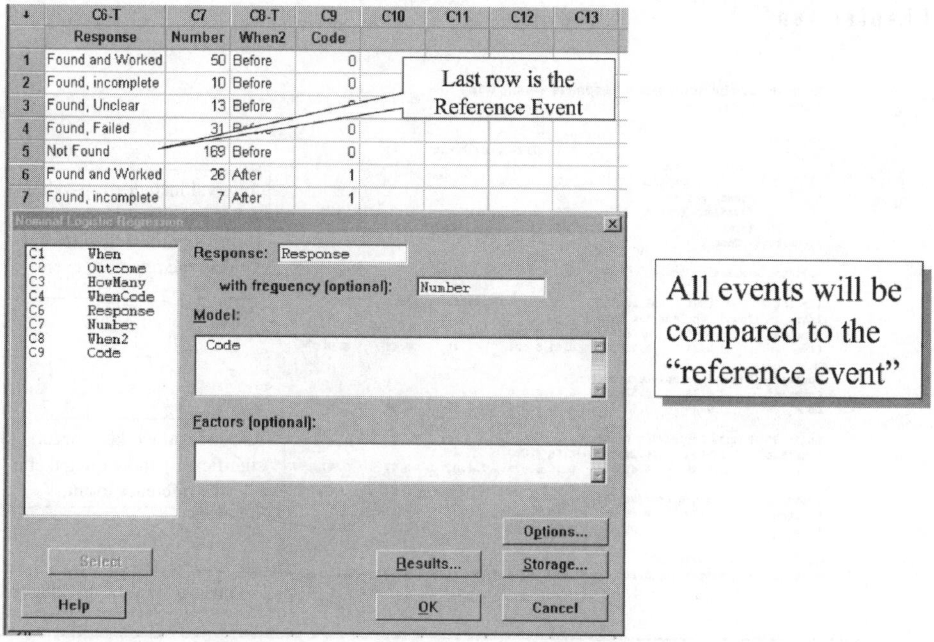

↓	C6-T	C7	C8-T	C9	C10	C11	C12	C13
	Response	Number	When2	Code				
1	Found and Worked	50	Before	0				
2	Found, incomplete	10	Before	0				
3	Found, Unclear	13	Before	0				
4	Found, Failed	31	Before	0				
5	Not Found	169	Before	0				
6	Found and Worked	26	After	1				
7	Found, incomplete	7	After	1				

Last row is the Reference Event

Nominal Logistic Regression

C1 When
C2 Outcome
C3 HowMany
C4 WhenCode
C6 Response
C7 Number
C8 When2
C9 Code

Response: Response

with frequency (optional): Number

Model:

Code

Factors (optional):

Select

Help

Options...

Results...

Storage...

OK

Cancel

All events will be compared to the "reference event"

FIGURE 10.39 Minitab's nominal logistic regression dialog box.

Nominal Logistic Regression: Response versus Code

```
Response Information

Variable  Value                 Count
Response  Not Found               215  (Reference Event)
          Found, Unclear           18
          Found, incomplete        17
          Found, Failed            40
          Found and Worked         76
          Total                   366
Frequency:  Number

Logistic Regression Table
                                                    Odds      95% CI
Predictor        Coef    SE Coef       Z      P    Ratio   Lower   Upper
Logit 1: (Found, Unclear/Not Found)
Constant      -2.5649     0.2878   -8.91  0.000
Code           0.3457     0.5519    0.63  0.531     1.41    0.48    4.17

Logit 2: (Found, incomplete/Not Found)
Constant      -2.8273     0.3254   -8.69  0.000
Code           0.9446     0.5201    1.82  0.069     2.57    0.93    7.13

Logit 3: (Found, Failed/Not Found)
Constant      -1.6959     0.1954   -8.68  0.000
Code           0.0645     0.4136    0.16  0.876     1.07    0.47    2.40

Logit 4: (Found and Worked/Not Found)
Constant      -1.2179     0.1610   -7.57  0.000
Code           0.6473     0.2935    2.21  0.027     1.91    1.07    3.40

Log-likelihood = -425.268
Test that all slopes are zero: G = 7.053, DF = 4, P-Value = 0.133
```

FIGURE 10.40 Minitab nominal logistic regression output.

Nominal Logistic Regression: Response versus Code

```
Response Information

Variable  Value               Count
Response  Not Found             215  (Reference Event)
          Found, Unclear         18
          Found, incomplete      17
          Found, Failed          40
          Found and Worked       76
          Total                 366
Frequency:  Number

Logistic Regression Table
                                                   Odds      95% CI
Predictor        Coef   SE Coef       Z      P    Ratio   Lower   Upper
Logit 1: (Found, Unclear/Not Found)
Constant      -2.5649    0.2878   -8.91  0.000
Code           0.3457    0.5519    0.63  0.531     1.41    0.48    4.17

Logit 2: (Found, incomplete/Not Found)
Constant      -2.8273    0.3254   -8.69  0.000
Code           0.9446    0.5201    1.82  0.069     2.57    0.93    7.13

Logit 3: (Found, Failed/Not Found)
Constant      -1.6959    0.1954   -8.68  0.000
Code           0.0645    0.4136    0.16  0.876     1.07    0.47    2.40

Logit 4: (Found and Worked/Not Found)
Constant      -1.2179    0.1610   -7.57  0.000
Code           0.6473    0.2935    2.21  0.027     1.91    1.07    3.40

Log-likelihood = -425.268
Test that all slopes are zero: G = 7.053, DF = 4, P-Value = 0.133
```

Interval includes 1, p-value > 0.05. I.e., "found but unclear" isn't significantly different than the reference event.

"Found and worked" occurs significantly more often than the reference event.

This is a "family" p-value

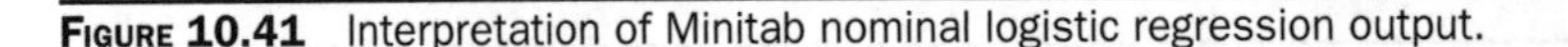

Figure 10.41 Interpretation of Minitab nominal logistic regression output.

Chi-Square Test: Before, After

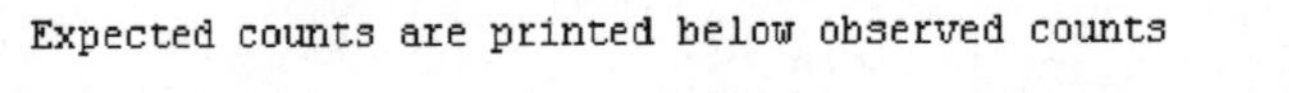

Expected counts are printed below observed counts

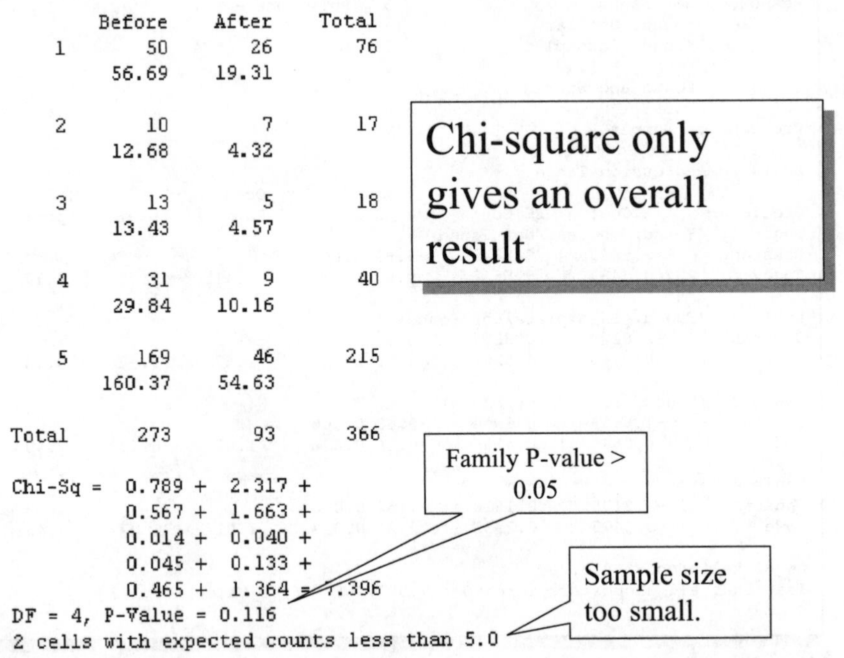

```
        Before    After    Total
  1         50       26       76
         56.69    19.31

  2         10        7       17
         12.68     4.32

  3         13        5       18
         13.43     4.57

  4         31        9       40
         29.84    10.16

  5        169       46      215
        160.37    54.63

Total      273       93      366

Chi-Sq =  0.789 +  2.317 +
          0.567 +  1.663 +
          0.014 +  0.040 +
          0.045 +  0.133 +
          0.465 +  1.364 = 7.396
DF = 4, P-Value = 0.116
2 cells with expected counts less than 5.0
```


Figure 10.42 Chi-square analysis of web design data.

Non-Parametric Methods

The most commonly used statistical tests (*t*-tests, Z-tests, ANOVA, etc.) are based on a number of assumptions (see testing assumptions above). Non-parametric tests, while not assumption-free, make no assumption of a specific *distribution* for the population. The qualifiers (assuming) for non-parametric tests are always much less restrictive than for their parametric counterparts. For example, classical ANOVA requires the assumptions of mutually independent random samples drawn from normal distributions that have equal variances, while the non-parametric counterparts require only the assumption that the samples come from any identical continuous distributions. Also, classical statistical methods are strictly valid only for data measured on interval or ratio scales, while non-parametric statistics apply to frequency or count data and to data measured on nominal or ordinal scales. Since interval and ratio data can be transformed to nominal or ordinal data, non-parametric methods are valid in all cases where classical methods are valid; the reverse is not true. Ordinal and nominal data are very common in Six Sigma work. Nearly all customer and employee surveys, product quality ratings, and many other activities produce ordinal and nominal data.

So if non-parametric methods are so great, why do we ever use parametric methods? When the assumptions hold, parametric tests will provide greater power than non-parametric tests. That is, the probability of rejecting H_0 when it is false is higher with parametric tests than with a non-parametric test using the same sample size. However, if the assumptions do not hold, then non-parametric tests may have considerably greater power than their parametric counterparts.

It should be noted that non-parametric tests perform comparisons using medians rather than means, ranks rather than measurements, and signs of difference rather than

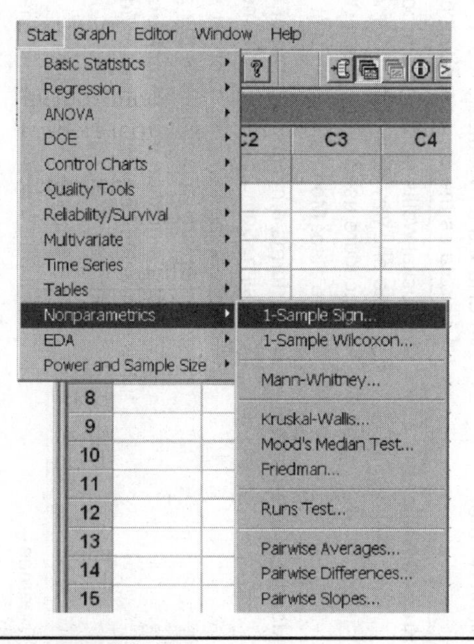

FIGURE 10.43 Minitab's non-parametric tests.

Minitab Non-Parametric Test	What It Does	ParametricAnalogs
1-sample sign	Performs a one-sample sign test of the median and calculates the corresponding point estimate and confidence interval.	• 1-sample *Z*-test • 1-sample *t*-test
1-sample Wilcoxon	Performs a one-sample Wilcoxon signed rank test of the median and calculates the corresponding point estimate and confidence interval.	• 1-sample *Z*-test • 1-sample *t*-test
Mann-Whitney	Performs a hypothesis test of the equality of two population medians and calculates the corresponding point estimate and confidence interval.	• 2-sample *t*-test
Kruskal-Wallis	Kruskal-Wallis performs a hypothesis test of the equality of population medians for a one-way design (two or more populations). This test is a generalization of the procedure used by the Mann-Whitney test. See also: Mood's median test.	• One-way ANOVA
Mood's median test	Performs a hypothesis test of the equality of population medians in a one-way design. Sometimes called a median test or sign scores test. Mood's median test is robust against outliers and errors in data, and is particularly appropriate in the preliminary stages of analysis. Mood's median test is more robust against outliers than the Kruskal-Wallis test, but is less powerful (the confidence interval is wider, on the average) for analyzing data from many distributions, including data from the normal distribution. See also: Kruskal-Wallis test.	• One-way ANOVA
Friedman	Performs a non-parametric analysis of a randomized block experiment. Randomized block experiments are a generalization of paired experiments. The Friedman test is a generalization of the paired sign test with a null hypothesis of treatments having no effect. This test requires exactly one observation per treatment-block combination.	• Two-way ANOVA • Paired sign test
Runs tests	Test whether or not the data order is random. Use Minitab's Stat > Quality Tools > Run Chart to generate a run chart.	• None
Pairwise averages	Pairwise averages calculates and stores the average for each possible pair of values in a single column, including each value with itself. Pairwise averages are also called Walsh averages. Pairwise averages are used, for example, for the Wilcoxon method.	• None

Pairwise differences	Pairwise differences calculates and stores the differences between all possible pairs of values formed from two columns. These differences are useful for non-parametric tests and confidence intervals. For example, the point estimate given by Mann-Whitney can be computed as the median of the differences.	• None
Pairwise slopes	Pairwise slopes calculates and stores the slope between all possible pairs of points, where a row in *y-x* columns defines a point in the plane. This procedure is useful for finding robust estimates of the slope of a line through the data.	• Simple linear regression
Levene's test	Test for equal variances. This method considers the distances of the observations from their sample median rather than their sample mean. Using the sample median rather than the sample mean makes the test more robust for smaller samples.	• Bartlett's test
Non-parametric Dist Analysis—Censored Data	Analyzes times-to-failure when no distribution can be found to fit the (censored) data. Tests for the equality of survival curves.	• Parametric Dist Analysis—Censored data
Hazard plots—non-parametric distribution analysis	If data are right censored, plots empirical hazard function or actuarial estimates. If data are arbitrarily censored, plots actuarial estimates.	• Hazard plots—parametric distribution analysis

Table 10.17 Applications for Minitab's Non-Parametric Tests*

measured differences. In addition to not requiring any distributional assumptions, these statistics are also more robust to outliers and extreme values.

The subject of non-parametric statistics is a big one and there are many entire books written about it. We can't hope to cover the entire subject in a book about Six Sigma. Instead, we briefly describe the non-parametric tests performed by Minitab (Fig. 10.43). Minitab's non-parametric tests cover a reasonably wide range of applications to Six Sigma work, as shown in Table 10.17.

Guidelines on When to Use Non-Parametric Tests

Use non-parametric analysis when *any* of the following are true (Gibbons, 1993):

- The data are counts or frequencies of different types of outcomes.
- The data are measured on a nominal scale.
- The data are measured on an ordinal scale.
- The assumptions required for the validity of the corresponding parametric procedure are not met or cannot be verified.
- The shape of the distribution from which the sample is drawn is unknown.
- The sample size is small.
- The measurements are imprecise.
- There are outliers and/or extreme values in the data, making the median more representative than the mean.

Use a parametric procedure when *both* of the following are true:

- The data are collected and analyzed using an interval or ratio scale of measurement.
- All of the assumptions required for the validity of that parametric procedure can be verified.

CHAPTER 11

The Improve/Design Phase

The primary objective of the Improve or Design stage of DMAIC/DMADV is to implement the new system. The first consideration is to prioritize the various opportunities, if more than one proposal exists. Once a preferred approach has been determined, the new process or product design is defined and optimal settings established. This new design can then be evaluated for risks and potential failure modes. If any of these steps require changes in prior assumptions, then steps must be repeated to properly evaluate the new proposal.

Using Customer Demands to Make Design and Improvement Decisions

Customer demands can easily be converted into design requirements and specifications. The term "translation" is used to describe this process because the activity literally involves interpreting the words from one language (the customer's) into those of another (the employee). For example, regarding the door of her automobile the customer might say "I want the door to close completely when I push it, but I don't want it swinging closed from just the wind or when I'm parked on a steep hill." The engineer working with this requirement must convert it into engineering terminology such as pounds of force required to move the door from an open to a closed position, the angle of the door when it's opened, and so on. Care must be taken to maintain the customers' intent throughout the development of internal requirements. The purpose of specifications is to transmit the voice of the customer throughout the organization.

In addition to the issue of maintaining the voice of the customer, there is the related issue of the importance assigned to each demand by the customer. Design of products and services always involves tradeoffs: gasoline economy suffers as vehicle weight increases, but safety improves as weight increases. The importance of each criterion must be determined by the customer. When different customers assign different importance to criteria, design decisions are further complicated.

It becomes difficult to choose from competing designs in the face of ambiguity and customer-to-customer variation. Add to this the differences between internal personnel and objectives—department versus department, designer versus designer cost versus quality, etc.—and the problem of choosing a design alternative quickly becomes complex. A rigorous process for deciding which alternative to settle on is helpful in dealing with the complexity.

Next, we must determine importance placed on each item by customers. There are a number of ways to do this:

- Have customers assign importance weights using a numerical scale (e.g., "How important is 'Easy self-help' on a scale between 1 and 10?").
- Have customers assign importance using a subjective scale (e.g., unimportant, important, very important, etc.).
- Have customers "spend" $100 by allocating it among the various items. In these cases it is generally easier for the customer to first allocate $100 to the major categories, then allocate another $100 to items within each subcategory. The subcategory weights are "local" in that they apply to the category. To calculate global weights for subcategory items, divide the subcategory weights by 100 and multiply them by the major category weight.
- Have customers evaluate a set of hypothetical product offerings and indicate their preference for each product by ranking the offerings, assigning a "likely to buy" rating, etc. The product offerings include a carefully selected mix of items chosen from the list of customer demands. The list is selected in such a way that the relative value the customer places *on each item in the offering* can be determined from the preference values. This is known as *conjoint analysis*, an advanced marketing technique that is described in courses on marketing statistics.
- Have customers evaluate the items in pairs, assigning a preference rating to one of the items in each pair, or deciding that both items in a pair are equally important. This is less tedious if the major categories are evaluated first, then the items within each category. The evaluation can use either numeric values or descriptive labels. The pairwise comparisons can be analyzed using a method known as the analytic hierarchical process (AHP; also sometimes referred to as a Prioritization Matrix) to determine the relative importance assigned to all of the items.

All of the above methods have their advantages and disadvantages. We will illustrate the use of AHP for our hypothetical software product. AHP is a powerful technique that has been proven in a wide variety of applications. In addition to its use in determining customer importance values, it is useful for decision making in general. Research has shown that people are better able to make one-on-one comparisons than to simultaneously compare several items.

Category Importance Weights

We begin our analysis by making pairwise comparison at the top level. The affinity diagram analysis identified five categories: easy to learn, easy to use quickly after I've learned it, Internet connectivity, works well with other software I own, and easy to maintain. Arrange these items in a matrix as shown in Fig. 11.1.

For our analysis we will assign verbal labels to our pairwise comparisons; the verbal responses will be converted into numerical values for analysis. All comparisons are made relative to the customer's goal of determining which product he will buy. The first cell in the matrix compares the "easy to learn" attribute and the "easy to use quickly after I've learned it" attribute. The customer must determine which is more important

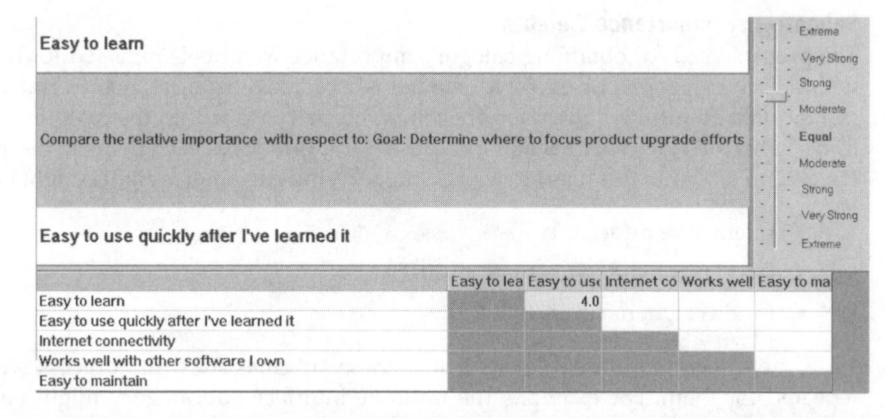

FIGURE 11.1 Matrix of categories for pairwise comparisons. (*Created using Expert Choice 2000 Software, www.expertchoice.com.**)

	Easy to lea	Easy to us(	Internet co	Works well	Easy to ma
Easy to learn		4.0	1.0	3.0	1.0
Easy to use quickly after I've learned it			0.2	0.33	0.25
Internet connectivity				3.0	3.0
Works well with other software I own					0.33
Easy to maintain	Incon: 0.05				

FIGURE 11.2 Completed top-level comparison matrix.

to him, or if the two attributes are of equal importance. In Fig. 11.1 this customer indicates that "easy to learn" is moderately to strongly preferred over "easy to use quickly after I've learned it" and the software has placed a + 4 in the cell comparing these two attributes. (The scale goes from −9 to +9, with "equal" being identified as a + 1.) The remaining attributes are compared one-by-one, resulting in the matrix shown in Fig. 11.2.

The shaded bars over the attribute labels provide a visual display of the relative importance of each major item to the customer. Numerically, the importance weights are:

- Easy to learn: 0.264 (26.4%)
- Easy to use quickly after I've learned it: 0.054 (5.4%)
- Internet connectivity: 0.358 (35.8%)
- Works well with other software I own: 0.105 (10.5%)
- Easy to maintain: 0.218 (21.8%)

These relative importance weights can be used in QFD and DFSS as well as in the AHP processes that we are illustrating here. In our allocation of effort, we will want to emphasize those attributes with high importance weights over those with lower weights.

---

*Although the analysis is easier with special software, you can obtain a good approximation using a spreadsheet. See Appendix 17 for details.

Subcategory Importance Weights

The process used for obtaining category importance weights is repeated for the items within each category. For example, the items interactive tutorial, good printed documentation, and intuitive interface are compared pairwise within the category "easy to learn." This provides weights that indicate the importance of each item on the category. For example, within the "easy to learn" category, the customer weights might be:

- Interactive tutorial: 11.7%
- Good printed documentation: 20.0%
- Intuitive interface: 68.3%

If there were additional levels below these subcategories, the process would be repeated for them. For example, the intuitive interface subcategory might be subdivided into "number of menus," "number of submenus," "menu items easily understood," etc. The greater the level of detail, the easier the translation of the customer's demands into internal specifications. The tradeoff is that the process quickly becomes tedious and may end up with the customer being asked for input he isn't qualified to provide. In the case of this example, we'd probably stop at the second level.

Global Importance Weights

The subcategory weights just obtained tell us how much importance the item has with respect to the category. Thus, they are often called local importance weights. However, they don't tell us about the impact of the item on the overall goal, which is called a global impact. This is easily determined by multiplying the subcategory item weight by the weight of the category the item belongs to. The global weights for our example are shown in Table 11.1 in descending order.

Using Weighted CTQs in Decision-Making

The first step in deciding upon a course of action is to identify the goal. For example, let's say you're the owner of the product development process for a company that sells software to help individuals manage their personal finances. The product, let's call it DollarWise, is dominant in its market and your company is well respected by its customers and competitors, in large part because of this product's reputation. The business is profitable and the leadership naturally wants to maintain this pleasant set of circumstances and to build on it for the future. The organization has committed itself to a strategy of keeping DollarWise the leader in its market segment so it can capitalize on its reputation by launching additional new products directed towards other financially oriented customer groups, such as small businesses. They have determined that product development is a core process for deploying this strategy.

As the process owner, or business process executive, you have control of the budget for product development, including the resources to upgrade the existing product. Although it is still considered the best personal financial software available, DollarWise is getting a little long in the tooth and the competition has steadily closed the technical gap. You believe that a major product upgrade is necessary and want to focus your resources on those things that matter most to customers. Thus, your goal is:

Choose the best product upgrade design concept

The global importance weights are most useful for the purpose of evaluating alternative routes to the overall goal. For our example, Internet connectivity obviously has a

Category	Subcategory	Local Weight	Global Weight
Easy to learn	Intuitive interface	68.3%	18.0%
Internet connectivity	Online billpay	43.4%	15.5%
Internet connectivity	Download statements	23.9%	8.6%
Internet connectivity	Download investment information	23.9%	8.6%
Works well with other software	Hotlinks to spreadsheet	75.0%	7.9%
Easy to maintain	Free Internet patches	35.7%	7.8%
Easy to maintain	Great, free self-help technical assistance on the Internet	30.8%	6.7%
Easy to learn	Good documentation	20.0%	5.3%
Easy to maintain	Reasonably priced advanced technical support	20.0%	4.4%
Internet connectivity	Works well at 56K	8.9%	3.2%
Easy to learn	Interactive tutorial	11.7%	3.1%
Easy to maintain	Automatic Internet upgrades	13.5%	2.9%
Works well with other software	Edit reports in word processor	25.0%	2.6%
Easy to use quickly after I've learned it	Savable frequently used reports	43.4%	2.3%
Easy to use quickly after I've learned it	Shortcut keys	23.9%	1.3%
Easy to use quickly after I've learned it	Short menus showing only frequently used commands	23.9%	1.3%
Easy to use quickly after I've learned it	Macro capability	8.9%	0.5%

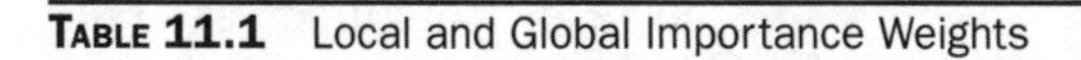

TABLE 11.1 Local and Global Importance Weights

huge customer impact. "Easy to use quickly after I've learned it" has a low impact. "Easy to learn" is dominated by one item: the user interface. These weights will be used to assess different proposed upgrade concepts. Each concept will be evaluated on each subcategory item and assigned a value depending on how well it addresses the item. The values will be multiplied by the global weights to arrive at an overall score for the concept. The scores can be rank-ordered to provide a list that you, the process owner, can use when making funding decisions. Or, more proactively, the information can be used to develop a concept that emphasizes the most important customer demands. Table 11.2 shows part of a table that assesses concepts using the global weights. The numerical rating used in the table is 0 = no impact, 1 = small impact, 3 = moderate impact, 5 = high impact. Since the global weights sum to 1 (100%), the highest possible

Item	Plan Customer Impact Score	Intuitive Interface	Online Billpay	Download Statements	Download Investment Information	Hotlinks to Spreadsheet	Free Internet Patches	Great, Free Self-Help Technical Assistance	Good Documen-tation	Reasonably Priced Advanced Technical Support	Works Well at 56K	Interactive Tutorial
GlobalWeight		18.0%	15.5%	8.6%	8.6%	7.9%	7.8%	6.7%	5.3%	4.4%	3.2%	3.1%
Concept A	3.57	3	5	1	1	3	3	4	5	5	5	5
Concept B	2.99	1	1	1	3	3	5	5	5	5	5	5
Concept C	4.15	5	5	5	5	5	5	3	3	1	3	3
Concept D	3.36	3	3	3	3	3	3	3	5	5	5	5
Concept E	2.30	5	0	0	0	5	5	1	1	0	1	1

Table 11.2 Example of Using Global Weights in Assessing Alternatives

score is 5. Of the five concepts evaluated, concept C has the highest score. It can be seen that concept C has a high impact on the six most important customer demands. It has at least a moderate impact on 10 of the top 11 items, with the exception of "reasonably priced advanced technical support." These items account for almost 90% of the customer demands.

The concept's customer impact score is, of course, just one input into the decision-making process. The rigor involved usually makes the score a very valuable piece of information. It is also possible to use the same process to incorporate other information, such as cost, timetable, feasibility, etc. into the final decision. The process owner would make pairwise comparisons of the different inputs (customer impact score, cost, feasibility, etc.) to assign weights to them, and use the weights to determine an overall concept score.

Pugh Concept Selection Method

The Pugh concept selection method is a simple alternative to the above approach of evaluating competing design concepts. The Pugh approach utilizes a simple matrix diagram to compare alternative concepts (Fig. 11.3). One concept is dubbed the "baseline" and all others are evaluated relative to the baseline in a simple Matrix diagram. In DMAIC the baseline is the current process. In DMADV, where there is no existing process or where the current process is deemed too bad to be salvaged, the baseline process is the one found "best" according to some criterion (e.g., fastest cycle time, lowest cost, fewest errors). If an alternative concept is better than the baseline with respect to a given criterion, it is given a "+" for that criterion. If it is worse it is given a "–." Otherwise it is considered to be the same and given an "S." Concept scores are found by summing the plus and minus signs, providing a count of pros and cons. This is only one input into the final choice of concepts, but the structure of the approach stimulates thought and discussion and usually proves to be very worthwhile.

Pugh Concept Selection Matrix Comparison Criteria	Baseline	Concept 1	Concept 2	Concept 3	Concept 4	Concept 5	Concept 6	Concept 7
Criterion 1	S	+	S					
Criterion 2	S	S	-					
Criterion *n*	S							
Total +'s	0	1	0					
Total –'s	0	0	1					

Compare current with selected alternatives

+ = Better Alternative, – = Worse Alternative, S = Same as Baseline

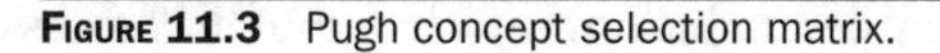

FIGURE 11.3 Pugh concept selection matrix.

Lean Techniques for Optimizing Flow

The key to value flow is to break the mental bonds of the batch-and-queue mindset. Batch and queue are everywhere. At your favorite restaurant where you are handed a little device to alert you when your table is ready. At the airport where you move from one line to another to another and show the same ID several times. At your physician's office where it's made clear to you that your time is less important than the doctor's time. On the phone where you find yourself on hold. On the waiting list for a surgical procedure. At home all day waiting for a cable installer who, we're told, will be there "sometime Wednesday."

Batch and queue are also endemic to our businesses. It's hard to imagine that at one point it was a fantastic innovation! Mass production is based on producing large lots of identical items to meet anticipated demand. This makes great efficiencies possible because the costs of setups, tooling, etc. are amortized over a very large number of units, making the per-unit costs very low. It also means inventory (queues for parts and materials), and longer cycle times due to the waiting. Choices are limited to those favored by the many. The term "customized," derived from the same root as customer, has no meaning. Production is to schedule, not to demand.

Flow focuses on the object of value. The product, design, service, order, etc., that is being created for the customer. The focus is *not* on the department, the supplier, the factory, the procedure, the tooling, the setup, the inventory or any other facet of the enterprise, or its operation. All work practices are carefully evaluated and rethought to eliminate stoppages of any kind so the object of value proceeds smoothly and continuously to the customer.

Tools to Help Improve Flow

Flow requires that the whole process be considered simultaneously. Generally the process begins with the order and ends with the customer receiving what was ordered. It requires, in effect, a customer-driven organization as described in Chap. 2. QFD is a useful tool in assuring that value is properly specified, designed, produced, and delivered to the customer. Other tools include:

- **5S**—5S is the starting point for Lean deployment. 5S stands for Sort, Set in order, Shine, Standardize, and Sustain. These terms are defined as follows:
  - **Sort**—Clearly distinguish what is necessary to do the job from what is not. Eliminate the unnecessary.
  - **Set in order**—Put needed items in their correct place to allow for easy accessibility and retrieval.
  - **Shine**—Keep the workplace clean and clear of clutter. This promotes safety as well as efficiency.
  - **Standardized cleanup**—Develop an approach to maintaining a clean and orderly work environment that works.
  - **Sustain**—Make a habit of maintaining your workplace.
- **Constraint management**—Constraints, or bottlenecks, require special attention. A process constraint is that step or part of the process that limits the throughput of the entire process. As such, they determine how much output the process can

produce. When a constraint isn't producing, the *process* isn't producing. Every effort needs to be focused on assuring that:

- The constraint has sufficient resources to keep running
- Every unit supplied to the constraint is of acceptable quality
- Every unit produced by the constraint is of acceptable quality
- The constraint is operated in as efficient a manner as is possible

- **Level loading**—Level loading is the process of generating a schedule that is level, stable, smooth, and responsive to the market. The goal of level loading is to make the same quantity of an item every day. It is driven by *Takt* time. A level loaded schedule can be obtained as follows:

$$\text{Calculate} = \frac{\text{daily work time}}{\text{daily quantity needed}} = \text{Take time} \tag{11.1}$$

  - For each part, list part name, part number, daily quantity needed, *Takt* time
  - Sort the list by quantity needed and *Takt* time. This is your level loaded schedule

- **Pull systems**—Traditional mass production is a *push* system. Push systems can be summarized as "Make a lot of stuff as cheaply as possible and hope people will buy it." Push systems minimize the number of setups and changeovers and use dedicated, specially designed equipment to produce identical units. Pull systems can be summarized as "Don't make anything until it is needed, then make it fast." A pull system controls the flow and quantity produced by replacing items when they are consumed. When I was in high school I worked in a supermarket that used a pull system. I'd walk down the aisles, note what was in short supply, then put more on the shelf. The storage area of a modern supermarket is very small compared to the retail floor area. In fact, supermarkets were the inspiration behind Taiichi Ohno's creating Lean at Toyota. Pull systems require level loading and flexible processes.
- **Flexible process**—Flexible processes are lightweight and maneuverable tools, and fixtures and equipment located and positioned to improve safety, ergonomics, quality, and productivity. They are the opposite of the big, heavy, permanently positioned counterparts traditionally used for mass production. A flexible shop can be quickly reconfigured to produce different items to meet changing customer demands. Flexible processes are related to level loading and pull. A completely flexible process would allow the factory to be instantly reconfigured to produce an item as soon as an order for it arrived. This ideal can't be met, but it can be approximated over some small time interval, such as a day.
- **Lot size reduction**—Lot size refers to the amount of an item that is ordered from the plant or supplier or issued as a standard quantity to the production process. The ideal lot size for flow is one. Larger lot sizes lead to larger quality problems due to delayed feedback, excessive inventory, obsolete inventory, etc. Of course, there are offsetting benefits such as quantity discounts, fewer setups, lower transportation costs, etc. In practice the costs and benefits must be balanced to achieve an optimum.

Putting all of these things together, the ideal scenario becomes: a customer orders an item or items (pull), the factory has the resources to produce the order (level loading), processes are configured to create the items ordered (flexible process), the order is produced and delivered to the customer exactly when he needs it.

Using Empirical Model Building to Optimize

Empirical model building is a statistical approach for determining optimal process or design settings. It uses a series of experimental designs to reduce the total possible process or product design space to hone in on the optimal settings with regard to one or more requirements.

If you are new to design of experiments and empirical model building, a metaphor may prove helpful. Imagine that you suddenly wake up in a strange wilderness. You don't know where you are, but you'd like to climb to the top of the nearest hill to see if there are any signs of civilization. What would you do?

A first step might be to take a good look around you. Is there anything you should know before starting out? You would probably pay particular attention to things that might be dangerous. If you are in a jungle these might be dangerous animals, quicksand, and other things to avoid. You'd also look for things that could be used for basic survival, such as food, shelter, and clothing. You may wish to establish a "base camp" where you can be ensured that all the basic necessities are available; a safe place to return to if things get a bit too exciting. In empirical modeling we also need to begin by becoming oriented with the way things are before we proceed to change them. We will call this *knowledge discovery* activity Phase 0.

Now that you have a feel for your current situation and you feel confident that you know something about where you are, you may begin planning your trip to the highest hill. Before starting out you will probably try to determine what you will need to make the trip. You are only interested in things that are truly important. However, since you are new to jungle travel, you decide to make a few short trips to be sure that you have what you need. For your first trip you pack up every conceivable item and set out. In all likelihood you will discover that you have more than you need. Those things that are not important you will leave at your camp. As part of your short excursions you also learn something about the local terrain close to your camp; not much, of course, but enough to identify which direction is uphill. This phase is equivalent to a *screening experiment*, which we call Phase I.

You now feel that you are ready to begin your journey. You take only those things you will need and head out into the jungle in the uphill direction. From time to time you stop to get your bearings and to be sure that you are still moving in the right direction. We call this hill-climbing *steepest ascent*, or Phase II.

At some point you notice that you are no longer moving uphill. You realize that this doesn't mean that you are at the highest point in your area of the jungle, only that you are no longer moving in the right direction. You decide to stop and make camp. The next morning you begin to explore the local area more carefully, making a few short excursions from your camp. The jungle is dense and you learn that the terrain in the immediate vicinity is irregular, sometimes steep, sometimes less steep. This is in contrast to the smooth and consistent uphill slope you were on during your ascent. We call this phase of your journey the *factorial experiment*, or Phase III.

Now you decide that a more organized approach will be needed to locate the nearby peak. You break out the heavy artillery, the GPS you've been carrying since the beginning! (one of those cheap ones that don't have built-in maps). You take several altitude readings from near your camp, and others at a carefully measured distance on all major compass headings. Each time you carefully record the altitude on a hand-drawn map. You use the map to draw contour lines of equal altitude and eventually a picture emerges that clearly shows the location of the top of the hill. This is the *composite design phase*, which we call Phase IV.

At last you reach the top of the hill. You climb to the top of a tree and are rewarded with a spectacular view, the best for miles around. You decide that you love the view so much, you will build your home on this hill and live there permanently. You make your home sturdy and strong, able to withstand the ravages of wind and weather that are sure to come to your little corner of the jungle. In other words, your home design is *robust*, or impervious to changes in its environment. We call the activity of building products and processes that are insensitive to changes in their operating parameters *robust product and process design*, which is Phase V of the journey.

Now that this little tale has been told, let's go on to the real thing, improving your products, processes, and services.

Phase 0: Getting Your Bearings

"Where Are We Anyway?"

Before any experimentation can begin the team should get an idea of what the major problems are, important measures of performance, costs, time and other resources available for experimentation, etc. Methods and techniques for conducting Phase 0 research are described in Chap. 10.

The central premise of the approach described in this section is that learning is, by its very nature, a sequential process. The experimenter, be it an individual or a team, begins with relatively little specific knowledge and proceeds to gain knowledge by conducting experiments on the process. As new knowledge is acquired, the learner is better able to determine which step is most appropriate to take next. In other words, experimentation always involves guesswork; but guesses become more educated as experimental data become available for analysis.

This approach is in contrast to the classical approach where an effort is made to answer all conceivably relevant questions in one large experiment. The classical approach to experimentation was developed primarily for agricultural experiments. Six Sigma applications are unlike agricultural applications in many ways, especially in that results become available quickly. The approach described here takes advantage of this to accelerate and direct learning.

We will use an example from electronic manufacturing. At the outset, a team of personnel involved in a soldering process received a mission from another team that had been evaluating problems for the factory as a whole. The factory team had learned that a leading reason for customer returns was solder problems. Another team discovered that the solder area spent more resources in terms of floor space than other areas; a major usage of floor space was for the storage of defective circuit boards and the repair of solder defects. Thus, the solder process improvement team was formed and asked to find ways to eliminate solder defects if possible, or to at least reduce them by a factor of 10. Team members included a Six Sigma technical leader, a process engineer, an inspector, a production operator, and a product engineer.

The team spent several meetings reviewing Pareto charts and problem reports. It also performed a process audit which uncovered several obvious problems. When the problems were repaired the team conducted a process capability study, which revealed a number of special causes of variation, which were investigated and corrected. Over a 4-month period, this preliminary work resulted in a 50% reduction in the number of solder defects, from about 160 defects per standard unit to the 70 to 80 defect range. The productivity of the solder area nearly doubled as a result of these efforts. While impressive, the results were still well short of the 10×minimum improvement the team was asked to deliver.

Phase I: The Screening Experiment

"What's Important Here?"

At this point the process was stable and the team was ready to move from the process control stage to the process improvement stage. This involved conducting designed experiments to measure important effects. The solder team decided to list as many items as possible that might be causing solder problems. Since many variables had already been studied as part of the Phase 0 work, the list was not unreasonably long. The team looked at ways to control the variables listed and was able to develop methods for eliminating the effects of many variables on their list. The remaining list included the following factors:

Variable	Low Level (−)	High Level (+)
A: Prebaking of boards in an oven	No	Yes
B: Preheat time	10 s	20 s
C: Preheat temperature	150°F	200°F
D: Distance from preheat element to board surface	25 cm	50 cm
E: Line speed	3 fpm	5 fpm
F: Solder temperature	495°F	505°F
G: Circuit density	Low	High
H: Was the board in a fixture?	No	Yes

This information was used to create an experimental design using a statistical software package. There are many packages on the market that perform similar analyses to the one shown here.

Since this is only to be a screening experiment, the team was not interested in obtaining estimates of factor interactions. The focus was to identify important main effects. The software allows selection from among several designs. The Black Belt decided upon the design which would estimate the main effects with the smallest number of test units. This design involved testing 16 units. The data matrix produced by the computer is shown in Table 11.3. The run order has been randomized by the computer. If the experiment cannot be conducted in that particular order, the computer software would allow the data to be run in blocks and it would adjust the analysis accordingly. The program also tells us that the design is of resolution IV,

Run	A	B	C	D	E	F	G	H	Response
1	+	−	−	−	−	+	+	+	65
2	+	−	+	+	−	+	−	−	85
3	+	+	−	−	+	+	−	−	58
4	−	+	−	−	+	−	+	+	57
5	−	−	−	−	−	−	−	−	63
6	+	+	+	+	+	+	+	+	75
7	−	+	−	+	−	+	+	−	77
8	−	+	+	−	−	+	−	+	60
9	+	−	+	−	+	−	−	+	67
10	+	+	+	−	−	−	+	−	56
11	−	−	+	−	+	+	+	−	63
12	−	−	−	+	+	+	−	+	81
13	+	+	−	+	−	−	−	+	73
14	+	−	−	+	+	−	+	−	87
15	−	+	+	+	+	−	−	−	75
16	−	−	+	+	−	−	+	+	84

TABLE 11.3 Screening Experiment Layout. Data matrix (Randomized)

which means that main effects are not confounded with each other or any two-factor interactions.

In Table 11.3 the "–" indicates that the variable is run at its low level, while a "+" sign indicates that it is to be run at its high level. For example, the unit for run #16 was processed as follows:

- Prebaking = No
- Preheat time = 10 sec
- Preheat temperature = 200°F
- Distance from preheat element to board surface = 50 cm
- Line speed = 3 fpm
- Solder temperature = 495°F
- Circuit density = High
- Fixture used = Yes
- Defects per standard unit = 84

Experimental data were collected using the randomized run order recommended by the software. The "response" column are data that were recorded in terms of defective solder joints per "standard unit," where a standard unit represented a

Estimated effects and coefficients for response (coded units)						
Term	*Effect*	*Coef*	*StDev coef*	*T*	*P*	
Constant		70.375	0.6597	106.67	0.000	
A	−0.750	−0.375	0.6597	−0.57	0.588	
B	8.000	4.000	0.6597	6.06	0.001	
C	−0.500	−0.250	0.6597	−0.38	0.716	
D	−18.500	−9.250	0.6597	−14.02	0.000	
E	0.000	0.000	0.6597	0.00	1.000	
F	−0.250	−0.125	0.6597	−0.19	0.855	
G	−0.250	−0.125	0.6597	−0.19	0.855	
H	0.250	0.125	0.6597	0.19	0.855	
ANOVA for defects (coded units)						
Source of variation	df	*Seq. SS*	*Adj. SS*	*Adj. MS*	*F*	*P-value*
Main effects	8	1629.00	1629.00	203.625	29.24	0.000
Residual error	7	48.75	48.75	6.964		
Total	15	1677.75				

TABLE 11.4 Results of Experimental Data Analysis. Fractional Factorial Fit

circuit board with a median number of solder joints.* The results are shown in Table 11.4.

A model that fits the data well would produce residuals that fall along a straight line. The Black Belt concluded that the fit of the model was adequate.

The analysis indicates that factors *B* (preheat time) and *D* (distance from preheat element to board surface) produce significant effects. Figure 11.4 shows a normal probability plot of the experimental effects. This figure plots the coefficients column from Table 11.4 on a normal probability scale. If the factor's effect was due to chance variation it would plot close to the line representing normal variation. In Fig. 11.5 the effects of *B* and *D* are shown to be further from the line than can be accounted for by random variation.

The effects of the significant factors are graphed in response units in Fig. 11.5.

Since the response is a defect count, the graph indicates that the low level of factor *D* gives better results, while the high level of factor *B* gives the better results. This can also be seen by examination of the coefficients for the variables. When *D* is low the

*Technically, a Poisson model would be the correct choice here. However, use of a normal model, which the analysis assumes, is reasonably accurate for defect counts of this magnitude. The team also evaluated the variance, more specifically, the log of the variance. The variances at each factor combination did not differ significantly and are not shown here.

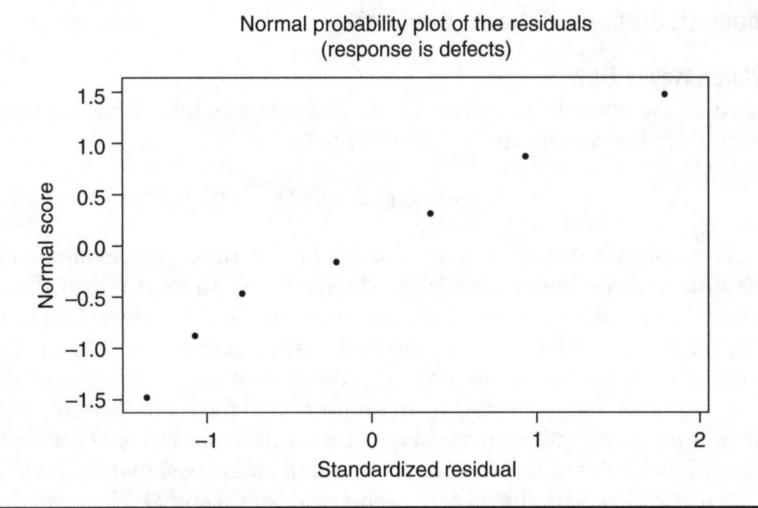

FIGURE 11.4 Residuals from experimental model.

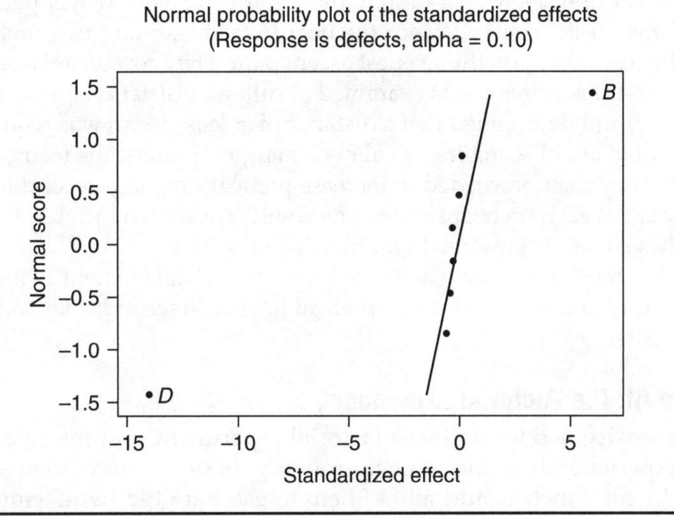

FIGURE 11.5 Significant factor effects.

average defect rate is 18.5 defects per unit better than when *D* is high; when *B* is high the average defect rate is 8 defects per unit better than when *B* is low.

The team met to discuss these results. They decided to set all factors that were not found to be statistically significant to the levels that cost the least to operate, and factors *B* and *D* at their midpoints. The process would be monitored at these settings for a while to determine that the results were similar to what the team expected based on the experimental analysis. While this was done, another series of experiments would be planned to further explore the significant effects uncovered by the screening experiment.

Phase II: Steepest Ascent (Descent)

"Which Way Is Up?"

Based on the screening experiment, the linear model for estimating the defect rate was found from the coefficients in Table 11.4 to be

$$\text{Defect rate} = 70.375 + 4B - 9.25D$$

The team wanted to conduct a series of experiments to evaluate factors B and D. The Phase I experiment reveals the direction and ratio at which B and D should be changed to get the most rapid improvement in the defect rate, that is, the direction of steepest ascent (where "ascent" means improvement in the measurement of interest). To calculate a series of points along the direction of steepest ascent, start at the center of the design and change the factors in proportion to the coefficients of the fitted equation; that is, for every 4 unit increase in factor B we decrease factor D 9.25 units. For the data at hand, the center of the experiment and unit sizes are shown in Table 11.5.

A test unit was produced at the center value of B and D. The team decided that they would reduce the preheat time (B) in increments of 5 seconds (1 unit), while lowering the distance from the heating element (D) by increments of $(9.25/4) \times 12.5$ cm = 28.9 cm. This resulted in a single experiment where $B = 20$ seconds, $D = 8.6$ cm. The result was 52 defects per unit. However, despite the improved solder defect performance, the team noted that at the short distance the board was beginning to scorch. This necessitated that the team abandon the steepest ascent path. They conducted a series of experiments where board scorching was examined at different distances to the preheating element (factor D) and determined that a distance of at least 15 cm was required to be confident they would avoid scorching. To allow a margin of safety, the team set the distance D at 20 cm. They then proceeded to increase preheat time in 5-second intervals, producing one board at each preheat setting. The results are shown in Table 11.6.

These data are presented graphically in Fig. 11.6.

With the distance fixed at 20 cm from the preheat element to the board surface, the best results were obtained with a preheat time of 40 seconds. Beyond that the defect rate was greater.

Phase III: The Factorial Experiment

The team decided to conduct a factorial experiment near the best settings to explore that experimental region more thoroughly. To do so, they decided to run a factorial experiment which would allow them to estimate the two-factor BD interaction as well as the main effects. They also wished to determine if there was any "curvature" in the area. This required that more than two levels be explored (only linear estimates are possible with two-level designs). Finally, the team wanted to obtain an

Factor	Unit Size	Center
B	5	15 s
D	12.5	37.5 cm

Table 11.5 Unit Sizes and Center of Experiment

Run	*B* (s)	*D* (cm)	Average Defects
1	15	37.5	70
2	20	8.75	52
3	25	20	51
4	30	20	31
5	35	20	18
6	40	20	12
7	45	20	10
8	50	20	13

TABLE 11.6 Data for Experiments on Path of Steepest Descent

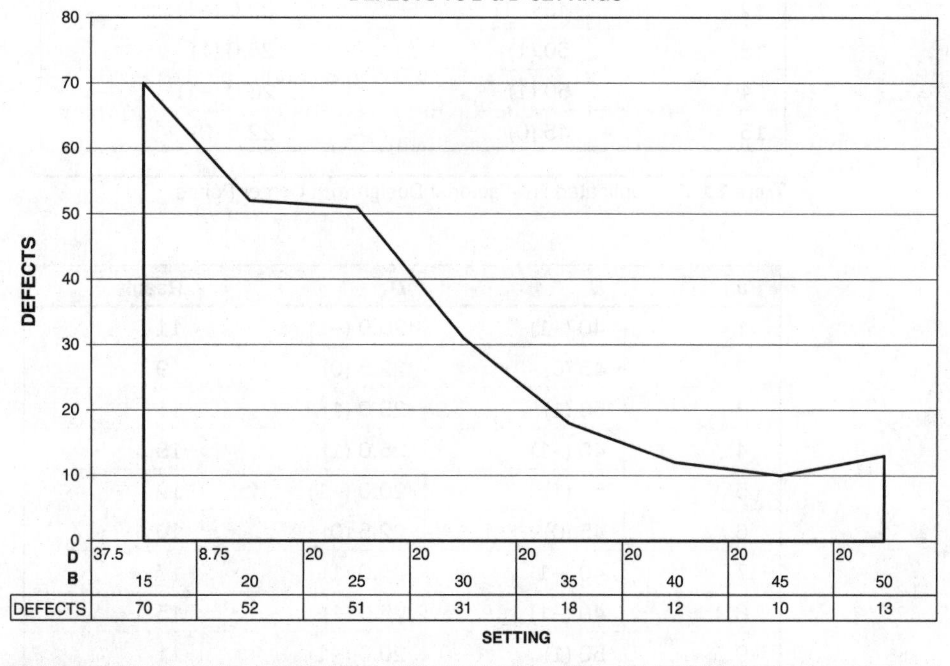

FIGURE 11.6 Steepest descent results.

estimate of the experimental error in the region; this required replicating the experiment. The design selected is shown in Table 11.7.

Code numbers used for the computer are shown in parentheses. The runs marked 0, 0 are center points. Note that each combination (i.e., set of plus and minus signs or zeros) is repeated three times. The team decided to center the design at the *B* value found to be steepest, $B = 45$ seconds. The interval for *D* was reduced to 2.5 cm and the experiment was centered one interval above $D = 20$ (i.e., at $D = 22.5$) (Table 11.8).

Run	*B*	*D*
1	40 (−1)	20.0 (−1)
2	45 (0)	22.5 (0)
3	50 (1)	25.0 (1)
4	40 (−1)	25.0 (1)
5	50 (1)	20.0 (−1)
6	45 (0)	22.5 (0)
7	40 (−1)	25.0 (1)
8	40 (−1)	25.0 (1)
9	50 (1)	20.0 (−1)
10	50 (1)	25.0 (1)
11	40 (−1)	20.0 (−1)
12	40 (−1)	20.0 (−1)
13	50 (1)	25.0 (1)
14	50 (1)	20.0 (−1)
15	45 (0)	22.5 (0)

TABLE 11.7 Replicated Full-Factorial Design with Center Points

Run	*B*	*D*	Result
1	40 (−1)	20.0 (−1)	11
2	45 (0)	22.5 (0)	9
3	50 (1)	25.0 (1)	11
4	40 (−1)	25.0 (1)	15
5	50 (1)	20.0 (−1)	12
6	45 (0)	22.5 (0)	10
7	40 (−1)	25.0 (1)	17
8	40 (−1)	25.0 (1)	15
9	50 (1)	20.0 (−1)	11
10	50 (1)	25.0 (1)	11
11	40 (−1)	20.0 (−1)	13
12	40 (−1)	20.0 (−1)	13
13	50 (1)	25.0 (1)	11
14	50 (1)	20.0 (−1)	11
15	45 (0)	22.5 (0)	10

TABLE 11.8 Results of Full Factorial Experiment with Center Points and Replicates

FRACTIONAL FACTORIAL FIT

Estimated effects and coefficients for defects (coded units)					
Term	Effect	Coef	StDev coef	T	P
Constant		12.583	0.2357	53.39	0.000
A	−2.833	−1.417	0.2357	−6.01	0.000
B	1.500	0.750	0.2357	3.18	0.010
A*B	−1.833	−0.917	0.2357	−3.89	0.003
C_t P_t		−2.917	0.5270	−5.53	0.000

Looking at the *P* column all terms in the model are significant (any *P* value below 0.05 indicates a significant effect). This analysis is confirmed by the ANOVA table (Table 11.9).

Looking at the *P* column of the ANOVA table, we see that main effects, the two-way interaction, and "curvature" are all significant ($P < 0.05$). Curvature is measured by comparing the average response at the center points with the responses at the corner points of the design. The fact that curvature is significant means that we are no longer experimenting in a linear region of the responses.

This means that our original coefficients, which were based on the linear model, are no longer adequate. Upon seeing these results, the Black Belt decided that it was necessary to proceed to Phase IV to better investigate the response region and to try to locate a stationary optimum.

Phase IV: The Composite Design

The Black Belt decided to try using a design known as a *composite design* or *central composite design* to obtain additional information on the region where the process

ANOVA for defects (coded units)						
Source of variation	*df*	*Seq. SS*	*Adj. SS*	*Adj. MS*	*F*	*P-value*
Main effects	2	30.833	30.833	15.4167	23.13	0.000
2-way interactions	1	10.083	10.083	10.0833	15.13	0.003
Curvature	1	20.417	20.417	20.4167	30.63	0.000
Residual error	10	6.667	6.667	0.6667		
Pure error	10	6.667	6.667	0.6667		
Total	14	68.000				

Table 11.9 ANOVA for Factorial Experiment with Center Points

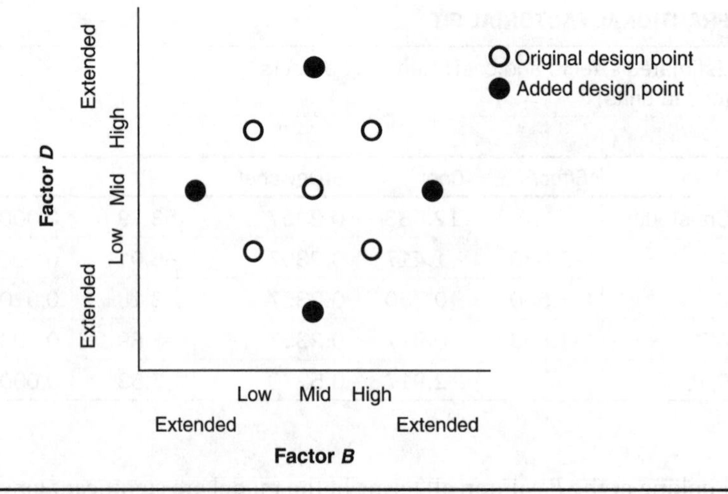

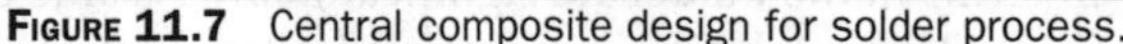

FIGURE 11.7 Central composite design for solder process.

was operating. This design involves augmenting the corner points and center point of the previous factorial experiment with additional points, as shown in Fig. 11.7. The points extend the design beyond the levels previously designed by the high and low values for each factor. The team decided that they could allow the distance to be decreased somewhat below the 20 cm "minimum" distance because they had added a 5 cm margin of safety. They also noted that they were now taking relatively small experimental steps compared to the large jumps they took during steepest ascent.

DOE software finds the coefficients of the equation that describes a complex region for the responses. The equation being fitted is:

$$y = \beta_0 + \beta_1 x_1 + \beta_2 x_2 + \beta_{11} x_1^2 + \beta_{22} x_2^2 + \beta_{12} x_1 x_2 + \varepsilon \tag{11.2}$$

The region described by this equation may contain a maximum, a minimum, or a "saddle point." At a maximum any movement away from the stationary point will cause the response to decrease. At the minimum any movement away from the stationary point will cause the response to increase. At a saddle point moving away from the stationary value of one variable will cause a decrease, while moving away from the stationary value of the other variable will cause an increase. Some DOE software will tell you the values of X and Y at the stationary point, and the nature of the stationary point (max, min, or saddle). Other DOE software display two-dimensional and three-dimensional drawings that graphically describe the region of experimentation. It is usually not difficult to interpret the response surface drawings.

The data collected by the team are shown in Table 11.10. Note that the data are shown in standard order, but the run order was random.

B	D	Defects
−1.41421	0.00000	16
1.00000	1.00000	11
0.00000	0.00000	9
0.00000	−1.41421	11
1.00000	−1.00000	9
1.41421	0.00000	4
0.00000	0.00000	10
0.00000	0.00000	10
0.00000	1.41421	15
0.00000	0.00000	9
0.00000	0.00000	10
−1.00000	1.00000	15
−1.00000	−1.00000	13

TABLE 11.10 Central Composite Design Experiment and Data

The computer analysis of these data is shown in Table 11.11.

The P-values indicate that all terms except the B^2 term and the interaction term are significant.

ANOVA for defects						
Source of variation	**Df**	**Seq. SS**	**Adj. SS**	**Adj. MS**	***F***	***P***
Regression	5	112.821	112.8211	22.5642	13.05	0.002
Linear	2	89.598	89.5980	44.7990	25.91	0.001
Square	2	23.223	23.2231	11.6115	6.72	0.024
Interaction	1	0.000	0.0000	0.0000	0.00	1.000
Residual Error	7	12.102	12.1020	1.7289		
Lack-of-Fit	3	10.902	10.9020	3.6340	12.11	0.018
Pure Error	4	1.200	1.2000	0.3000		
Total	12	124.923				
Unusual observations for defects						
Observation	**Defects**	**Fit**	**StDev Fit**	**Residual**	**St Resid**	
6	4.000	5.836	1.039	−1.836	−2.28R	

Estimated regression coefficients for defects						
Term	Coef	StDev coef	T	P		
Constant	9.600	0.5880	16.326	0.000		
B	−3.121	0.4649	−6.714	0.000		
D	1.207	0.4649	2.597	0.036		
B*B	0.325	0.4985	0.652	0.535		
D*D	1.825	0.4985	3.661	0.008		
B*D	0.000	0.6574	0.000	1.000		
S = 1.315		R−Sq = 90.3%		R−Sq(adj) = 83.4%		

TABLE 11.11 Analysis of Central Composite Design. Estimate Regression Coefficients for *Y*

R denotes an observation with a large standardized residual. The team confirmed the defect count for observation 6.

The ANOVA indicates that the lack of fit is significantly greater than pure error. However, the Black Belt felt the magnitude of the lack of fit was tolerable. It also indicates that the interaction term is not significant and could be removed from the model, gaining a degree of freedom for estimating the error.

The response surface 3D and contour plots are shown in Figs. 11.8(a) and (b).

The analysis could become somewhat more advanced if the Black Belt chose to perform a *canonical analysis* to investigate the nature of the response surface in greater detail. Canonical analysis involves finding a *stationary point S* and performing a coordinate system transformation to eliminate the cross-product and first order terms. The techniques for performing this analysis are described in a number of advanced texts

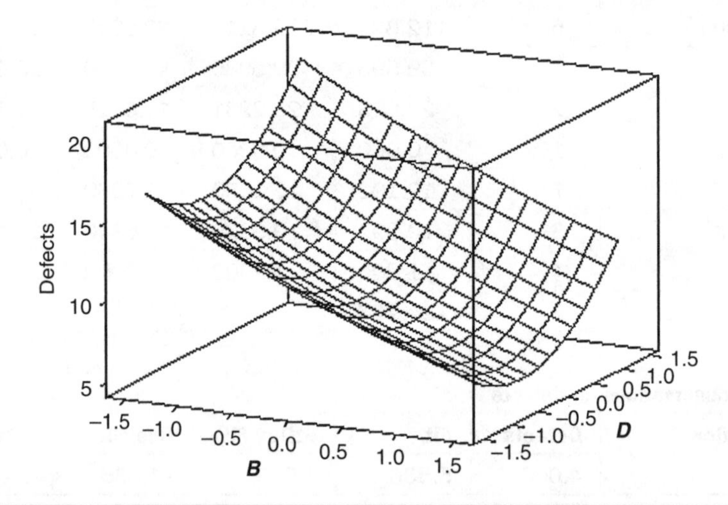

FIGURE 11.8 (a) Response surface plot for defect data.

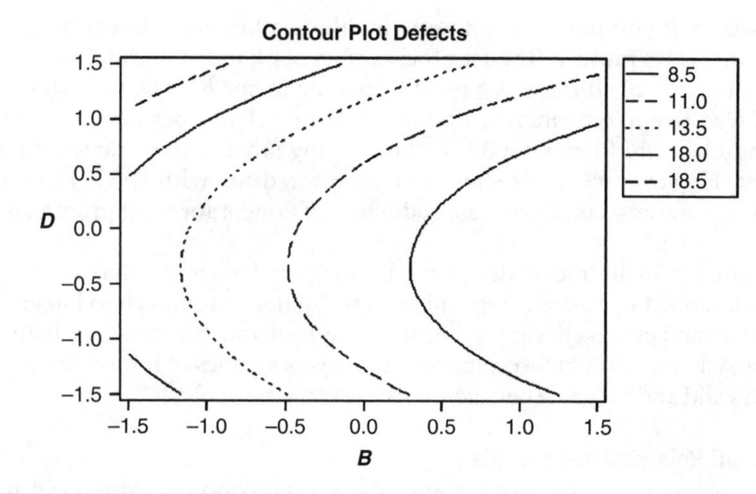

FIGURE 11.8 (b) Contour plot for defect data.

(Box and Draper, 1987; Meyers and Montgomery, 1995). However, it is obvious from the contour plot and the 3D response surface plot that there may still be some room to improve by holding *D* constant and gradually increasing *B*.

At this point the team decided that they had reached a point of diminishing returns as far as the current process is concerned. The data indicate that the existing process for wave solder can, if properly controlled, manage to produce 10 or fewer defects per standard unit at the center of the last design. This is about 16 times better than the process was producing at the outset of the project and about 7 times better than the average result of the first experiment.

The team, guided by the Black Belt, decided to set the process at the center point of the last experiment ($B = 0, D = 0$) and to implement Evolutionary Operation (EVOP) to pursue further optimization. EVOP involves running a series of designed experiments on production units, with operating personnel making small changes (Box and Draper, 1969). By restricting EVOP to small process changes the risk of producing scrap is reduced. Although the movements in process settings are small, the cumulative improvement in performance can be substantial. The apparent gradual slope of the defect rate in the $+B$ direction also made it unlikely that the process would "fall off of a cliff" during EVOP.

The Black Belt helped set up EVOP on the process and train supervisory and hourly personnel in its use. She also agreed to provide ongoing support in the form of periodic visits and availability should questions arise. The team decided that after turning over process improvement to operating personnel, they would look at ways of maintaining their gains, while simultaneously investigating basic process and product design changes to obtain further improvement.

Phase V: Robust Product and Process Design

Maintaining gains involves, among other things, creating processes and products that operate close to their optimum conditions even when changes occur. Robust design can begin with careful study of the contour plot. Note that if you start at $B = D = 0$ and move from along a line from left to right the response changes relatively slowly.

However, if you move from the center along a line from lower to upper, the defect rate increases rapidly. Robust process control planning should take such nonlinearity into account. If there is a need to change factor *B* or *D*, they should be changed in a way that avoids increasing the defect rate. This does *not* mean that all changes should be forbidden; after all, without change there can be no learning or improvement. However, changes should be monitored (as with EVOP) to provide a filter between the customer and the production of nonconforming product that may occur during the learning process.

More formally, robust design can be integrated into experimental design. The methods described by Genichi Taguchi are a well-known approach to integrating DOE and product and process design. While there has been much criticism of Taguchi's statistical approach, there is a broad consensus that his principles of robust parameter design are both valid and valuable contributions to Six Sigma analysis.

Taguchi Robustness Concepts

This section will introduce some of the special concepts introduced by Dr. Genichi Taguchi of Japan. A complete discussion of Taguchi's approach to designed experiments is beyond the scope of this book. However, many of Taguchi's ideas are useful, such as his concept of designing processes and products to be robust to the potential sources of variation that could influence them.

Introduction Quality is defined as the loss imparted to the society from the time a product is shipped (Taguchi, 1986). Taguchi divides quality control efforts into two categories: online quality control and off-line quality control. Online quality control—involves diagnosis and adjusting of the process, forecasting and correction of problems, inspection and disposition of product, and follow-up on defectives shipped to the customer. Off-line quality control—quality and cost control activities conducted at the product and the process design stages in the product development cycle. There are three major aspects to off-line quality control:

1. System design is the process of applying scientific and engineering knowledge to produce a basic functional prototype design. The prototype model defines the initial settings of product or process design characteristics.
2. Parameter design is an investigation conducted to identify settings that minimize (or at least reduce) the performance variation. A product or a process can perform its intended function at many settings of its design characteristics. However, variation in the performance characteristics may change with different settings. This variation increases both product manufacturing and lifetime costs. The term *parameter design* comes from an engineering tradition of referring to product characteristics as product parameters. An exercise to identify optimal parameter settings is therefore called *parameter design*.
3. Tolerance design is a method for determining tolerances that minimize the sum of product manufacturing and lifetime costs. The final step in specifying product and process designs is to determine tolerances around the nominal settings identified by parameter design. It is still a common practice in industry to assign tolerances by convention rather than scientifically. Tolerances that are too narrow increase manufacturing costs, and tolerances that are too wide increase performance variation and the lifetime cost of the product.

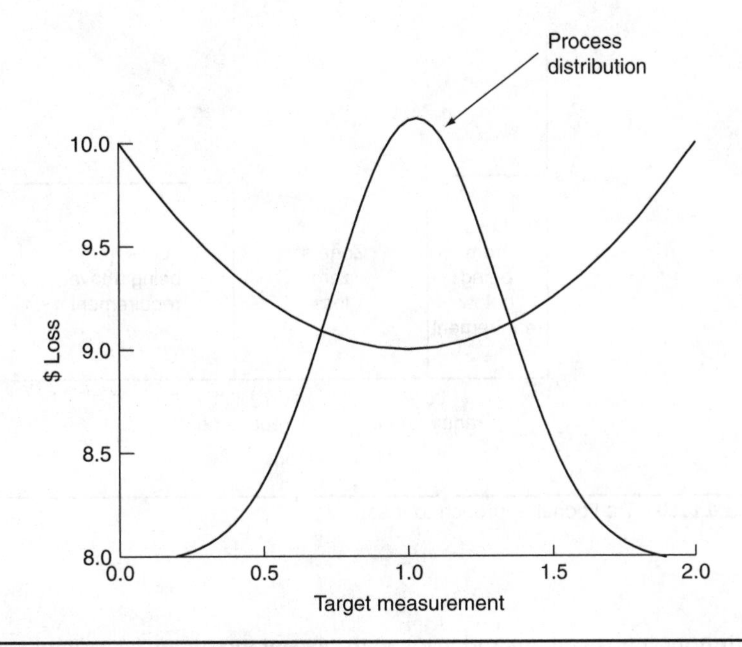

FIGURE 11.9 Taguchi's quadratic loss function.

Expected loss the monetary losses an arbitrary user of the product is likely to suffer at an arbitrary time during the product's life span due to performance variation. Taguchi advocates modeling the loss function so the issue of parameter design can be made more concrete. The most often-used model of loss is the quadratic loss function illustrated in Fig. 11.9. Note that the loss from operating the process is found by integrating the process pdf over the dollar-loss function. Under this model there is always a benefit to

1. Moving the process mean closer to the target value
2. Reducing variation in the process

Of course, there is often a cost associated with these two activities.

Weighing the cost/benefit ratio is possible when viewed from this perspective.

Note the contrast between the quadratic loss function and the conceptual loss function implicit in the traditional management view. The traditional management approach to loss is illustrated in Fig. 11.10.

Interpretation of Fig. 11.10: there is no loss as long as a product or service meets requirements. There is no "target" or "optimum": just barely meeting requirements is as good as operating anywhere else within the zone of zero loss. Deviating a great deal from requirements incurs the same loss as being just barely outside the prescribed range. The process distribution is irrelevant as long as it meets the requirements.

Note that under this model of loss there is no incentive for improving a process that meets the requirements since there is no benefit, that is, the loss is zero. Thus, cost > benefit for any process that meets requirements. This effectively destroys the idea of

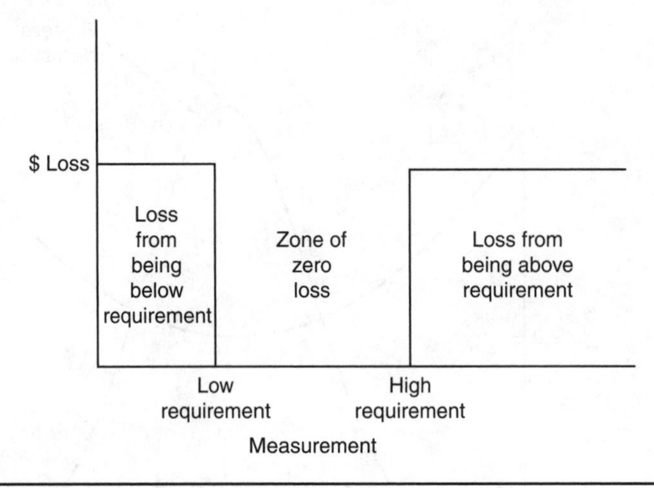

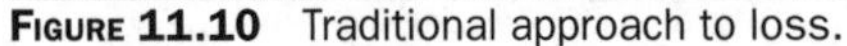
FIGURE 11.10 Traditional approach to loss.

continuous improvement and leads to the acceptance of an "acceptable quality level" as an operating standard.

Noise—the term used to describe all those variables, except design parameters, that cause performance variation during a product's life span and across different units of the product. Sources of noise are classified as either external sources or internal sources. External sources of noise—variables external to a product that affect the product's performance. Internal sources of noise—the deviations of the actual characteristics of a manufactured product from the corresponding nominal settings.

Performance statistics—estimate the effect of noise factors on the performance characteristics. Performance statistics are chosen so that maximizing the performance measure will minimize expected loss. Many performance statistics used by Taguchi use "signal to noise ratios" which account jointly for the levels of the parameters and the variation of the parameters.

Summary of the Taguchi Method

The Taguchi method for identifying settings of design parameters that maximize a performance statistic is summarized by Kackar (1985):

- Identify initial and competing settings of the design parameters, and identify important noise factors and their ranges.
- Construct the design and noise matrices, and plan the parameter design experiment.
- Conduct the parameter design experiment and evaluate the performance statistic for each test run of the design matrix.
- Use the values of the performance statistic to predict new settings of the design parameters.
- Confirm that the new settings do indeed improve the performance statistic.

Data Mining, Artificial Neural Networks, and Virtual Process Mapping

As beneficial and productive as design of experiments can be, the process of conducting them has its drawbacks. The workplace, be it a factory, a retail establishment or an office, is designed around a routine. The routine is the "real work" that must be done to generate the sales which, in turn, produce the revenues that keep the enterprise in existence. By its very nature, experimenting means disrupting the routine. Important things are changed to determine what effect they have on various important metrics. Often, these effects are unpleasant; that's why they weren't changed in the first place! The routine was often established to steer a comfortable course that avoids the disruption and waste that results from making changes.

The problem is, without change things can never improve. Six Sigma generates as much improvement by changing things as it does by reducing variability.

In this section we present a way of conducting "virtual" experiments using existing data and artificial neural network (neural net) software. Neural nets are popular because they have a proven track record in many data mining and decision-support applications. Neural nets are a class of very powerful, general purpose tools readily applied to prediction, classification, and clustering. They have been applied across a broad range of industries from predicting financial series to diagnosing medical conditions, from identifying clusters of valuable customers to identifying fraudulent credit card transactions, from recognizing numbers written on checks to predicting failure rates of engines (Berry and Linoff, 1997). In this section we explore only the application of neural nets to design of experiments for Six Sigma, but this merely scratches the surface of the potential applications of neural nets for quality and performance improvement.

Neural networks use a digital computer to model the neural connections in human brains. When used in well-defined domains, their ability to generalize and learn from data mimics our ability to learn from experience. However, there is a drawback. Unlike a well-planned and executed DOE, a neural network does not provide a mathematical model of the process.* For the most part, neural networks must be approached as black boxes with mysterious internal workings, much like the mystery of the human mind it is designed to imitate.

All companies record important data, some in well-designed data warehouses, some in file drawers. These data represent potential value to the Six Sigma team. They contain information that can be used to evaluate process performance. If the data include information on process settings, for example, they may be matched up to identify possible cause and effect relationships and point the direction for improvement. The activity of sifting through a database for useful information is known as *data mining*. The process works as follows:

1. Create a detailed inventory of data available throughout the organization.
2. Determine the variables which apply to the process being improved.

*It is possible, however, to include various transformed variables to "help" the neural net if one has a model in mind. For example, in addition to feeding the neural net X1 and X2 raw data, one could include higher-order polynomial and interaction terms as inputs to the neural network.

3. Using a subset of the data which include the most extreme values, train the neural net to recognize relationships between patterns in the independent variables and patterns in the dependent variables.
4. Validate the neural net's predictive capacity with the remaining data.
5. Perform experimental designs as described in the section above entitled "Using Empirical Model Building to Optimize." However, instead of making changes to the actual process, make changes to the "virtual process" as modeled by the neural net.
6. Once Phase IV has been completed, use the settings from the neural net as a starting point for conducting experiments on the actual process. In other words, begin experimenting at Phase I with a screening experiment.

It can be seen that the entire soft experimentation process is part of Phase 0 in the empirical model building process. It helps answer the question "Where are we?" It is important to recognize that neural net experiments are not the same as live experiments. However, the cost of doing them is minimal compared with live experiments and the process of identifying input and output variables, deciding at which levels to test these variable, etc. will bear fruit when the team moves on to the real thing. Also, soft experiments allow a great deal more "what if?" analysis, which may stimulate creative thinking from team members.

Example of Neural Net Models

The data in Table 11.12 are from the solder process described above. Data were not gathered for a designed experiment, but were merely collected during the operation of the process. The data were used to train and validate a neural net.

The neural net model is shown in Fig. 11.11.

The model was trained using the above data, producing the process map shown in Fig. 11.12.

You can see that the surface described by the neural net is similar to the one modeled earlier using DOE. Both models direct the B and D settings to similar levels and both make similar predictions for the defect rate.

The neural net software also allows "what if" analysis. Since these data are from the region where the team ran its last phase of experiments, they could be used to conduct virtual DOE. The neural net's What If? contour plot dialog box is shown in Fig. 11.13.

The virtual DOE values are entered in the What If? dialog box and the neural net's predictions are used in the experimental design just as if they had been obtained using data from a real experiment. If you have data covering the entire region of interest, the neural net may bring you very close to the optimum settings even before you do your first actual experiment.

Optimization Using Simulation

Simulation is a means of experimenting with a detailed model of a real system to determine how the system will respond to changes in its structure, environment, or underlying assumptions. A system is defined as a combination of elements that interact to

PH_Time	PH_Distance	Defects
38	22.5	15
40	20	13
40	25	16
45	17.5	15
45	22.5	5
45	26	11
50	20	12
42	22.5	10
50	25	3
42	22	11
46	22	4
55	25	4
55	21	17
55	25	15
50	24	3
49	25	3
57	37	10
35	25	20
45	37.5	17
30	20	27
30	22.5	33
30	25	37
30	27.5	50
30	37.5	57
50	20	13
50	22.5	5
50	25	3
50	30	5
50	14	12
50	37.5	14
50	45	16
50	50	40
60	20	35
60	25	18
60	37.5	12

TABLE 11.12 Solder Process Data for Virtual Process Mapping

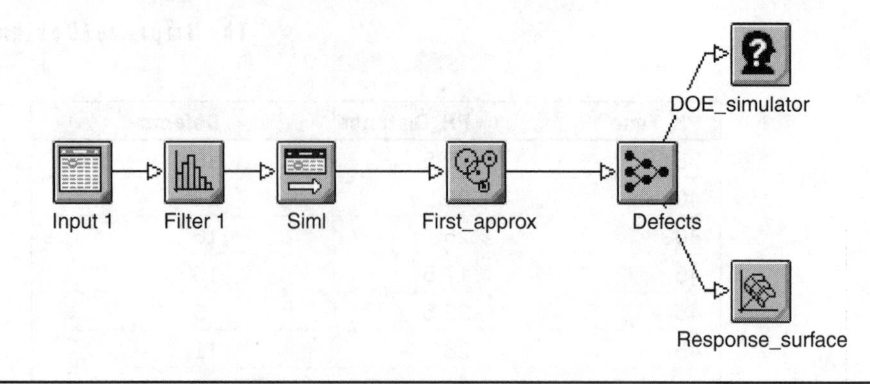

FIGURE 11.11 Neural net model for solder defects.

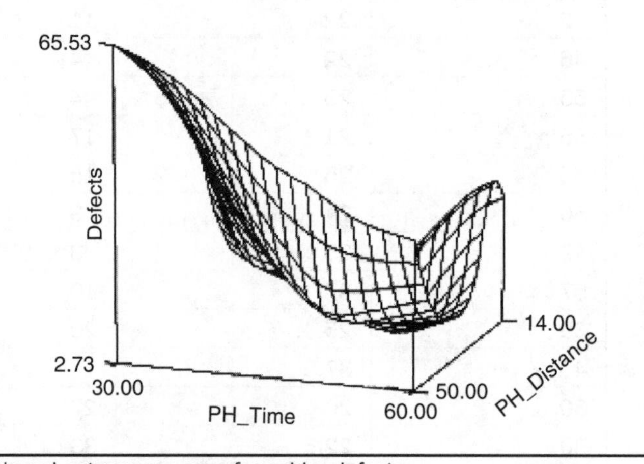

FIGURE 11.12 Neural net process map for solder defects.

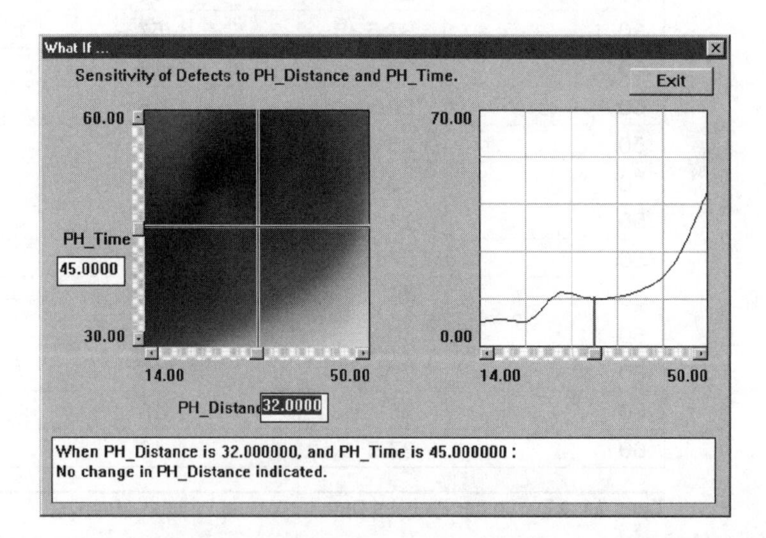

FIGURE 11.13 What If? contour plot dialog box.

accomplish a specific objective. A group of machines performing related manufacturing operations would constitute a system. These machines may be considered, as a group, an element in a larger production system. The production system may be an element in a larger system involving design, delivery, etc.

Simulations allow the system designer to solve problems. To the extent that the computer model behaves as the real world system it models, the simulation can help answer important questions. Care should be taken to prevent the model from becoming the focus of attention. If important questions can be answered more easily without the model, then the model should not be used

The modeler must specify the scope of the model and the level of detail to include in the model. Only those factors which have a significant impact on the model's ability to serve its stated purpose should be included. The level of detail must be consistent with the purpose. The idea is to create, as economically as possible, a replica of the real-world system that can provide answers to important questions. This is usually possible at a reasonable level of detail.

Well designed simulations provide data on a wide variety of systems metrics, such as throughput, resource utilization, queue times, and production requirements. While useful in modeling and understanding existing systems, they are even better suited to evaluating proposed process *changes*. In essence, simulation is a tool for rapidly generating and evaluating ideas for process improvement. By applying this technology to the creativity process, Six Sigma improvements can be greatly accelerated.

Predicting CTQ Performance

A key consideration for any design concept is the CTQ that would result from deploying the design. It is often very difficult to determine the overall result of a series of process steps, but relatively easy to study each step individually. Software can then be used to simulate the process a number of times and calculate the CTQ at the end of the series.

Example of Predicting CTQ Performance

An order fulfillment process design concept has been studied by benchmarking each step. Some of the steps were observed internally, while others were operating in companies considered best in class for the particular step. The overall CTQ target was to ship within 24 hours of receiving the customer's call. The distribution of the time to complete each of the individual steps is shown in Fig. 11.14.

An Excel based computer program* was used to simulate 10,000 orders going through the process, producing the results shown in Fig. 11.15. The simulation indicates that the CTQ Total Cycle Time will be met 99.9% of the time, for a process sigma level of 4.6. Since the process goal is Six Sigma, there is a gap of 1.4 sigma (about 997 PPM) to be addressed. The distribution of the individual steps provides valuable clues about where to focus our attention. Analysis of Fig. 11.15 indicates that the problem is a long tail to the right, so we should look at steps where this is also the case. "Ship order" and "enter order" are both prime suspects and candidates for improvement. A new concept design would then be resimulated.

*Crystal Ball Professional, www.decisioneering.com.

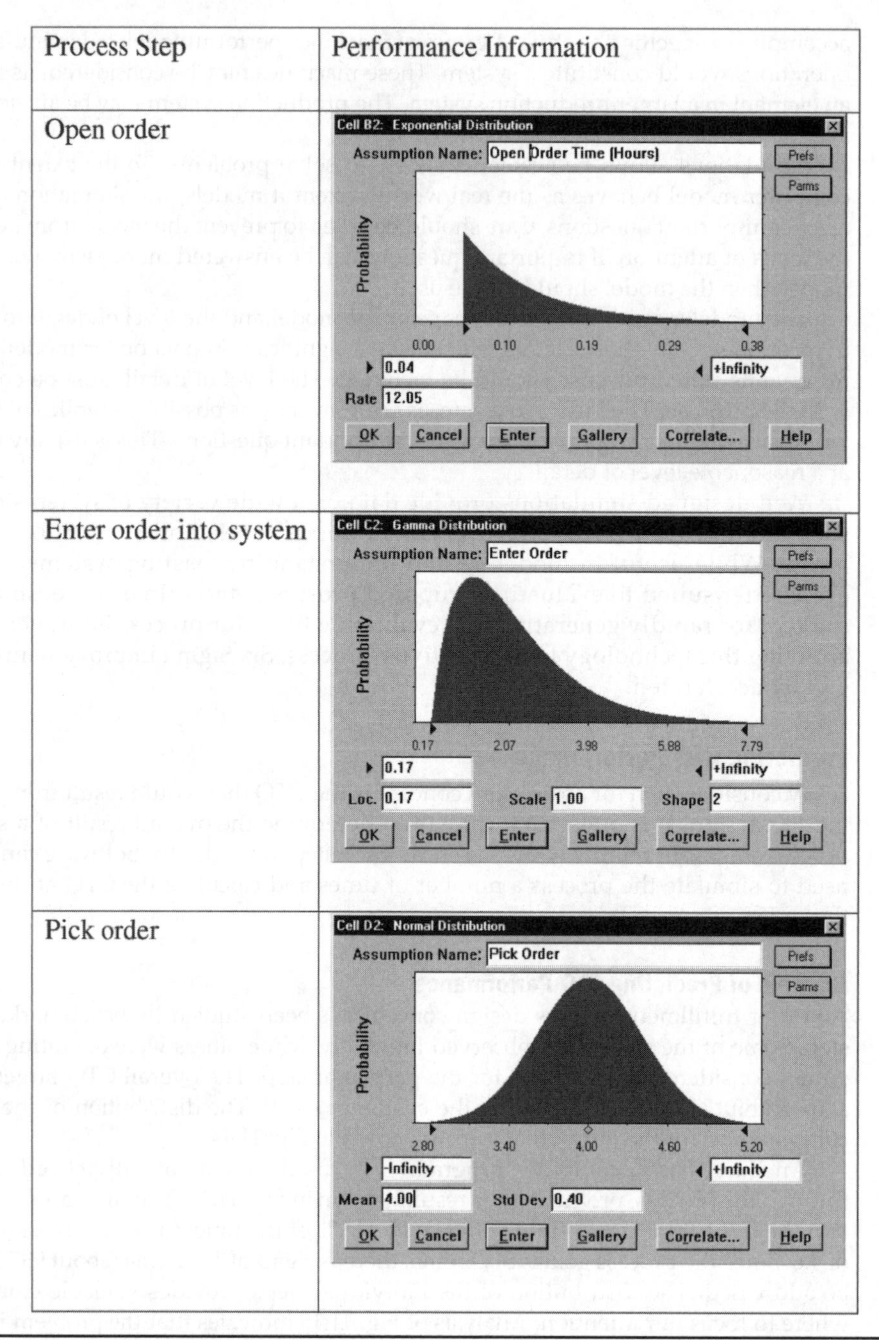

Process Step	Performance Information
Open order	
Enter order into system	
Pick order	

FIGURE 11.14 Order fulfillment process.

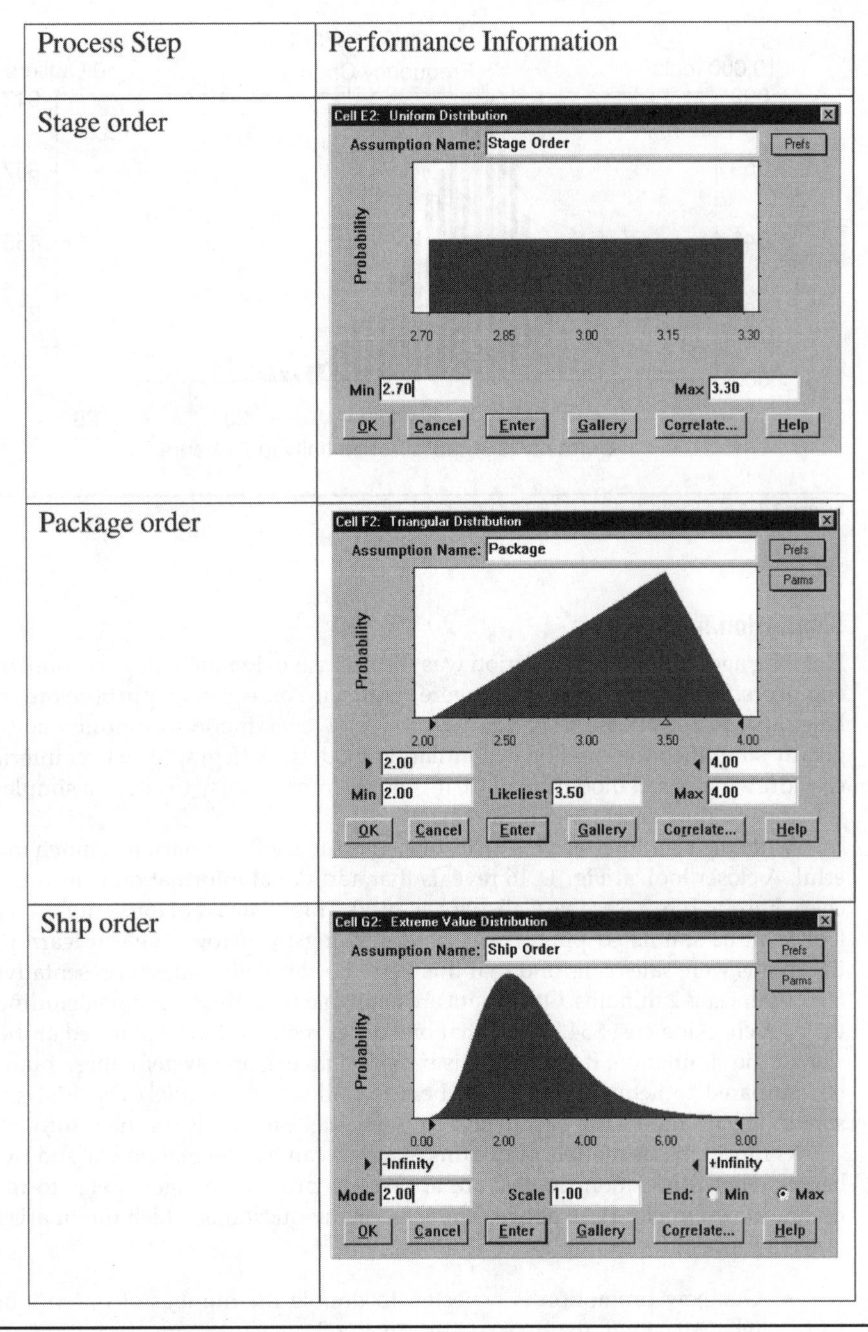

Figure 11.14 *(Continued)*

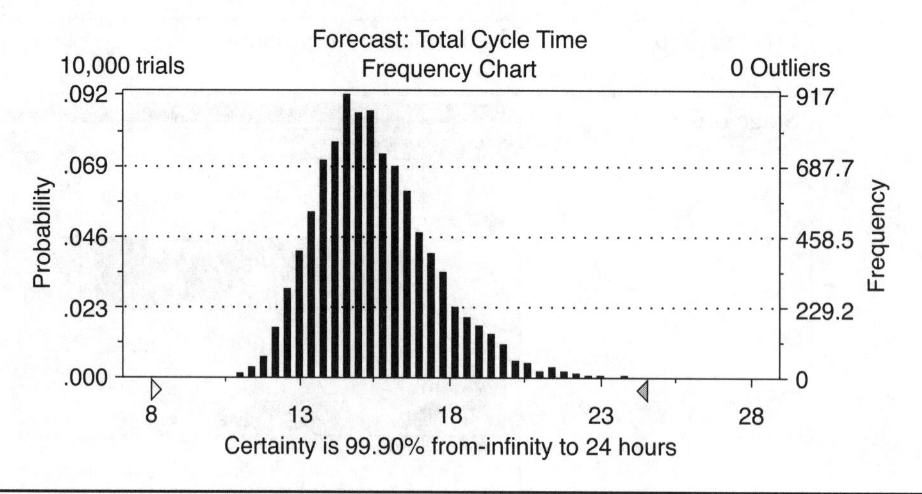

FIGURE 11.15 Cycle time for 10,000 orders.

Simulation Tools

Not long ago, computer simulation was the exclusive domain of highly trained systems engineers. These early simulations were written in some general purpose programming language, such as FORTRAN, Pascal, or C. However, modern computer software has greatly simplified the creation of simulation models. With graphical user interfaces and easy drawing-based model creation, it is now almost as easy to create a simulation as it is to draw a flow chart (see Fig. 11.16).

While the user interface may *look* like an ordinary flow chart, it is much more powerful. A closer look at Fig. 11.16 reveals that additional information is recorded on the chart. Next to box 2: Receive Call, we see In: 60. This means that the simulation program looked at 60 simulated phone calls. By following the arrows we can learn that 20 of these calls were sales calls and that this kept the simulated sales representatives active for 2 hours and 2 minutes. Other data are available from the diagram, including the fact that QA checking cost $34.83 and that one order remained to be shipped at the conclusion of the simulation. If the model is based on an existing system, these numbers will be compared to actual process numbers to validate the model. The first simulation should always model the past to ensure the model's forecasts for the future.

As intriguing as models of existing systems can be, the real power and excitement begins when simulation models are applied to process changes. Refer to the simple model shown in Fig. 11.16 above. There are many questions which might arise regarding this process, for example,

- Our new promotion is expected to double the number of orders phoned in, what effect will that have on production?
- If the QA check was performed by production personnel, what effect would that have on QA cost? Total cost? Production throughput?

In general, the model lets us determine what happens to Y_i if we change X. Changes often create *unanticipated consequences* throughout a complex system due to their effects

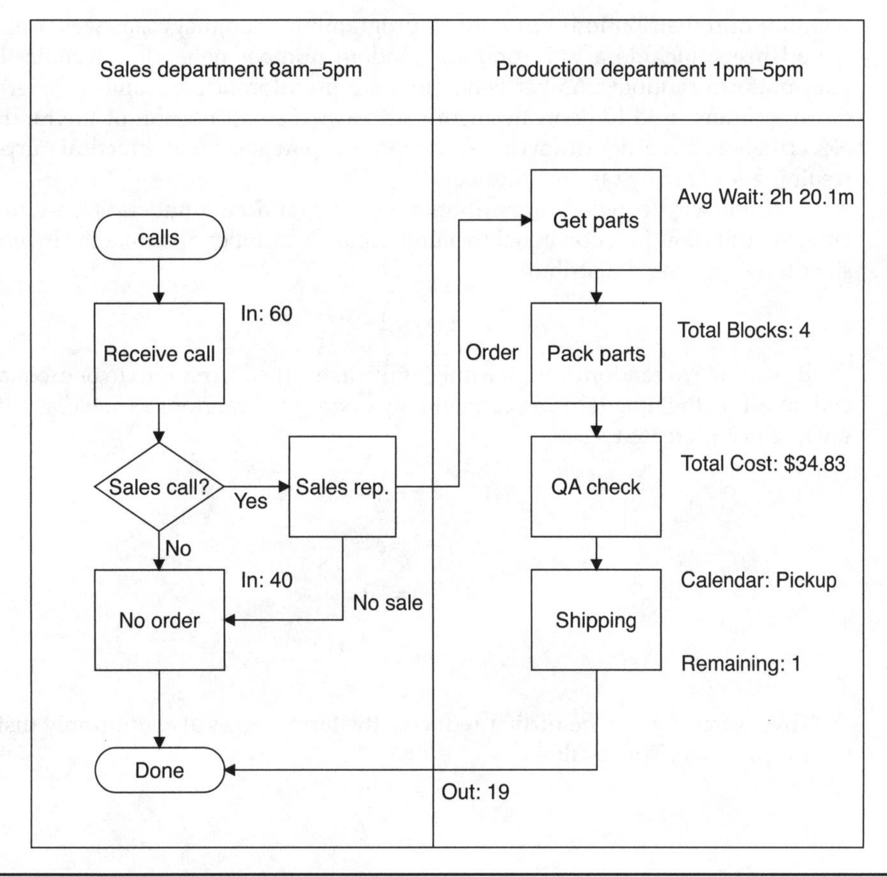

Figure 11.16 Simulation software interface.

on interrelated processes. For example, changing the volume of calls might cause an increase in the idle time of the QA checker because it increases the delay time in the "Get Parts" bottleneck process. Once this fact has been revealed by the simulation, the manager can deal with it. Furthermore, the manager's proposed solution can also be tested by simulation before it's tried in the real world. For example, the manager might propose to cross-train the QA person to be able to help Get Parts. This would theoretically reduce the wait at the Get Parts step while simultaneously increasing the utilization of the QA person. The simulation would allow this hypothesis to be tested before it's tried. Perhaps it will show that the result is merely to move the bottleneck from one process step to another, rather than eliminating it. Anyone who has spent any length of time in the working world is familiar with these "hydraulic models" where managers' attempts to fix one problem only result in the creation of new problems. By discovering this before trying it, money is saved and morale improved.

Random Number Generators

The heart of any simulation is the generation of random numbers. Random numbers from specific distributions are generated by transforming random numbers from the unit, uniform distribution. Virtually all programming languages, as well as electronic spreadsheets, include a unit, uniform random number generator. Technically, these unit, uniform random number generators are pseudorandom number generators, as the algorithms used to generate them take away a small portion of the randomness. Nevertheless, these algorithms are extremely efficient and for all practical purposes the result is a set of truly random numbers.

A simple way to generate distribution-specific random numbers is to set the cumulative distribution function equal to a unit, random number and take the inverse. Consider the exponential distribution

$$F(x) = 1 - e^{-\lambda x} \tag{11.3}$$

By setting r, a random variable uniformly distributed from zero to one, equal to $F(x)$ and inverting the function, an exponentially distributed random variable, x, with a failure rate of λ is created.

$$\begin{aligned} r &= 1 - e^{-\lambda x} \\ 1 - r &= e^{-\lambda x} \\ \ln(1 - r) &= -\lambda x \\ x &= -\frac{\ln(1 - r)}{\lambda} \end{aligned} \tag{11.4}$$

This expression can be further reduced; the term $1 - r$ is also uniformly distributed from zero to one. The result is

$$x = -\frac{\ln r}{\lambda} \tag{11.5}$$

Table 11.13 contains some common random number generators.

After the desired random number generator(s) have been constructed, the next step is to mathematically model the situation under study. After completing the model, it is important to *validate* and *verify* the model. A valid model is a reasonable representation of the situation being studied A model is verified by determining that the mathematical and computer model created represents the intended conceptual model.

Enough iterations should be included in the simulation to provide a *steady-state* solution, which is reached when the output of the simulation from one iteration to the next changes negligibly. When calculating means and variances, 1,000 iterations is usually sufficient. If calculating confidence limits, many more iterations are required; after all, for 99% confidence limits the sample size for the number of random deviates exceeding the confidence limit is 1/100th the number of iterations.

Example: A Simulation of Receiving Inspection*

This example describes a simulation model of a complex inspection operation at a factory of a large unionized defense contractor. The plant receives four types of parts: electrical, simple mechanical, complex mechanical, and parts or materials that require nondestructive testing (NDT). Union regulations required four different inspector grades. The plant

*Pyzdek, 1992b.

Distributors	Probability Density Function	Random Number Generators[§]
Uniform	$f(x)=\frac{1}{b-a}, a \le x \le b$	$x = a + (b - a)r$
Exponential	$f(x) = \lambda e^{-\lambda x},\ 0 < x < \infty$	$x=-\frac{1}{\lambda}\ln r$
Normal	$f(x)=\frac{1}{\alpha\sqrt{2\pi}}\exp\left[-\frac{1}{2}\left(\frac{x-\mu}{\sigma}\right)^2\right], -\infty < x < \infty$	$x_1 = \left[\sqrt{-2\ln r_1}\ \cos(2\pi r_2)\right]\sigma+\mu$ $x_2 = \left[\sqrt{-2\ln r_1}\ \sin(2\pi r_2)\right]\sigma+\mu^{\dagger}$
Lognormal	$f(x)=\frac{1}{\sigma x\sqrt{2\pi}}\exp\left[-\frac{1}{2}\left(\frac{\ln x-\mu}{\sigma}\right)^2\right], x>0$	$x_1 = \exp\left[\sqrt{-2\ln r_1}\ \cos(2\pi r_2)\right]\sigma+\mu$ $x_2 = \left[\sqrt{-2\ln r_1}\ \sin(2\pi r_2)\right]\sigma+\mu^{\dagger}$
Weibull	$f(x)=\frac{\beta x^{\beta-1}}{\theta^{\beta}}\exp\left(\frac{x}{\theta}\right)^{\beta}, x>0$	$x = \theta(-\ln r)^{1/\beta}$
Poisson	$f(x)=\frac{e^{-\lambda t}(\lambda t)^x}{x!}, x=0,1,2,\ldots,\infty$	$x=\begin{cases} 0, -\frac{1}{\lambda}\ln r > t^{\ddagger} \\ x, \sum_{i=1}^{x} -\frac{1}{\lambda}\ln r_i < t < \sum_{i=1}^{x+1} -\frac{1}{\lambda}\ln r_i \end{cases}$
Chi-square	$f(x)=\frac{1}{2^{v/2}\Gamma(v/2)}x^{(v/2-1)}e^{-x/2}, x>0$	$x=\sum_{i=1}^{v} Z_i^2$ z_i is a standard normal random deviate.
Beta	$f(x)=\frac{1}{B(q,p)}x^{p-1}(1-x)^{q-1},$ $0 \le x \le 1, p>0, q>0$	$x=\frac{r^{1/p}}{r^{1/p}+r^{1/q}}$

[†]Two uniform random numbers must be generated, with the result being two normally distributed random numbers.
[‡]Increase the value of x until the inequality is satisfied.
[§]Statistical Software, such as MINITAB, have these functions built-in.

TABLE 11.13 Random Number Generators

Distributors	Probability Density Function	Random Number Generators§
Gamma	$f(x)=\frac{\lambda^n}{\Gamma(\eta)}x(\eta-1)e^{-\lambda t}$, $x?0,\ \eta?0,\ \lambda?0$	1. η is a non-integer shape parameter. 2. Let η_1, = the truncated integer root of η 3. Let $q=-\ln \prod_{j=1}^{\eta_i} r_j$. 4. Let $A=\eta-\eta_1$, and $B=1-A$. 5. Generate a random number and let $y_t=r_j^{t/A}$. 6. Generate a random number and let $y_2=r_{i+1}^{t/B}$. 7. If $y_1+y_2\le 1$ go to 9. 8. Let $i=i+2$ and go to 5. 9. Let $z=y_1/+(y_1+y_2)$. 10. Generate a random number, r_n. 11. Let $W=-\ln r_n$. 12. $x=(q+zW)\lambda$.
Binomial	$P(x)=(n/x)p^x(1-p)^{n-x}, x=0,1,\ldots,n$	$x=\sum_{i=1}^{n} y_i,\ y_i=\begin{cases}0, r_i>p\\1, r_i\le p\end{cases}$
Geometric	$P(x)=p(1-p)^{x-1}$, $x=1, 2, 3,\ldots$	$\frac{\ln(1-r)}{\ln(1-p)}\le x\le\frac{\ln(1-r)}{\ln(1-p)}+1^{\dagger}$
Student's t	$f(x)=\frac{\Gamma[(v-1)/2]}{\Gamma(v/2)\sqrt{\pi v}}\left(1+\frac{x^2}{v}\right)^{-(v+1)/2}, -\infty<x<\infty$	$x=\frac{z_t}{\left(\sum_{i=2}^{v+1}\frac{z_i^2}{v}\right)^{1/2}}$ z_i is a standard normal random deviate.
F	$f(x)=\left(\frac{\Gamma[(v_1+v_2)/2](v_1/v_2)^{v_1/2}}{\Gamma(v_1/2)\Gamma(v_2/2)}\right)\times\left(\frac{x^{v_1/2-1}}{(1+v_1x/v_2)^{(v_1+v_2)/2}}\right), x>0.$	$x=\frac{v_2\sum_{i=1}^{v_1} z_i^2}{v_1\sum_{i=v_1+1}^{v_1+v_2} z_i^2}$ z_i is a standard normal random deviate.

Table 11.13 Random Number Generators (*Continued*)

is experiencing a growing backlog of orders awaiting inspection. The backlog is having an adverse effect on production scheduling, including frequent missile assembly stoppages. A computer simulation will be conducted to answer the following questions:

1. Is the backlog a chance event that will eventually correct itself without intervention, or is a permanent change of process required?
2. If additional personnel are hired, will they be adequately utilized?
3. Which types of job skills are required?
4. Will additional automated or semi-automated inspection equipment alleviate the problem?

Model Development

The first phase of the project is to develop an accurate model of the Receiving Inspection process. One element to evaluate is the distribution of arrivals of the various parts. Figure 11.17 compares the empirical distribution of the electrical lots with the predictions of an exponential arrival time model Data were gathered from a recent work- month.

Similar "eyeball fits" were obtained from the arrival distributions of the other three part types. The exponential model seems to provide adequate representation of the data

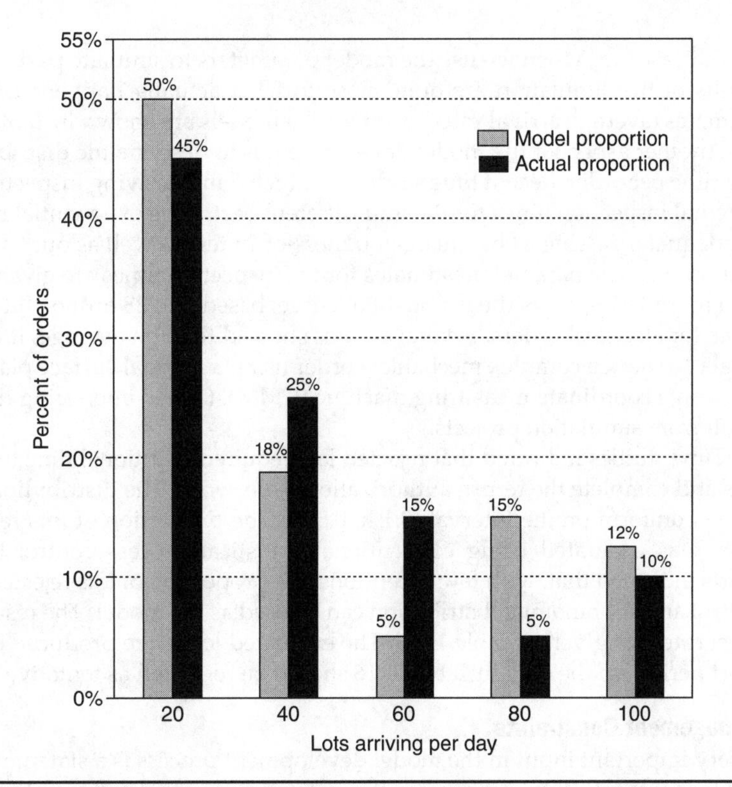

FIGURE 11.17 Electrical order arrivals predicted versus actual.

Order Type	MeanArrival Rate (Orders-Per-Hour)
Electrical	4.292
Simple mechanical	6.849
Complex mechanical	1.541
Nondestructive test	0.630

TABLE 11.14 Average Arrival Rates

Order Type	Average Inspection Time (Orders-Per-Hour)
Electrical	1.681
Simple mechanical	2.500
Complex mechanical	0.597
Nondestructive test	0.570

TABLE 11.15 Average Inspection Times

in each case (i.e., when we use the model parameters to simulate past performance, the results of the simulation are quite close to what actually happened). The parameter estimates (average arrival rates) used for the models are shown in Table 11.14.

Another aspect of the model development is to describe the distribution of inspection time per order. Recent time studies conducted in Receiving Inspection provide data of actual inspection times for the four different parts. The exponential model proved to be adequate, passing a chi-square goodness-of-fit test as well as our "simulation of the past" check. The parameter estimates for the inspection times are given in Table 11.15.

Figure 11.18 shows the exponential curve, based on 228 orders, fitted to inspection times for electrical orders. Several studies showed that, on average, it takes four times longer to check a complex mechanical order using a manual surface plate layout than it takes on a coordinate measuring machine (CMM). (These interesting discoveries often result from simulation projects.)

Time studies indicated that rejected lots required additional time to fill out rejected tags and complete the return authorization paperwork. The distribution of this process time is uniform on the interval [0.1 h, 0.5 h]. The proportion of lots rejected, by order type, was evaluated using conventional statistical process control techniques. The charts indicated that, with few exceptions, the proportion of lots rejected is in statistical control and the binomial distribution can be used as the model. The resulting estimated reject rates are given in Table 11.16. The evaluated lots were produced over a relatively short period, so the data in Table 11.16 should be regarded as tentative.

Management Constraints

A very important input in the model development process is a statement of constraints by management. With this project, the constraints involve a description of permissible

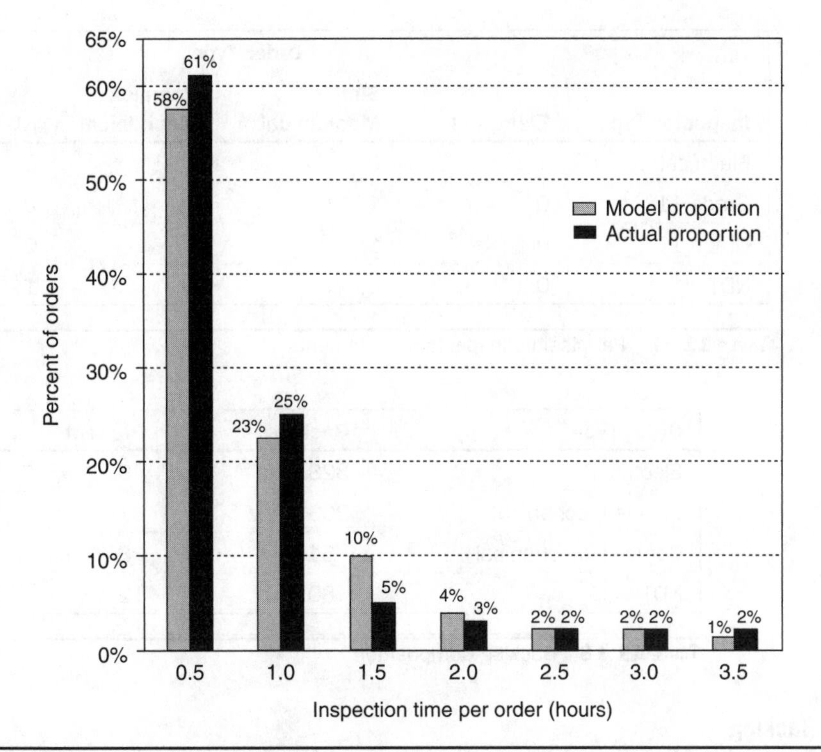

FIGURE 11.18 Electrical order inspection times.

Order Type	% Lots Rejected	Lots Evaluated
Electrical	2.2	762
Simple mechanical	1.1	936
Complex mechanical	25.0	188
Nondestructive test	0.5	410

TABLE 11.16 Average Reject Rates

job assignments for each inspector classification, the overtime policy, the queue discipline, and the priority discipline for assigning jobs to inspectors as well as to the CMMs. Table 11.17 summarizes the permissible job assignments. A "0" indicates that the assignment is not permitted, while a "1" indicates a permissible assignment.

The simulation will be run under the assumption that overtime is not permitted; 40 hours is considered a complete work week. Preference is given to the lower paid Grade 11 inspectors when a simple mechanical order is assigned. The CMM is to be used for complex mechanical parts when it is available. Inspection is conducted using a first-in first-out (FIFO) protocol.

Inspector Type	Order Type			
	Electrical	Simple Mechanical	Complex Mechanical	NDT
Electrical	1	0	0	0
Grade 11	0	1	0	0
Grade 19	0	1	1	0
NDT	0	0	0	1

TABLE 11.17 Permissible Inspector Assignments

Order Type	Backlog	Percent
Electrical	328	49
Simple mechanical	203	31
Complex mechanical	51	8
NDT	80	12

TABLE 11.18 Backlog Composition

Backlog

The backlog is a composite of all part types. Information on when the particular order entered the backlog is not available. At the time the simulation was proposed, the backlog stood at 662 orders with the composition shown in Table 11.18.

By the time the computer program was completed 4 weeks later, the backlog had dwindled to 200 orders. The assumption was made that the percentage of each order type remained constant and the simulation was run with a 200 order backlog.

The Simulation

The first simulation mimics the current system so the decision maker can determine if the backlog is just a chance event that will work itself out. The simulation begins with the current staff, facilities, and backlog and runs 4 simulated regular weeks of 40 hours per week. This is done 6 times and the following statistics computed:

1. Average delay awaiting inspection
2. Maximum delay awaiting inspection
3. Average backlog
4. Maximum backlog
5. Utilization of the inspectors
6. Utilization of the CMM

Statistics 1 through 4 will be computed for each part type; statistic 5 will be computed for each inspector type.

Modified Systems Simulations provide an ideal way of evaluating the impact of proposed management changes. Such changes might include inspection labor and the number of CMMs; therefore, these were programmed as input variables. In discussions with management, the following heuristic rules were established:

$$\text{If } U_i < (n_i - 1)/n_i,\ \text{i} = 1; 2; 3; 4, \text{ then let } n_\text{i} = n_\text{i} - 1$$

where U_i = Utilization of inspector type *i* and n_i = Number of inspectors of type *i*

For example, suppose there are three electrical inspectors (i.e., $n_i = 3$), and the utilization of electrical inspectors is 40% ($U_i = 0.4$). The heuristic rule would recommend eliminating an electrical inspector since 0:4 < (3 – 1)/3 = 0.67.

A decision was made that the reductions would take place only if the backlog was under control for a given order type. The author interpreted this to mean that a two sigma interval about the average change in backlog should either contain zero backlog growth, or be entirely negative.

Results of Simulations The first simulation was based on the existing system, coded 5–2–5–2–1, meaning

- 5 electrical inspectors
- 2 grade 11 inspectors
- 5 grade 19 inspectors
- 2 NDT inspectors
- 1 CMM

The results of this simulation are shown in Table 11.19.

After 6 simulated weeks:

Job Type	Average Utilization	Avearge Change in Backlog	Std. Dev. of Change in Backlog
Electrical	0.598	–96.333	6.3140
Mech-simple	0.726	–64.000	8.4617
Mech-complex	0.575	–14.500	3.5637
NDT	0.640	–22.500	3.7283

The heuristic rule describes the direction to go with staffing, but not how far. Based solely on the author's intuition, the following configuration was selected for the next simulation:

- 3 electrical inspectors
- 2 grade 11 inspectors
- 3 grade 19 inspectors
- 2 NDT inspectors
- 1 CMM

The results of simulating this 3–2–3–2–1 system are given in Table 11.20. All average utilization values pass the heuristic rule and the backlog growth is still, on the average,

Type Inspection	Inspectors	Inspector Utilization	Backlog		CMM	
			Avg.	Max.	Number	Utilization
Run 1						
Electrical	5	0.577	8.5	98		
Mech-simple	2	0.704	1.6	61		
Mech-complex	5	0.545	0.7	16	1	0.526
NDT	2	0.622	4.3	25		
Run 2						
Electrical	5	0.623	7.5	97		
Mech-simple	2	0.752	1.9	68		
Mech-complex	5	0.621	0.6	11	1	0.501
NDT	2	0.685	5.0	24		
Run 3						
Electrical	5	0.613	8.3	107		
Mech-simple	2	0.732	1.5	51		
Mech-complex	5	0.596	2.0	30	1	0.495
NDT	2	0.541	3.5	23		
Run 4						
Electrical	5	0.608	4.9	93		
Mech-simple	2	0.726	1.5	67		
Mech-complex	5	0.551	0.8	14	1	0.413
NDT	2	0.665	3.5	28		
Run 5						
Electrical	5	0.567	6.8	91		
Mech-simple	2	0.684	2.9	77		
Mech-complex	5	0.554	0.6	13	1	0.506
NDT	2	0.592	2.1	21		
Run 6						
Electrical	5	0.598	6.6	96		
Mech-simple	2	0.755	2.4	65		
Mech-complex	5	0.584	1.6	19	1	0.493
NDT	2	0.735	5.0	22		

Table 11.19 Current System 5–2–5–2–1 Simulation Results

Type Inspection	Inspectors	Inspector Utilization	Backlog		CMM	
			Avg.	Max.	Number	Utilization
Run 1						
Electrical	3	0.935	49.4	101		
Mech-simple	2	0.847	7.5	61		
Mech-complex	3	0.811	2.0	16	1	0.595
NDT	2	0.637	8.2	28		
Run 2						
Electrical	3	0.998	81.7	114		
Mech-simple	2	0.866	8.2	70		
Mech-complex	3	0.863	2.5	16	1	0.629
NDT	2	0.631	3.5	22		
Run 3						
Electrical	3	0.994	74.3	109		
Mech-simple	2	0.889	12.0	73		
Mech-complex	3	0.891	6.2	32	1	0.623
NDT	2	0.679	6.4	27		
Run 4						
Electrical	3	0.879	31.2	109		
Mech-simple	2	0.927	7.2	52		
Mech-complex	3	0.924	5.6	26	1	0.632
NDT	2	0.715	3.8	25		
Run 5						
Electrical	3	0.992	45.6	117		
Mech-simple	2	0.791	3.7	43		
Mech-complex	3	0.761	1.8	18	1	0.537
NDT	2	0.673	2.3	24		
Run 6						
Electrical	3	0.990	39.9	95		
Mech-simple	2	0.844	6.9	63		
Mech-complex	3	0.800	1.7	18	1	0.606
NDT	2	0.716	4.2	24		

TABLE 11.20 3–2–3–2–1 System Simulation Results

comfortably negative. However, the electrical order backlog reduction is considerably more erratic when the inspection staff is reduced.

After 6 simulations:

Job Type	Average Utilization	Average Change in Backlog	Std. Dev. of Change in Backlog
Electrical	0.965	–91.833	20.5856
Mech-simple	0.861	–54.667	8.7331
Mech-complex	0.842	–15.833	1.3292
NDT	0.676	–23.500	1.3784

While this configuration was acceptable, the author believed that additional trials might allow replacement of one or more of the highly paid grade 19 inspectors with the lower paid grade 11 inspectors. A number of combinations were tried, resulting in the 3–3–1–2–1 system shown in Table 11.21.

After 6 simulations:

Job Type	Average Utilization	Average Change in Backlog	Std. Dev. of Change in Backlog
Electrical	0.965	–93.667	6.9762
Mech-simple	0.908	–57.500	5.8224
Mech-complex	0.963	–5.500	18.1411
NDT	0.704	–25.500	2.7386

The 3–3–1–2–1 system complies with all management constraints relating to resource utilization and backlog control. It is recommended to management with the caution that they carefully monitor the backlog of complex mechanical orders. For this type of order, the simulation indicates negative backlog growth on average, but with periods of positive backlog growth being possible.

Conclusion The simulation allowed the receiving inspection process to be "changed" without actually disrupting operations. In the computer, inspectors can be added, removed, or reassigned without the tremendous impact on morale and operations that would result from making these changes in the real world. It is a simple matter to add additional CMMs which would cost six figures in the real world. It is just as easy to try different job assignment protocols, examine the impact of a proposed new product line, look at new work area layouts, see if we can solve a temporary problem by working overtime or hiring temporary workers, etc. The effect of these changes can be evaluated in a few days, rather than waiting several months to learn that the problem was not resolved.

Virtual Doe Using Simulation Software

Modern simulation software can interface with statistical analysis software to allow more detailed analysis of proposed new products and processes. In this section I'll demonstrate this capability with iGrafx Process for Six Sigma and Minitab. However, these capabilities are also incorporated into other software packages.

Type Inspection	Inspectors	Inspector Utilization	Backlog Avg.	Max.	CMM Number	Utilization
Run 1						
Electrical	3	0.937	37.0	110		
Mech-simple	3	0.885	13.1	61		
Mech-complex	1	0.967	7.4	21	1	0.718
NDT	2	0.604	3.4	25		
Run 2						
Electrical	3	0.932	26.8	100		
Mech-simple	3	0.888	7.9	58		
Mech-complex	1	0.925	17.8	49	1	0.722
NDT	2	0.607	4.0	27		
Run 3						
Electrical	3	0.997	74.1	119		
Mech-simple	3	0.915	14.6	58		
Mech-complex	1	0.957	20.6	40	1	0.807
NDT	2	0.762	7.1	22		
Run 4						
Electrical	3	0.995	42.2	96		
Mech-simple	3	0.976	38.4	79		
Mech-complex	1	0.997	23.8	56	1	0.865
NDT	2	0.758	4.8	30		
Run 5						
Electrical	3	0.996	61.3	121		
Mech-simple	3	0.913	7.7	50		
Mech-complex	1	0.996	21.7	52	1	0.909
NDT	2	0.820	7.4	30		
Run 6						
Electrical	3	0.933	35.3	101		
Mech-simple	3	0.867	5.7	59		
Mech-complex	1	0.938	17.8	49	1	0.736
NDT	2	0.674	8.8	33		

TABLE 11.21 3–3–1–2–1 System Simulation Results

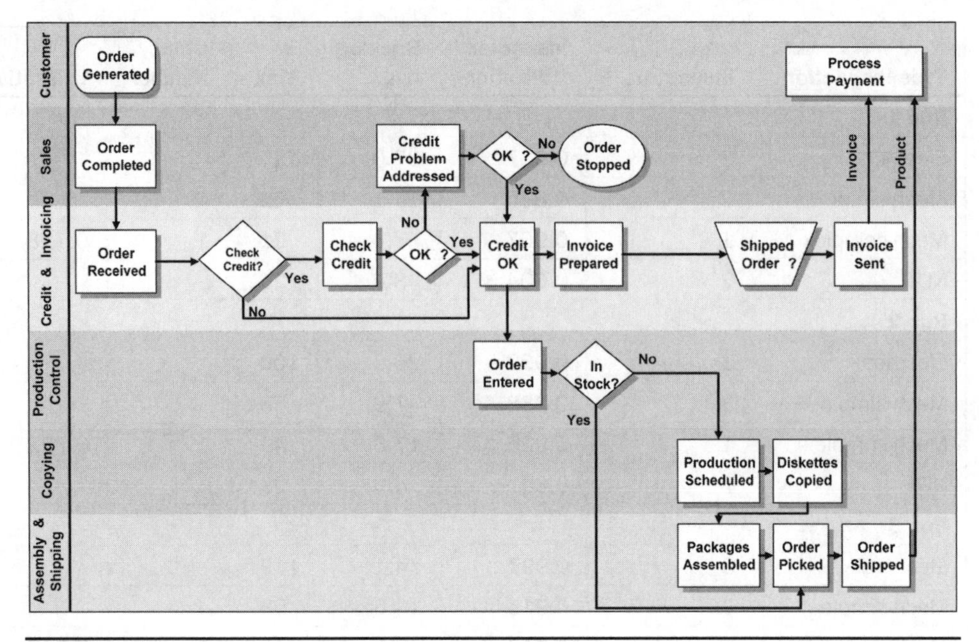

Figure 11.19 Process to be evaluated using virtual DOE.

Example

A Six Sigma team has developed the process shown in Fig. 11.19. The CTQs for the process are the cycle times required for processing transactions for new and existing customers. They want to recommend staff levels that produce good CTQ results for both customer types during both normal and busy workload times. They will determine the recommended staff levels by performing simulations and analyzing the results using DOE techniques.

Figure 11.20 shows the dialog box for the iGrafx "RapiDOE" procedure. The top part of the box displays the available factors. The team wants to evaluate their two CTQs as the interarrival rate varies, and for different staff levels for six different types of workers. The middle of the box indicates that the experiment will be replicated four times. The bottom of the dialog box shows the CTQs the Black Belt has selected for evaluation.

Figure 11.21 shows the RapiDOE Minitab options dialog box. The Black Belt will use a fractional factorial design with 64 runs. Minitab's display of Available Factorial Designs (Fig. 11.22) indicates that this half-fraction seven-factor, 64 run design is resolution VII. This will allow the estimation of all main effects, two-factor interactions, and three-factor interactions.

In just a few minutes the 256 simulated experimental runs are completed (Fig. 11.23). The analysis of these results proceeds in exactly the same way as it would with the results from real-world experiments. Of course, the conclusions are not as trustworthy as real-world experimentation would provide. However, they are certainly a lot cheaper and faster to obtain and they provide a great deal of insight into the process bottlenecks, areas of potential improvement and other important factors. Virtual DOE also allows trial-and-error without disrupting operations or impacting customers.

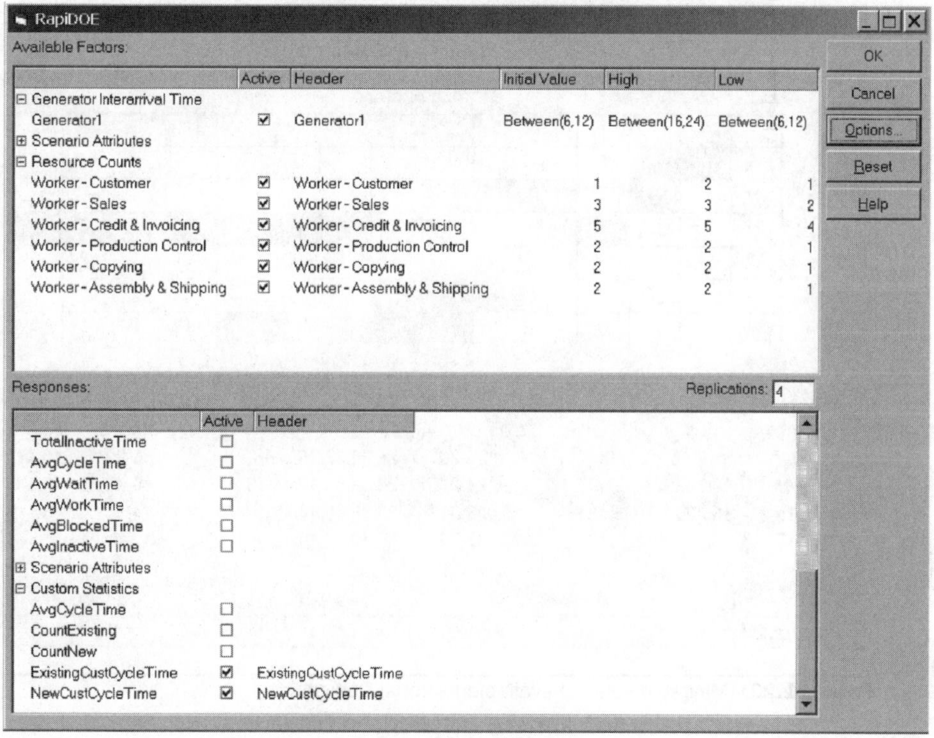

RapiDOE

Available Factors:

	Active	Header	Initial Value	High	Low
⊟ Generator Interarrival Time					
Generator1	☑	Generator1	Between(6,12)	Between(16,24)	Between(6,12)
⊞ Scenario Attributes					
⊟ Resource Counts					
Worker - Customer	☑	Worker - Customer	1	2	1
Worker - Sales	☑	Worker - Sales	3	3	2
Worker - Credit & Invoicing	☑	Worker - Credit & Invoicing	5	5	4
Worker - Production Control	☑	Worker - Production Control	2	2	1
Worker - Copying	☑	Worker - Copying	2	2	1
Worker - Assembly & Shipping	☑	Worker - Assembly & Shipping	2	2	1

OK
Cancel
Options...
Reset
Help

Responses:

Replications: 4

	Active	Header
TotalInactiveTime	☐	
AvgCycleTime	☐	
AvgWaitTime	☐	
AvgWorkTime	☐	
AvgBlockedTime	☐	
AvgInactiveTime	☐	
⊞ Scenario Attributes		
⊟ Custom Statistics		
AvgCycleTime	☐	
CountExisting	☐	
CountNew	☐	
ExistingCustCycleTime	☑	ExistingCustCycleTime
NewCustCycleTime	☑	NewCustCycleTime

FIGURE 11.20 iGrafx Process for Six Sigma RapiDOE display.

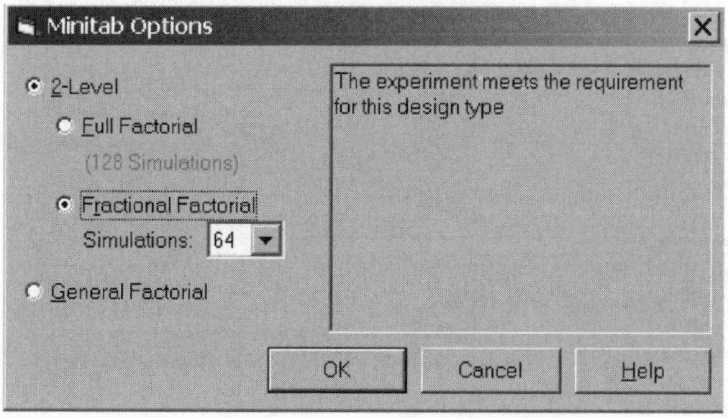

Minitab Options

◉ 2-Level
○ Full Factorial
(128 Simulations)
◉ Fractional Factorial
Simulations: 64
○ General Factorial

The experiment meets the requirement for this design type

OK
Cancel
Help

FIGURE 11.21 iGrafx process for Six Sigma RapiDOE Minitab options.

Create Factorial Design - Display Available Designs

Available Factorial Designs (with Resolution)

Factors

Runs	2	3	4	5	6	7	8	9	10	11	12	13	14	15
4	Full	III												
8		Full	IV	III	III	III								
16			Full	V	IV	IV	IV	III	III	III	III	III	III	III
32				Full	VI	IV	IV	IV	IV	IV	IV	IV	IV	IV
64					Full	VII	V	IV	IV	IV	IV	IV	IV	IV
128						Full	VIII	VI	V	V	IV	IV	IV	IV

Available Resolution III Plackett-Burman

Factors	Runs	Factors	Runs	Factors	Runs
2-7	8,12,16,20,...,48	20-23	24,28,32,36,...,48	36-39	40,44,48
8-11	12,16,20,24,...,48	24-27	28,32,36,40,44,48	40-43	44,48
12-15	16,20,24,28,...,48	28-31	32,36,40,44,48	44-47	48
16-19	20,24,28,32,...,48	32-35	36,40,44,48		

Help OK

Figure 11.22 Minitab display of available Factorial Designs.

↓	C4-T	C5	C6	C7	C8	C9	C10	C11	C12
	Generator1	Customer	Sales Worker	Credit	Production	Copying	Assembly &	ExistingCust	NewCustCycle
236	Between(6,12)	1	3	4	2	1	1	436.792	455.372
237	Between(16,24)	2	2	4	2	1	2	126.808	130.433
238	Between(6,12)	2	2	4	2	1	1	285.695	305.960
239	Between(16,24)	1	2	4	2	1	1	145.640	132.390
240	Between(6,12)	1	2	4	2	1	2	134.134	129.493
241	Between(16,24)	2	3	5	1	1	1	142.785	139.555
242	Between(6,12)	2	3	5	1	1	2	126.208	132.606
243	Between(16,24)	1	3	5	1	1	2	124.875	129.400
244	Between(6,12)	1	3	5	1	1	1	524.238	550.081
245	Between(16,24)	2	2	5	1	1	2	134.932	129.397
246	Between(6,12)	2	2	5	1	1	1	529.054	548.717
247	Between(16,24)	1	2	5	1	1	1	141.672	136.337
248	Between(6,12)	1	2	5	1	1	2	127.931	131.428
249	Between(16,24)	2	3	4	1	1	2	133.150	126.706
250	Between(6,12)	2	3	4	1	1	1	362.809	376.929
251	Between(16,24)	1	3	4	1	1	1	136.857	134.455
252	Between(6,12)	1	3	4	1	1	2	135.973	131.701
253	Between(16,24)	2	2	4	1	1	1	137.652	144.981
254	Between(6,12)	2	2	4	1	1	2	132.612	132.871
255	Between(16,24)	1	2	4	1	1	2	128.793	126.469
256	Between(6,12)	1	2	4	1	1	1	489.439	456.210

Figure 11.23 Partial display of results from virtual DOE.

Risk Assessment Tools

While reliability prediction is a valuable activity, it is even more important to design reliable systems in the first place. Proposed designs must be evaluated to detect potential failures prior to building the system. Some failures are more important than others, and the assessment should highlight those failures most deserving of attention and scarce resources. Once failures have been identified and prioritized, a system can be designed that is robust, that is, it is insensitive to most conditions that might lead to problems.

Design Review

Design reviews are conducted by specialists, usually working on teams. Designs are, of course, reviewed on an ongoing basis as part of the routine work of a great number of people. However, the term as used here refers to the formal design review process. The purposes of formal design review are threefold:

1. Determine if the product will actually work as desired and meet the customer's requirements.
2. Determine if the new design is producible and inspectable.
3. Determine if the new design is maintainable and repairable.

Design review should be conducted at various points in the design and production process. Review should take place on preliminary design sketches, after prototypes have been designed, and after prototypes have been built and tested, as developmental designs are released, etc. Designs subject to review should include parts, subassemblies, and assemblies.

Fault-Tree Analysis

While FMEA (see section "Failure Mode and Effect Analysis") is a bottom-up approach to reliability analysis, FTA is a top-down approach. FTA provides a graphical representation of the events that might lead to failure. Some of the symbols used in construction of fault trees are shown in Table 11.22.

In general, FTA follows these steps:

1. Define the top event, sometimes called the *primary event*. This is the failure condition under study.
2. Establish the boundaries of the FTA.
3. Examine the system to understand how the various elements relate to one another and to the top event.
4. Construct the fault tree, starting at the top event and working downward.
5. Analyze the fault tree to identify ways of eliminating events that lead to failure.
6. Prepare a corrective action plan for preventing failures and a contingency plan for dealing with failures when they occur.
7. Implement the plans.
8. Return to step 1 for the new design.

Figure 11.24 illustrates an FTA for an electric motor.

Gate Symbol	Gate Name	Causal Relations
	AND gate	Output event occurs if all the input events occur simultaneously
	OR gate	Output event occurs if any one of the input events occurs
	Inhibit gate	Input produces output when conditional event occurs
	Priority AND gate	Output event occurs if all input events occur in the order from left to right
	Exclusive OR gate	Output event occurs if one, but not both, of the input events occur
m n inputs	*m*-out-of-*n* gate (voting or sample gate)	Output event occurs if *m*-out-of-*n* input events occur
Event Symbol	**Meaning**	
rectangle	Event represented by a gate	
circle	Basic event with sufficient data	
diamond	Undeveloped event	
switch or house	Either occurring or not occurring	
oval	Conditional event used with inhibit gate	
triangles	Transfer symbol	

Source: *Handbook of Reliability Engineering and Management*, McGraw-Hill, reprinted with permission of the publisher.

TABLE 11.22 Fault-Tree Symbols

Safety Analysis

Safety and reliability are closely related. A safety problem is created when a critical failure occurs, which reliability theory addresses explicitly with such tools as FMEA and FTA. The modern evaluation of safety/reliability takes into account the probabilistic nature of failures. With the traditional approach a safety factor would be defined using Eq. (11.6).

$$SF = \frac{\text{average strength}}{\text{worst expected stress}} \tag{11.6}$$

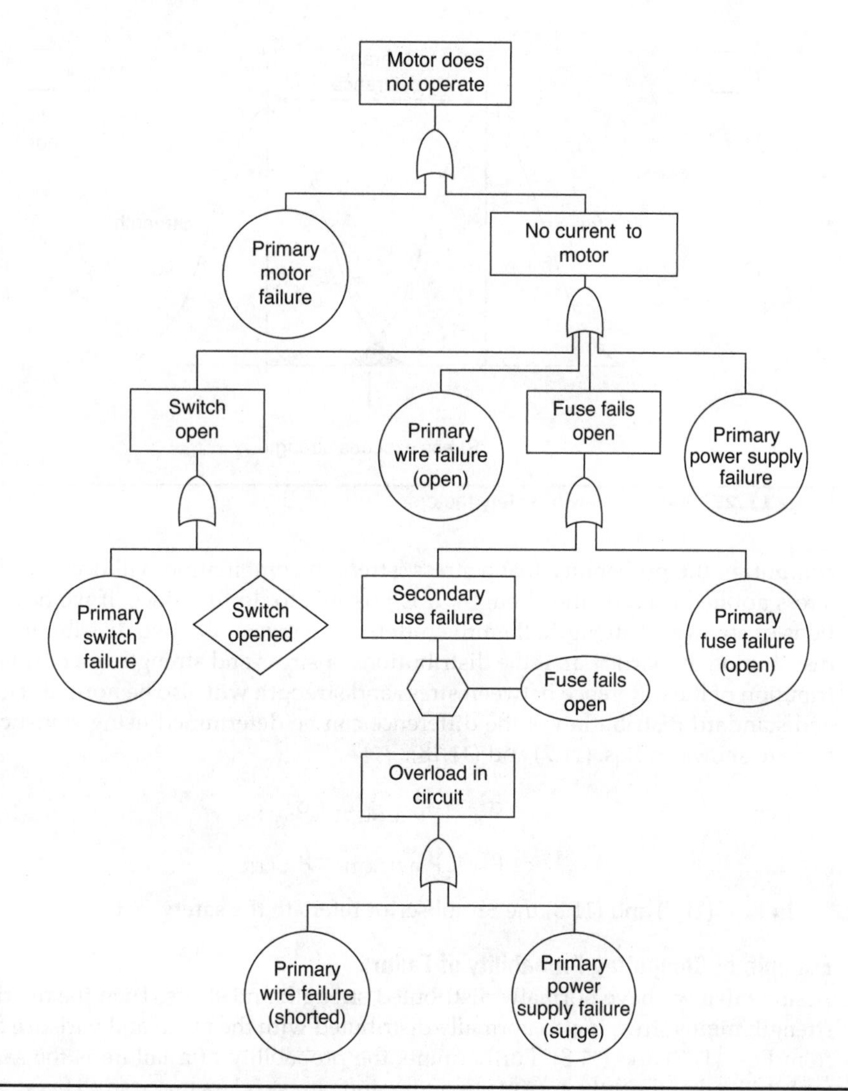

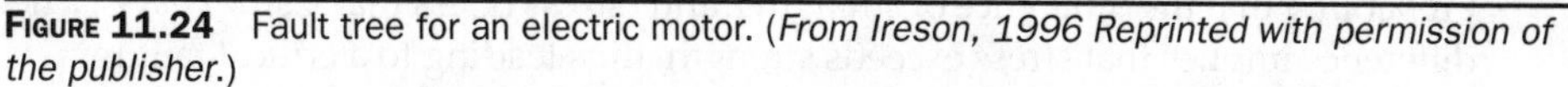

FIGURE 11.24 Fault tree for an electric motor. (*From Ireson, 1996 Reprinted with permission of the publisher.*)

The problem with this approach is quite simple: it doesn't account for variation in either stress or strength. The fact of the matter is that both strength and stress will vary over time, and unless this variation is dealt with explicitly we have no idea what the "safety factor" really is. The modern view is that a safety factor is the difference between an improbably high stress (the maximum expected stress, or "reliability boundary") and an improbably low strength (the minimum expected strength). Figure 11.25 illustrates the modern view of safety factors. The figure shows two *distributions*, one for stress and one for strength.

Since any strength or stress is theoretically possible, the traditional concept of a safety factor becomes vague at best and misleading at worst. To deal intelligently with this situation, we must consider *probabilities* instead of *possibilities*. This is done by

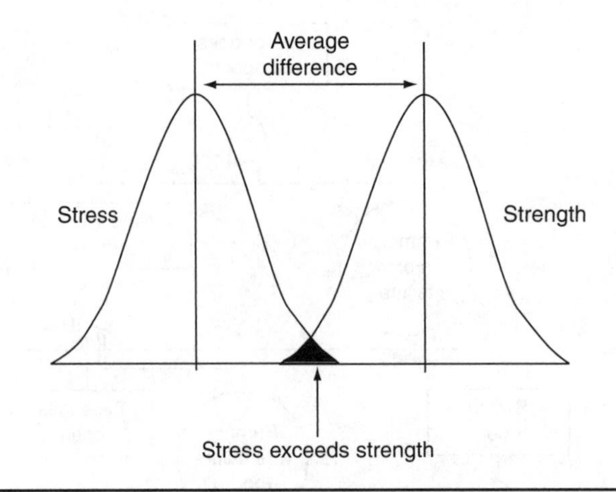

FIGURE 11.25 Modern view of safety factors.

computing the probability that a stress/strength combination will occur such that the stress applied exceeds the strength. It is possible to do this since, if we have distributions of stress and strength, then the difference between the two distributions is also a distribution. In particular, if the distributions of stress and strength are normal, the distribution of the difference between stress and strength will also be normal. The average and standard distribution of the difference can be determined using statistical theory, and are shown in Eqs. (11.7) and (11.18).

$$\sigma^2_{SF} = \sigma^2_{STRENGTH} + \sigma^2_{STRESS} \tag{11.7}$$

$$\mu_{SF} = \mu_{STRENGTH} - \mu_{STRESS} \tag{11.8}$$

In Eqs. (11.7) and (11.8) the SF subscript refers to the safety factor.

Example of Computing Probability of Failure

Assume that we have normally distributed strength and stress. Then the distribution of strength minus stress is also normally distributed with the mean and variance computed from Eqs. (11.7) and (11.8). Furthermore, the probability of a failure is the same as the probability that the difference of strength minus stress is less than zero. That is, a negative difference implies that stress exceeds strength, thus leading to a critical failure.

Assume that the strength of a steel rod is normally distributed with $\mu = 50{,}000^{\#}$ and $\sigma = 5{,}000^{\#}$. The steel rod is to be used as an undertruss on a conveyor system. The stress observed in the actual application was measured by strain gages and it was found to have a normal distribution with $\mu = 30{,}000^{\#}$ and $\sigma = 3{,}000^{\#}$. What is the expected reliability of this system?

Solution The mean variance and standard deviation of the difference is first computed using Eqs. (11.7) and (11.18), giving

$$\sigma^2_{DIFFERENCE} = \sigma^2_{STRENGTH} = \sigma^2_{STRESS} = 5{,}000^2 + 3{,}000^2 = 34{,}000{,}000$$

$$\sigma = \sqrt{34{,}000{,}000} = 5.831^{\#}$$

$$\mu_{DIFFERENCE} = \mu_{STRENGTH} - \mu_{STRESS} = 50{,}000^{\#} - 30{,}000^{\#} = 20{,}000$$

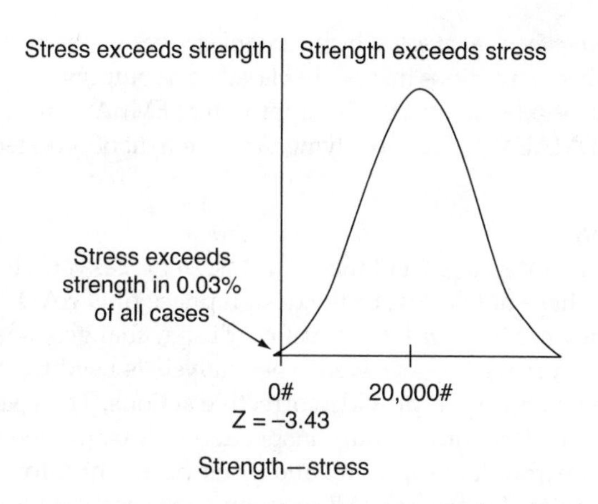

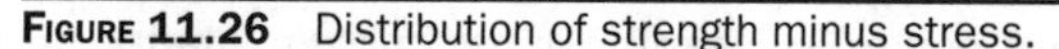

FIGURE 11.26 Distribution of strength minus stress.

We now compute Z which transforms this normal distribution to a standard normal distribution (see the section entitled "Normal Distribution" in Chap. 8).

$$z = \frac{0^{\#} - 20{,}000^{\#}}{5{,}831^{\#}} = -3.43$$

Using a normal table (Appendix 2), we now look up the probability associated with this Z value and find it is 0.0003. This is the probability of failure, about 3 chances in 10,000. The reliability is found by subtracting this probability from 1, giving 0.9997. Thus, the reliability of this system (and safety for this particular failure mode) is 99.97%. This example is summarized in Fig. 11.26.

Failure Mode and Effect Analysis

Failure mode and effect analysis, or FMEA, is an attempt to delineate all possible failures, their effect on the system, the likelihood of occurrence, and the probability that the failure will go undetected. FMEA provides an excellent basis for classification of characteristics, that is, for identifying CTQs and other critical variables. As with Pareto analysis, one objective of FMEA is to direct the available resources toward the most promising opportunities. An extremely unlikely failure, even a failure with serious consequences, may not be the best place to concentrate preventative efforts. FMEA can be combined with decision analysis methods such as AHP and QFD to help guide preventive action planning.

FMEA came into existence on the space program in the 1960s. Later it was incorporated into military standards, in particular Mil-Std-1629A.* There are two primary approaches for accomplishing an FMEA:

1. *The hardware approach* which lists individual hardware items and analyzes their possible failure modes. This FMEA approach is sometimes used in product DFSS projects.

*Mil-Std-1629A actually calls the approach FMECA, which stands for Failure mode, effect, and criticality analysis, but over time the "C" has been dropped from common usage. However, criticality analysis is still very much a part of FMEA.

2. *The functional approach* which recognizes that every item is designed to perform a number of functions that can be classified as outputs. The outputs are listed and their failure modes analyzed. This approach to FMEA is most common on both DMAIC and DMADV projects involving improvement of processes or complex systems.

FMEA Process

The FMEA is an integral part of the early design process and it should take place during the improve phase of DMAIC or the design phase of DMADV. FMEAs are living documents and they must be updated to reflect design changes, which makes them useful in the control or verify phases as well. The analysis is used to assess high risk items and the activities underway to provide corrective actions. The FMEA is also used to define special test considerations, quality inspection points, preventive maintenance actions, operational constraints, useful life, and other pertinent information and activities necessary to minimize failure risk. All recommended actions which result from the FMEA must be evaluated and formally dispositioned by appropriate implementation or documented rationale for no action. The following steps are used in performing an FMEA:

a. Define the system to be analyzed. Complete system definition includes identification of internal and interface functions, expected performance at all system levels, system restraints, and failure definitions. Functional narratives of the system should include descriptions of each goal in terms of functions which identify tasks to be performed for each goal and operational mode. Narratives should describe the environmental profiles, expected cycle times and equipment utilization, and the functions and outputs of each item.

b. Construct process maps which illustrate the operation, interrelationships, and interdependencies of functional entities.

c. Conduct SIPOC analysis for each subprocess in the system. All process and system interfaces should be indicated.

d. List the intended function of each step in the process or subprocess.

e. For each process step, identify all potential item and interface failure modes and define the effect on the immediate function or item, on the system, and on the mission to be performed for the customer.

f. Evaluate each failure mode in terms of the worst potential consequences which may result and assign a severity classification category, or SEV (see Table 11.23).

g. Determine the likelihood of occurrence of each failure mode and assign an occurrence risk category, or OCC (see Table 11.23).

h. Identify failure detection methods and assign a detectability risk category, or DET (see Table 11.23).

i. Calculate the risk priority number (RPN) for the current system. $RPN = SEV \times OCC \times DET$.

j. Determine compensating provisions for each failure mode.

k. Identify corrective design or other actions required to eliminate failure or control the risk. Assign responsibility and due dates for corrective actions.

l. Identify effects of corrective actions on other system attributes.

Rating	Severity (SEV)	Occurrence (OCC)	Detectability (DET)
Rating	How significant is this failure's effect to the customer?	How likely is the cause of this failure to occur?	How likely is it that the existing system will detect the cause, if it occurs? Note: p is the estimated probability of failure *not* being detected.
1	Minor. Customer won't notice the effect or will consider it insignificant.	Not likely.	Nearly certain to detect before reaching the customer. ($p \approx 0$)
2	Customer will notice the effect.	Documented low failure rate.	Extremely low probability of reaching the customer without detection. ($0 < p \leq 0.01$)
3	Customer will become irritated at reduced performance.	Undocumented low failure rate.	Low probability of reaching the customer without detection. ($0.01 < p \leq 0.05$)
4	Marginal. Customer dissatisfaction due to reduced performance.	Failures occur from time-to-time.	Likely to be detected before reaching the customer. ($0.05 < p \leq 0.20$)
5	Customer's productivity is reduced.	Documented moderate failure rate.	Might be detected before reaching the customer. ($0.20 < p \leq 0.50$)
6	Customer will complain. Repair or return likely. Increased internal costs (scrap, rework, etc.).	Undocumented moderate failure rate.	Unlikely to be detected before reaching the customer. ($0.50 < p \leq 0.70$)
7	Critical. Reduced customer loyalty. Internal operations adversely impacted.	Documented high failure rate.	Highly unlikely to detect before reaching the customer. ($0.70 < p \leq 0.90$)
8	Complete loss of customer goodwill. Internal operations disrupted.	Undocumented high failure rate.	Poor chance of detection. ($0.90 < p \leq 0.95$)
9	Customer or employee safety compromised. Regulatory compliance questionable.	Failures common.	Extremely poor chance of detection. ($0.95 < p \leq 0.99$)
10	Catastrophic. Customer or employee endangered without warning. Violation of law or regulation.	Failures nearly always occur.	Nearly certain that failure won't be detected. ($p \approx 1$)

TABLE 11.23 FMEA Severity, Likelihood, Detectability Rating Guidelines

m. Identify severity, occurrence, and detectability risks after the corrective action and calculate the "after" RPN.

n. Document the analysis and summarize the problems which could not be corrected and identify the special controls which are necessary to reduce failure risk.

RPNs are useful in setting priorities, with larger RPNs receiving greater attention than smaller RPNs. Some organizations have guidelines requiring action based on the absolute value of the RPN. For example, Boeing recommends that action be required if the RPN > 120.

A worksheet similar to worksheet 1 can be used to document and guide the team in conducting an FMEA. FMEA is incorporated into software packages, including some that perform QFD. There are numerous resources available on the web to assist you with FMEA, including spreadsheets, real-world examples of FMEA, and much more.*

Defining New Performance Standards Using Statistical Tolerancing

For our discussion of statistical tolerancing we will use the definitions of limits proposed by Juran and Gryna (1993), which are shown in Table 11.24.

In manufacturing it is common that parts interact with one another. A pin fits through a hole, an assembly consists of several parts bonded together, etc. Figure 11.27 illustrates one example of interacting parts.

Suppose that all three layers of this assembly were manufactured to the specifications indicated in Fig. 11.27. A logical specification on the overall stack height would be found by adding the nominal dimensions and tolerances for each layer; for example, 0.175 inch ± 0.0035 inch, giving limits of 0.1715 inch and 0.1785 inch. The lower specification is equivalent to a stack where all three layers are at their minimums, the upper specification is equivalent to a stack where all three layers are at their maximums, as shown in Table 11.25.

Adding part tolerances is the usual way of arriving at assembly tolerances, but it is usually too conservative, especially when manufacturing processes are both capable and in a state of statistical control. For example, assume that the probability of getting any particular layer below its low specification was 1 in 100 (which is a conservative estimate for a controlled, capable process). Then the probability that a particular stack would be below the lower limit of 0.1715 inch is

$$\frac{1}{100} \times \frac{1}{100} \times \frac{1}{100} = \frac{1}{1{,}000{,}000}$$

Similarly, the probability of getting a stack that is too thick would be 1 in a million. Thus, setting component and assembly tolerances by simple addition is extremely conservative, and often costly.

The statistical approach to tolerancing is based on the relationship between the variances of a number of independent causes and the variance of the dependent or overall result. The equation is:

$$\sigma_{\text{result}} = \sqrt{\sigma^2_{\text{cause A}} + \sigma^2_{\text{cause B}} + \sigma^2_{\text{cause C}} + \cdots} \tag{11.9}$$

*http://www.fmeainfocentre.com/

Name of Limit	Meaning
Tolerance	Set by the engineering design function to define the minimum and maximum values allowable for the product to work properly
Statistical tolerance	Calculated from process data to define the amount of variation that the process exhibits; these limits will contain a specified proportion of the total population
Prediction	Calculated from process data to define the limits which will contain all of k future observations
Confidence	Calculated from data to define an interval within which a population parameter lies
Control	Calculated from process data to define the limits of chance (random) variation around some central value

TABLE 11.24 Definitions of Limits

Minimum	Maximum
0.0240	0.0260
0.0995	0.1005
0.0480	0.0520
0.1715	0.1785

TABLE 11.25 Minimum and Maximum Multilayer Assemblies

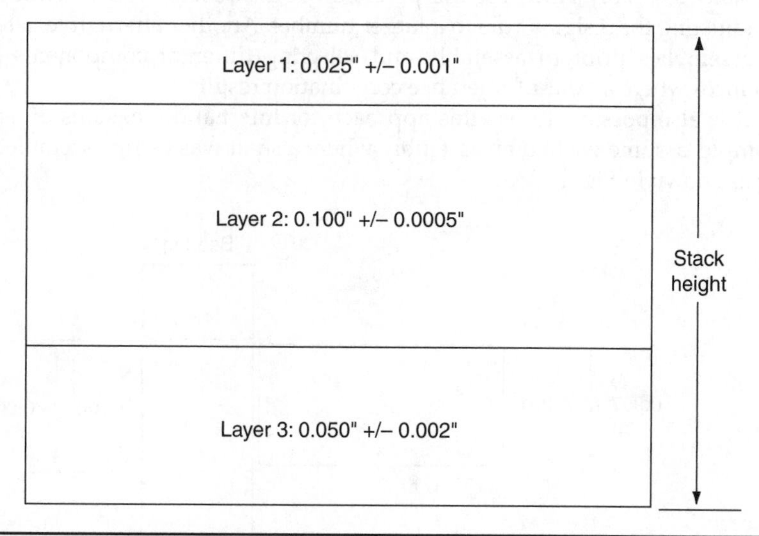

FIGURE 11.27 A multilevel circuit board assembly.

For our example, the equation is

$$\sigma_{\text{stack}} = \sqrt{\sigma^2_{\text{layer 1}} + \sigma^2_{\text{layer 2}} + \sigma^2_{\text{layer 3}}} \tag{11.10}$$

Of course, engineering tolerances are usually set without knowing which manufacturing process will be used to manufacture the part, so the actual variances are not known. However, a worst-case scenario would be where the process was just barely able to meet the engineering requirement. In Chap. 6 (Process capability topic in CTQ section) we learned that this situation occurs when the engineering tolerance is 6 standard deviations wide (±3 standard deviations). Thus, we can write Eq. (11.11) as

$$\frac{T}{3} = \sqrt{\left(\frac{T_A}{3}\right)^2 + \left(\frac{T_B}{3}\right)^2 + \left(\frac{T_C}{3}\right)^2} \tag{11.11}$$

or

$$T_{\text{stack}} = \sqrt{T^2_{\text{layer 1}} + T^2_{\text{layer 2}} + T^2_{\text{layer 3}}}$$

In other words, instead of simple addition of tolerances, the squares of the tolerances are added to determine the square of the tolerance for the overall result.

The result of the statistical approach is a dramatic *increase* in the allowable tolerances for the individual piece-parts. For our example, allowing each layer a tolerance of ±0.002 inch would result in the same stack tolerance of 0.0035 inch. This amounts to doubling the tolerance for layer 1 and quadrupling the tolerance for layer 3, without changing the tolerance for the overall stack assembly. There are many other combinations of layer tolerances that would yield the same stack assembly result, which allows a great deal of flexibility for considering such factors as process capability and costs.

The penalty associated with this approach is a slight probability of an out-of-tolerance assembly. However, this probability can be set to as small a number as needed by adjusting the 3 sigma rule to a larger number. Another alternative is to measure the subassemblies prior to assembly and selecting different components in those rare instances where an out-of-tolerance combination results.

It is also possible to use this approach for internal dimensions of assemblies. For example, assume we had an assembly where a shaft was being assembled with a bearing as shown in Fig. 11.28.

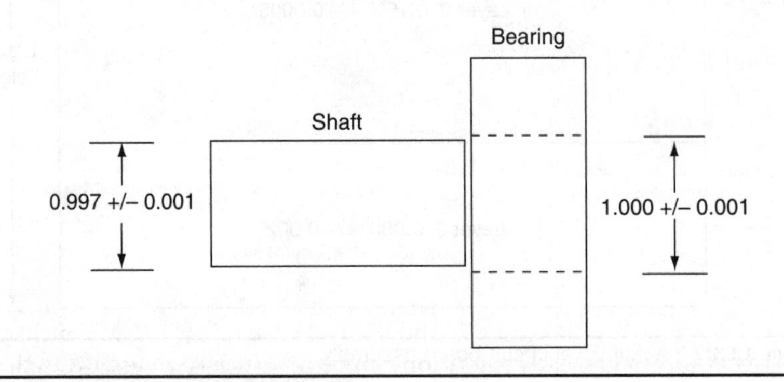

FIGURE 11.28 A bearing and shaft assembly.

The clearance between the bearing and the shaft can be computed as

$$\text{Clearance} = \text{bearing inside diameter} - \text{shaft outside diameter} \tag{11.12}$$

The minimum clearance will exist when the bearing inside diameter is at its smallest allowed and the shaft outside diameter is at its largest allowed. Thus,

$$\text{Minimum clearance} = 0.999 \text{ in} - 0.998 \text{ in} = 0.001 \text{ in}$$

The maximum clearance will exist when the bearing inside diameter is at its largest allowed and the shaft outside diameter is at its smallest allowed,

$$\text{Maximum clearance} = 1.001 \text{ in} - 0.996 \text{ in} = 0.005 \text{ in}$$

Thus, the assembly tolerance can be computed as

$$T_{\text{assembly}} = 0.005 \text{ in} - 0.001 \text{ in} = 0.004 \text{ in}$$

The statistical tolerancing approach is used here in the same way as it was used above. Namely,

$$\frac{T}{3} = \sqrt{\left(\frac{T_A}{3}\right)^2 + \left(\frac{T_B}{3}\right)^2} \tag{11.13}$$

or

$$T_{\text{assembly}} = \sqrt{T^2_{\text{bearing}} + T^2_{\text{shaft}}}$$

For our example we get

$$T_{\text{assembly}} = 0.004'' = \sqrt{T^2_{\text{bearing}} + T^2_{\text{shaft}}}$$

If we assume equal tolerances for the bearing and the shaft the tolerance for each becomes

$$(0.004)^2 = T^2_{\text{bearing}} + T^2_{\text{shaft}} = 2T^2$$

$$T = \sqrt{\frac{(0.004)^2}{2}} = \pm 0.0028$$

Which nearly triples the tolerance for each part.

Assumptions of Formula

The formula is based on several assumptions:

- The component dimensions are independent and the components are assembled randomly. This assumption is usually met in practice.
- Each component dimension should be approximately normally distributed.
- The actual average for each component is equal to the nominal value stated in the specification. For the multilayer circuit board assembly example, the averages for layers 1, 2, and 3 must be 0.025 inch, 0.100 inch, and 0.050 inch respectively. This condition can be met by applying SPC to the manufacturing processes.

Reasonable departures from these assumptions are acceptable. The author's experience suggests that few problems will appear as long as the subassembly manufacturing processes are kept in a state of statistical control.

Tolerance Intervals

We have found that confidence limits may be determined so that the interval between these limits will cover a population parameter with a specified confidence, that is, a specified proportion of the time. Sometimes it is desirable to obtain an interval which will cover a fixed portion of the population distribution with a specified confidence. These intervals are called *tolerance intervals*, and the end points of such intervals are called *tolerance limits*. For example, a manufacturer may wish to estimate what proportion of product will have dimensions that meet the engineering requirement. In Six Sigma, tolerance intervals are typically of the form $\bar{X} \pm Ks$, where K is determined, so that the interval will cover a proportion P of the population with confidence γ. Confidence limits for μ are also of the form $\bar{X} \pm Ks$. However, we determine K so that the confidence interval would cover the population mean μ a certain proportion of the time. It is obvious that the interval must be longer to cover a large portion of the distribution than to cover just the single value μ. Table 8 in the Appendix gives K for $P = 0.90$, 0.95, 0.99, 0.999 and $\gamma = 0.90$, 0.95, 0.99 and for many different sample sizes n.

Example of Calculating a Tolerance Interval

Assume that a sample of $n = 20$ from a stable process produced the following results: $\bar{X} = 20$, $s = 1.5$. We can estimate that the interval $\bar{X} \pm Ks = 20 \pm 3.615(1.5) = 20 \pm 5.4225$, the interval from 14.5775 to 25.4225 will contain 99% of the population with confidence 95%. The K values in the table assume normally distributed populations.

CHAPTER 12

Control/Verify Phase

The main objectives of the Control/Verify stage is to:

- Statistically validate that the new process or design meets the objectives and benefits sought through the project
- Develop and implement a control plan to institutionalize the new process or design
- Document lessons learned and project findings, as discussed in the "Tracking Six Sigma Project Results" section of Chap. 4

Validating the New Process or Product Design

Although the design was thoroughly evaluated, there is no substitute for doing. The team should ensure that their operating procedures, operator training, materials, information systems, etc. actually produce the predicted results. The pilot run consists of a small-scale, limited time run of the new design under the careful watch of the process expert. Metrics are collected and analyzed using SPC analysis to determine if the CTQ predictions are reasonably accurate under real-world conditions Actual customers are served by the new design and their reactions closely monitored. Of course, the results of the pilot are analyzed bearing in mind that proficiency will improve with practice. Still, unanticipated problems are nearly always discovered during pilot runs and they should not be overlooked.

Full-scale operations are to the pilot run as the pilot run is to a simulation. The handoff should be gradual, with redesign options open until enough time has passed to ensure that the new design is stable. Process owners are the primary decision maker when it comes to declaring the handoff complete. The transition should be planned as a subproject, with tasks, due dates, and responsibility assigned.

Business Process Control Planning

The project has finished successfully! Or has it? You've met the project's goals and the customer and sponsor have accepted the deliverables. But don't be too hasty to declare victory. The last battle is yet to be fought. The battle against creeping disorder, the battle against entropy. That battle to ensure that the gains you made are permanent.

Maintaining Gains

All organizations have systems designed to ensure stability and to protect against undesirable change. Often these systems also make it more difficult to make beneficial change; perhaps you encountered an example or two while pursuing your Six Sigma project! Still, once you've created an improved business system these "anti-change" systems can be your friend. Here are some suggested ways to protect your hard-won gains.

- Policy changes—Which corporate policies should be changed as a result of the project? Have some policies been rendered obsolete? Are new policies needed?
- New standards—Did the project bring the organization into compliance with a standard (e.g., ISO 9000, environmental standards, product safety standards)? If so, having the company adopt the standard might prevent backsliding. Are there any industry standards which, if adopted, would help maintain the benefits of the project? Customer standards? ANSI, SAE, JCAHO, NCQA, ASTM, ASQ or any other standard-making organization standards? Government standards? Don't forget that compliance with accepted standards is often an effective marketing tool; ask your marketing people if this is the case and, if so, get their help in adopting the standard.
- Modify procedures—Procedures describe the way things are supposed to be done. Since the project produced better (different) results, presumably some things are being done differently. Be sure these differences are incorporated into formal procedures.
- Modify quality appraisal and audit criteria—The quality control activity in an organization exists to ensure conformance to requirements. This will work for you by ensuring that the changes made to documentation will result in changes in the way the work is done. In many cases, where the Six Sigma project results in fundamental improvements to a process, the existing control schemes may provide an unnecessarily stringent inspection or control regimen that may be eased given the improved performance. A project that built sufficient stakeholder buy-in with documented analysis will pay dividends in this case.
- Update prices and contract bid models—The way product is priced for sale is directly related to profit, loss, and business success. Because of this, project improvements that are embedded in bid models and price models will be institutionalized by being indirectly integrated into an array of accounting and information systems.
- Change engineering drawings—Many Six Sigma projects create engineering change requests as part of their problem solution. For example, when a Six Sigma project evaluates process capability it is common to discover that the engineering requirements are excessively tight. Perhaps designers are using worst-case tolerancing instead of statistical tolerancing. The project team should ensure that these discoveries result in actual changes to engineering drawings.
- Change manufacturing planning—An organization's manufacturing plans describe in detail how product is to be processed and produced. Often the Six Sigma project team will discover better ways of doing things. If manufacturing plans are not changed the new and improved approach is likely to be lost due to personnel turnovers, etc. For those organizations that have no manufacturing

plans, the Six Sigma project team should develop them, at least for products and processes developed as part of the project. Note: this should not be considered scope creep or scope drift because it is directly related to the team's goals. However, it will be better still if the team can obtain a permanent policy change to make manufacturing planning a matter of policy (see "Policy changes" above).

- Revise accounting systems—Six Sigma projects take a value stream perspective of business systems, that is, a global approach. However, many accounting systems (such as activity based costing) look at local activities in isolation from their place in the overall scheme of things. If kept in place, these accounting systems produce perverse incentives that will eventually undo all of the good the team has done by breaking the integrated value delivery process into a series of competing fiefdoms. Consider changing to throughput accounting or other accounting systems better aligned with a process and systems perspective.
- Revise budgets—Improvements mean that more can be done with less. Budgets should be adjusted accordingly. However, the general rule of free markets should also be kept in mind: capital flows to the most efficient.
- Revise manpower forecasts—Toyota's Taiichi Ohno says that he isn't interested in labor savings, only in manpower savings. In other words, if as a result of a Six Sigma project the same number of units can be produced with fewer people, this should be reflected in staffing requirements. I hasten to point out, however, that research shows that Six Sigma and total quality firms *increase* employment at roughly triple the rate of non-Six Sigma firms. Greater efficiency, higher quality, and faster cycle times allow firms to create more value for customers, thus generating more sales. Investors, employees and other stakeholders benefit.
- Modify training—Personnel need to become familiar with the new way of doing things. Be sure all current employees are retrained, and new employees receive the proper indoctrination. Evaluate existing training materials and revise them as necessary.
- Change information systems—For example, Manufacturing Resource Planning MRP, inventory requirements, etc. Much of what occurs in the organization is not touched by humans. For example:
  - A purchase order might be issued automatically when inventories for a part reach a certain level. However, a Six Sigma project may have eliminated the need for safety stock.
  - MRP may generate a schedule based on cycle times rendered obsolete by improvements in cycle times.

When Six Sigma projects change the underlying relationships on which the automated information systems are based, programs should be modified to reflect this.

Tools and Techniques Useful for Control Planning

- Project planning—Many of the Six Sigma tools and techniques used during the Define, Measure, Analyze and Improve phases can also be used to develop a control plan. Perhaps most important is to keep in mind that control planning is a (sub) project. The deliverable is an effective and implemented control system.

The activities, responsibilities, durations and due dates necessary to produce the deliverable should be carefully listed. If the process changes are extensive, the control subproject may require another sponsor to take ownership of the control process after the team disbands and the main project sponsor accepts the new system. A detailed Business Process Change control plan should be prepared and kept up to date until the Black Belt, sponsor, and process owner are confident that the improvements are permanent.

- Brainstorming—The Six Sigma team should brainstorm to expand the list presented above with ideas from their own organization.
- Force-field diagram—A force-field diagram can be very useful at this point. Show the forces that will push to undo the changes, and create counterforces that will maintain them. The ideas obtained should be used to develop a process control plan that will ensure that the organization continues to enjoy the benefits of the Six Sigma project.
- Process decision program chart—The PDPC is a useful tool in developing a contingency plan.
- Failure mode and effect analysis—Using FMEA in the improve phase was discussed in detail in Chap. 11. Its output provides necessary input for the control plan.

Using SPC for Ongoing Control

Assuming that the organization's leadership has created an environment where open and honest communication can flourish, SPC implementation becomes a matter of (1) selecting processes for applying the SPC approach and (2) selecting variables within each process as part of a detailed process control plan.

Preparing the Process Control Plan

Process control plans should be prepared for each key process. The plans should be prepared by teams of people who understand the process. The team should begin by creating a flow chart of the process using the process elements determined in creating the house of quality (see the QFD discussion in Chap. 2). The flow chart will show how the process elements relate to one another and it will help in the selection of control points. It will also show the point of delivery to the customer, which is usually an important control point. Note that the customer may be an internal customer.

For any given process there are a number of different types of process elements. Some process elements are *internal* to the process, others *external*. The rotation speed of a drill is an internal process element, while the humidity in the building is external. Some process elements, while important, are easy to hold constant at a given value so that they do not change unless deliberate action is taken. We will call these *fixed* elements. Other process elements vary of their own accord and must be watched; we call these *variable* elements. The drill rotation speed can be set in advance, but the line voltage for the drill press may vary, which causes the drill speed to change in spite of its initial setting (a good example of how a correlation matrix might be useful). Figure 12.1 provides a planning guide based on the internal/external and fixed/variable classification scheme. Of course, other classification schemes may be more suitable on a given project and the analyst is encouraged to develop the approach

	Internal	External
	I	II
Fixed	• Setup approval • Periodic audits • Preventive maintenance	• Audit • Certification
	III	IV
Variable	• Control charts • Mistake-proof product • Mistake-proof process • Sort the output	• Supplier SPC • Receiving inspection • Supplier sorting • Mistake-proof product

FIGURE 12.1 Guide to selecting and controlling process variables.

that best serves his or her needs. For convenience, each class is identified with a Roman numeral; I = fixed–internal, II = fixed–external, III = variable–internal, and IV = variable–external.

In selecting the appropriate method of control for each process element, pay particular attention to those process elements which received high importance rankings in the house of quality analysis. In some cases an important process element is very expensive to control. When this happens, look at the QFD correlation matrix or the statistical correlation matrix for possible assistance. The process element may be correlated with other process elements that are less costly to control. Either correlation matrix will also help you to minimize the number of control charts. It is usually unnecessary to keep control charts on several variables that are correlated with one another. In these cases, it may be possible to select the process element that is least expensive (or most sensitive) to monitor as the control variable.

As Fig. 12.1 indicates, control charts are not always the best method of controlling a given process element. In fact, control charts are seldom the method of choice. When process elements are important we would prefer that they *not vary at all!* Only when this cannot be accomplished economically should the analyst resort to the use of control charts to monitor the element's variation. Control charts may be thought of as a control mechanism of last resort. Control charts are useful only when the element being monitored can be expected to exhibit measurable and "random-looking" variation when the process is properly controlled. A process element that always checks "10" if everything is okay is not a good candidate for control charting. Nor is one that checks "10" or "12," but never anything else. Ideally, the measurements being monitored with variables control charts will be capable of taking on any value, that is, the data will be continuous. Discrete measurement data can be used if it's not too discrete; indeed, all real-world data are somewhat discrete. As a rule of thumb, at least ten different values should appear in the data set and no one value should comprise more than 20% of the data set. When the measurement data become too discrete for SPC, monitor them with checksheets or simple time-ordered plots.

Of course, the above discussion applies to measurement data. Attribute control charts can be used to monitor process elements that are discrete counts.

Any process control plan must include instructions on the action to be taken if problems appear. This is particularly important where control charts are being used for process control. Unlike process control procedures such as audits or setup approvals, it is not always apparent just what is wrong when a control chart indicates a problem. The investigation of special causes of variation usually consists of a number of predetermined actions (such as checking the fixture or checking a cutting tool) followed by notifying someone if the items checked don't reveal the source of the problem. Also verify that the arithmetic was done correctly and that the point was plotted in the correct position on the control chart.

The reader may have noticed that Fig. 12.1 includes "sort the output" as part of the process control plan. Sorting the output implies that the process is not capable of meeting the customer's requirements, as determined by a process capability study and the application of Deming's all-or-none rules. However, even if sorting is taking place, SPC is still advisable. SPC will help ensure that things don't get any worse. SPC will also reveal improvements that may otherwise be overlooked. The improvements may result in a process that is good enough to eliminate the need for sorting.

Process Control Planning for Short and Small Runs

A starting place for understanding SPC for short and small runs is to define our terms. The question "what is a short run?" will be answered for our purposes as an environment that has a large number of jobs per operator in a production cycle, each job involving different product. A production cycle is typically a week or a month. *A small run* is a situation where only a very few products of the same type are to be produced. An extreme case of a small run is the one-of-a-kind product, such as the Hubble Space Telescope. Short runs need not be small runs; a can manufacturing line can produce over 100,000 cans in an hour or two. Likewise small runs are not necessarily short runs; the Hubble Space Telescope took over 15 years to get into orbit (and even longer to get into orbit and working properly)! However, it is possible to have runs that are both short and small. Programs such as Just-In-Time (JIT) inventory control are making this situation more common all of the time.

Process control for either small or short runs involves similar strategies. Both situations involve markedly different approaches than those used in the classical mass-production environment. Thus, this section will treat both the small run and the short run situations simultaneously. You should, however, select the SPC tool that best fits your particular situation.

Strategies for Short and Small Runs

Juran's famous trilogy separates quality activities into three distinct phases (Juran and Gryna, 1988):

- Planning
- Control
- Improvement

Figure 12.2 provides a graphic portrayal of the Juran trilogy.

When faced with small or short runs the emphasis should be placed in the planning phase. As much as possible needs to be done *before* any product is made, because it simply isn't possible to waste time or materials "learning from mistakes" made during production. It is also helpful to realize that the Juran trilogy is usually applied to *products*, while SPC applies to *processes*. It is quite possible that the element being monitored

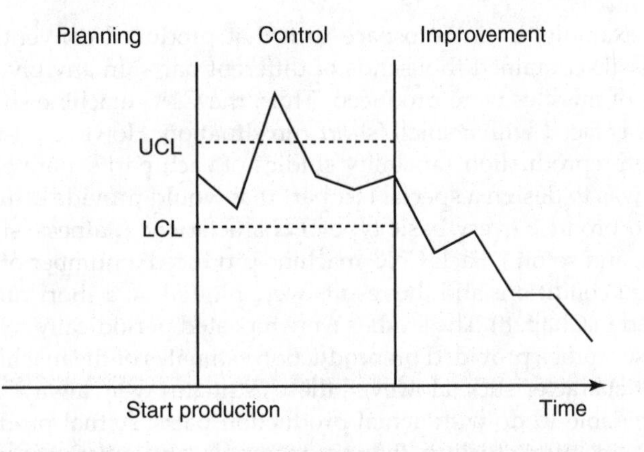

FIGURE 12.2 Juran's trilogy.

with SPC is a process element and not a product feature at all. In this case there really is no "short run," despite appearances to the contrary.

A common problem with application of SPC to short/small runs is that people fail to realize the limitations of SPC in this application. Even the use of SPC to *long production runs* will benefit from a greater emphasis on preproduction planning. In the best of all worlds, SPC will merely confirm that the correct process has been selected and controlled in such a way that it consistently produces well-designed parts at very close to the desired target values for every dimension.

Preparing the Short Run Process Control Plan

Plans for short runs require a great deal of up-front attention. The objective is to create a list of as many potential sources of variation as possible and to take action to deal with them *before* going into production. One of the first steps to be taken is to identify which processes may be used to produce a given part; this is called the "Approved Process List." Analogously, parts that can be produced by a given process should also be identified; this is called the "Approved Parts List." These determinations are made based on process capability studies (Pyzdek, 1992a). The approach described in this guide uses process capability indices, specifically C_{PK} (the number of standard deviations between the mean and the nearest specification limit). The use of this capability index depends on a number of assumptions, such as normality of the data etc.; Pyzdek (1992b) describes the proper use, and some common abuses, of capability indices.

Because short runs usually involve less than the recommended number of pieces the acceptability criteria are usually modified. When less than 50 observations are used to determine the capability I recommend that the capability indices be modified by using a $\pm 4\sigma$ minimum acceptable process width (instead of $\pm 3\sigma$) and a minimum acceptable C_{PK} of 1.5 (instead of 1.33). Don't bother making formal capability estimates until you have at least 20 observations. (You can see in Chap. 8 that these observations need not always be from 20 separate parts.)

When preparing for short runs it often happens that actual production parts are not available in sufficient quantity for process capability studies. One way of dealing with this situation is to study process elements separately and to then sum the variances from all of the known elements to obtain an estimate of the best overall variance a given process will be able to produce.

For example, in an aerospace firm that produced conventional guided missiles, each missile contained thousands of different parts. In any given month only a small number of missiles were produced. Thus, the CNC machine shop (and the rest of the plant) was faced with a small/short run situation. However, it was not possible to do separate preproduction capability studies of each part separately. The approach used instead was to design a special test part that would provide estimates of the machine's ability to produce every basic type of characteristic (flatness, straightness, angularity, location, and so on.). Each CNC machine produced a number of these test parts under controlled conditions and the results were plotted on a short run X R chart (these are described in Chap. 8). The studies were repeated periodically for each machine.

These studies provided preproduction estimates of the machine's ability to produce different characteristics. However, these estimates were always *better* than the process would be able to do with actual production parts. Actual production would involve different operators, tooling, fixtures, materials, and other common and special causes not evaluated by the *machine capability study*. Preliminary Approved Parts lists and Preliminary Approved Process lists were created from the capability analysis using the more stringent acceptability criteria described earlier (C_{PK} at least 1.5 based on a $\pm 4\sigma$ process spread). When production commenced the actual results of the production runs were used instead of the estimates based on special runs. Once sufficient data were available, the parts were removed from the preliminary lists and placed on the appropriate permanent lists.

When creating Approved Parts and Approved Process lists always use the most stringent product requirements to determine *the process requirement*. For example, if a process will be used to drill holes in 100 different parts with hole location tolerances ranging from 0.001 inch to 0.030 inch, the process requirement is 0.001 inch. The process capability estimate is based on its ability to hold the 0.001 inch tolerance.

The approach used is summarized as follows:

1. Get the process into statistical control.
2. Set the control limits *without regard to the requirement.*
3. Based on the calculated process capability, determine if the most stringent product requirement can be met.

Process Audits

The requirements for all processes should be documented. A process audit checklist should be prepared and used to determine the condition of the process prior to production. The audit can be performed by the operator himself, but the results should be documented. The audit should cover known or suspected sources of variation. These include such things as the production plan, condition of fixtures, gage calibration, the resolution of the gaging being used, obvious problems with materials or equipment, operator changes, and so on.

SPC can be used to monitor the results of the process audits over time. For example, an audit score can be computed and tracked using an individuals control chart.

Selecting Process Control Elements

Many short run SPC programs bog down because the number of control charts being used grows like Topsy. Before anyone knows what is happening they find the walls plastered with charts that few understand and no one uses. The operators and inspectors wind up spending more time filling out paperwork than they spend on true value-added work. Eventually the entire SPC program collapses under its own weight.

One reason for this is that people tend to focus their attention on the *product* rather than on the *process*. Control elements are erroneously selected because they are functionally important. A great fear is that an important product feature will be produced out of specification and that it will slip by unnoticed. This is a misunderstanding of the purpose of SPC, which is to provide a means of *process* control; SPC is not intended to be a substitute for inspection or testing. The guiding rule of selecting control items for SPC is:

SPC control items should be selected to provide a maximum amount of information regarding the state of the process at a minimum cost.

Fortunately most process elements are correlated with one another. Because of this one process element may provide information not only about itself, but about several others as well. This means that a small number of process control elements will often explain a large portion of the process variance.

Although sophisticated statistical methods exist to help determine which groups of process elements explain the most variance, common sense and knowledge of the process can often do as well, if not better. The key is to think about the process carefully. What are the "generic process elements" that affect all parts? How do the process elements combine to affect the product? Do several process elements affect a single product feature? Do changes in one process element automatically cause changes in some other process elements? What process elements or product features are most sensitive to unplanned changes?

Example One

The CNC machines mentioned earlier were extremely complex. A typical machine had dozens of different tools and produced hundreds of different parts with thousands of characteristics. However, the SPC team reasoned that the machines themselves involved only a small number of "generic operations": select a tool, position the tool, remove metal, and so on. Further study revealed that nearly all of the problems encountered after the initial setup involved only the ability of the machine to position the tool precisely. A control plan was created that called for monitoring no more than one variable for each axis of movement. The features selected were those farthest from the machine's "home position" and involving the most difficult to control operations. Often a single feature provided control of more than one axis of movement, for example, the location of a single hole provides information on the location of the tool in both the X and Y directions.

As a result of this system no part had more than four features monitored with control charts, even though many parts had thousands of features. Subsequent sophisticated multivariate evaluation of the accumulated data by a statistician revealed that the choices made by the team explained over 90% of the process variance.

Example Two

A wave solder machine was used to solder printed circuit boards for a manufacturer of electronic test equipment. After several months of applying SPC the SPC team evaluated the data and decided that they needed only a single measure of product quality for SPC purposes: defects per 1,000 solder joints. A single control chart was used for dozens of different circuit boards. They also determined that most of the process variables being checked could be eliminated. The only process variables monitored in the future would be flux density, solder chemistry (provided by the vendor), solder temperature, and final rinse contamination. Historic data showed that one of these variables was nearly always out of control when process problems were encountered. Other variables were monitored with periodic audits using checksheets, but they were not charted.

Notice that in both of these examples all of the variables being monitored were related to the *process*, even though some of them were product features. The terms "short run" and "small run" refer to the product variables only; the process is in continuous operation so its run size and duration is neither small nor short.

The Single Part Process

The ultimate small run is the single part. A great deal can be learned by studying single pieces, even if your situation involves more than one part.

The application of SPC to single pieces may seem incongruous. Yet when we consider that the "P" in SPC stands for *process* and not product, perhaps it is possible after all. Even the company producing one-of-a-kind product usually does so with the same equipment, employees, facilities, etc. In other words, they use the same *process* to produce different *products*. Also, they usually produce products that are similar, even though not identical. This is also to be expected. It would be odd indeed to find a company fabricating microchips one day and baking bread the next. The processes are too dissimilar. The company assets are, at least to a degree, product-specific.

This discussion implies that the key to controlling the quality of single parts is to concentrate on the process elements rather than on the product features. This is the same rule we applied earlier to larger runs. In fact, it's a good rule to apply to all SPC applications, regardless of the number of parts being produced!

Consider a company manufacturing communications satellites. The company produces a satellite every year or two. The design and complexity of each satellite is quite different than any other. How can SPC be applied at this company?

A close look at a satellite will reveal immense complexity. The satellite will have thousands of terminals, silicon solar cells, solder joints, fasteners, and so on. Hundreds, even thousands of people are involved in the design, fabrication, testing, and assembly. In other words, there are *processes* that involve massive amounts of repetition. The processes include engineering (errors per engineering drawing); terminal manufacture (size, defect rates); solar cell manufacture (yields, electrical properties); soldering (defects per 1,000 joints, strength); fastener installation quality (torque) and so on.

Another example of a single-piece run is software development. The "part" in this case is the working copy of the software delivered to the customer. Only a singe unit of product is involved. How can we use SPC here?

Again, the answer comes when we direct our attention to the underlying process. Any marketable software product will consist of thousands, perhaps millions of bytes of finished machine code. This code will be compiled from thousands of lines of source code. The source code will be arranged in modules; the modules will contain procedures; the procedures will contain functions; and so on. Computer science has developed a number of ways of measuring the quality of computer code. The resulting numbers, called computer metrics, can be analyzed using SPC tools just like any other numbers.

The processes that produced the code can thus be measured, controlled and improved. If the process is in statistical control, the process elements, such as programmer selection and training, coding style, planning, procedures, etc. must be examined. If the process is not in statistical control, the special cause of the problem must be identified.

As discussed earlier, although the single part process is a small run, it isn't necessarily a short run. By examining the process rather than the part, improvement possibilities will begin to suggest themselves. The key is to find the process, to define its elements so they may be measured, controlled, and improved.

Other Elements of the Process Control Plan

In addition to the selection of process control elements, the PCP should also provide information on the method of inspection, dates and results of measurement error studies, dates and results of process capability studies, subgroup sizes and methods of selecting subgroups, sampling frequency, required operator certifications, pre-production checklists, notes and suggestions regarding previous problems, etc. In short, the PCP provides a complete, detailed road-map that describes how process integrity will be measured and maintained. By preparing a PCP the *inputs* to the process are controlled, thus assuring that the *outputs* from the process will be consistently acceptable.

Pre-Control

The Pre-Control method was originally developed by Dorian Shainin in the 1950s. According to Shainin, Pre-Control is a simple algorithm for controlling a process based on the tolerances. It assumes the process is producing product with a measurable and adjustable quality characteristic which varies according to some distribution. It makes no assumptions concerning the actual shape and stability of the distribution. Cautionary zones are designated just inside each tolerance extreme. A new process is qualified by taking consecutive samples of individual measurements until five in a row fall within the central zone before two in a row fall into the cautionary zones. To simplify the application, Pre-Control charts are often color-coded. On such charts the central zone is colored green, the cautionary zones yellow, and the zone outside of the tolerance red. Pre-Control is not equivalent to SPC. SPC is designed to identify special causes of variation; Pre-Control starts with a process that is known to be capable of meeting the tolerance and ensures that it does so. SPC and process capability analysis should always be used before Pre-Control is applied.*

Once the process is qualified, it is monitored by taking periodic samples consisting of two individuals each (called the A, B pair). Action is taken only if both A and B are in the cautionary zone. Processes must be requalified after any action is taken.

Setting up Pre-Control

Figure 12.3 illustrates the Pre-Control zones for a two-sided tolerance (i.e., a tolerance with both a lower specification limit and an upper specification limit).

Figure 12.4 illustrates the Pre-Control zones for a one-sided tolerance (i.e., a tolerance with only a lower specification limit or only an upper specification limit). Examples of this situation are flatness, concentricity, runout, and other total indicator reading type features.

*The reader should keep in mind that Pre-Control should not be considered a *replacement* for SPC.

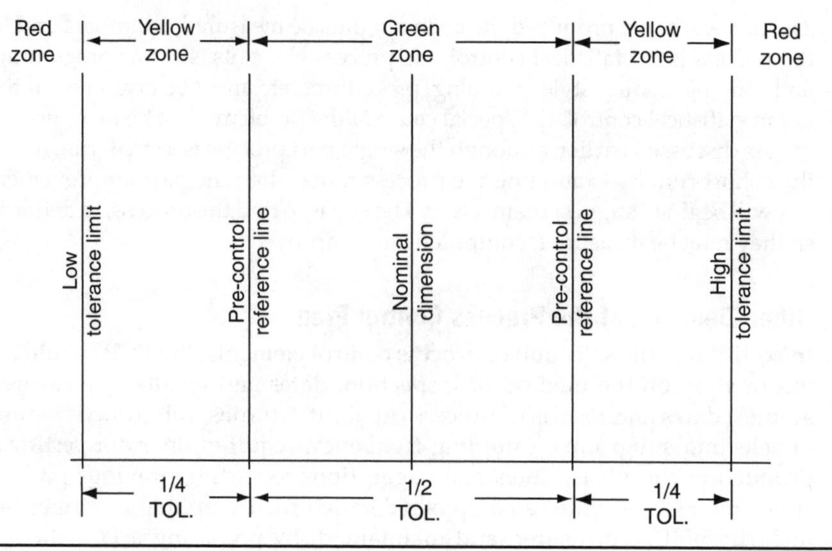

FIGURE 12.3 Pre-Control zones (two-sided tolerance).

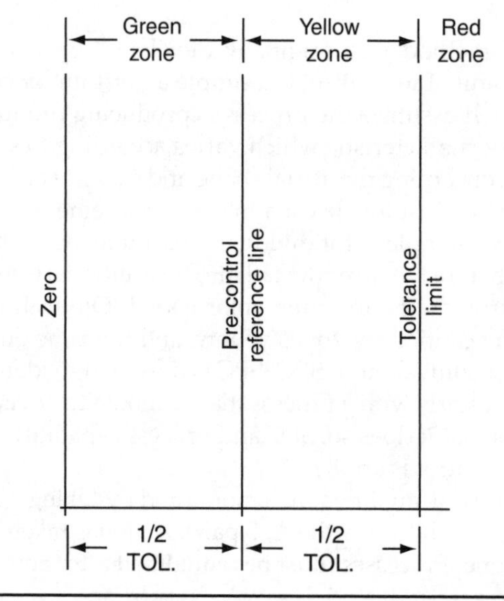

FIGURE 12.4 Pre-Control zones (one-sided tolerance).

Figure 12.5 illustrates the Pre-Control zones for characteristics with minimum or maximum specification limits. Examples of this situation are tensile strength, contamination levels, etc. In this situation place one reference line a quarter of the way from the tolerance limit toward the best sample produced during past operations.

Using Pre-Control

The first step is *setup qualification*. To begin, measure every piece produced until you obtain five greens in a row. If one yellow is encountered, restart the count. If two yellows

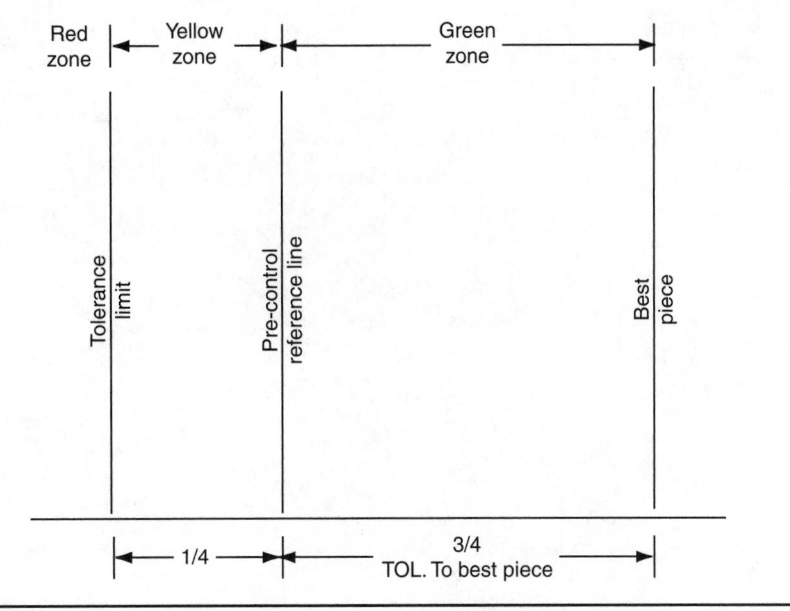

Figure 12.5 PRE-Control zones (minimum/maximum specifications).

in a row or any reds are encountered, adjust the process and restart the count. This step replaces first-piece inspection.

After setup qualification you will enter the *run* phase. Measure two consecutive pieces periodically (the A, B pair). If both are yellow on the same side, adjust. If yellow on opposite sides, call for help to reduce the variability of the process. If either is red, adjust. In the case of two yellows, the adjustment must be made immediately to prevent nonconforming work. In the case of red, stop; nonconforming work is already being produced. Segregate all nonconforming product according to established procedures.

Shainin and Shainin (1988) recommend adjusting the inspection frequency such that six A, B pairs are measured on average between each process adjustment. A simple formula for this is shown in Eq. (12.1).

$$\text{Minutes between measurements} = \text{hours between adjustments} \times 10 \qquad (12.1)$$

APPENDIX 1

Glossary of Basic Statistical Terms*

Acceptable quality level—The maximum percentage or proportion of variant units in a lot or batch that, for the purposes of acceptance sampling, can be considered satisfactory as a process average.

Analysis of Variance (ANOVA)—A technique which subdivides the total variation of a set of data into meaningful component parts associated with specific sources of variation for the purpose of testing some hypothesis on the parameters of the model or estimating variance components.

Assignable cause—A factor which contributes to variation and which is feasible to detect and identify.

Average Outgoing Quality (AOQ)—The expected quality of outgoing product following the use of an acceptance sampling plan for a given value of incoming product quality.

Average Outgoing Quality Limit (AOQL)—For a given acceptance sampling plan, the maximum AOQ over all possible levels of incoming quality.

Chance causes—Factors, generally numerous and individually of relatively small importance, which contribute to variation, but which are not feasible to detect or identify.

Coefficient of determination—A measure of the part of the variance for one variable that can be explained by its linear relationship with a second variable. Designated by ρ^2 or r^2.

Coefficient of multiple correlation—A number between 0 and 1 that indicates the degree of the combined linear relationship of several predictor variables, $X_1, X_2, \ldots, X_p$ to the response variable Y. It is the simple correlation coefficient between predicted and observed values of the response variable.

Coefficient of variation—A measure of relative dispersion that is the standard deviation divided by the mean and multiplied by 100 to give a percentage value. This measure cannot be used when the data take both negative and positive values or when it has been coded in such a way that the value $X = 0$ does not coincide with the origin.

Confidence limits—The end points of the interval about the sample statistic that is believed, with a specified confidence coefficient, to include the population parameter.

Consumer's risk (β)—For a given sampling plan, the probability of acceptance of a lot, the quality of which has a designated numerical value representing a level which it is seldom desired to accept. Usually the designated value will be the Limiting Quality Level (LQL).

Correlation coefficient—A number between –1 and 1 that indicates the degree of linear relationship between two sets of numbers:

$$r_{xy} = \frac{s_{xy}}{s_x s_y} = \frac{n\sum XY - \sum X \sum Y}{\sqrt{\left[n\sum X^2 - \left(\sum X\right)^2\right]\left[n\sum Y^2 - \left(\sum Y\right)^2\right]}}$$

Defect—A departure of a quality characteristic from its intended level or state that occurs with a severity sufficient to cause an associated product or service not to satisfy intended normal, or reasonably foreseeable, usage requirements.

Defective—A unit of product or service containing at least one defect, or having several imperfections that in combination cause the unit not to satisfy intended normal, or reasonably foreseeable, usage requirements. The word *defective* is appropriate for use when a unit of product or service is evaluated in terms of usage (as contrasted to conformance to specifications).

Double sampling—Sampling inspection in which the inspection of the first sample of size n_1, leads to a decision to accept a lot, not to accept it, or to take a second sample of size n_2, and the inspection of the second sample then leads to a decision to accept or not to accept the lot.

Experiment design—The arrangement in which an experimental program is to be conducted, and the selection of the versions (levels) of one or more factors or factor combinations to be included in the experiment.

Factor—An assignable cause which may affect the responses (test results) and of which different versions (levels) are included in the experiment.

Factorial experiments—Experiments in which all possible treatment combinations formed from two or more factors, each being studied at two or more versions (levels), are examined so that interactions (differential effects) as well as main effects can be estimated.

Frequency distribution—A set of all the various values that individual observations may have and the frequency of their occurrence in the sample or population.

Histogram—A plot of the frequency distribution in the form of rectangles whose bases are equal to the cell interval and whose areas are proportional to the frequencies.

Hypothesis, alternative—The hypothesis that is accepted if the null hypothesis is disproved. The choice of alternative hypothesis will determine whether "one-tail" or "two-tail" tests are appropriate.

Hypothesis, null—The hypothesis tested in tests of significance is that there is no difference (null) between the population of the sample and specified population (or between the populations associated with each sample). The null hypothesis can never be proved true. It can, however, be shown, with specified risks of error, to be untrue; that is, a difference can be shown to exist between the populations. If it is not disproved, one usually acts on the assumption that there is no adequate reason to doubt that it is true. (It may be that there is insufficient power to prove the existence of a difference rather than that there is no difference; that is, the sample size may be too small. By specifying the minimum difference that one wants to detect and β, the risk of failing to detect a difference of this size, the actual sample size required, however, can be determined.)

In-control process—A process in which the statistical measure(s) being evaluated are in a "state of statistical control."

Kurtosis—A measure of the shape of a distribution. A positive value indicates that the distribution has longer tails than the normal distribution (platykurtosis); while a negative value indicates that the distribution has shorter tails (leptokurtosis). For the normal distribution, the kurtosis is 0.

Mean, standard error of—The standard deviation of the average of a sample of size *n*.

$$s_{\bar{x}} = \frac{s_x}{\sqrt{n}}$$

Mean—A measure of the location of a distribution. The centroid.

Median—The middle measurement when an odd number of units are arranged in order of size; for an ordered set $X_1, X_2, \ldots, X_{2k-1}$

$$\text{Median} = X_k$$

When an even number are so arranged, the median is the average of the two middle units; for an ordered set $X_1, X_2, \ldots, X_{2k}$

$$\text{Median} = \frac{X_k + X_{k+1}}{2}$$

Mode—The most frequent value of the variable.

Multiple sampling—Sampling inspection in which, after each sample is inspected, the decision is made to accept a lot, not to accept, it or to take another sample to reach the decision. There may be a prescribed maximum number of samples, after which a decision to accept or not to accept must be reached.

Operating Characteristics curve (OC curve)—

1. For isolated or unique lots or a lot from an isolated sequence: a curve showing, for a given sampling plan, the probability of accepting a lot as a function of the lot quality. (Type A)
2. For a continuous stream of lots: a curve showing, for a given sampling plan, the probability of accepting a lot as a function of the process average. (Type B)
3. For continuous sampling plans: a curve showing the proportion of submitted product over the long run accepted during the sampling phases of the plan as a function of the product quality.
4. For special plans: a curve showing, for a given sampling plan, the probability of continuing to permit the process to continue without adjustment as a function of the process quality.

Parameter—A constant or coefficient that describes some characteristic of a population (e.g., standard deviation, average, regression coefficient).

Population—The totality of items or units of material under consideration.

NOTE: The items may be units or measurements, and the population may be real or conceptual. Thus *population* may refer to all the items actually produced in a given day or all that might be produced if the process were to continue *in-control*.

Power curve—The curve showing the relation between the probability (1 – β) of rejecting the hypothesis that a sample belongs to a given population with a given characteristic(s) and the actual population value of that characteristic(s). NOTE: if β is used instead of (1 – β), the curve is called an operating characteristic curve (OC curve) (used mainly in sampling plans for quality control).

Process capability—The limits within which a tool or process operate based upon minimum variability as governed by the prevailing circumstances.
NOTE: The phrase "by the prevailing circumstances" indicates that the definition of inherent variability of a process involving only one operator, one source of raw material, etc., differs from one involving multiple operators, and many sources of raw material, etc. If the measure of inherent variability is made within very restricted circumstances, it is necessary to add components for frequently occurring assignable sources of variation that cannot economically be eliminated.

Producer's risk (α)—For a given sampling plan, the probability of not accepting a lot the quality of which has a designated numerical value representing a level which it is generally desired to accept. Usually the designated value will be the **Acceptable Quality Level (AQL).**

Quality—The totality of features and characteristics of a product or service that bear on its ability to satisfy given needs.

Quality assurance—All those planned or systematic actions necessary to provide adequate confidence that a product or service will satisfy given needs.

Quality control—The operational techniques and the activities which sustain a quality of product or service that will satisfy given needs; also the use of such techniques and activities.

Random sampling—The process of selecting units for a sample of size *n* in such a manner that all combinations of *n* units under consideration have an equal or ascertainable chance of being selected as the sample.

***R* (range)**—A measure of dispersion which is the difference between the largest observed value and the smallest observed value in a given sample. While the range is a measure of dispersion in its own right, it is sometimes used to estimate the population standard deviation, but is a biased estimator unless multiplied by the factor $(1/d_2)$ appropriate to the sample size.

Replication—The repetition of the set of all the treatment combinations to be compared in an experiment. Each of the repetitions is called a *replicate*.

Sample—A group of units, portion of material, or observations taken from a larger collection of units, quantity of material, or observations that serves to provide information that may be used as a basis for making a decision concerning the larger quantity.

Single sampling—Sampling inspection in which the decision to accept or not to accept a lot is based on the inspection of a single sample of size *n*.

Skewness—A measure of the symmetry of a distribution. A positive value indicates that the distribution has a greater tendency to tail to the right (positively skewed or skewed to the right), and a negative value indicates a greater tendency of the distribution to tail to the left (negatively skewed or skewed to the left). Skewness is 0 for a normal distribution.

Standard deviation—

1. σ—population standard deviation. A measure of variability (dispersion) of observations that is the positive square root of the population variance.

2. *s*—sample standard deviation. A measure of variability (dispersion) that is the positive square root of the sample variance.

$$\sqrt{\frac{1}{n}\sum (X_i - \bar{X})^2}$$

Statistic—A quantity calculated from a sample of observations, most often to form an estimate of some population parameter.

Type I error (acceptance control sense)—The incorrect decision that a process is unacceptable when, in fact, perfect information would reveal that it is located within the "zone of acceptable processes."

Type II error (acceptance control sense)—The incorrect decision that a process is acceptable when, in fact, perfect information would reveal that it is located within the "zone of rejectable processes."

Variance—

1. σ^2—population variance. A measure of variability (dispersion) of observations based upon the mean of the squared deviation from the arithmetic mean.
2. s^2—sample variance. A measure of variability (dispersion) of observations in a sample based upon the squared deviations from the arithmetic average divided by the degrees of freedom.

APPENDIX 2

Area Under the Standard Normal Curve

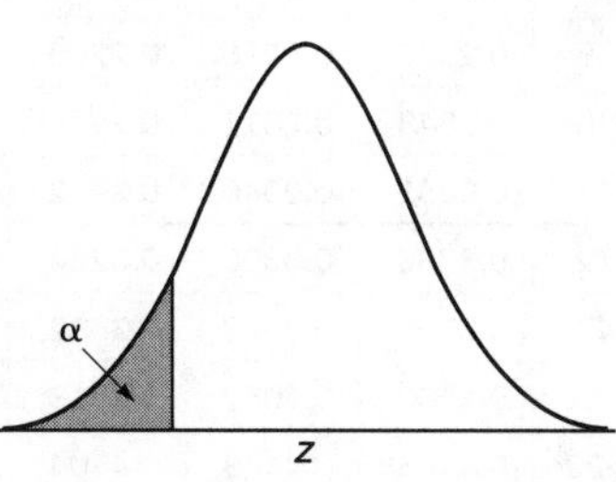

z	0.00	0.01	0.02	0.03	0.04	0.05	0.06	0.07	0.08	0.09
–3.4	0.0003	0.0003	0.0003	0.0003	0.0003	0.0003	0.0003	0.0003	0.0003	0.0002
–3.3	0.0005	0.0005	0.0005	0.0004	0.0004	0.0004	0.0004	0.0004	0.0004	0.0003
–3.2	0.0007	0.0007	0.0006	0.0006	0.0006	0.0006	0.0006	0.0005	0.0005	0.0005
–3.1	0.0010	0.0009	0.0009	0.0009	0.0008	0.0008	0.0008	0.0008	0.0007	0.0007
–3.0	0.0013	0.0013	0.0013	0.0012	0.0012	0.0011	0.0011	0.0011	0.0010	0.0010
–2.9	0.0019	0.0018	0.0018	0.0017	0.0016	0.0016	0.0015	0.0015	0.0014	0.0014
–2.8	0.0026	0.0025	0.0024	0.0023	0.0023	0.0022	0.0021	0.0021	0.0020	0.0019
–2.7	0.0035	0.0034	0.0033	0.0032	0.0031	0.0030	0.0029	0.0028	0.0027	0.0026
–2.6	0.0047	0.0045	0.0044	0.0043	0.0041	0.0040	0.0039	0.0038	0.0037	0.0036
–2.5	0.0062	0.0060	0.0059	0.0057	0.0055	0.0054	0.0052	0.0051	0.0049	0.0048
–2.4	0.0082	0.0080	0.0078	0.0075	0.0073	0.0071	0.0069	0.0068	0.0066	0.0064
–2.3	0.0107	0.0104	0.0102	0.0099	0.0096	0.0094	0.0091	0.0089	0.0087	0.0084
–2.2	0.0139	0.0136	0.0132	0.0129	0.0125	0.0122	0.0119	0.0116	0.0113	0.0110
–2.1	0.0179	0.0174	0.0170	0.0166	0.0162	0.0158	0.0154	0.0150	0.0146	0.0143
–2.0	0.0228	0.0222	0.0217	0.0212	0.0207	0.0202	0.0197	0.0192	0.0188	0.0183
–1.9	0.0287	0.0281	0.0274	0.0268	0.0262	0.0256	0.0250	0.0244	0.0239	0.0233

z	0.00	0.01	0.02	0.03	0.04	0.05	0.06	0.07	0.08	0.09
–1.8	0.0359	0.0351	0.0344	0.0336	0.0329	0.0322	0.0314	0.0307	0.0301	0.0294
–1.7	0.0446	0.0436	0.0427	0.0418	0.0409	0.0401	0.0392	0.0384	0.0375	0.0367
–1.6	0.0548	0.0537	0.0526	0.0516	0.0505	0.0495	0.0485	0.0475	0.0465	0.0455
–1.5	0.0668	0.0655	0.0643	0.0630	0.0618	0.0606	0.0594	0.0582	0.0571	0.0559
–1.4	0.0808	0.0793	0.0778	0.0764	0.0749	0.0735	0.0721	0.0708	0.0694	0.0681
–1.3	0.0968	0.0951	0.0934	0.0918	0.0901	0.0885	0.0869	0.0853	0.0838	0.0823
–1.2	0.1151	0.1131	0.1112	0.1093	0.1075	0.1056	0.1038	0.1020	0.1003	0.0985
–1.1	0.1357	0.1335	0.1314	0.1292	0.1271	0.1251	0.1230	0.1210	0.1190	0.1170
–1.0	0.1587	0.1562	0.1539	0.1515	0.1492	0.1469	0.1446	0.1423	0.1401	0.1379
–0.9	0.1841	0.1814	0.1788	0.1762	0.1736	0.1711	0.1685	0.1660	0.1635	0.1611
–0.8	0.2119	0.2090	0.2061	0.2033	0.2005	0.1977	0.1949	0.1922	0.1894	0.1867
–0.7	0.2420	0.2389	0.2358	0.2327	0.2296	0.2266	0.2236	0.2206	0.2177	0.2148
–0.6	0.2743	0.2709	0.2676	0.2643	0.2611	0.2578	0.2546	0.2514	0.2483	0.2451
–0.5	0.3085	0.3050	0.3015	0.2981	0.2946	0.2912	0.2877	0.2843	0.2810	0.2776
–0.4	0.3446	0.3409	0.3372	0.3336	0.3300	0.3264	0.3228	0.3192	0.3156	0.3121
–0.3	0.3821	0.3783	0.3745	0.3707	0.3669	0.3632	0.3594	0.3557	0.3520	0.3483
–0.2	0.4207	0.4168	0.4129	0.4090	0.4052	0.4013	0.3974	0.3936	0.3897	0.3859
–0.1	0.4602	0.4562	0. 4522	0.4483	0.4443	0.4404	0.4364	0.4325	0.4286	0.4247
–0.0	0.5000	0.4960	0.4920	0.4880	0.4840	0.4801	0.4761	0.4721	0.4681	0.4641
0.0	0.5000	0.5040	0.5080	0.5120	0.5160	0.5199	0.5239	0.5279	0.5319	0.5359
0.1	0.5398	0.5438	0.5478	0.5517	0.5557	0.5596	0.5636	0.5675	0.5714	0.5753
0.2	0.5793	0.5832	0.5871	0.5910	0.5948	0.5987	0.6026	0.6064	0.6103	0.6141
0.3	0.6179	0.6217	0.6255	0.6293	0.6331	0.6368	0.6406	0.6443	0.6480	0.6517
0.4	0.6554	0.6591	0.6628	0.6664	0.6700	0.6736	0.6772	0.6808	0.6844	0.6879
0.5	0.6915	0.6950	0.6985	0.7019	0.7054	0.7088	0.7123	0.7157	0.7190	0.7224
0.6	0.7257	0.7291	0.7324	0.7357	0.7389	0.7422	0.7454	0.7486	0.7517	0.7549
0.7	0.7580	0.7611	0.7642	0.7673	0.7704	0.7734	0.7764	0.7794	0.7823	0.7852
0.8	0.7881	0.7910	0.7939	0.7967	0.7995	0.8023	0.8051	0.8078	0.8106	0.8133
0.9	0.8159	0.8186	0.8212	0.8238	0.8264	0.8289	0.8315	0.8340	0.8365	0.8389
1.0	0.8413	0.8438	0.8461	0.8485	0.8508	0.8531	0.8554	0.8577	0.8599	0.8621
1.1	0.8643	0.8665	0.8686	0.8708	0.8729	0.8749	0.8770	0.8790	0.8810	0.8830
1.2	0.8849	0.8869	0.8888	0.8907	0.8925	0.8944	0.8962	0.8980	0.8997	0.9015
1.3	0.9032	0.9049	0.9066	0.9082	0.9099	0.9115	0.9131	0.9147	0.9162	0.9177
1.4	0.9192	0.9207	0.9222	0.9236	0.9251	0.9265	0.9279	0.9292	0.9306	0.9319
1.5	0.9332	0.9345	0.9357	0.9370	0.9382	0.9394	0.9406	0.9418	0.9429	0.9441
1.6	0.9452	0.9463	0.9474	0.9484	0.9495	0.9505	0.9515	0.9525	0.9535	0.9545
1.7	0.9554	0.9564	0.9573	0.9582	0.9591	0.9599	0.9608	0.9616	0.9625	0.9633
1.8	0.9641	0.9649	0.9656	0.9664	0.9671	0.9678	0.9686	0.9693	0.9699	0.9706
1.9	0.9713	0.9719	0.9726	0.9732	0.9738	0.9744	0.9750	0.9756	0.9761	0.9767

z	0.00	0.01	0.02	0.03	0.04	0.05	0.06	0.07	0.08	0.09
2.0	0.9772	0.9778	0.9783	0.9788	0.9793	0.9798	0.9803	0.9808	0.9812	0.9817
2.1	0.9821	0.9826	0.9830	0.9834	0.9838	0.9842	0.9846	0.9850	0.9854	0.9857
2.2	0.9861	0.9864	0.9868	0.9871	0.9875	0.9878	0.9881	0.9884	0.9887	0.9890
2.3	0.9893	0.9896	0.9898	0.9901	0.9904	0.9906	0.9909	0.9911	0.9913	0.9916
2.4	0.9918	0.9920	0.9922	0.9925	0.9927	0.9929	0.9931	0.9932	0.9934	0.9936
2.5	0.9938	0.9940	0.9941	0.9943	0.9945	0.9946	0.9948	0.9949	0.9951	0.9952
2.6	0.9953	0.9955	0.9956	0.9957	0.9959	0.9960	0.9961	0.9962	0.9963	0.9964
2.7	0.9965	0.9966	0.9967	0.9968	0.9969	0.9970	0.9971	0.9972	0.9973	0.9974
2.8	0.9974	0.9975	0.9976	0.9977	0.9977	0.9978	0.9979	0.9979	0.9980	0.9981
2.9	0.9981	0.9982	0.9982	0.9983	0.9984	0.9984	0.9985	0.9985	0.9986	0.9986
3.0	0.9987	0.9987	0.9987	0.9988	0.9988	0.9989	0.9989	0.9989	0.9990	0.9990
3.1	0.9990	0.9991	0.9991	0.9991	0.9992	0.9992	0.9992	0.9992	0.9993	0.9993
3.2	0.9993	0.9993	0.9994	0.9994	0.9994	0.9994	0.9994	0.9995	0.9995	0.9995
3.3	0.9995	0.9995	0.9995	0.9996	0.9996	0.9996	0.9996	0.9996	0.9996	0.9997
3.4	0.9997	0.9997	0.9997	0.9997	0.9997	0.9997	0.9997	0.9997	0.9997	0.9998

APPENDIX 3

Critical Values of the t-Distribution

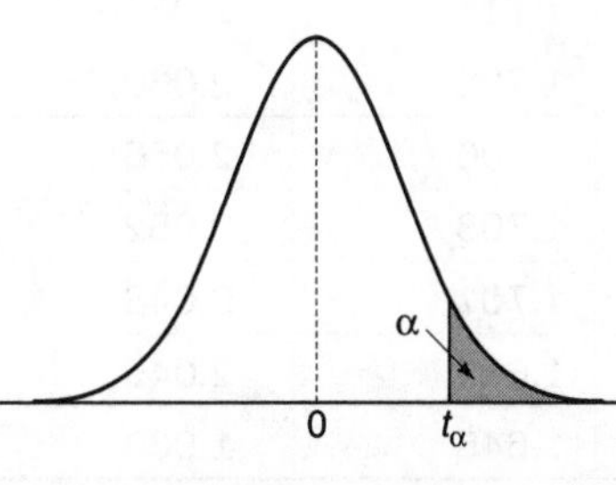

	α				
df	0.1	0.05	0.025	0.01	0.005
1	3.078	6.314	12.706	31.821	63.657
2	1.886	2.920	4.303	6.965	9.925
3	1.638	2.353	3.182	4.541	5.841
4	1.533	2.132	2.776	3.747	4.604
5	1.476	2.015	2.571	3.365	4.032
6	1.440	1.943	2.447	3.143	3.707
7	1.415	1.895	2.365	2.998	3.499
8	1.397	1.860	2.306	2.896	3.355
9	1.383	1.833	2.262	2.821	3.250
10	1.372	1.812	2.228	2.764	3.169
11	1.363	1.796	2.201	2.718	3.106
12	1.356	1.782	2.179	2.681	3.055
13	1.350	1.771	2.160	2.650	3.012
14	1.345	1.761	2.145	2.624	2.977
15	1.341	1.753	2.131	2.602	2.947

	α				
df	**0.1**	**0.05**	**0.025**	**0.01**	**0.005**
16	1.337	1.746	2.120	2.583	2.921
17	1.333	1.740	2.110	2.567	2.898
18	1.330	1.734	2.101	2.552	2.878
19	1.328	1.729	2.093	2.539	2.861
20	1.325	1.725	2.086	2.528	2.845
21	1.323	1.721	2.080	2.518	2.831
22	1.321	1.717	2.074	2.508	2.819
23	1.319	1.714	2.069	2.500	2.807
24	1.318	1.711	2.064	2.492	2.797
25	1.316	1.708	2.060	2.485	2.787
26	1.315	1.706	2.056	2.479	2.779
27	1.314	1.703	2.052	2.473	2.771
28	1.313	1.701	2.048	2.467	2.763
29	1.311	1.699	2.045	2.462	2.756
∞	1.282	1.645	1.960	2.326	2.576

APPENDIX 4
Chi-Square Distribution

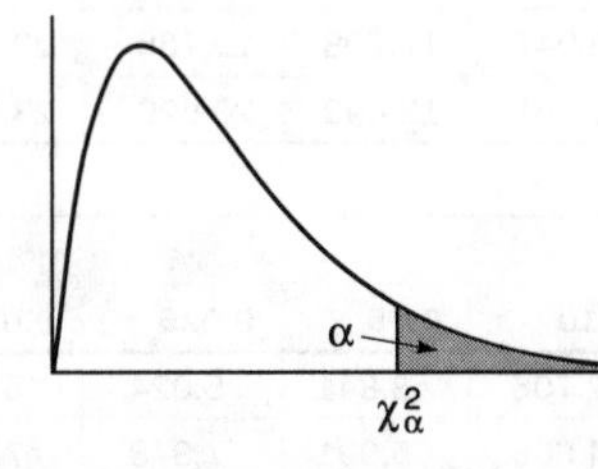

	α									
γ	**0.995**	**0.99**	**0.98**	**0.975**	**0.95**	**0.90**	**0.80**	**0.75**	**0.70**	**0.50**
	0.00004	0.000	0.001	0.001	0.004	0.016	0.064	0.102	0.148	0.455
2	0.0100	0.020	0.040	0.051	0.103	0.211	0.446	0.575	0.713	0.1386
3	0.0717	0.115	0.185	0.216	0.352	0.584	1.005	1.213	1.424	2.366
4	0.207	0.297	0.429	0.484	0.711	1.064	1.649	1.923	2.195	3.357
5	0.412	0.554	0.752	0.831	1.145	1.610	2.343	2.675	3.000	4.351
6	0.676	0.872	1.134	1.237	1.635	2.204	3.070	3.455	3.828	5.348
7	0.989	1.239	1.564	1.690	2.167	2.833	3.822	4.255	4.671	6.346
8	1.344	1.646	2.032	2.180	2.733	3.490	4.594	5.071	5.527	7.344
9	1.735	2.088	2.532	2.700	3.325	4.168	5.380	5.899	6.393	8.343
10	2.156	2.558	3.059	3.247	3.940	4.865	6.179	6.737	7.267	9.342
11	2.603	3.053	3.609	3.816	4.575	5.578	6.989	7.584	8.148	10.341
12	3.074	3.571	4.178	4.404	5.226	6.304	7.807	8.438	9.034	11.340
13	3.565	4.107	4.765	5.009	5.892	7.042	8.634	9.299	9.926	12.340
14	4.075	4.660	5.368	5.629	6.571	7.790	9.467	10.165	10.821	13.339
15	4.601	5.229	5.985	6.262	7.261	8.547	10.307	11.037	11.721	14.339
16	5.142	5.812	6.614	6.908	7.962	9.312	11.152	11.912	12.624	15.338
17	5.697	6.408	7.255	7.564	8.672	10.085	12.002	12.792	13.531	16.338
18	6.265	7.015	7.906	8.231	9.390	10.865	12.857	13.675	14.440	17.338
19	6.844	7.633	8.567	8.907	10.117	11.651	13.716	14.562	15.352	18.338
20	7.434	8.260	9.237	9.591	10.851	12.443	14.578	15.452	16.266	19.337

	α									
γ	0.995	0.99	0.98	0.975	0.95	0.90	0.80	0.75	0.70	0.50
21	8.034	8.897	9.915	10.283	11.591	13.240	15.445	16.344	17.182	20.337
22	8.643	9.542	10.600	10.982	12.338	14.041	16.314	17.240	18.101	21.337
23	9.260	10.196	11.293	11.689	13.091	14.848	17.187	18.137	19.021	22.337
24	9.886	10.856	11.992	12.401	13.848	15.659	18.062	19.037	19.943	23.337
25	10.520	11.524	12.697	13.120	14.611	16.473	18.940	19.939	20.867	24.337
26	11.160	12.198	13.409	13.844	15.379	17.292	19.820	20.843	21.792	25.336
27	11.808	12.879	14.125	14.573	16.151	18.114	20.703	21.749	22.719	26.336
28	12.461	13.565	14.847	15.308	16.928	18.939	21.588	22.657	23.647	27.336
29	13.121	14.256	15.574	16.047	17.708	19.768	22.475	23.567	24.577	28.336
30	13.787	14.953	16.306	16.791	18.493	20.599	23.364	24.478	25.508	29.336

	α									
γ	0.30	0.25	0.20	0.10	0.05	0.025	0.02	0.01	0.005	0.001
1	1.074	1.323	1.642	2.706	3.841	5.024	5.412	6.635	7.879	10.828
2	2.408	2.773	3.219	4.605	5.991	7.378	7.824	9.210	10.597	13.816
3	3.665	4.108	4.642	6.251	7.815	9.348	9.837	11.345	12.838	16.266
4	4.878	5.385	5.989	7.779	9.488	11.143	11.668	13.277	14.860	18.467
5	6.064	6.626	7.289	9.236	11.070	12.833	13.388	15.086	16.750	20.515
6	7.231	7.841	8.558	10.645	12.592	14.449	15.033	16.812	18.548	22.458
7	8.383	9.037	9.803	12.017	14.067	16.013	16.622	18.475	20.278	24.322
8	9.524	10.219	11.030	13.362	15.507	17.535	18.168	20.090	21.955	26.124
9	10.656	11.389	12.242	14.684	16.919	19.023	19.679	21.666	23.589	27.877
10	11.781	12.549	13.442	15.987	18.307	20.483	21.161	23.209	25.188	29.588
11	12.899	13.701	14.631	17.275	19.675	21.920	22.618	24.725	26.757	31.264
12	14.011	14.845	15.812	18.549	21.026	23.337	24.054	26.217	28.300	32.909
13	15.119	15.984	16.985	19.812	22.362	24.736	25.472	27.688	29.819	34.528
14	16.222	17.117	18.151	21.064	23.685	26.119	26.873	29.141	31.319	36.123
15	17.322	18.245	19.311	22.307	24.996	27.488	28.259	30.578	32.801	37.697
16	18.418	19.369	20.465	23.542	26.296	28.845	29.633	32.000	34.267	39.252
17	19.511	20.489	21.615	24.769	27.587	30.191	30.995	33.409	35.718	40.790
18	20.601	21.605	22.760	25.989	28.869	31.526	32.346	34.805	37.156	42.312
19	21.689	22.718	23.900	27.204	30.144	32.852	33.687	36.191	38.582	43.820
20	22.775	23.828	25.038	28.412	31.410	34.170	35.020	37.566	39.997	45.315
21	23.858	24.935	26.171	29.615	32.671	35.479	36.343	38.932	41.401	46.797
22	24.939	26.039	27.301	30.813	33.924	36.781	37.659	40.289	42.796	48.268
23	26.018	27.141	28.429	32.007	35.172	38.076	38.968	41.638	44.181	49.728
24	27.096	28.241	29.553	33.196	36.415	39.364	40.270	42.980	45.559	51.179
25	28.172	29.339	30.675	34.382	37.652	40.646	41.566	44.314	46.928	52.620
26	29.246	30.435	31.795	35.563	38.885	41.923	42.856	45.642	48.290	54.052
27	30.319	31.528	32.912	36.741	40.113	43.195	44.140	46.963	49.645	55.476
28	31.391	32.620	34.027	37.916	41.337	44.461	45.419	48.278	50.993	56.892
29	32.461	33.711	35.139	39.087	42.557	45.722	46.693	49.588	52.336	58.301
30	33.530	34.800	36.250	40.256	43.773	46.979	47.962	50.892	53.672	59.703

APPENDIX 5

F Distribution ($\alpha = 1\%$)

$F_{.99}(n_1, n_2)$

n_1 = degrees of freedom for numerator

n_2 = degrees of freedom for denominator

n_2 \ n_1	1	2	3	4	5	6	7	8	9	10
1	4052	4999.5	5403	5625	5764	5859	5928	5982	6022	6056
2	98.50	99.00	99.17	99.25	99.30	99.33	99.36	99.37	99.39	99.40
3	34.12	30.82	29.46	28.71	28.24	27.91	27.67	27.49	27.35	27.23
4	21.20	18.00	16.69	15.98	15.52	15.21	14.98	14.80	14.66	14.55
5	16.26	13.27	12.06	11.39	10.97	10.67	10.46	10.29	10.16	10.05
6	13.75	10.92	9.78	9.15	8.75	8.47	8.26	8.10	7.98	7.87
7	12.25	9.55	8.45	7.85	7.46	7.19	6.99	6.84	6.72	6.62
8	11.26	8.65	7.59	7.01	6.63	6.37	6.18	6.03	5.91	5.81
9	10.56	8.02	6.99	6.42	6.06	5.80	5.61	5.47	5.35	5.26
10	10.04	7.56	6.55	5.99	5.64	5.39	5.20	5.06	4.94	4.85
11	9.65	7.21	6.22	5.67	5.32	5.07	4.89	4.74	4.63	4.54
12	9.33	6.93	5.95	5.41	5.06	4.82	4.64	4.50	4.39	4.30
13	9.07	6.70	5.74	5.21	4.86	4.62	4.44	4.30	4.19	4.10
14	8.86	6.51	5.56	5.04	4.69	4.46	4.28	4.14	4.03	3.94
15	8.68	6.36	5.42	4.89	4.56	4.32	4.14	4.00	3.89	3.80
16	8.53	6.23	5.29	4.77	4.44	4.20	4.03	3.89	3.78	3.69
17	8.40	6.11	5.18	4.67	4.34	4.10	3.93	3.79	3.68	3.59
18	8.29	6.01	5.09	4.58	4.25	4.01	3.84	3.71	3.60	3.51
19	8.18	5.93	5.01	4.50	4.17	3.94	3.77	3.63	3.52	3.43
20	8.10	5.85	4.94	4.43	4.10	3.87	3.70	3.56	3.46	3.37
21	8.02	5.78	4.87	4.37	4.04	3.81	3.64	3.51	3.40	3.31
22	7.95	5.72	4.82	4.31	3.99	3.76	3.59	3.45	3.35	3.26
23	7.88	5.66	4.76	4.26	3.94	3.71	3.54	3.41	3.30	3.21
24	7.82	5.61	4.72	4.22	3.90	3.67	3.50	3.36	3.26	3.17
25	7.77	5.57	4.68	4.18	3.85	3.63	3.46	3.32	3.22	3.13
26	7.72	5.53	4.64	4.14	3.82	3.59	3.42	3.29	3.18	3.09
27	7.68	5.49	4.60	4.11	3.78	3.56	3.39	3.26	3.15	3.06
28	7.64	5.45	4.57	4.07	3.75	3.53	3.36	3.23	3.12	3.03
29	7.60	5.42	4.54	4.04	3.73	3.50	3.33	3.20	3.09	3.00
30	7.56	5.39	4.51	4.02	3.70	3.47	3.30	3.17	3.07	2.98
40	7.31	5.18	4.31	3.83	3.51	3.29	3.12	2.99	2.89	2.80
60	7.08	4.98	4.13	3.65	3.34	3.12	2.95	2.82	2.72	2.63
120	6.85	4.79	3.95	3.48	3.17	2.96	2.79	2.66	2.56	2.47
∞	6.63	4.61	3.78	3.32	3.02	2.80	2.64	2.51	2.41	2.32

n_2 \ n_1	12	15	20	24	30	40	60	120	∞
1	6106	6157	6209	6235	6261	6287	6313	6339	6366
2	99.42	99.43	99.45	99.46	99.47	99.47	99.48	99.49	99.50
3	27.05	26.87	26.69	26.60	26.50	26.41	26.32	26.22	26.13
4	14.37	14.20	14.02	13.93	13.84	13.75	13.65	13.56	13.46
5	9.89	9.72	9.55	9.47	9.38	9.29	9.20	9.11	9.02
6	7.72	7.56	7.40	7.31	7.23	7.14	7.06	6.97	6.88
7	6.47	6.31	6.16	6.07	5.99	5.91	5.82	5.74	5.65
8	5.67	5.52	5.36	5.28	5.20	5.12	5.03	4.95	4.86
9	5.11	4.96	4.81	4.73	4.65	4.57	4.48	4.40	4.31
10	4.71	4.56	4.41	4.33	4.25	4.17	4.08	4.00	3.91
11	4.40	4.25	4.10	4.02	3.94	3.86	3.78	3.69	3.60
12	4.16	4.01	3.86	3.78	3.70	3.62	3.54	3.45	3.36
13	3.96	3.82	3.66	3.59	3.51	3.43	3.34	3.25	3.17
14	3.80	3.66	3.51	3.43	3.35	3.27	3.18	3.09	3.00
15	3.67	3.52	3.37	3.29	3.21	3.13	3.05	2.96	2.87
16	3.55	3.41	3.26	3.18	3.10	3.02	2.93	2.84	2.75
17	3.46	3.31	3.16	3.08	3.00	2.92	2.83	2.75	2.65
18	3.37	3.23	3.08	3.00	2.92	2.84	2.75	2.66	2.57
19	3.30	3.15	3.00	2.92	2.84	2.76	2.67	2.58	2.49
20	3.23	3.09	2.94	2.86	2.78	2.69	2.61	2.52	2.42
21	3.17	3.03	2.88	2.80	2.72	2.64	2.55	2.46	2.36
22	3.12	2.98	2.83	2.75	2.67	2.58	2.50	2.40	2.31
23	3.07	2.93	2.78	2.70	2.62	2.54	2.45	2.35	2.26
24	3.03	2.89	2.74	2.66	2.58	2.49	2.40	2.31	2.21
25	2.99	2.85	2.70	2.62	2.54	2.45	2.36	2.27	2.17
26	2.96	2.81	2.66	2.58	2.50	2.42	2.33	2.23	2.13
27	2.93	2.78	2.63	2.55	2.47	2.38	2.29	2.20	2.10
28	2.90	2.75	2.60	2.52	2.44	2.35	2.26	2.17	2.06
29	2.87	2.73	2.57	2.49	2.41	2.33	2.23	2.14	2.03
30	2.84	2.70	2.55	2.47	2.39	2.30	2.21	2.11	2.01
40	2.66	2.52	2.37	2.29	2.20	2.11	2.02	1.92	1.80
60	2.50	2.35	2.20	2.12	2.03	1.94	1.84	1.73	1.60
120	2.34	2.19	2.03	1.95	1.86	1.76	1.66	1.53	1.38
∞	2.18	2.04	1.88	1.79	1.70	1.59	1.47	1.32	1.00

APPENDIX 6

F Distribution ($\alpha = 5\%$)

$F_{.95}(n_1, n_2)$

n_1 = degrees of freedom for numerator

n_2 = degrees of freedom for denominator

n_2 \ n_1	1	2	3	4	5	6	7	8	9	10
1	161.4	199.5	215.7	224.6	230.2	234.0	236.8	238.9	240.5	241.9
2	18.51	19.00	19.16	19.25	19.30	19.33	19.35	19.37	19.38	19.40
3	10.13	9.55	9.28	9.12	9.01	8.94	8.89	8.85	8.81	8.79
4	7.71	6.94	6.59	6.39	6.26	6.16	6.09	6.04	6.00	5.96
5	6.61	5.79	5.41	5.19	5.05	4.95	4.88	4.82	4.77	4.74
6	5.99	5.14	4.76	4.53	4.39	4.28	4.21	4.15	4.10	4.06
7	5.59	4.47	4.35	4.12	3.97	3.87	3.79	3.73	3.68	3.64
8	5.32	4.46	4.07	3.84	3.69	3.58	3.50	3.44	3.39	3.35
9	5.12	4.26	3.86	3.63	3.48	3.37	3.29	3.23	3.18	3.14
10	4.96	4.10	3.71	3.48	3.33	3.22	3.14	3.07	3.02	2.98
11	4.84	3.98	3.59	3.36	3.20	3.09	3.01	2.95	2.90	2.85
12	4.75	3.89	3.49	3.26	3.11	3.00	2.91	2.85	2.80	2.75
13	4.67	3.81	3.41	3.18	3.03	2.92	2.83	2.77	2.71	2.67
14	4.60	3.74	3.34	3.11	2.96	2.85	2.76	2.70	2.65	2.60
15	4.54	3.68	3.29	3.06	2.90	2.79	2.71	2.64	2.59	2.54
16	4.49	3.63	3.24	3.01	2.85	2.74	2.66	2.59	2.54	2.49
17	4.45	3.59	3.20	2.96	2.81	2.70	2.61	2.55	2.49	2.45
18	4.41	3.55	3.16	2.93	2.77	2.66	2.58	2.51	2.46	2.41
19	4.38	3.52	3.13	2.90	2.74	2.63	2.54	2.48	2.42	2.38
20	4.35	3.49	3.10	2.87	2.71	2.60	2.51	2.45	2.39	2.35
21	4.32	3.47	3.07	2.84	2.68	2.57	2.49	2.42	2.37	2.32
22	4.30	3.44	3.05	2.82	2.66	2.55	2.46	2.40	2.34	2.30
23	4.28	3.42	3.03	2.80	2.64	2.53	2.44	2.37	2.32	2.27
24	4.26	3.40	3.01	2.78	2.62	2.51	2.42	2.36	2.30	2.25
25	4.24	3.39	2.99	2.76	2.60	2.49	2.40	2.34	2.28	2.24
26	4.23	3.37	2.98	2.74	2.59	2.47	2.39	2.32	2.27	2.22
27	4.21	3.35	2.96	2.73	2.57	2.46	2.37	2.31	2.25	2.20
28	4.20	3.34	2.95	2.71	2.56	2.45	2.36	2.29	2.24	2.19
29	4.18	3.33	2.93	2.70	2.55	2.43	2.35	2.28	2.22	2.18
30	4.17	3.32	2.92	2.69	2.53	2.42	2.33	2.27	2.21	2.16
40	4.08	3.23	2.84	2.61	2.45	2.34	2.25	2.18	2.12	2.08
60	4.00	3.15	2.76	2.53	2.37	2.25	2.17	2.10	2.04	1.99
120	3.92	3.07	2.68	2.45	2.29	2.17	2.09	2.02	1.96	1.91
∞	3.84	3.00	2.60	2.37	2.21	2.10	2.01	1.94	1.88	1.83

n_1 = degrees of freedom for numerator

n_2 = degrees of freedom for denominator

n_2 \ n_1	12	15	20	24	30	40	60	120	∞
1	243.9	245.9	248.0	249.1	250.1	251.1	252.2	253.3	254.3
2	19.41	19.43	19.45	19.45	19.46	19.47	19.48	19.49	19.50
3	8.74	8.70	8.66	8.64	8.62	8.59	8.57	8.55	8.53
4	5.91	5.86	5.80	5.77	5.75	5.72	5.69	5.66	5.63
5	4.68	4.62	4.56	4.53	4.50	4.46	4.43	4.40	4.36
6	4.00	3.94	3.87	3.84	3.81	3.77	3.74	3.70	3.67
7	3.57	3.51	3.44	3.41	3.38	3.34	3.30	3.27	3.23
8	3.28	3.22	3.15	3.12	3.08	3.04	3.01	2.97	2.93
9	3.07	3.01	2.94	2.90	2.86	2.83	2.79	2.75	2.71
10	2.91	2.85	2.77	2.74	2.70	2.66	2.62	2.58	2.54
11	2.79	2.72	2.65	2.61	2.57	2.53	2.49	2.45	2.40
12	2.69	2.62	2.54	2.51	2.47	2.43	2.38	2.34	2.30
13	2.60	2.53	2.46	2.42	2.38	2.34	2.30	2.25	2.21
14	2.53	2.46	2.39	2.35	2.31	2.27	2.22	2.18	2.13
15	2.48	2.40	2.33	2.29	2.25	2.20	2.16	2.11	2.07
16	2.42	2.35	2.28	2.24	2.19	2.15	2.11	2.06	2.01
17	2.38	2.31	2.23	2.19	2.15	2.10	2.06	2.01	1.96
18	2.34	2.27	2.19	2.15	2.11	2.06	2.02	1.97	1.92
19	2.31	2.23	2.16	2.11	2.07	2.03	1.98	1.93	1.88
20	2.28	2.20	2.12	2.08	2.04	1.99	1.95	1.90	1.84
21	2.25	2.18	2.10	2.05	2.01	1.96	1.92	1.87	1.81
22	2.23	2.15	2.07	2.03	1.98	1.94	1.89	1.84	1.78
23	2.20	2.13	2.05	2.01	1.96	1.91	1.86	1.81	1.76
24	2.18	2.11	2.03	1.98	1.94	1.89	1.84	1.79	1.73
25	2.16	2.09	2.01	1.96	1.92	1.87	1.82	1.77	1.71
26	2.15	2.07	1.99	1.95	1.90	1.85	1.80	1.75	1.69
27	2.13	2.06	1.97	1.93	1.88	1.84	1.79	1.73	1.67
28	2.12	2.04	1.96	1.91	1.87	1.82	1.77	1.71	1.65
29	2.10	2.03	1.94	1.90	1.85	1.81	1.75	1.70	1.64
30	2.09	2.01	1.93	1.89	1.84	1.79	1.74	1.68	1.62
40	2.00	1.92	1.84	1.79	1.74	1.69	1.64	1.58	1.51
60	1.92	1.84	1.75	1.70	1.65	1.59	1.53	1.47	1.39
120	1.83	1.75	1.66	1.61	1.55	1.50	1.43	1.35	1.25
∞	1.75	1.67	1.57	1.52	1.46	1.39	1.32	1.22	1.00

APPENDIX 7

Poisson Probability Sums

$$\sum_{x=0}^{r} p(x;\mu)$$

r	μ								
	0.1	**0.2**	**0.3**	**0.4**	**0.5**	**0.6**	**0.7**	**0.8**	**0.9**
0	0.9048	0.8187	0.7408	0.6703	0.6065	0.5488	0.4966	0.4493	0.4066
1	0.9953	0.9825	0.9631	0.9384	0.9098	0.8781	0.8442	0.8088	0.7725
2	0.9998	0.9989	0.9964	0.9921	0.9856	0.9769	0.9659	0.9526	0.9371
3	1.0000	0.9999	0.9997	0.9992	0.9982	0.9966	0.9942	0.9909	0.9865
4	1.0000	1.0000	1.0000	0.9999	0.9998	0.9996	0.9992	0.9986	0.9977
5	1.0000	1.0000	1.0000	1.0000	1.0000	1.0000	0.9999	0.9998	0.9997
6	1.0000	1.0000	1.0000	1.0000	1.0000	1.0000	1.0000	1.0000	1.0000

r	μ								
	1.0	**1.5**	**2.0**	**2.5**	**3.0**	**3.5**	**4.0**	**4.5**	**5.0**
0	0.3679	0.2231	0.1353	0.0821	0.0498	0.0302	0.0183	0.0111	0.0067
1	0.7358	0.5578	0.4060	0.2873	0.1991	0.1359	0.0916	0.0611	0.0404
2	0.9197	0.8088	0.6767	0.5438	0.4232	0.3208	0.2381	0.1736	0.1247
3	0.9810	0.9344	0.8571	0.7576	0.6472	0.5366	0.4335	0.3423	0.2650
4	0.9963	0.9814	0.9473	0.8912	0.8153	0.7254	0.6288	0.5321	0.4405
5	0.9994	0.9955	0.9834	0.9580	0.9161	0.8576	0.7851	0.7029	0.6160
6	0.9999	0.9991	0.9955	0.9858	0.9665	0.9347	0.8893	0.8311	0.7622
7	1.0000	0.9998	0.9989	0.9958	0.9881	0.9733	0.9489	0.9134	0.8666
8	1.0000	1.0000	0.9998	0.9989	0.9962	0.9901	0.9786	0.9597	0.9319
9	1.0000	1.0000	1.0000	0.9997	0.9989	0.9967	0.9919	0.9829	0.9682
10	1.0000	1.0000	1.0000	0.9999	0.9997	0.9990	0.9972	0.9933	0.9863
11	1.0000	1.0000	1.0000	1.0000	0.9999	0.9997	0.9991	0.9976	0.9945
12	1.0000	1.0000	1.0000	1.0000	1.0000	0.9999	0.9997	0.9992	0.9980
13	1.0000	1.0000	1.0000	1.0000	1.0000	1.0000	0.9999	0.9997	0.9993
14	1.0000	1.0000	1.0000	1.0000	1.0000	1.0000	1.0000	0.9999	0.9998
15	1.0000	1.0000	1.0000	1.0000	1.0000	1.0000	1.0000	1.0000	0.9999
16	1.0000	1.0000	1.0000	1.0000	1.0000	1.0000	1.0000	1.0000	1.0000

r	μ								
	5.5	6.0	6.5	7.0	7.5	8.0	8.5	9.0	9.5
0	0.0041	0.0025	0.0015	0.0009	0.0006	0.0003	0.0002	0.0001	0.0001
1	0.0266	0.0174	0.0113	0.0073	0.0047	0.0030	0.0019	0.0012	0.0008
2	0.0884	0.0620	0.0430	0.0296	0.0203	0.0138	0.0093	0.0062	0.0042
3	0.2017	0.1512	0.1118	0.0818	0.0591	0.0424	0.0301	0.0212	0.0149
4	0.3575	0.2851	0.2237	0.1730	0.1321	0.0996	0.0744	0.0550	0.0403
5	0.5289	0.4457	0.3690	0.3007	0.2414	0.1912	0.1496	0.1157	0.0885
6	0.6860	0.6063	0.5265	0.4497	0.3782	0.3134	0.2562	0.2068	0.1649
7	0.8095	0.7440	0.6728	0.5987	0.5246	0.4530	0.3856	0.3239	0.2687
8	0.8944	0.8472	0.7916	0.7291	0.6620	0.5925	0.5231	0.4557	0.3918
9	0.9462	0.9161	0.8774	0.8305	0.7764	0.7166	0.6530	0.5874	0.5218
10	0.9747	0.9574	0.9332	0.9015	0.8622	0.8159	0.7634	0.7060	0.6453
11	0.9890	0.9799	0.9661	0.9467	0.9208	0.8881	0.8487	0.8030	0.7520
12	0.9955	0.9912	0.9840	0.9730	0.9573	0.9362	0.9091	0.8758	0.8364
13	0.9983	0.9964	0.9929	0.9872	0.9784	0.9658	0.9486	0.9261	0.8981
14	0.9994	0.9986	0.9970	0.9943	0.9897	0.9827	0.9726	0.9585	0.9400
15	0.9998	0.9995	0.9988	0.9976	0.9954	0.9918	0.9862	0.9780	0.9665
16	0.9999	0.9998	0.9996	0.9990	0.9980	0.9963	0.9934	0.9889	0.9823
17	1.0000	0.9999	0.9998	0.9996	0.9992	0.9984	0.9970	0.9947	0.9911
18	1.0000	1.0000	0.9999	0.9999	0.9997	0.9993	0.9987	0.9976	0.9957
19	1.0000	1.0000	1.0000	1.0000	0.9999	0.9997	0.9995	0.9989	0.9980
20	1.0000	1.0000	1.0000	1.0000	1.0000	0.9999	0.9998	0.9996	0.9991
21	1.0000	1.0000	1.0000	1.0000	1.0000	1.0000	0.9999	0.9998	0.9996
22	1.0000	1.0000	1.0000	1.0000	1.0000	1.0000	1.0000	0.9999	0.9999
23	1.0000	1.0000	1.0000	1.0000	1.0000	1.0000	1.0000	1.0000	0.9999
24	1.0000	1.0000	1.0000	1.0000	1.0000	1.0000	1.0000	1.0000	1.0000

r	μ								
	10.0	11.0	12.0	13.0	14.0	15.0	16.0	17.0	18.0
0	0.0000	0.0000	0.0000	0.0000	0.0000	0.0000	0.0000	0.0000	0.0000
1	0.0005	0.0002	0.0001	0.0000	0.0000	0.0000	0.0000	0.0000	0.0000
2	0.0028	0.0012	0.0005	0.0002	0.0001	0.0000	0.0000	0.0000	0.0000
3	0.0103	0.0049	0.0023	0.0011	0.0005	0.0002	0.0001	0.0000	0.0000
4	0.0293	0.0151	0.0076	0.0037	0.0018	0.0009	0.0004	0.0002	0.0001
5	0.0671	0.0375	0.0203	0.0107	0.0055	0.0028	0.0014	0.0007	0.0003
6	0.1301	0.0786	0.0458	0.0259	0.0142	0.0076	0.0040	0.0021	0.0010
7	0.2202	0.1432	0.0895	0.0540	0.0316	0.0180	0.0100	0.0054	0.0029
8	0.3328	0.2320	0.1550	0.0998	0.0621	0.0374	0.0220	0.0126	0.0071
9	0.4579	0.3405	0.2424	0.1658	0.1094	0.0699	0.0433	0.0261	0.0154
10	0.5830	0.4599	0.3472	0.2517	0.1757	0.1185	0.0774	0.0491	0.0304
11	0.6968	0.5793	0.4616	0.3532	0.2600	0.1848	0.1270	0.0847	0.0549
12	0.7916	0.6887	0.5760	0.4631	0.3585	0.2676	0.1931	0.1350	0.0917
13	0.8645	0.7813	0.6815	0.5730	0.4644	0.3632	0.2745	0.2009	0.1426
14	0.9165	0.8540	0.7720	0.6751	0.5704	0.4657	0.3675	0.2808	0.2081

	μ								
r	**10.0**	**11.0**	**12.0**	**13.0**	**14.0**	**15.0**	**16.0**	**17.0**	**18.0**
15	0.9513	0.9074	0.8444	0.7636	0.6694	0.5681	0.4667	0.3715	0.2867
16	0.9730	0.9441	0.8987	0.8355	0.7559	0.6641	0.5660	0.4677	0.3751
17	0.9857	0.9678	0.9370	0.8905	0.8272	0.7489	0.6593	0.5640	0.4686
18	0.9928	0.9823	0.9626	0.9302	0.8826	0.8195	0.7423	0.6550	0.5622
19	0.9965	0.9907	0.9787	0.9573	0.9235	0.8752	0.8122	0.7363	0.6509
20	0.9984	0.9953	0.9884	0.9750	0.9521	0.9170	0.8682	0.8055	0.7307
21	0.9993	0.9977	0.9939	0.9859	0.9712	0.9469	0.9108	0.8615	0.7991
22	0.9997	0.9990	0.9970	0.9924	0.9833	0.9673	0.9418	0.9047	0.8551
23	0.9999	0.9995	0.9985	0.9960	0.9907	0.9805	0.9633	0.9637	0.8989
24	1.0000	0.9998	0.9993	0.9980	0.9950	0.9888	0.9777	0.9594	0.9317
25	1.0000	0.9999	0.9997	0. 9990	0.9974	0.9938	0.9869	0.9748	0.9554
26	1.0000	1.0000	0.9999	0.9995	0.9987	0.9967	0.9925	0.9848	0.9718
27	1.0000	1.0000	0.9999	0.9998	0.9994	0.9983	0.9959	0.9912	0.9827
28	1.0000	1.0000	1.0000	0.9999	0.9997	0.9991	0.9978	0.9950	0.9897
29	1.0000	1.0000	1.0000	1.0000	0.9999	0.9996	0.9989	0.9973	0.9941
30	1.0000	1.0000	1.0000	1.0000	0.9999	0.9998	0.9994	0.9986	0.9967
31	1.0000	1.0000	1.0000	1.0000	1.0000	0.9999	0.9997	0.9993	0.9982
32	1.0000	1.0000	1.0000	1.0000	1.0000	1.0000	0.9999	0.9996	0.9990
33	1.0000	1.0000	1.0000	1.0000	1.0000	1.0000	0.9999	0.9998	0.9995
34	1.0000	1.0000	1.0000	1.0000	1.0000	1.0000	1.0000	0.9999	0.9998
35	1.0000	1.0000	1.0000	1.0000	1.0000	1.0000	1.0000	1.0000	0.9999
36	1.0000	1.0000	1.0000	1.0000	1.0000	1.0000	1.0000	1.0000	0.9999
37	1.0000	1.0000	1.0000	1.0000	1.0000	1.0000	1.0000	1.0000	1.0000

APPENDIX 8

Tolerance Interval Factors

	$\gamma = 0.90$				$\gamma = 0.95$				$\gamma = 0.99$			
n	$p =$ 0.90	$p =$ 0.95	$p =$ 0.99	$p =$ 0.999	$p =$ 0.90	$p =$ 0.95	$p =$ 0.99	$p =$ 0.999	$p =$ 0.90	$p =$ 0.95	$p =$ 0.99	$p =$ 0.999
2	15.978	18.800	24.167	30.227	32.019	37.674	48.430	60.573	160.193	188.491	242.300	303.054
3	5.847	6.919	8.974	11.309	8.380	9.916	12.861	16.208	18.930	22.401	29.055	36.616
4	4.166	4.943	6.440	8.149	5.369	6.370	8.299	10.502	9.398	11.150	14.527	18.383
5	3.494	4.152	5.423	6.879	4.275	5.079	6.634	8.415	6.612	7.855	10.260	13.015
6	3.131	3.723	4.870	6.188	3.712	4.414	5.775	7.337	5.337	6.345	8.301	10.548
7	2.902	3.452	4.521	5.750	3.369	4.007	5.248	6.676	4.613	5.488	7.187	9.142
8	2.743	3.264	4.278	5.446	3.316	3.732	4.891	6.226	4.147	4.936	6.468	8.234
9	2.626	3.125	4.098	5.220	2.967	3.532	4.631	5.899	3.822	4.550	5.966	7.600
10	2.535	3.018	3.959	5.046	2.839	3.379	4.433	5.649	3.582	4.265	5.594	7.129
11	2.463	2.933	3.849	4.906	2.737	3.259	4.277	5.452	3.397	4.045	5.308	6.766
12	2.404	2.863	3.758	4.792	2.655	3.162	4.150	5.291	3.250	3.870	5.079	6.477
13	2.355	2.805	3.682	4.697	2.587	3.081	4.044	5.158	3.130	3.727	4.893	6.240
14	2.314	2.756	3.618	4.615	2.529	3.012	3.955	5.045	3.029	3.608	4.737	6.043
15	2.278	2.713	3.562	4.545	2.480	2.954	3.878	4.949	2.945	3.507	4.605	5.876
16	2.246	2.676	3.514	4.484	2.437	2.903	3.812	4.865	2.872	3.421	4.492	5.732
17	2.219	2.643	3.471	4.430	2.400	2.858	3.754	4.791	2.808	3.345	4.393	5.607
18	2.194	2.614	3.433	4.382	2.366	2.819	3.702	4.725	2.753	3.279	4.307	5.497
19	2.172	2.588	3.399	4.339	2.337	2.784	3.656	4.667	2.703	3.221	4.230	5.399
20	2.152	2.564	3.368	4.300	2.310	2.752	3.615	4.614	2.659	3.168	4.161	5.312
21	2.135	2.543	3.340	4.264	2.286	2.723	3.577	4.567	2.620	3.121	4.100	5.234
22	2.118	2.524	3.315	4.232	2.264	2.697	3.543	4.523	2.584	3.078	4.044	5.163
23	2.103	2.506	3.292	4.203	2.244	2.673	3.512	4.484	2.551	3.040	3.993	5.098
24	2.089	2.480	3.270	4.176	2.225	2.651	3.483	4.447	2.522	3.004	3.947	5.039
25	2.077	2.474	3.251	4.151	2.208	2.631	3.457	4.413	2.494	2.972	3.904	4.985
30	2.025	2.413	3.170	4.049	2.140	2.549	3.350	4.278	2.385	2.841	3.733	4.768
35	1.988	2.368	3.112	3.974	2.090	2.490	3.272	4.179	2.306	2.748	3.611	4.611
40	1.959	2.334	3.066	3.917	2.052	2.445	3.213	4.104	2.247	2.677	3.518	4.493
45	1.935	2.306	3.030	3.871	2.021	2.408	3.165	4.042	2.200	2.621	3.444	4.399
50	1.916	2.284	3.001	3.833	1.996	2.379	3.126	3.993	2.162	2.576	3.385	4.323

TABLE A8-1 Values of k for Two-Sided Limits

	$\gamma = 0.90$				$\gamma = 0.95$				$\gamma = 0.99$			
n	$p =$ 0.90	$p =$ 0.95	$p =$ 0.99	$p =$ 0.999	$p =$ 0.90	$p =$ 0.95	$p =$ 0.99	$p =$ 0.999	$p = 0.90$	$p =$ 0.95	$p = 0.99$	$p =$ 0.999
3	4.258	5.310	7.340	9.651	6.158	7.655	10.552	13.857	–	–	–	–
4	3.187	3.957	5.437	7.128	4.163	5.145	7.042	9.215	–	–	–	–
5	2.742	3.400	4.666	6.112	3.407	4.202	5.741	7.501	–	–	–	–
6	2.494	3.091	4.242	5.556	3.006	3.707	50.62	6.612	4.408	5.409	7.334	9.540
7	2.333	2.894	3.972	5.201	2.755	3.399	4.641	6.061	3.856	4.730	6.411	8.348
8	2.219	2.755	3.783	4.955	2.582	3.188	4.353	5.686	3.496	4.287	5.811	7.566
9	2.133	2.649	3.641	4.772	2.454	3.031	4.143	5.414	3.242	3.971	5.389	7.014
10	2.065	2.568	3.532	4.629	2.355	2.911	3.981	5.203	3.048	3.739	5.075	6.603
11	2.012	2.503	3.444	4.515	2.275	2.815	3.852	5.036	2.897	3.557	4.828	6.284
12	1.966	2.448	3.371	4.420	2.210	2.736	3.747	4.900	2.773	3.410	4.633	6.032
13	1.928	2.403	3.310	4.341	2.155	2.670	3.659	4.787	2.677	3.290	4.472	5.826
14	1.895	2.363	3.257	4.274	2.108	2.614	3.585	4.690	2.592	3.189	4.336	5.651
15	1.866	2.329	3.212	4.215	2.068	2.566	3.520	4.607	2.521	3.102	4.224	5.507
16	1.842	2.299	3.172	4.146	2.032	2.523	3.463	4.534	2.458	3.028	4.124	5.374
17	1.820	2.272	3.136	4.118	2.001	2.468	3.415	4.471	2.405	2.962	4.038	5.268
18	1.800	2.249	3.106	4.078	1.974	2.453	3.370	4.415	2.357	2.906	3.961	5.167
19	1.781	2.228	3.078	4.041	1.949	2.423	3.331	4.364	2.315	2.855	3.893	5.078
20	1.765	2.208	3.052	4.009	1.926	2.396	3.295	4.319	2.275	2.807	3.832	5.003
21	1.750	2.190	3.028	3.979	1.905	2.371	3.262	4.276	2.241	2.768	3.776	4.932
22	1.736	2.174	3.007	3.952	1.887	2.350	3.233	4.238	2.208	2.729	3.727	4.866
23	1.724	2.159	2.987	3.927	1.869	2.329	3.206	4.204	2.179	2.693	3.680	4.806
24	1.712	2.145	2.969	3.904	1.853	2.309	3.181	4.171	2.154	2.663	3.638	4.755
25	1.702	2.132	2.952	3.882	1.838	2.292	3.158	4.143	2.129	2.632	3.601	4.706
30	1.657	2.080	2.884	3.794	1.778	2.220	3.064	4.022	2.029	2.516	3.446	4.508
35	1.623	2.041	2.833	3.730	1.732	2.166	2.994	3.934	1.957	2.431	3.334	4.364
40	1.598	2.010	2.793	3.679	1.697	2.126	2.941	3.866	1.902	2.365	3.250	4.255
45	1.577	1.986	2.762	3.638	1.669	2.092	2.897	3.811	1.857	2.313	3.181	4.168
50	1.560	1.965	2.735	3.604	1.646	2.065	2.963	3.766	1.821	2.296	3.124	4.096

Table A8-2 Values of k for One-Sided Limits

n	$\gamma = 0.90$	$\gamma = 0.95$	$\gamma = 0.99$	$\gamma = 0.995$
2	0.052	0.026	0.006	0.003
4	0.321	0.249	0.141	0.111
6	0.490	0.419	0.295	0.254
10	0.664	0.606	0.496	0.456
20	0.820	0.784	0.712	0.683
40	0.907	0.887	0.846	0.829
60	0.937	0.924	0.895	0.883
80	0.953	0.943	0.920	0.911
100	0.962	0.954	0.936	0.929
150	0.975	0.969	0.957	0.952
200	0.981	0.977	0.968	0.961
500	0.993	0.991	0.987	0.986
1000	0.997	0.996	0.994	0.993

TABLE A8-3 Proportion of Population Covered with γ% Confidence and Sample Size n

α	$\gamma = 0.90$	$\gamma = 0.95$	$\gamma = 0.99$	$\gamma = 0.995$
0.005	777	947	1325	1483
0.01	388	473	662	740
0.05	77	93	130	146
0.01	38	46	64	72
0.15	25	30	42	47
0.20	18	22	31	34
0.25	15	18	24	27
0.30	12	14	20	22
0.40	6	10	14	16
0.50	7	8	11	12

TABLE A8-4 Sample Size Required to Cover $(1 - \alpha)$% of the Population with γ% Confidence

APPENDIX 9

Control Chart Constants

Observations in Sample, n	Chart for Averages			Chart for Standard Deviations					
	Factors for Control Limits			Factors for Central Line		Factors for Control Limits			
	A	A_2	A_3	c_4	$1/c_4$	B_3	B_4	B_5	B_6
2	2.121	1.880	2.659	0.7979	1.2533	0	3.267	0	2.606
3	1.732	1.023	1.954	0.8862	1.1284	0	2.568	0	2.276
4	1.500	0.729	1.628	0.9213	1.0854	0	2.266	0	2.088
5	1.342	0.577	1.427	0.9400	1.0638	0	2.089	0	1.964
6	1.225	0.483	1.287	0.9515	1.0510	0.030	1.970	0.029	1.874
7	1.134	0.419	1.182	0.9594	1.0423	0.118	1.882	0.113	1.806
8	1.061	0.373	1.099	0.9650	1.0363	0.185	1.815	0.179	1.751
9	1.000	0.337	1.032	0.9693	1.0317	0.239	1.761	0.232	1.707
10	0.949	0.308	0.975	0.9727	1.0281	0.284	1.716	0.276	1.669
11	0.905	0.285	0.927	0.9754	1.0252	0.321	1.679	0.313	1.637
12	0.866	0.266	0.886	0.9776	1.0229	0.354	1.646	0.346	1.610
13	0.832	0.249	0.850	0.9794	1.0210	0.382	1.618	0.374	1.585
14	0.802	0.235	0.817	0.9810	1.0194	0.406	1.594	0.399	1.563
15	0.775	0.223	0.789	0.9823	1.0180	0.428	1.572	0.421	1.544
16	0.750	0.212	0.763	0.9835	1.0168	0.448	1.552	0.440	1.526
17	0.728	0.203	0.739	0.9845	1.0157	0.466	1.534	0.458	1.511
18	0.707	0.194	0.718	0.9854	1.0148	0.482	1.518	0.475	1.496
19	0.688	0.187	0.698	0.9862	1.0140	0.497	1.503	0.490	1.483
20	0.671	0.180	0.680	0.9869	1.0133	0.510	1.490	0.504	1.470
21	0.655	0.173	0.663	0.9876	1.0126	0.523	1.477	0.516	1.459
22	0.640	0.167	0.647	0.9882	1.0119	0.534	1.466	0.528	1.448
23	0.626	0.162	0.633	0.9887	1.0114	0.545	1.455	0.539	1.438
24	0.612	0.157	0.619	0.9892	1.0109	0.555	1.445	0.549	1.429
25	0.600	0.153	0.606	0.9896	1.0105	0.565	1.435	0.559	1.420

Observations in Sample, n	Chart for Ranges							x Charts
	Factors for Central Line			Factors for Control Limits				
	d_2	$1/d_2$	d_3	D_1	D_2	D_3	D_4	E_2
2	1.128	0.8865	0.853	0	3.686	0	3.267	2.660
3	1.693	0.5907	0.888	0	4.358	0	2.574	1.772
4	2.059	0.4857	0.880	0	4.698	0	2.282	1.457
5	2.326	0.4299	0.864	0	4.918	0	2.114	1.290
6	2.534	0.3946	0.848	0	5.078	0	2.004	1.184
7	2.704	0.3698	0.833	0.204	5.204	0.076	1.924	1.109
8	2.847	0.3512	0.820	0.388	5.306	0.136	1.864	1.054
9	2.970	0.3367	0.808	0.547	5.393	0.184	1.816	1.010
10	3.078	0.3249	0.797	0.687	5.469	0.223	1.777	0.975
11	3.173	0.3152	0.787	0.811	5.535	0.256	1.744	0.945
12	3.258	0.3069	0.778	0.922	5.594	0.283	1.717	0.921
13	3.336	0.2998	0.770	1.025	5.647	0.307	1.693	0.899
14	3.407	0.2935	0.763	1.118	5.696	0.328	1.672	0.881
15	3.472	0.2880	0.756	1.203	5.741	0.347	1.653	0.864
16	3.532	0.2831	0.750	1.282	5.782	0.363	1.637	0.849
17	3.588	0.2787	0.744	1.356	5.820	0.378	1.622	0.836
18	3.640	0.2747	0.739	1.424	5.856	0.391	1.608	0.824
19	3.689	0.2711	0.734	1.487	5.891	0.403	1.597	0.813
20	3.735	0.2677	0.729	1.549	5.921	0.415	1.585	0.803
21	3.778	0.2647	0.724	1.605	5.951	0.425	1.575	0.794
22	3.819	0.2618	0.720	1.659	5.979	0.434	1.566	0.786
23	3.858	0.2592	0.716	1.710	6.006	0.443	1.557	0.778
24	3.895	0.2567	0.712	1.759	6.031	0.451	1.548	0.770
25	3.931	0.2544	0.708	1.806	6.056	0.459	1.541	0.763

APPENDIX 10

Control Chart Equations

	***np* Chart**	***p* Chart**
LCL	$\text{LCL} = n\bar{p} - \sqrt[3]{n\bar{p}\left(1 - \frac{n\bar{p}}{n}\right)}$ or 0 if LCL is negative	$\text{LCL} = \bar{p} - 3\sqrt{\frac{\bar{p}(1-\bar{p})}{n}}$ or 0 if LCL is negative
Center Line	$n\bar{p} = \frac{\text{Sum of items with problems}}{\text{Number of subgroups}}$	$\bar{p} = \frac{\text{Sum of items with problems}}{\text{Number of items in all subgroups}}$
UCL	$\text{UCL} = n\bar{p} + 3\sqrt{n\bar{p}\left(1 - \frac{n\bar{p}}{n}\right)}$ or *n* if UCL is greater than n	$\text{UCL} = \bar{p} + 3\sqrt{\frac{\bar{p}(1-\bar{p})}{n}}$ or 1 if UCL is greater than 1

	***c* Chart**	***u* Chart**
LCL	$\text{LCL} = \bar{c} - 3\sqrt{\bar{c}}$ or 0 if LCL is negative	$\text{LCL} = \bar{u} - 3\sqrt{\frac{\bar{u}}{n}}$ or 0 if LCL is negative
Center Line	$\bar{c} = \frac{\text{Sum of problems}}{\text{Number of subgroups}}$	$\bar{u} = \frac{\text{Sum of problems}}{\text{Number of units in all subgroups}}$
UCL	$\text{UCL} = \bar{c} + 3\sqrt{\bar{c}}$	$\text{UCL} = \bar{u} + 3\sqrt{\frac{\bar{u}}{n}}$

	x Chart	$\bar{X}$ Chart
LCL	$LCL = \bar{X} - 2.66(M\bar{R})$	$LCL = \bar{\bar{X}} - A_2\bar{R}$
Center Line	$\bar{X} = \frac{\text{Sum of measurements}}{\text{Number of measurements}}$	$\bar{\bar{X}} = \frac{\text{Sum of subgroup averages}}{\text{Number of averages}}$
UCL	$UCL = \bar{X} + 2.66(M\bar{R})$	$UCL = \bar{\bar{X}} + A_2\bar{R}$

	R Chart	
LCL	$LCL = D_3\bar{R}$	
Center Line	$\bar{R} = \frac{\text{Sum of ranges}}{\text{Number of ranges}}$	
UCL	$UCL = D_4\bar{R}$	

APPENDIX 11

Table of d_2^* Values

		m = Repeat Readings Taken						
		2	**3**	**4**	**5**	**6**	**7**	**8**
g = # parts × # inspectors	1	1.41	1.91	2.24	2.48	2.67	2.83	2.96
	2	1.28	1.81	2.15	2.40	2.60	2.77	2.91
	3	1.23	1.77	2.12	2.38	2.58	2.75	2.89
	4	1.21	1.75	2.11	2.37	2.57	2.74	2.88
	5	1.19	1.74	2.10	2.36	2.56	2.73	2.87
	6	1.18	1.73	2.09	2.35	2.56	2.73	2.87
	7	1.17	1.73	2.09	2.35	2.55	2.72	2.87
	8	1.17	1.72	2.08	2.35	2.55	2.72	2.87
	9	1.16	1.72	2.08	2.34	2.55	5.72	2.86
	10	1.16	1.72	2.08	2.34	2.55	2.72	2.86
	11	1.16	1.71	2.08	2.34	2.55	2.72	2.86
	12	1.15	1.71	2.07	2.34	2.55	2.72	2.85
	13	1.15	1.71	2.07	2.34	2.55	2.71	2.85
	14	1.15	1.71	2.07	2.34	2.54	2.71	2.85
	15	1.15	1.71	2.07	2.34	2.54	2.71	2.85
	>15	1.128	1.693	2.059	2.326	2.534	2.704	2.847

		m = Repeat Readings Taken						
		9	**10**	**11**	**12**	**13**	**14**	**15**
g = # parts × # inspectors	1	3.08	3.18	3.27	3.35	3.42	3.49	3.55
	2	3.02	3.13	3.22	3.30	3.38	3.45	3.51
	3	3.01	3.11	3.21	3.29	3.37	3.43	3.50
	4	3.00	3.10	3.20	3.28	3.36	3.43	3.49
	5	2.99	3.10	3.19	3.28	3.35	3.42	3.49
	6	2.99	3.10	3.19	3.27	3.35	3.42	3.49
	7	2.99	3.10	3.19	3.27	3.35	3.42	3.48
	8	2.98	3.09	3.19	3.27	3.35	3.42	3.48
	9	2.98	3.09	3.18	3.27	3.35	3.42	3.48
	10	2.98	3.09	3.18	3.27	3.34	3.42	3.48
	11	2.98	3.09	3.18	3.27	3.34	3.41	3.48
	12	2.98	3.09	3.18	3.27	3.34	3.41	3.48
	13	2.98	3.09	3.18	3.27	3.34	3.41	3.48
	14	2.98	3.08	3.18	3.27	3.34	3.41	3.48
	15	2.98	3.08	3.18	3.26	3.34	3.41	3.48
	>15	2.970	3.078	3.173	3.258	3.336	3.407	3.472

APPENDIX 12

Factors for Short Run Control Charts for Individuals, *x*-bar, and *R* Charts

	Subgroup Size											
	1 (*R* Based on Moving Range of 2)				2				3			
g	A_{2F}	D_{4F}	A_{2S}	D_{4S}	A_{2F}	D_{4F}	A_{2S}	D_{4S}	A_{2F}	D_{4F}	A_{2S}	D_{4S}
1	NA	NA	236.5	128	NA	NA	167	128	NA	NA	8.21	14
2	12.0	2.0	20.8	16.0	8.49	2.0	15.70	15.6	1.57	1.9	2.72	7.1
3	6.8	2.7	9.6	15.0	4.78	2.7	6.76	14.7	1.35	2.3	1.90	4.5
4	5.1	3.3	6.6	8.1	3.62	3.3	4.68	8.1	1.26	2.4	1.62	3.7
5	4.4	3.3	5.4	6.3	3.12	3.3	3.82	6.3	1.20	2.4	1.47	3.4
6	4.0	3.3	4.7	5.4	2.83	3.3	3.34	5.4	1.17	2.5	1.39	3.3
7	3.7	3.3	4.3	5.0	2.65	3.3	3.06	5.0	1.14	2.5	1.32	3.2
8	3.6	3.3	4.1	4.7	2.53	3.3	2.87	4.7	1.13	2.5	1.28	3.1
9	3.5	3.3	3.9	4.5	2.45	3.3	2.74	4.5	1.12	2.5	1.25	3.0
10	3.3	3.3	3.7	4.5	2.37	3.3	2.62	4.5	1.10	2.5	1.22	3.0
15	3.1	3.5	3.3	4.1	2.18	3.5	2.33	4.1	1.08	2.5	1.15	2.9
20	3.0	3.5	3.1	4.0	2.11	3.5	2.21	4.0	1.07	2.6	1.12	2.8
25	2.9	3.5	3.0	3.8	2.05	3.5	2.14	3.8	1.06	2.6	1.10	2.7

Numbers enclosed in bold boxes represent the recommended minimum number of subgroups for starting a control chart.

	Subgroup Size							
	4				5			
g	A_{2F}	D_{4F}	A_{2S}	D_{4S}	A_{2F}	D_{4F}	A_{2S}	D_{4S}
1	NA	NA	3.05	13	NA	NA	1.8	5.1
2	0.83	1.9	1.44	3.5	0.58	1.7	1.0	3.2
3	0.81	1.9	1.14	3.2	0.59	1.8	0.83	2.8
4	0.79	2.1	1.01	2.9	0.59	1.9	0.76	2.6
5	0.78	2.1	0.95	2.8	0.59	2.0	0.72	2.5
6	0.77	2.2	0.91	2.7	0.59	2.0	0.70	2.4
7	0.76	2.2	0.88	2.6	0.59	2.0	0.68	2.4
8	0.76	2.2	0.86	2.6	0.59	2.0	0.66	2.3
9	0.76	2.2	0.85	2.5	0.59	2.0	0.65	2.3
10	0.75	2.2	0.83	2.5	0.58	2.0	0.65	2.3
15	0.75	2.3	0.80	2.4	0.58	2.1	0.62	2.2
20	0.74	2.3	0.78	2.4	0.58	2.1	0.61	2.2
25	0.74	2.3	0.77	2.4	0.58	2.1	0.60	2.2

Numbers enclosed in bold boxes represent the recommended minimum number of subgroups for starting a control chart.

APPENDIX 13

Sample Customer Survey

Taken from *How did we do?*, a patient satisfaction survey for the XXX Community Hospital. (3/15/94)

For each of the following statements, please check the appropriate box. Mark the NA box if you had no opportunity to judge that aspect of care during your stay at XXX Community Hospital.	**Strongly agree**	**Agree**	**Neither agree nor disagree**	**Disagree**	**Strongly disagree**	**NA**
I received my medication on time	☐	☐	☐	☐	☐	☐
The menu offered foods I liked	☐	☐	☐	☐	☐	☐
My doctor kept me informed	☐	☐	☐	☐	☐	☐
My room was clean	☐	☐	☐	☐	☐	☐
The discharge process was smooth	☐	☐	☐	☐	☐	☐
My doctor was available	☐	☐	☐	☐	☐	☐
The hospital was well supplied	☐	☐	☐	☐	☐	☐
I received the foods I selected from the menu	☐	☐	☐	☐	☐	☐
The staff answered my call light quickly	☐	☐	☐	☐	☐	☐
The food looked good	☐	☐	☐	☐	☐	☐
I was informed of what I should do after discharge	☐	☐	☐	☐	☐	☐
My bed was comfortable	☐	☐	☐	☐	☐	☐
The hospital staff took good care of me	☐	☐	☐	☐	☐	☐
I knew my doctor's name	☐	☐	☐	☐	☐	☐
The staff treated one another with respect	☐	☐	☐	☐	☐	☐
The hospital was well maintained	☐	☐	☐	☐	☐	☐
The food tasted good	☐	☐	☐	☐	☐	☐
My medications were ready when I was ready to go	☐	☐	☐	☐	☐	☐
The billing procedures were explained to me	☐	☐	☐	☐	☐	☐
I was served the right amount of food	☐	☐	☐	☐	☐	☐
The nurse checked on me frequently	☐	☐	☐	☐	☐	☐
I had assistance making plans to leave the hospital	☐	☐	☐	☐	☐	☐
My doctor told me when I was going home	☐	☐	☐	☐	☐	☐
The food servers were pleasant	☐	☐	☐	☐	☐	☐

For each of the following statements, please check the appropriate box. Mark the NA box if you had no opportunity to judge that aspect of care during your stay at XXX Community Hospital.	Strongly agree	Agree	Neither agree nor disagree	Disagree	Strongly disagree	NA
The hospital was clean	☐	☐	☐	☐	☐	☐
Overall, the hospital staff treated me with respect	☐	☐	☐	☐	☐	☐
My room was quiet	☐	☐	☐	☐	☐	☐
The staff met my special needs	☐	☐	☐	☐	☐	☐
The attitude of the staff was nice	☐	☐	☐	☐	☐	☐
I was escorted out of the hospital at discharge	☐	☐	☐	☐	☐	☐
My room was comfortable	☐	☐	☐	☐	☐	☐
My diet was what the doctor ordered	☐	☐	☐	☐	☐	☐
The staff kept me informed about my care	☐	☐	☐	☐	☐	☐
I was satisfied with my doctor(s)	☐	☐	☐	☐	☐	☐
Meals were served on time	☐	☐	☐	☐	☐	☐
The staff were helpful	☐	☐	☐	☐	☐	☐
The discharge process was speedy	☐	☐	☐	☐	☐	☐
My doctor knew who I was	☐	☐	☐	☐	☐	☐
My medications/wound care/equipment were explained to me	☐	☐	☐	☐	☐	☐
I was treated well	☐	☐	☐	☐	☐	☐
I was prepared to go home	☐	☐	☐	☐	☐	☐
The staff were attentive to my needs	☐	☐	☐	☐	☐	☐
I had the same doctor(s) throughout my hospitalization	☐	☐	☐	☐	☐	☐
The nurses acted in a professional manner	☐	☐	☐	☐	☐	☐
The staff knew what care I needed	☐	☐	☐	☐	☐	☐
I would refer a family member to XXX Community Hospital	☐	☐	☐	☐	☐	☐
I would choose to come back to XXX Community Hospital	☐	☐	☐	☐	☐	☐

Were there any incidents you remember from your stay that were especially PLEASANT?

Were there any incidents you remember from your stay that were especially UNPLEASANT?

We welcome any other suggestions you have to offer.

Thank you for your assistance!

APPENDIX 14

Process σ Levels and Equivalent PPM Quality Levels

Based on the assumption that in the long term the process could drift by plus or minus 1.5σ.

Process σ Level	Process PPM
6.27	1
6.12	2
6.0	3.4
5.97	4
5.91	5
5.88	6
5.84	7
5.82	8
5.78	9
5.77	10
5.61	20
5.51	30
5.44	40
5.39	50
5.35	60
5.31	70
5.27	80

Process σ Level	Process PPM
5.25	90
5.22	100
5.04	200
4.93	300
4.85	400
4.79	500
4.74	600
4.69	700
4.66	800
4.62	900
4.59	1,000
4.38	2,000
4.25	3,000
4.15	4,000
4.08	5,000
4.01	6,000
3.96	7,000

Process σ Level	Process PPM
3.91	8,000
3.87	9,000
3.83	10,000
3.55	20,000
3.38	30,000
3.25	40,000
3.14	50,000
3.05	60,000
2.98	70,000
2.91	80,000
2.84	90,000
2.78	100,000
2.34	200,000
2.02	300,000
1.75	400,000
1.50	500,000

APPENDIX 15
Black Belt Effectiveness Certification

Black Belt Certification Recommendation

Name ______________________________ (as it will appear on the certificate)
Address ____
City _________________________ State ________ Zip ______________
Social Security Number ____
We the undersigned, on behalf of ____________________, the Six Sigma organization, certify the above named individual as a Six Sigma Black Belt within [COMPANY].

Printed or typed Board member name	Signature	Date Signed

[COMPANY] Black Belt Skill Set Certification Process

Introduction

This document describes the process and provides the minimum acceptable criteria for certifying an individual as a [COMPANY] Six Sigma Black Belt. [COMPANY] certification involves recognition by the [COMPANY] and his or her peers, and should not be construed as a professional license.

Process

The [COMPANY] determines recognition as a [COMPANY] Six Sigma Black Belt. [COMPANY] certification requires that the applicant pass the [COMPANY]'s Black Belt examination. The exam covers the core skill set of the Black Belt Body of Knowledge (BOK) as defined by the [COMPANY]. The [COMPANY] will score the candidate and determine if his or her score meets the [COMPANY]'s minimum passing score for each section of the BOK, as well as for the overall score. The [COMPANY] also provides criteria for assessing the candidate's effectiveness by evaluating his or her

- Ability to achieve significant, tangible results by applying the Six Sigma approach.
- Ability to lead organizational change as demonstrated by the candidate's leadership, teamwork, project management, and communication skills.

The exam will be administered by the Six Sigma organization. The Six Sigma organization is responsible for assuring the integrity of the exam, verifying the identity of the candidate sitting for the exam, and enforcing time limits. The Six Sigma organization will evaluate the candidate's effectiveness using the [COMPANY] requirements and will notify the [COMPANY] when a candidate who has passed the [COMPANY] BOK exam has met the effectiveness requirements.

[COMPANY] Black Belt Effectiveness Certification Criteria

This section describes the criteria for certifying that a [COMPANY] Black Belt candidate is "effective" in applying the Six Sigma approach. Effectiveness means that the candidate has demonstrated the ability to lead the change process by successfully applying Six Sigma methodologies on more than one significant project. Success is demonstrated by achieving documented substantial, sustained, and tangible results. Examples of results are cost savings or cost avoidance validated by finance and accounting experts, improved customer satisfaction, reduced cycle time, increased revenues and profits, reduced accident rates, improved morale, reduction of critical to customer defects, etc. Merely demonstrating the use of Six Sigma tools is *not* sufficient. Nor is the delivery of intermediate "products" such as Pareto diagrams or process maps.

In addition on passing the [COMPANY] BOK exam, certification requires the following:

1. Acceptable completion of a Black Belt training curriculum approved by the Six Sigma organization.
2. Demonstration of clear and rational thought process.
   a. Ability to analyze a problem following a logical sequence.
   b. Usage of facts and data to guide decisions and action.
3. Ability to clearly explain Six Sigma and the DMAIC project cycle in layman's terms.
4. Ability to achieve tangible results, for example,
   a. Completed two or more projects which employed the Six Sigma approach (DMAIC or equivalent).
      i. Projects reviewed by appropriate personnel.
      ii. Deliverables accepted by the project sponsor.

iii. Projects documented in the manner prescribed by the Six Sigma organization.

iv. Projects used the Six Sigma approach and correctly employed a significant subset of basic, intermediate, and advanced Six Sigma tools and techniques (see page 517 for a listing).

b. Ability to perform benefit/cost analysis.

c. Ability to quantify deliverables in terms meaningful to the organization, for example, cost, quality, cycle time, safety improvement, etc.

d. Ability to identify and overcome obstacles to progress.

e. Ability to work within time, budget, and operational constraints.

5. Demonstrated ability to explain the tools of Six Sigma to others.
6. Demonstrated interpersonal and leadership skills necessary to be an effective change agent within the organization.

[COMPANY] Black Belt Certification Board

The [COMPANY] recommends that each area of effectiveness be rated by at least two qualified individuals. Table 15.1 provides guidelines for identifying members of the [COMPANY] Black Belt Certification Board.

Assessment Subject Area	Board Member
Change agent skills	Supervisor, project sponsor(s), Six Sigma champion, mentor, process owner, Green Belt
Application of tools and techniques	Black Belt instructor, Master Black Belt, [COMPANY] Certified Master Black Belt consultant
Ability to achieve results	Project sponsor, process owner, team members, Green Belt, Six Sigma champion, [COMPANY] Certified Master Black Belt consultant

Table 15.1 [COMPANY] Black Belt Certification Board Member Selection Guide

Effectiveness Questionnaire

The [COMPANY] provides questionnaires to assist [COMPANY] Certification Board members with their assessment. It is strongly recommended that the candidate perform a self-assessment using the [COMPANY]'s questionnaire prior to applying for certification. The candidate should provide the Six Sigma champion with a list of potential members of his or her Certification Board.

The effectiveness questionnaire includes a set of assessment questions for each subject area. The results of the questionnaires can be summarized and used as input into the Six Sigma organization's certification process. A form for this is provided below. The scoring summary sheet summarizes the evaluator's scores by category. Worksheet items scored in the top 3 boxes are considered to be acceptable. Particular attention should be directed to any worksheet item scored in the lower 4 boxes. Since there are 10 choices for each item, any score below 5 indicates that the evaluator disagreed with the survey item. Survey items are worded in such a way that evaluators should agree with them for qualified Black Belt candidates. Disagreement indicates an area for improvement. The scores are, of course, not the only input. The [COMPANY] Certification Board must also consider any other relevant factors before reaching their decision.

The Scoring Summary and Assessment Worksheets may be reproduced as necessary.

[COMPANY] Black Belt Notebook and Oral Review

[COMPANY] Black Belt candidates should provide Certification Board members with written documentation of their on the job applications of the Six Sigma approach. These "notebooks" should include all relevant information, including project charters, demonstrations of tool usage, samples of data used, excerpts of presentations to sponsors or leaders, team member names, project schedules and performance to these schedules, financial and other business results, etc. The notebooks can be distributed to Certification Board members as either soft copies or hard copies, at their discretion.

Even with the best documentation, it is difficult to assess effectiveness properly without providing the candidate the opportunity to present his or her work and respond to questions. Six Sigma organizations should require that [COMPANY] Black Belt candidates deliver an oral presentation to the Certification Board. The oral review will also provide the Certification Board with a firsthand demonstration of the candidate's communication skills.

Change Agent Skills Assessment Worksheet			
Black Belt Candidate		Date of Assessment	
Certification Board Member		Role	

1. The candidate effectively identifies and recruits Six Sigma team members.

   Strongly Disagree Strongly Agree

   ☐ ☐ ☐ ☐ ☐ ☐ ☐ ☐ ☐ ☐

2. The candidate effectively develops Six Sigma team dynamics and motivates participants.

   Strongly Disagree Strongly Agree

   ☐ ☐ ☐ ☐ ☐ ☐ ☐ ☐ ☐ ☐

3. The candidate is able to apply conflict resolution techniques.

Strongly Disagree Strongly Agree

☐ ☐ ☐ ☐ ☐ ☐ ☐ ☐ ☐ ☐

4. The candidate is able to overcome obstacles to change.

Strongly Disagree Strongly Agree

☐ ☐ ☐ ☐ ☐ ☐ ☐ ☐ ☐ ☐

5. The candidate utilizes a logical approach to problem solving.

Strongly Disagree Strongly Agree

☐ ☐ ☐ ☐ ☐ ☐ ☐ ☐ ☐ ☐

6. The candidate effectively facilitates group discussions and meetings.

Strongly Disagree Strongly Agree

☐ ☐ ☐ ☐ ☐ ☐ ☐ ☐ ☐ ☐

7. The candidate's presentations are well organized and easy to understand.

Strongly Disagree Strongly Agree

☐ ☐ ☐ ☐ ☐ ☐ ☐ ☐ ☐ ☐

8. The candidate identifies and mobilizes sponsors for change.

Strongly Disagree Strongly Agree

☐ ☐ ☐ ☐ ☐ ☐ ☐ ☐ ☐ ☐

9. The candidate builds a shared vision of the desired state with champions and sponsors.

Strongly Disagree Strongly Agree

☐ ☐ ☐ ☐ ☐ ☐ ☐ ☐ ☐ ☐

10. The candidate effectively communicates with and obtains support from all levels of management.

Strongly Disagree Strongly Agree

☐ ☐ ☐ ☐ ☐ ☐ ☐ ☐ ☐ ☐

11. The candidate identifies gaps between as-is and desired performance.

Strongly Disagree Strongly Agree

☐ ☐ ☐ ☐ ☐ ☐ ☐ ☐ ☐ ☐

12. The candidate identifies and obtains support from all key stakeholders.

Strongly Disagree Strongly Agree

☐ ☐ ☐ ☐ ☐ ☐ ☐ ☐ ☐ ☐

Application of Tools and Techniques Assessment Worksheet			
Black Belt Candidate		Date of Assessment	
Certification Board Member		Role	

1. The candidate uses an appropriate mix of basic, intermediate, and advanced Six Sigma tools.*

   Strongly Disagree — Strongly Agree

   ☐ ☐ ☐ ☐ ☐ ☐ ☐ ☐ ☐ ☐

2. The candidate uses the tools of Six Sigma properly.

   Strongly Disagree — Strongly Agree

   ☐ ☐ ☐ ☐ ☐ ☐ ☐ ☐ ☐ ☐

3. The candidate applies the correct Six Sigma tools at the proper point in the project.

   Strongly Disagree — Strongly Agree

   ☐ ☐ ☐ ☐ ☐ ☐ ☐ ☐ ☐ ☐

4. The candidate asks for help with Six Sigma tools when necessary.

   Strongly Disagree — Strongly Agree

   ☐ ☐ ☐ ☐ ☐ ☐ ☐ ☐ ☐ ☐

5. The candidate has a working knowledge of word processors, spreadsheets, and presentation software.

   Strongly Disagree — Strongly Agree

   ☐ ☐ ☐ ☐ ☐ ☐ ☐ ☐ ☐ ☐

6. The candidate has a working knowledge of a full-featured statistical software package.

   Strongly Disagree — Strongly Agree

   ☐ ☐ ☐ ☐ ☐ ☐ ☐ ☐ ☐ ☐

7. The candidate understands the limitations as well as the strengths of quantitative methods.

   Strongly Disagree — Strongly Agree

   ☐ ☐ ☐ ☐ ☐ ☐ ☐ ☐ ☐ ☐

*See page 517 for examples of these tools.

Ability to Achieve Results Assessment Worksheet			
Black Belt Candidate		Date of Assessment	
Certification Board Member		Role	

1. The candidate has completed more than one Six Sigma project which produced tangible results.

   Strongly Disagree Strongly Agree

   ☐ ☐ ☐ ☐ ☐ ☐ ☐ ☐ ☐ ☐

2. The candidate's projects had an acceptable project charter, including sponsorship, problem statement, business case, etc.

   Strongly Disagree Strongly Agree

   ☐ ☐ ☐ ☐ ☐ ☐ ☐ ☐ ☐ ☐

3. The projects employed the Six Sigma approach (DMAIC or equivalent).

   Strongly Disagree Strongly Agree

   ☐ ☐ ☐ ☐ ☐ ☐ ☐ ☐ ☐ ☐

4. The projects' deliverables were clearly defined in tangible terms.

   Strongly Disagree Strongly Agree

   ☐ ☐ ☐ ☐ ☐ ☐ ☐ ☐ ☐ ☐

5. The projects produced significant improvements to an important business process.

   Strongly Disagree Strongly Agree

   ☐ ☐ ☐ ☐ ☐ ☐ ☐ ☐ ☐ ☐

6. The current baseline sigma level was determined using valid data.

   Strongly Disagree Strongly Agree

   ☐ ☐ ☐ ☐ ☐ ☐ ☐ ☐ ☐ ☐

7. The final sigma level was calculated using valid data and showed improvements that were both statistically significant and important to the organization.

   Strongly Disagree Strongly Agree

   ☐ ☐ ☐ ☐ ☐ ☐ ☐ ☐ ☐ ☐

8. An acceptable control plan has been implemented to assure that improvements are maintained.

Strongly Disagree Strongly Agree

☐ ☐ ☐ ☐ ☐ ☐ ☐ ☐ ☐ ☐

9. The projects' financial benefits were validated by experts in accounting or finance.

Strongly Disagree Strongly Agree

☐ ☐ ☐ ☐ ☐ ☐ ☐ ☐ ☐ ☐

10. Key customers were identified and their critical requirements defined.

Strongly Disagree Strongly Agree

☐ ☐ ☐ ☐ ☐ ☐ ☐ ☐ ☐ ☐

11. Project sponsors are satisfied with their project's deliverables.

Strongly Disagree Strongly Agree

☐ ☐ ☐ ☐ ☐ ☐ ☐ ☐ ☐ ☐

12. Projects identified and corrected root causes, not symptoms.

Strongly Disagree Strongly Agree

☐ ☐ ☐ ☐ ☐ ☐ ☐ ☐ ☐ ☐

13. All key stakeholders were kept informed of project status and are aware of final outcomes.

Strongly Disagree Strongly Agree

☐ ☐ ☐ ☐ ☐ ☐ ☐ ☐ ☐ ☐

14. Projects were completed on time.

Strongly Disagree Strongly Agree

☐ ☐ ☐ ☐ ☐ ☐ ☐ ☐ ☐ ☐

15. Projects were completed within budget.

Strongly Disagree Strongly Agree

☐ ☐ ☐ ☐ ☐ ☐ ☐ ☐ ☐ ☐

16. Projects were conducted in a manner that minimized disruptions to normal work.

Strongly Disagree Strongly Agree

☐ ☐ ☐ ☐ ☐ ☐ ☐ ☐ ☐ ☐

Assessment Comments	
Assessment Subject Area	**Comments**
Change agent skills	
Applications of tools and techniques	
Ability to achieve results	

Scoring Summary				
Evaluator	**Subject Area**	**Items Scored 4 or Less**	**% in Top 3 Boxes**	**Comment**
	Change agent skills			
	Application of tools and techniques			
	Ability to achieve results			
	Change agent skills			
	Application of tools and techniques			
	Ability to achieve results			
	Change agent skills			
	Application of tools and techniques			
	Ability to achieve results			
	Change agent skills			
	Application of tools and techniques			
	Ability to achieve results			
	Change agent skills			
	Application of tools and techniques			
	Ability to achieve results			

Examples of Six Sigma Tools and Analytical Concepts		
Basic	**Intermediate**	**Advanced**
☐ DMAIC	☐ Control charts for measurements	☐ Exponentially weighted moving average control charts
☐ SIPOC	☐ Control charts for attributes	☐ Short run SPC
☐ DPMO	☐ Process capability	☐ Design and analysis of experiments
☐ Computer skills	☐ Yield analysis (e.g., first pass yield, rolled throughput yield)	☐ ANOVA, MANOVA and other general linear models
☐ Scales of measurement	☐ Measurement error analysis (gage R&R)	☐ Multiple linear regression
☐ Pareto analysis	☐ Correlation analysis	☐ Basic reliability analysis
☐ Process mapping, flowcharts	☐ Simple linear regression	☐ Design for Six Sigma
☐ Check sheets	☐ Chi-square	☐ Simulation and modeling
☐ Cause-and-effect diagrams	☐ Type I and Type II errors	☐ Statistical tolerancing
☐ Scatter plots	☐ Confidence interval interpretation	☐ Response surface methods
☐ Run charts	☐ Hypothesis tests	☐ Robust design concepts
☐ Histograms	☐ Normality assessment and transformations	☐ Design, validation and analysis of customer surveys
☐ Ogives	☐ *Z* transformations	☐ Logistic regression
☐ Descriptive statistics (e.g., mean, standard deviation, skewness)	☐ Process sigma calculations	
☐ Enumerative vs. analytic statistics		
☐ Stem-and-leaf, boxplots		
☐ Basic probability concepts		
☐ Discrete probability distributions (binomial, Poisson, hypergeometric)		
☐ Continuous probability distributions (normal, exponential, etc.)		
☐ 7M tools		
☐ FMEA		
☐ Sampling		
☐ CTx identification		

APPENDIX 16

Green Belt Effectiveness Certification

Green Belt Certification Recommendation

Name ______________________ (as it will appear on the Certificate)

Payroll Number ______________________

Org Code ______________________ Date ______________________

We the undersigned, on behalf of [COMPANY] certify the above named individual as a Six Sigma Green Belt.

Printed or Type Board Member Name	Signature	Date Signed

Green Belt Skill Set Certification Process

Introduction

This document describes the process and provides the minimum criteria for certifying an individual as a Six Sigma Green Belt. Certification involves recognition by [COMPANY], and should not be construed as a professional license.

Green Belt Effectiveness Certification Criteria

To become a Certified Green Belt, the candidate must demonstrate:

1. Ability to lead organizational change as demonstrated by the candidate's leadership, teamwork, project management, communication, and technical skills.
2. Ability to achieve tangible results that have a significant impact by applying the Six Sigma approach.

This section describes the criteria for certifying that a Green Belt candidate is "effective" in applying the Six Sigma approach. Effectiveness means that the candidate has demonstrated the ability to lead the change process by successfully applying Six Sigma methodologies on a significant project. Success is demonstrated by achieving documented substantial, tangible and sustained results. Examples of results are cost savings or cost avoidance validated by finance and accounting experts, improved customer satisfaction, reduced cycle time, increased revenues and profits, reduced accident rates, improved employee morale, reduction of critical to customer defects, etc. Merely demonstrating the use of Six Sigma tools is *not* sufficient. Nor is the delivery of intermediate "products" such as Pareto diagrams or process maps.

Certification as a Green Belt requires the following:

1. Acceptable completion of a Green Belt training curriculum approved by the Six Sigma organization.
2. Demonstration of clear and rational thought process.
   a. Ability to analyze a problem following a logical sequence.
   b. Usage of facts and data to guide decisions and action.
3. Ability to clearly explain Six Sigma and the DMAIC project cycle in layman's terms.
4. Ability to achieve tangible results, for example,
   a. Completed one or more projects that employed the Six Sigma approach (DMAIC or equivalent).
      i. Projects reviewed by appropriate personnel.
      ii. Deliverables accepted by the project sponsor.
      iii. Projects documented in a Green Belt notebook arranged in the DMAIC or equivalent format.
      iv. Projects used the Six Sigma approach and correctly employed a significant subset of basic tools and at least some intermediate Six Sigma tools and techniques (see page 529) for a listing).
   b. Ability to perform benefit/cost analysis.
   c. Ability to quantify deliverables in terms meaningful to the organization, for example, cost, quality, cycle time, safety improvement, etc.
   d. Ability to identify and overcome obstacles to progress.
   e. Ability to work within time, budget, and operational constraints.

5. Demonstrated ability to explain the tools of Six Sigma to others in ordinary language.
6. Demonstrated interpersonal and leadership skills necessary to be an effective change agent within the organization.

Green Belt Certification Board

Effectiveness must be determined by qualified individuals familiar with the candidate's performance in the given effectiveness area. Table A16.1 provides guidelines for identifying prospective members of the Green Belt Certification Board. It is the Green Belt's responsibility to assist with the selection of their Certification Board.

Effectiveness Questionnaire

It is strongly recommended that the candidate perform a self-assessment prior to applying for certification.

Certification Board members are encouraged to use the following questionnaires to assist them with their assessment. The candidate should provide the Six Sigma champion with a list of potential members of his or her Certification Board. When questionnaires are completed by someone other than a Certification Board member, they should be sent directly to a Certification Board member.

Scoring Guidelines

The effectiveness questionnaire includes a set of assessment questions for each subject area. The results of the questionnaires can be summarized and used as input into the certification process. A form for this is provided below. The scoring summary sheet summarizes the evaluator's scores by category. Worksheet items scored in the top 3 boxes are considered to be acceptable. Particular attention should be directed to any worksheet item scored in the lower 4 boxes. Since there are 10 choices for each item, any score below 5 indicates that the evaluator disagreed with the survey item. Survey items are worded in such a way that evaluators should agree with them for qualified Green Belt candidates; that is, higher scores are always better. Disagreement (low scores) in a few areas does not necessarily disqualify a candidate for certification. However, it indicates areas which need improvement and it is recommended that certification be granted only if the candidate agrees to a program for addressing these areas. The scores

Assessment Subject Area	Board Member
Change agent skills	Supervisor, project sponsor(s), Six Sigma champion, mentor, process owner, Black Belt
Application of tools and techniques	Green Belt instructor, Master Black Belt, qualified Certified Master Black Belt Six Sigma consultant
Ability to achieve results	Project sponsor, process owner, team members, Green Belt, Six Sigma champion, Certified Master Black Belt, qualified Six Sigma consultant

TABLE A16.1 Green Belt Certification Board Member Selection Guide

are, of course, not the only input. Ultimately each Certification Board member must exercise his or her own judgment and consider any other relevant factors before reaching a decision.

The Scoring Summary and Assessment Worksheets may be reproduced as necessary.

Green Belt notebook

Green Belt candidates should provide Certification Board members with written documentation of their on the job applications of the Six Sigma approach. These "Green Belt notebooks" should include all relevant information, including project characters, demonstrations of tool usage, samples of data used, excerpts of presentations to sponsors or leaders, team member names, project schedules and performance to these schedules, financial and other business results, etc. The notebooks can be distributed to Certification Board members as either soft copies or hard copies, at the candidate's discretion.

Change Agent Skills Assessment Worksheet			
Green Belt Candidate		Date of Assessment	
Certification Board Member		Role	

1. The candidate effectively identifies and recruits Six Sigma team members.

   Strongly Disagree ☐ ☐ ☐ ☐ ☐ ☐ ☐ ☐ ☐ ☐ Strongly Agree

2. The candidate effectively develops Six Sigma team dynamics and motivates participants.

   Strongly Disagree ☐ ☐ ☐ ☐ ☐ ☐ ☐ ☐ ☐ ☐ Strongly Agree

3. The candidate is able to apply conflict resolution techniques.

   Strongly Disagree ☐ ☐ ☐ ☐ ☐ ☐ ☐ ☐ ☐ ☐ Strongly Agree

4. The candidate is able to overcome obstacles to change.

   Strongly Disagree ☐ ☐ ☐ ☐ ☐ ☐ ☐ ☐ ☐ ☐ Strongly Agree

5. The candidate utilizes a logical approach to problem solving.

   Strongly Disagree ☐ ☐ ☐ ☐ ☐ ☐ ☐ ☐ ☐ ☐ Strongly Agree

6. The candidate effectively facilitates group discussions and meetings.

Strongly Disagree Strongly Agree

☐ ☐ ☐ ☐ ☐ ☐ ☐ ☐ ☐ ☐

7. The candidate's presentations are well organized and easy to understand.

Strongly Disagree Strongly Agree

☐ ☐ ☐ ☐ ☐ ☐ ☐ ☐ ☐ ☐

8. The candidate identifies and mobilizes sponsors for change.

Strongly Disagree Strongly Agree

☐ ☐ ☐ ☐ ☐ ☐ ☐ ☐ ☐ ☐

9. The candidate builds a shared vision of the desired state with champions and sponsors.

Strongly Disagree Strongly Agree

☐ ☐ ☐ ☐ ☐ ☐ ☐ ☐ ☐ ☐

10. The candidate identifies gaps between as-is and desired performance.

Strongly Disagree Strongly Agree

☐ ☐ ☐ ☐ ☐ ☐ ☐ ☐ ☐ ☐

11. The candidate identifies all key stakeholders and obtains support for the project.

Strongly Disagree Strongly Agree

☐ ☐ ☐ ☐ ☐ ☐ ☐ ☐ ☐ ☐

Application of Tools and Techniques Assessment Worksheet			
Green Belt Candidate		Date of Assessment	
Certification Board Member		Role	

1. The candidate uses an appropriate mix of basic and intermediate Six Sigma tools.*

Strongly Disagree Strongly Agree

☐ ☐ ☐ ☐ ☐ ☐ ☐ ☐ ☐ ☐

*See page 529 for a partial listing of these tools.

2. The candidate uses the tools of Six Sigma properly.

Strongly Disagree Strongly Agree

☐ ☐ ☐ ☐ ☐ ☐ ☐ ☐ ☐ ☐

3. The candidate applies the correct Six Sigma tools at the proper point in the project.

Strongly Disagree Strongly Agree

☐ ☐ ☐ ☐ ☐ ☐ ☐ ☐ ☐ ☐

4. The candidate asks for help with Six Sigma tools when necessary.

Strongly Disagree Strongly Agree

☐ ☐ ☐ ☐ ☐ ☐ ☐ ☐ ☐ ☐

5. The candidate can clearly explain all of the Six Sigma tools used on their projects in ordinary language. Note: candidates are not required to be able to *perform* all of the analyses without assistance, but they are required to understand basic or intermediate tools used for their projects.

Strongly Disagree Strongly Agree

☐ ☐ ☐ ☐ ☐ ☐ ☐ ☐ ☐ ☐

6. The candidate understand the limitations as well as the strengths of quantitative methods.

Strongly Disagree Strongly Agree

☐ ☐ ☐ ☐ ☐ ☐ ☐ ☐ ☐ ☐

Ability to Achieve Results Assessment Worksheet			
Green Belt Candidate		Date of Assessment	
Certification Board Member		Role	

1. The candidate has successfully completed at least one Six Sigma project which produced tangible results.

Strongly Disagree Strongly Agree

☐ ☐ ☐ ☐ ☐ ☐ ☐ ☐ ☐ ☐

2. The candidate's project(s) had an acceptable project charter, including sponsorship, problem statement, business case, etc.

Strongly Disagree Strongly Agree

☐ ☐ ☐ ☐ ☐ ☐ ☐ ☐ ☐ ☐

3. The projects employed the Six Sigma approach (DMAIC or equivalent).

   Strongly Disagree Strongly Agree

   ☐ ☐ ☐ ☐ ☐ ☐ ☐ ☐ ☐ ☐

4. The projects' deliverables were clearly defined in tangible terms.

   Strongly Disagree Strongly Agree

   ☐ ☐ ☐ ☐ ☐ ☐ ☐ ☐ ☐ ☐

5. The projects produced significant improvements to an important business process.

   Strongly Disagree Strongly Agree

   ☐ ☐ ☐ ☐ ☐ ☐ ☐ ☐ ☐ ☐

6. The baseline performance level was determined using valid data.

   Strongly Disagree Strongly Agree

   ☐ ☐ ☐ ☐ ☐ ☐ ☐ ☐ ☐ ☐

7. The final performance level was calculated using valid data and showed improvements that were both statistically significant and important to the organization.

   Strongly Disagree Strongly Agree

   ☐ ☐ ☐ ☐ ☐ ☐ ☐ ☐ ☐ ☐

8. An acceptable control plan has been implemented to assure that improvements are maintained.

   Strongly Disagree Strongly Agree

   ☐ ☐ ☐ ☐ ☐ ☐ ☐ ☐ ☐ ☐

9. The projects' financial benefits were validated by experts in accounting or finance.

   Strongly Disagree Strongly Agree

   ☐ ☐ ☐ ☐ ☐ ☐ ☐ ☐ ☐ ☐

10. Key customers were identified and their critical requirements defined.

    Strongly Disagree Strongly Agree

    ☐ ☐ ☐ ☐ ☐ ☐ ☐ ☐ ☐ ☐

11. Project sponsors are satisfied with their project's deliverables.

    Strongly Disagree Strongly Agree

    ☐ ☐ ☐ ☐ ☐ ☐ ☐ ☐ ☐ ☐

12. Project identified and corrected root causes, not symptoms.

Strongly Disagree Strongly Agree

☐ ☐ ☐ ☐ ☐ ☐ ☐ ☐ ☐ ☐

13. All key stakeholders were kept informed of project status and are aware of final outcomes.

Strongly Disagree Strongly Agree

☐ ☐ ☐ ☐ ☐ ☐ ☐ ☐ ☐ ☐

14. Projects were completed on time.

Strongly Disagree Strongly Agree

☐ ☐ ☐ ☐ ☐ ☐ ☐ ☐ ☐ ☐

15. Projects were completed within budget.

Strongly Disagree Strongly Agree

☐ ☐ ☐ ☐ ☐ ☐ ☐ ☐ ☐ ☐

16. Projects were conducted in a manner that minimized disruptions to normal work.

Strongly Disagree Strongly Agree

☐ ☐ ☐ ☐ ☐ ☐ ☐ ☐ ☐ ☐

Assessment Comments	
Assessment Subject Area	**Comments**
Change agent skills	
Application of tools and techniques	
Ability to achieve results	

Scoring Summary				
Evaluator	**Subject Area**	**Items scored 4 or less**	**% in top 3 boxes**	**Comment**
	Change agent skills			
	Application of tools and techniques			
	Ability to achieve results			
	Change agent skills			
	Application of tools and techniques			
	Ability to achieve results			
	Change agent skills			
	Application of tools and techniques			
	Ability to achieve results			
	Change agent skills			
	Application of tools and techniques			
	Ability to achieve results			
	Change agent skills			
	Application of tools and techniques			
	Ability to achieve results			

Examples of Six Sigma Tools and Analytical Concepts	
Basic	**Intermediate**
☐ DMAIC	☐ Control charts for measurements
☐ SIPOC	☐ Control charts for attributes
☐ DPMO	☐ Process capability
☐ Computer skills	☐ Yield analysis (e.g., first pass yield, rolled throughput yield)
☐ Scales of measurement	☐ Measurement error analysis (gage R&R)
☐ Pareto analysis	☐ Correlation analysis
☐ Process mapping, flowcharts	☐ Simple linear regression
☐ Check sheets	☐ Chi-square
☐ Cause-and-effect diagrams	☐ Type I and Type II errors
☐ Scatter plots	☐ Confidence interval interpretation
☐ Run charts	☐ Hypothesis tests
☐ Histograms	☐ Normality assessment and transformations
☐ Ogives	☐ *Z*-transformations
☐ Descriptive statistics (e.g., mean, standard deviation, skewness)	☐ Process sigma calculations
☐ Enumerative vs. analytic statistics	
☐ Stem-and-leaf, boxplots	
☐ Basic probability concepts	
☐ Discrete probability distributions (binominal, Poisson, hypergeometric)	
☐ Continuous probability distributions (normal, exponential, etc.)	
☐ 7M tools	
☐ FMEA	
☐ Sampling	
☐ CTx identification	

APPENDIX 17

AHP Using Microsoft Excel™

The analytic hierarchical process (AHP) is a powerful technique for decision making. It is also quite elaborate and if you wish to obtain exact results you will probably want to acquire specialized software, such as *Expert Choice 2000* (www.expertchoice.com). However, if all you need is a good approximation, and if you are willing to forgo some of the bells and whistles, you can use a spreadsheet to perform the analysis. To demonstrate this, we will use Microsoft Excel to repeat the analysis we performed in Chap. 2.

Example

In Chap. 2 we analyzed the high-level requirements for a software development process and obtained this matrix of pairwise comparisons from our customers.

	Easy to Learn	Easy to Use	Internet Connectivity	Works Well	Easy to Manage
Easy to learn		4.0	1.0	3.0	1.0
Easy to use quickly after I've learned it			5.0	3.0	4.0
Internet connectivity				3.0	3.0
Works well with other software I own					3.0
Easy to maintain	Incon: 0.05				

The meaning of the numbers is described in Chap. 2. The Excel equivalent of this is

	A	B	C	D	E	F
1	Attribute	A	B	C	D	E
2	A-Easy to learn	0.00	4.00	1.00	3.00	1.00
3	B-Easy to use	0.25	0.00	0.20	0.33	0.25
4	C-Connectivity	1.00	5.00	0.00	3.00	3.00
5	D-Compatible	0.33	3.00	0.33	0.00	0.33
6	E-Easy to maintain	1.00	4.00	0.33	3.00	0.00

Note that the paler numbers in the original matrix have become reciprocals, for example, the pale 5.0 is now 0.20, or 1/5. Also note that the numbers on the diagonal are zeros, that is, the comparison of an attribute with itself has no meaning. Finally, the numbers below the diagonals are the reciprocals of the corresponding comparison above the diagonal. For example, the cell C2 has a 4.00, indicating that attribute A is preferred over attribute B; so the cell B3 must contain $1/4 = 0.25$ to show the same thing.

To calculate the weight for each item, we must obtain the grand total for the entire matrix, then divide the row totals by the grand total. This is shown below:

	A	B	C	D	E	F		
1	Attribute	A	B	C	D	E	Total	Weight
2	A-Easy to learn	0.00	4.00	1.00	3.00	1.00	9.00	26.2%
3	B-Easy to use	0.25	0.00	0.20	0.33	0.25	1.03	3.0%
4	C-Connectivity	1.00	5.00	0.00	3.00	3.00	12.00	34.9%
5	D-Compatible	0.33	3.00	0.33	0.00	0.33	4.00	11.6%
6	E-Easy to maintain	1.00	4.00	0.33	3.00	0.00	8.33	24.2%
7	Grand total						34.37	

These results are shown in the figure below.

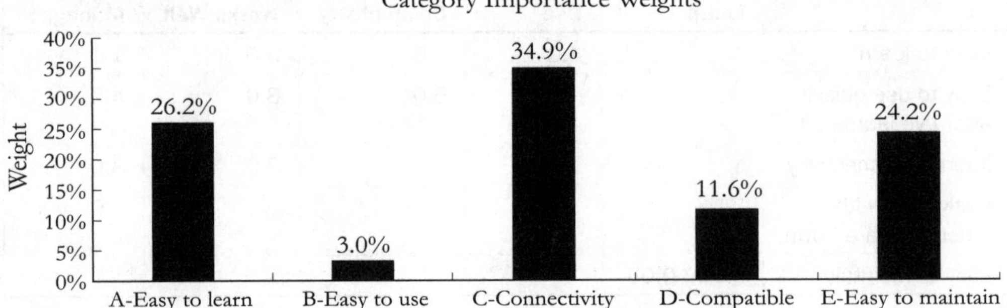

Compare these weights to those obtained by the exact analysis obtained using Expert Choice 2000.

Category	Exact Weight	Spreadsheet Weight
Easy to learn	26.4%	26.2%
Easy to use quickly after I've learned it	5.4%	3.0%
Internet connectivity	35.8%	34.9%
Works well with other software I own	10.5%	11.6%
Easy to maintain	21.8%	24.2%

The conclusions are essentially the same for both analyses.

References

Abraham, B. and Whitney, J. B. (1990). "Applications of EWMA Charts to Data from Continuous Processess," *Annual Quality Congress Transactions*, Milwaukee, WI: ASQ Quality Press.

Akao, Y., Editor (1990). *Quality Function Deployment: Integrating Customer Requirements into Product Design*, Cambridge, MA: Productivity Press.

Alwan, L. C. and Roberts, H. V. (1989). "Time Series Modeling for Statistical Process Control," in Keats, J. B. and Hubel, N. F., Editors, *Statistical Process Control in Automated Manufacturing*, New York: Marcel Dekker.

ASQ Statistics Division. (1983). *Glossary and Tables for Statistical Quality Control*, Milwaukee, WI: ASQ Quality Press.

Aubrey, C. A. and Felkins, P. K. (1988). *Teamwork: Involving People in Quality and Productivity Improvement*, Milwaukee, WI: ASQ Quality Press.

Berry, Michael J. A. and Linoff, Gordon. (1997). *Data Mining Techniques for Marketing, Sales, and Customer Support*, New York: John Wiley & Sons, pp. 369–370.

Blauth, Chris (AchieveGlobal 2008). *Web presentation: Managing Change: Keys to Success*, June 12.

Box, G. E. P. and Draper, N. R. (1969). *Evolutionary Operation: A Statistical Method for Process Improvement*, New York: John Wiley & Sons.

Box, G .E. P. and Draper, N. R. (1987). *Empirical Model-Building and Response Surfaces*, New York: John Wiley & Sons.

Box, G. E. P., Hunter, W. G. and Hunter, J. S. (1978). *Statistics for Experimenters*, New York: John Wiley & Sons.

Brassard, M. (1989). *The Memory Jogger Plus +*, Methuen, MA: GOAL/QPC.

Buffett, Warren E. (1996). *An Owners Manual*, Omaha, NE: Berkshire Hathaway.

Burke, E., Kloeber, J. M. and Deckro, R. F. (2002). "Using and Abusing QFD Scores," *Quality Engineering*, vol. 15, no. l, pp. 23–36.

Burr, I. W. (1976). *Statistical Quality Control Methods, Statistics: Textbooks and Monographs*, vol. 16, New York: Marcel-Dekker, Inc.

Camp, R. C. (1989). *Benchmarking: The Search for Industry Best Practices That Lead to Superior Performance*, Milwaukee, WI: ASQ Quality Press and White Plains, NY: Quality Resources.

Campanella, J., Editor. (1990). *Principles of Quality Costs*, 2nd Edition, Milwaukee, WI: ASQ Quality Press.

Campanella, J., Editor. (1999). *Principles of Quality Costs*, 3rd Edition, Milwaukee, WI: ASQ Quality Press.

Carder, B. and Clark, J. D. (1992). "The Theory and Practice of Employee Recognition," *Quality Progress*, December.

Deming, W. E. (1975). "On Probability as a Basis for Action," *The American Statistician*, vol. 29, no. 4, pp. 146–152.

Deming, W. E. (1986). *Out of the Crisis*, Cambridge, MA: MIT Center for Advanced Engineering Study.

Deming, W. E. (1993). *The New Economics for Industry, Government, Education*, Cambridge, MA: MIT Center for Advanced Engineering Study.

DeToro, I. (1995). "The 10 Pitfalls of Benchmarking," *Quality Progress*, January, pp. 61–63.

Dillman, D. A. (1983). "Mail and Other Self-Administered Questionnaires," in Rossi, P., Wright, J., and Anderson, A., Editors, *Handbook of Survey Research*, New York: Academic Press, Inc., pp. 359–377.

Draper, N. and Smith, H. (1981). *Applied Regression Analysis*, 2nd Edition, New York: John Wiley & Sons.

Duncan, A. J. (1974). *Quality Control and Industrial Statistics*, 4th Edition, Homewood, IL: Irwin.

Eades, Keith M. (2004). *Solution Selling*, New York: McGraw-Hill.

Edosomwan, J. A. (1993). *Customer and Market-Driven Quality Management*, Milwaukee, WI: ASQ Quality Press.

Efron, B. (1982). *The Jackknife, the Bootstrap, and Other Resampling Plans*, Philadelphia, PA: Society for Industrial and Applied Mathematics.

Fivars, G. (1980). "The Critical Incident Technique: a Bibliography," Research and Publication Service, American Institutes for Research ED 195 681.

Flanagan, J.C. (1954). "The Critical Incident Technique," *Psychological Bulletin*, vol. 51, no. 4, July, pp. 327–358.

Forsha, H. J. (1992). *The Pursuit of Quality through Personal Change*, Milwaukee, WI: ASQ Quality Press.

Forum Corporation. (1996). Annual Report.

Galloway, D. (1994). *Mapping Work Processes*, Milwaukee, WI: ASQ Quality Press.

GAO (1986). "Developing and Using Questionnaires—Transfer Paper 7," Washington, DC: United States General Accounting Office.

George, M. L. (2002). *Lean Six Sigma*, New York: McGraw-Hill.

Gibbons, Jean Dickinson. (1993). *Nonparametric Statistics: An Introduction*, New York: Sage Publications, p. 63.

Goldratt, Eliyahu M. (1990). *The Haystack Syndrome: Sifting Information Out of the Data Ocean*, Great Barrington, MA: North River Press, pp. 59–63.

Hammer, M. and Champy, J. (1993). *Reengineering the Corporation: A Manifesto for Business Revolution*, New York: HarperCollins Publishers.

Hammer, Michael and Stanton, Steve. (1999). "How Process Enterprises *Really*," *Harvard Business Review*, November-December, pp. 108–118.

Harrington, H. J. (1992). "Probability and Statistics," in Pyzdek, T. and Berger, R. W., Editors, *Quality Engineering Handbook*, Milwaukee, WI: ASQ Quality Press, pp. 513–577.

Hayes, Bob E. (1992). *Measuring Customer Satisfaction: Development and Use of Questionnaires*, Milwaukee, WI: ASQ Quality Press.

Hicks, Charles R. (1973). *Fundamental Concepts in the Design of Experiments*, 2nd Edition, New York: Holt, Rinehart, Winston, pp. 31–38.

Hicks, Charles R. (1993). *Fundamental Concepts in the Design of Experiments*, 3rd Edition, New York: Holt, Rinehart, Winston.

Hillier, F. S. (1969). "X-bar and R-chart Control Limits Based on a Small Number of Subgroups," *Journal of Quality Technology*, vol. l, no. l, January, pp. 17–26.

Holland, John H. (1996). *Hidden Order: How Adaptation Builds Complexity*, Reading, MA: Perseus Books, p. 56.

Hunter, J. S. (1986). "The Exponentially Weighted Moving Average," *Journal of Quality Technology*, vol. 18, pp. 203–210.

Hunter, J. S. (1989). "A One Point Plot Equivalent to the Shewhart Chart with Western Electric Rules," *Quality Engineering*, vol. 2, pp. 13–19.

Hurley, H. and Loew, C. (1996). "A Quality Change for New Product Development," *Quality Observer*, January, pp. 10–13.

Hutton, D. W. (1994). *The Change Agent's Handbook: A Survival Guide for Quality Improvement Champions*, Milwaukee, WI: ASQ Quality Press.

Imai, M. (l986). *Kaizen*, New York: Random House.

Ireson, W., Coombs, C., Moss, R. (1996). *Handbook of Reliability Engineering and Management*, New York: McGraw-Hill.

Ishikawa, K. (1985). *What is Total Quality Control the Japanese Way?*, Englewood Cliffs, NJ: Prentice-Hall, Inc.

Johnson, D. W. and Johnson, F. P. (1999). *Joining Together: Group Theory and Group Skills*, Englewood Cliffs, NJ: Prentice-Hall.

Johnson, R. S. (1993). *TQM: The Mechanics of Quality Processes*, Milwaukee, WI: ASQ Quality Press.

Joiner, B. L. (1994). *Fourth Generation Management: The New Business Consciousness*, New York: McGraw-Hill, Inc.

Juran, J. M. and Gryna, F. M. (1988). *Juran's Quality Control Handbook*, 4th Edition, New York: McGraw-Hill.

Juran, J. M. and Gryna, F. M. (1993). *Quality Planning and Analysis*, 3rd Edition, New York: McGraw-Hill.

Kackar, R. N. (1985). "Off-line Quality Control, Parameter Design, and the Taguchi Method" *Journal of Quality Technology*, vol. 17, pp. 176–188.

Kaplan, Robert S. and Norton, David P. (1992). "The Balanced Scorecard—Measures that Drive Performance," *Harvard Business Review*, January–February, pp. 71–79.

Keller, Paul A. (2005). *Six Sigma Demystified*, New York: McGraw-Hill, Inc.

King, B. (1987). *Better Designs in Half the Time: Implementing QFD in America*, Methuen, MA: Goal/QPC.

Kirkpatrick, D. L. (1996). "Evaluation," in Craig, R.L., Editor-in-chief, *The ASTD Training and Development Handbook: A Guide to Human Resources Development*, New York: McGraw-Hill, pp. 294–312.

Koch, G. G. and Gillings, D. B. (1983). "Statistical Inference, Part I," in Kotz, Samuel and Johnson, Norman L, Editors-in-chief, *Encyclopedia of Statistical Sciences*, New York: John Wiley & Sons vol. 4, pp. 84–88.

Kohn, A. (1993). *Punished by Rewards: The Trouble with Gold Stars, Incentive Plans, A's, Praise and Other Bribes*, New York: Houghton Mifflin Company.

Kotler, P. (1991). *Marketing Management: Analysis, Planning, Implementation, and Control*, 7th Edition, Englewood Cliffs, NJ: Prentice-Hall.

Krisnamoorthi, K.S. (1991). "On Assignable Causes that Cannot be Eliminated—An Example from a Foundry," *Quality Engineering*, vol. 3, pp. 41– 47.

Levinson, Jay Conrad; Godin, Seth (Contributor); and Rubin, Charles. (1995). *The Guerrilla Marketing Handbook*, Boston, MA: Houghton Mifflin Co.

Main, Jeremy (1994). *Quality Wars: The Triumphs and Defeats of American Business*, New York: The Free Press, p. 173.

Meyers, Raymond H. and Montgomery, Douglas C. (1995). *Response Surface Methodology: Process and Product Optimization Using Designed Experiments*, New York: John Wiley & Sons.

Mizuno, S., Editor. (1988). *Management for Quality Improvement: The 7 New QC Tools*, Cambridge, MA: Productivity Press.

Montgomery, D. C. (1984). *Design and Analysis of Experiments*, 2nd Edition, New York: John Wiley & Sons.

Montgomery, Douglas C. (1996). *Design and Analysis of Experiments*, New York: John Wiley & Sons.

Natrella, M.G. (1963). *Experimental Statistics: NBS Handbook 91*, Washington, DC: US Government Printing Office.

Nelson, L. S. (1984). "The Shewhart Control Chart—Tests for Special Causes," *Journal of Quality Technology*, vol. 16, no. 4, October 1984, pp. 237–239.

Ohno, Taiichi. (1988). *Toyota Production System: Beyond Large-Scale Production*, Portland, OR: Productivity Press.

Palm, A. C. (1990). "SPC versus Automatic Process Control," *Annual Quality Congress Transactions*, Milwaukee, WI: ASQ Quality Press.

Phillips, J. J. (1996). "Measuring the Results of Training," in Craig, R.L. Editor-in-chief, *The ASTD Training and Development Handbook: A Guide to Human Resources Development*, New York: McGraw-Hill, pp. 313–341.

Proschan, F. and Savage, I. R. (1960). "Starting a Control Chart," *Industrial Quality Control*, 17(3), September, pp. 12–13.

Provost, L. P. (1988). "Interpretation of Results of Analytic Studies," paper presented at 1989 NYU Deming Seminar for Statisticians, March 13, New York: New York University.

Pyzdek, T. (1976). "The Impact of Quality Cost Reduction on Profits," *Quality Progress*, May, pp. 14–15.

Pyzdek, T. (1985). "A Ten-step Plan for Statistical Process Control Studies," *Quality Progress*, April, pp. 77–81, and July, pp. 18–20.

Pyzdek, T. (1989a), *What Every Engineer Should Know About Quality Control*, New York: Marcel-Dekker, Inc.

Pyzdek, T. (1990). *Pyzdek's Guide to SPC Volume One*, Tucson, AZ: Quality Publishing, Inc.

Pyzdek, T. (1992a). *Pyzdek's Guide to SPC Volume Two—Applications and Special Topics*, Tucson, AZ: Quality Publishing, Inc.

Pyzdek, T. (1992b). "A Simulation of Receiving Inspection," *Quality Engineering*, 4(1), pp. 9–19.

Pyzdek, T. (1992c). *Turn-key SPC Instructor's Manual*, Tucson, AZ: Quality Publishing.

Reichheld, F. F. (1996). *The Loyalty Effect: The Hidden Force Behind Growth, Profits, and Lasting Value*, New York: McGraw-Hill.

Rose, K. H. (1995). "A Performance Measurement Model," *Quality Progress*, February, pp. 63–66.

Ruskin, A. M. and Estes, W. E. (1995). *What Every Engineer Should Know About Project Management*, 2nd Edition, New York: Marcel-Dekker.

Saaty, T. L. (1988). *Decision Making for Leaders: The Analytic Hierarchy Process for Decisions in a Complex World*, Pittsburgh, PA: RWS Publications.

Scholtes, P. R. (1988). *The Team Handbook: How to Use Teams to Improve Quality*, Madison, WI: Joiner Associates, Inc.

Schuman, S. P. (1996). "What to Look for in a Group Facilitator," *Quality Progress*, June, pp. 69–72.

Shainin, D. and Shainin, P. D. (1988). "Statistical Process Control," in Juran, J. M., and Gryna, F. M., *Juran's Quality Control Handbook*, 4th Edition, New York: McGraw-Hill.

Sheridan, B. M. (1993). *Policy Deployment: The TQM Approach to Long-Range Planning*, Milwaukee, WI: ASQ Quality Press.

Shewhart, W. A. (1939, 1986). *Statistical Method from the Viewpoint of Quality Control*, New York: Dover Publications.

Shewhart, W. A. (1931, 1980). *Economic Control of Quality of Manufacturing*, Milwaukee, WI: ASQ Quality Press.
Simon, J. L. (1992). *Resampling: the New Statistics*, Arlington, VA: Resampling Stats, Inc.
Spencer, Mary. (1999). "DFE and Lean Principles," *The Monitor*, 6(2), Fall, pp. 15–16.
Stewart, T.A. (1995). "After All You've Done for Your Customers, Why are They Still NOT HAPPY?" *Fortune*, December 11.
Suminski, L. T., Jr. (1994). "Measuring the Cost of Quality," *Quality Digest*, March, pp. 26–32.
Taguchi, G. (1986). *Introduction to Quality Engineering: Designing Quality into Products and Processes*, White Plains, NY: Quality Resources.
Tuckman, B. W. (1965). "Development Sequence in Small Groups," *Psychological Bulletin*, 63, June, pp. 384–399.
Tufte, E. R. (2001). *The Visual Display of Quantitative Information*, Cheshire, CT: Graphics Press.
Tukey, J. W. (1977). *Exploratory Data Analysis*, Reading, MA: Addison-Wesley.
Vaziri, H. K. (1992). "Using Competitive Benchmarking to Set Goals," *Quality Progress*, November, pp. 81–85.
Waddington, P. (1996). *Dying for Information: An Investigation of Information Overload in the UK and Worldwide*, London: Reuters Business Information.
Wheeler, D. J. (1991). "Shewhart's Charts: Myths, Facts, and Competitors," *Annual Quality Congress Transactions*, ASQ, Milwaukee, WI.
Womack, J. P. and Jones, D. T. (1996). *Lean Thinking: Banish Waste and Create Wealth in Your Corporation*, New York: Simon & Shuster.
Zaltman, Berald. (2002). "Hidden Minds," *Harvard Business Review*, June, pp. 26–27.

Index

Note: Page numbers referencing figures are followed by an "*f*, "page numbers referencing tables are italicized followed by a "*t*," and page numbers referring to footnotes are followed by an "*n*."

D

E

F

T